Mass spectrometry is a versatile analytical technique in widespread use in industry, laboratories and universities. Its ability to identify and quantify materials quickly and, if necessary, in minute amounts, makes it an ideal method for solving analytical problems in a huge variety of fields, such as chemistry, biochemistry, space research, health, archaeology, forensic science, pharmaceutical chemistry, industrial research and environmental science. This book introduces the basis of mass spectrometry, covers the full range of commonly used techniques and describes a wide variety of applications. The authors adopt an instructional approach and make use of recent examples to illustrate important points.

The second edition updates the first by removing much of the material on older techniques where those methods have been superseded. It includes an impressive and extended range of applications of mass spectrometry that have developed in the last ten years, in particular the analysis of once difficult polar compounds and substances with very large masses such as proteins. The increased role of computers has been included and powerful methods in which mass spectrometry is combined with newer separation techniques are covered. Sections on ionization and mass analysis of ions have been extended to address these new methods.

The book, requiring no previous knowledge of mass spectrometry, is an ideal teaching text in particular for students of chemistry and biochemistry, at both undergraduate and postgraduate level. It will also be of considerable interest to researchers in universities, analytical laboratories and industry because of the wide range of topics covered and the development in depth of the commoner techniques.

Mass spectrometry for chemists and biochemists

SECOND EDITION

Mass spectrometry for chemists and biochemists

SECOND EDITION

Robert A.W. Johnstone *University of Liverpool*

and Malcolm E. Rose *The Open University*

CAMBRIDGE
UNIVERSITY PRESS

Published by the Press Syndicate of the University of Cambridge

The Pitt Building, Trumpington Street, Cambridge CB2 1RP

40 West 20th Street, New York, NY 10011-4211, USA

10 Stamford Road, Oakleigh, Melbourne 3166, Australia

First published 1982

Reprinted with amendments 1987

Second edition 1996

A catalogue record for this book is available from the British Library

Library of Congress cataloguing in publication data

Johnstone, R. A. W. (Robert Alexander Walker)

Mass spectrometry for chemists and biochemists / Robert A. W.

Johnstone and Malcolm E. Rose. – 2nd ed.

   p.   cm.

Rose's name appears first on the earlier edition.

Includes bibliographical references (p. –) and index.

ISBN 0 521 41466 0 (hardcover : alk. paper). – ISBN 0 521 42497 6

(pbk. : alk. paper)

1. Mass spectrometry. I. Rose, M. E. II. Title.

QD96.M3R67   1996

543′.0873–dc20   95–40173 CIP

ISBN 0 521 41466 0 hardback

ISBN 0 521 42497 6 paperback

Transferred to digital printing 2002

Dedicated to Barbara and Colin   M.E.R.

Dedicated to Christine, Steven and Fiona   R.A.W.J.

Contents

Introduction		xiii
Acknowledgments		xvii
Table of quantities		xviii
List of abbreviations		xx
1	**The mass spectrum**	**1**
1.1	Formation of ions	1
1.2	Formation of the mass spectrum following electron ionization	1
1.3	Multiply charged ions	9
1.4	Isotopes	10
1.5	Metastable ions	14
1.6	Elemental compositions of ions	15
1.7	Appearance of the mass spectrum	17
1.8	Formation of the mass spectrum following ionization by other methods	18
2	**Instrument design**	**22**
2.1	Introduction	22
2.2	Inlet systems	23
2.3	Ion sources	27
2.4	Analysers (separation of ions)	37
2.5	Detection and recording of spectra	55
2.6	A total system	58
3	**Methods of ionization**	**60**
3.1	Introduction	60
3.2	Electron ionization	64

3.3	Chemical ionization	66
3.4	Negative-ion chemical ionization	78
3.5	Pulsed positive-ion/negative-ion chemical ionization	82
3.6	Modified electron and chemical ionization	82
3.7	Field ionization	85
3.8	Field desorption	89
3.9	Spraying and ion evaporation	91
3.10	High-speed particles	100
3.11	Lasers	108
3.12	Plasma desorption	110
4	**Computers in mass spectrometry: data systems**	**113**
4.1	Introduction	113
4.2	Data acquisition and mass calibration	114
4.3	Control of the mass spectrometer	121
4.4	Data processing	123
4.5	Data display units	141
5	**Combined chromatography and mass spectrometry**	**142**
5.1	Introduction	142
5.2	Gas chromatography/mass spectrometry	144
5.3	Liquid chromatography/mass spectrometry	158
5.4	Some applications of liquid chromatography/mass spectrometry	166
5.5	Capillary electrophoresis/mass spectrometry	172
5.6	Supercritical fluid chromatography/mass spectrometry	176
5.7	Thin-layer chromatography/mass spectrometry	180
6	**Uses of derivatization**	**183**
6.1	Introduction	183
6.2	Imparting volatility and thermal stability	186
6.3	Modelling the molecule for analysis	188
6.4	Inorganic compounds	203
7	**Quantitative mass spectrometry**	**205**
7.1	Introduction and principles	205
7.2	Calibration and internal standards	208

7.3   Selected ion monitoring   213
7.4   Applications based on gas chromatography/mass
       spectrometry   219
7.5   Applications based on direct inlet   224
7.6   Applications based on liquid chromatography/mass
       spectrometry   227
7.7   Selected reaction monitoring   228

**8   Metastable ions and mass spectrometry/mass
        spectrometry   232**
8.1   The origin of metastable ions   232
8.2   The usefulness of metastable ions   236
8.3   Metastable ions in conventional mass spectrometers   238
8.4   Metastable ions in mass spectrometers of reversed geometry   256
8.5   Triple quadrupole analysers   259
8.6   Triple-focussing magnetic-sector mass spectrometers   260
8.7   Activation of normal ions   261
8.8   Mass spectrometry/mass spectrometry (tandem MS)   270
8.9   Summary and miscellany of MS/MS experiments   281

9   Theory of mass spectrometry   289
9.1   Introduction   289
9.2   Energy states resulting from electron ionization at low gas
       pressures   291
9.3   Energy states resulting from ionization at higher gas
       pressures   298
9.4   Energy states resulting from ionization in condensed
       (liquid) media   300
9.5   Energy states resulting from collisional and other activation 301
9.6   Formation of ions   303
9.7   Theories of fragmentation rates   310
9.8   Thermochemical arguments   313
9.9   Ion lifetimes   315
9.10   Qualitative theories   316

10   Structure elucidation   325
10.1. Classifications of mass spectra   325

10.2.	Examination of the mass spectrum	329
10.3.	Modification of mass spectra through instrumental parameters	337
10.4.	Postulation of ion structures	344
10.5.	Fragmentation of hydrocarbons	346
10.6.	Primary fragmentations of aliphatic heteroatomic compounds	353
10.7.	Primary fragmentations of aromatic heteroatomic compounds	373
10.8.	Subsequent decomposition of primary fragment ions	382
10.9.	Rearrangement accompanying fragmentation	387
10.10.	Fragmentation following other methods of ionization	393
10.11	A suggested scheme for interpretation of mass spectra	395
11	**Examples of structure elucidation by mass spectrometry**	**397**
11.1	Introduction	397
11.2	Example A	401
11.3	Example B	406
11.4	Example C	408
11.5	Example D	411
11.6	Example E	413
11.7	Example F	415
11.8	Example G	418
11.9	Example H	420
12	**Further discussion of selected topics**	**427**
12.1	Ionization and appearance energies	427
12.2	Isotope analysis and labelling	431
12.3	Ion cyclotron resonance spectroscopy	444
12.4	Pyrolysis/mass spectrometry and pyrolysis/chromatography/mass spectrometry	455
12.5	Further literature on mass spectrometry	473
	References	478
	Index	496

Introduction

'Chaos, rudis indegestaque moles'   Ovid, *Metamorphoses, i7*

The basis of mass spectrometry is the production of ions from neutral compounds and, particularly in chemistry and biochemistry, the examination of the subsequent decomposition of those ions. In mass spectrometry, a substance can be characterized by investigating the *chemistry* of ions resulting from that substance. Because the technique involves a chemical reaction, the sample being investigated is not recoverable; however, only very small quantities of material are required for the analysis. Most other physical methods of analysis deal with a narrowly defined property of a molecule, but this is not so in mass spectrometry. As with any chemical reaction, the precise outcome, the mass spectrum, is dependent on a considerable number of factors, such as temperature, concentration, effects of the medium and so on. It is this uncertainty which lends mass spectrometry its versatility, intricacy and charm.

Chemical reactions in solids, liquids and gases are usually discussed in terms of isolated molecules whereas their actual behaviour is the result of 'group' effects arising from collisional or other activation of molecules. Most mass spectra are measured at low pressures when collisions between ions and molecules are so rare that interpretation of mass spectra in terms of isolated species is more satisfactory. For practical purposes, these low concentrations of ions impose severe restraints on methods of investigating them. The gas densities of ions are usually so low that other common physical methods such as ultraviolet, infrared and nuclear magnetic resonance spectroscopy cannot be used directly to elucidate the structures and internal energies of the ions. At best, postulated ion structures can only be described as consistent with indirect evidence based on isotopic labelling, thermochemical arguments, reasoning by analogy and so on. As its name implies, a mass spectrometer measures mass (or strictly mass-to-charge ratio) and gives no direct information on ion structures. Because of this lack of knowledge of structures, fundamental investigation of the mechanisms of ion decomposition has been severely hampered. Nevertheless, it needs to be stressed that, even without any sound basic understanding of ion structures and mechanisms of mass spectrometric fragmentations, a great deal of analytical information may be extracted by use of empirical rules and concepts. At this stage, it is worth reminding the reader that mass spectrometry alone seldom gives a unique solution to a problem and it is best used in conjunction with information obtained by other means.

Following the construction of the first mass spectrometers many years ago, there followed a period of modest interest by physicists and chemists until it was realized by organic chemists that here was an instrument which could be of immense value to their work. When *Mass Spectrometry for Organic Chemists* first appeared (1972), the technique already held a prime position in chemistry, especially in organic chemistry, for elucidation of molecular structures. Both the applications and development of mass spectrometry continued to increase rapidly. By the time the second edition appeared in 1982, mass spectrometry had been applied with advantage in many other, diverse fields such that it became necessary to re-name it *Mass Spectrometry for Chemists and Biochemists*. Notably, its use spread to biochemistry, medicine and toxicology. Many of these wider applications were due to improved instrumentation and could be attributed to three interconnected factors: (a) the linking of mass spectrometers to computers, (b) improvements permitting ready combination of mass spectrometers with separation devices such as gas and liquid chromatographs and (c) development of sophisticated means of producing ions from neutral species. These three topics have continued to be exploited since 1982, together with new instrumental techniques and improved limits of detection. Additionally, innovations in instrument design have seen the introduction of the 'bench top' mass spectrometer for widespread use.

The present text was conceived as a simple up-dating of *Mass Spectrometry for Chemists and Biochemists*, but the increasing applications of the technique begged a wider scope for the book and the newer developments have necessitated considerable rewriting and expansion. Despite all these changes, it is appropriate that the title of the text should remain as it is because the great majority of users of the techniques covered are schooled in chemistry and biochemistry. To write a simple introductory text to this subject is a somewhat uncomfortable task since it becomes necessary to make statements which may be arguable, but the authors are of the opinion that simplicity and accuracy are not mutually exclusive in this context. The layout for this latest edition has been changed considerably. After an introductory chapter, instrumentation is described and there follows a fresh chapter on ionization methods, including new techniques developed since the last edition appeared in 1982. The increased usage of computers for control of instruments and acquisition and manipulation of data was thought to merit its own new chapter 4. Combined chromatography/mass spectrometry, being a routine and important tool in many different fields, is described in some detail in the ensuing chapter. Technologically, combined liquid chromatography/mass spectrometry is almost as advanced as the older gas chromatography/mass spectrometry and the two methods are covered in depth in the same chapter, rather than separately. Other combinations of mass spectrometry, with thin-layer chromatography, supercritical fluid chromatography and capillary electrophoresis, are also covered in chapter 5. The important topic of derivatization as an adjunct to mass spectrometric investigations merits its own chapter. One trend discernible for some time in mass spectrometry is away from simple identification of unknown substances and towards quantification of known compounds. Chapter 7 is devoted to these aspects and will be of most use to those interested in life sciences and analytical chemistry. Chapter 8 is completely new and covers developments in tandem mass spectrometry, or as it is also called, mass spectrometry/mass spectrometry. It was believed recently by some that tandem mass spectrometry would supplant combined chromatography/mass spectrometry but this has not occurred and the two techniques co-exist with their own qualities. Metastable ions are discussed conveniently in this same chapter. In this introductory text, it has been felt necessary to include some theoretical aspects of mass spectrometry as well as its empirical applications because, by understanding fundamentals, an empiricist may make better use of the technique. Such an approach is embodied in chapter 9. The next sections (chapters 10 and 11) deal with empirical application of mass spectrometry to structural elucidation, which forms a large part of routine mass spectrometry for a typical user. The examples of structural elucidation were chosen to illustrate many of the modern methods of mass spectrometric analysis. Miscellaneous topics which would not be in context elsewhere in the book are included in chapter 12.

The book is suitable for those without prior knowledge of mass spectrometry and, whilst being an introductory text, it does provide an approach to the more advanced topics for the interested reader. As part of this approach, an extensive bibliography is included, facilitating consultation of further, more specialized literature. For the convenience of the reader, a simple table of quantities (conversion factors) appears at the front of the book.

Acknowledgments

The authors gratefully acknowledge the kind help and photographs provided by Micromass UK Ltd, Finnigan MAT Ltd and Horizon Instruments.

Table of quantities

Because of the variety of some units of quantity and pressure used in mass spectrometry in the past in contrast to the preferred standard SI units now in place, the relationships between some of the more frequently found units are defined below for the convenience of the mass spectrometrist.

Energy
1 kcal = 4.184 kilojoules
1 electron volt (eV) = 1.602×10^{-22} kJ
$$= 96.44 \text{ kJ mol}^{-1}$$
$$= 23.05 \text{ kcal mol}^{-1}$$

Pressure
1 Pascal (Pa) = 1 Newton per square metre (N m^{-2})
1 Torr = 133.3 N m^{-2} = 133.3 Pa
$$= 1 \text{ mm Hg (at } 0°C)$$
1 bar = 10^5 Pascals (N m^{-2}) = 0.987 atmospheres
1 atmosphere = 101.3 kN m^{-2}
$$= 1.013 \text{ bar}$$
$$= 760 \text{ Torr}$$

Gas constant
$R = 8.314 \text{ J K}^{-1} \text{ mol}^{-1}$

Sub-units of the gramme

1 milligramme (mg)	$= 10^{-3}$ g
1 microgramme (μg)	$= 10^{-6}$ g
1 nanogramme (ng)	$= 10^{-9}$ g
1 picogramme (pg)	$= 10^{-12}$ g
1 femtogramme (fg)	$= 10^{-15}$ g
1 attogramme (ag)	$= 10^{-18}$ g

Sub-units of the mole use the same prefixes (e.g., 1 pmol $= 10^{-12}$ mol)

List of abbreviations

For the convenience of the reader, some of the more commonly used mass spectrometric abbreviations are listed below.

ADC, analogue-to-digital converter

APCI, atmospheric pressure chemical ionization

API, atmospheric pressure ionization

B, used to represent a magnetic field (extended to mean a magnetic sector in a mass spectrometer)

B/E, a mode of linked scanning combining magnetic and electric fields

B^2/E, a mode of linked scanning using the square of a magnetic field with an electric field

CE, capillary electrophoresis

CEM, channel electron multiplier

CE/MS, combined capillary electrophoresis/mass spectrometry

CI, chemical ionization

CIA, collisionally induced activation

CID, collision-induced dissociation or decomposition

DAC, digital-to-analogue converter

DCI, direct or desorption chemical ionization

DEI, direct or desorption electron ionization

E, used to represent an electric field (extended to mean an electric sector in a mass spectrometer)

E^2/V, a mode of linked scanning using the square of an electric field with an accelerating voltage

e, represents the unit charge on an electron

E_{CM}, collision energy referenced to the centre of mass of the colliding species

E_{LAB}, collision energy referenced to the laboratory frame of reference

EI, electron ionization (formerly, electron impact)

ESI, electrospray ionization

FAB, fast atom bombardment

FD, field desorption

FI, field ionization

FIB, fast ion bombardment

FTICR, Fourier transform ion cyclotron mass spectrometry

GC/MS, combined gas chromatography/mass spectrometry

ICP/MS, inductively coupled plasma/mass spectrometry

ICR, ion cyclotron resonance

IKES, ion kinetic energy spectroscopy

ITMS, ion trap mass spectrometry

LC/MS, combined liquid chromatography/mass spectrometry

LSIMS, secondary ion mass spectrometry in the liquid phase

MALDI, matrix-assisted laser desorption ionization

MIKES, mass-analysed ion kinetic energy spectroscopy

MIPS, million instructions per second

MPI, multiphoton ionization

MS/MS, mass spectrometry/mass spectrometry or tandem mass spectrometry

MS^n, general designation of mass spectrometry to the nth degree (e.g., $MS/MS = MS^2$)

m/z, the ratio between the mass (m) of an ion and the number (z) of electronic charges on it. In older publications, this is referred to as m/e but e should be reserved for the *charge* on an electron and not the *number* of charges on an ion. The total electronic charge on an ion is ze.

3-NOBA, 3-nitrobenzyl alcohol

PFK, perfluorokerosene

PID, photon-induced dissociation or decomposition

PPNICI, pulsed positive-ion/negative-ion chemical ionization

Py/GC/MS, pyrolysis/gas chromatography/mass spectrometry

Py/MS, pyrolysis/mass spectrometry

Q, quadrupolar field (extended to mean a quadrupole mass filter)

QET, quasi-equilibrium theory

QQQ, triple quadrupole analyser

%RA, percentage relative abundance

REMPI, resonance-enhanced multiphoton ionization

%RIC, percentage reconstructed ion current

RISC, reduced instruction set computer or computation

SFC/MS, combined supercritical fluid chromatography/mass
        spectrometry

SID, surface-induced dissociation or decomposition

SIM, selected (or single) ion monitoring

SIMS, secondary ion mass spectrometry

SIR, selected (or single) ion recording (equivalent to SIM)

SRM, selected reaction monitoring

SSMS, spark source mass spectrometry

%TIC, percentage total ion current

TG/MS, combined thermogravimetry/mass spectrometry

TLC/MS, combined thin-layer chromatography/mass spectrometry

TMS, trimethylsilyl

TOF, time-of-flight mass spectrometry

V, used to represent a voltage difference

VDU, visual display unit (workstation monitor)

z, represents the number of charges on an ion

1 The mass spectrum

1.1. FORMATION OF IONS

A mass spectrometer works with electrically charged particles and, before a mass spectrum can be obtained, the substance under examination must be ionized if it is not already ionic. Most research into mass spectrometry has been carried out on positive ions and these are discussed in detail. Negative ions are discussed as the occasion demands

A molecule (M) can be ionized by removal or addition of an electron to give species ($M^{+\cdot}$ and $M^{-\cdot}$ respectively) having a mass which, for practical purposes, is identical to that of the original molecule, the mass of an electron being so small. Optionally, the molecule may be ionized by addition or subtraction of other charged species (X^+) to give ions ($[M + X]^+$ or $[M - X]^-$); in these cases, the resulting ions contain all or most of the original molecule but have masses that are different from that of the original so they are called *quasi-molecular* ions. Finally, the original substance may be a salt (M^+X^-), in which instance, it is ionized already and the oppositely charged species need only be separated prior to mass spectrometry.

Both currently and historically, the most widespread means of ionization has been the removal of an electron from a molecule to give $M^{+\cdot}$ through use of another electron (*electron ionization,* EI). Therefore, this discussion on the formation of a mass spectrum begins with EI, the other methods being introduced later.

1.2. FORMATION OF THE MASS SPECTRUM FOLLOWING
ELECTRON IONIZATION

When a molecule is ionized by removal of an electron, a *molecular ion* ($M^{+\cdot}$, a cation radical) is produced and this may contain sufficient excess of internal energy to fragment by ejection of a neutral particle (N) with the formation of a *fragment ion* ($A^{+\bullet}$ or A^+). A neutral molecule gives a radical-cation as the molecular ion, and the fragment ion may be either a cation or a radical-cation. The ejected neutral particle (N) may be a radical or neutral molecule.

$$M - e^- \longrightarrow M^{+\bullet} \longrightarrow A^+ + N^\bullet$$

or

$$M - e^- \longrightarrow M^{+\bullet} \longrightarrow A^{+\bullet} + N$$

If the fragment ion (e.g. A^+) has sufficient excess of internal energy, then further decomposition may occur with the formation of new fragment ions (B^+, C^+, etc.) until there is insufficient excess of internal energy in any one ion for further reaction.

$$M^{+\bullet} \xrightarrow{\quad N^\bullet \quad} A^+ \xrightarrow{\quad N_a \quad} B^+ \xrightarrow{\quad N_b \quad} C^+ \longrightarrow \text{etc.}$$

Such a series of decompositions when elucidated from a mass spectrum is a *fragmentation pathway.* The molecular ion ($M^{+\bullet}$) and any of the fragment ions (A^+, B^+, C^+, etc.) may decompose by more than one pathway. The various fragmentation pathways together compose a *fragmentation pattern* character-istic of the compound under investigation. At one extreme, the fragmenta-tion pattern might consist of only one pathway and result in a very simple mass spectrum. At the other extreme, the fragmentation pattern contains many, often interlocking pathways producing a complex spectrum. The extent to which fragmentation takes place along the individual pathways is determined by the excess of internal energy imparted to the molecular ion ($M^{+\bullet}$), its structure and the time allowed between ion formation and detec-tion. Hence, the *mass spectrum* is not simply the fragmentation pattern but is the appearance of the fragmentation pattern at specified energies and times.

A mass spectrometer is designed to separate and measure the masses of ions by making use of their mass-to-charge (m/z) ratios. An ion is usually formed with a single positive charge ($z = 1$) so that m/z is then equivalent to m and gives the mass of the ion directly. Formerly, e and not z was used to represent the charge on the ion. Accordingly, many existing publications represent the mass-to-charge ratio as m/e. The term m/z is used here to conform with IUPAC recommendations.

During any one interval of time in the spectrometer, molecular ions will be produced with various excesses of internal energy. Each molecular ion will fragment at a rate determined by its initial energy because, at the low pressures normally obtaining in an ion source, ion/ion and ion/molecule collisions are rare so that collisional equilibration of internal energies does not occur (except in techniques like chemical ionization). Some molecular ions may have insufficient energy to fragment whereas others have so much energy that decomposition proceeds right through a fragmentation pathway. Because of the initial range of internal energies in the molecular ions, a short period of time after ionization would see the presence of ions $M^{+\cdot}$, A^+, B^+, C^+, etc. in amounts determined by their individual rates of formation and decomposition and the initial energy imparted to M. If a sample of these ions in the ion source is obtained and their relative amounts measured, the results can be displayed with m/z values as abscissae and *ion abundances* as ordinates (figure 1.1). Such a picture of a sample of the ions in the ion source is the basis of the mass spectrum, which may be recorded on a cathode ray tube, paper chart, photographic plate or via a computer. An example of part of a mass spectrum recorded as peaks on a paper chart is shown in figure 1.2.

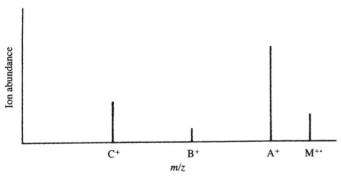

Figure 1.1. Abundances of ions $M^{+\cdot}$, A^+, B^+ and C^+ in a sample of ions drawn from an ion source.

As a standard practice, it is usual to make a record of a mass spectrum in either a *normalized* or *percentage relative abundance* (%RA) form or as a *percentage of total ion current* (%TIC). In a normalized record, the biggest peak in the spectrum is called the *base peak* and its height is put equal to 100 units; the relative heights of all other peaks are referred to this base peak and lie between 0 and 100 units (note that the base peak is not necessarily

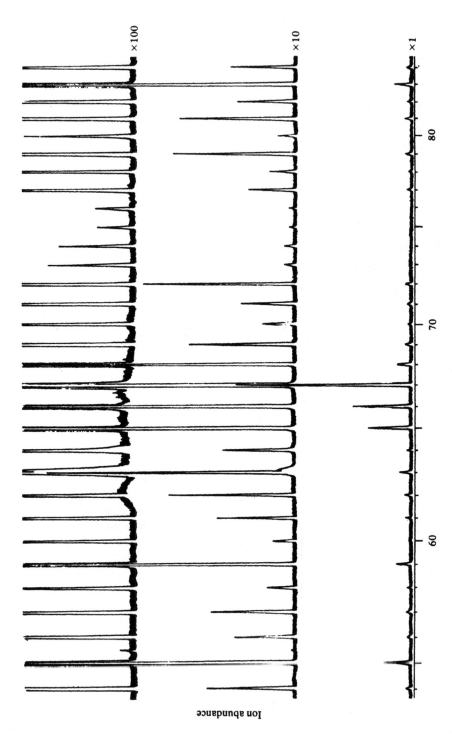

Figure 1.2. Part of a mass spectrum recorded on photographic paper (the three traces record the spectrum at increasing sensitivities from the lower to the upper trace).

Table 1.1. *Normalized abundances[a] in the mass spectrum of 1,3-dimethylbenzene.*

m/z	Abundance	m/z	Abundance	m/z	Abundance
38	1.2	65	5.5	102	1.1
39	7.1	74	1.1	103	6.1
50	2.8	77	18.3	104	2.6
51	9.1	78	5.1	105	28.7
52	5.0	79	6.5	106	61.7
54	2.2	89	1.9	107	5.4
62	1.6	91	100.0		
63	4.2	92	7.8		

Notes:

[a] Ion abundances less than 1 per cent of the base peak have been omitted.

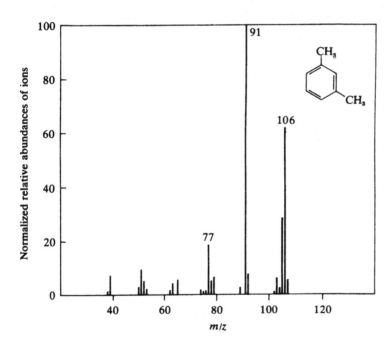

Figure 1.3. A line diagram showing normalized relative abundances of ions in the mass spectrum of 1,3-dimethylbenzene. Note that the base peak at m/z 91 is made equal to 100 units.

the molecular ion peak, although it may be). The height of a peak represents the abundance of ions at that particular m/z value. Table 1.1 illustrates part of the mass spectrum of 1,3-dimethylbenzene recorded in this way and figure 1.3 shows the same spectrum as a normalized line diagram. This form of output is typical of that provided by a computer. Very often not all the peaks in a mass spectrum appear in the normalized line diagram since peaks of less than 1 per cent of the size of the base peak are frequently arbitrarily omitted as unimportant. Care must be exercised in this respect because, even though ions may not be abundant, they can be important for elucidating structures from mass spectra. For example, the molecular ions of a compound undergoing extensive fragmentation will have low abundance but they are still a very important feature of the spectrum. Fragment ions may decompose at about the same rate as they are formed, when again their abundance would be low but they could be important for unravelling a fragmentation pathway. For compounds of high molecular mass, the low-mass ions (e.g. below m/z 40) may be numerous and abundant but also may have little value for the interpretation of a spectrum; such low-mass ions are frequently omitted from the normalized line diagram.

To obtain a record of a mass spectrum by the percentage of total ion current method, abundances of all ions giving peaks of significant size, from the molecular ion down to a suitably chosen low mass (often about m/z 40 for electron ionization and possibly m/z 100 for other ionization methods) are added together as a measure of the total ion current (TIC). The relative contribution of each ion to this total is then calculated as a percentage (%TIC). Table 1.2 records part of the mass spectrum of 1,3-dimethylbenzene from m/z 38 to the molecular ion, with abundances of ions as %TIC. This last method of recording a mass spectrum is used less frequently than the normalization method but has some advantages in emphasizing the relative importance of an ion in the whole spectrum. It is helpful to know this relative importance when the mass spectrometer is used not to record a whole mass spectrum but to monitor ions of only one mass or a small number of masses in a mass spectrum (see later, chapter 7). Also, the percentage ion current method finds some favour in comparing the spectra of isomers, when changes in the relative contributions of certain ions to the total ion current may be sufficient to distinguish between the isomers. The difference between the two ways of recording spectra in tabular form is more apparent than real since the relative abundance of a peak in a normalized spectrum is simply related numerically to its value as a percentage of total ion current. However, if the spectrum has been processed to remove 'background' ions of constant, unwanted impurities such as chemical ionization reactant gases or decomposition products of the stationary phases of gas chromatography/mass spectrometry systems (section 4.4.2), this simple relationship may no longer exist. If such

Table 1.2. *Ion abundances[a] in the mass spectrum of 1,3-dimethylbenzene as %TIC.*

m/z	Abundance	m/z	Abundance	m/z	Abundance
39	2.5	65	1.9	92	2.7
50	1.0	77	6.4	103	2.1
51	3.2	78	1.8	105	10.0
52	1.7	79	2.2	106	21.6
63	1.5	91	35.1	107	1.9

Notes:

[a] Ion abundances of less than 1 per cent are omitted

processing has been performed then the %TIC method is often replaced by an equivalent method: the percentage reconstructed ion current (%RIC). This latter sums the ion current from all consequential ions remaining after subtraction of 'background' ions.

It should be noted that the normalization method is likely to be the less accurate because a small irregularity in recording or measuring the base peak will markedly affect all the other ion abundances normalized against it. The measurement of the height of the base peak is prone to error, caused for example by the recording system becoming saturated. Usually, the percentage of total ion current method is virtually unaffected by such anomalies and is to be preferred, particularly if the mass spectral data are acquired and processed by a computer when the greater numerical difficulty of the method is unimportant. Most data systems allow either or both methods to be selected, with presentation of the spectrum in tabular or diagrammatic form. Table 1.3 shows a computer output listing of the mass spectrum of decahydronaphthalene in both %RA and %TIC forms.

Two main facets of mass spectrometry will be apparent from the above description. (i) In contrast with what happens in most other common methods of physicochemical spectroscopic analysis, some or all of the sample is consumed in mass spectrometry and is not recoverable. However, mass spectrometers are very sensitive instruments and spectra may be obtained from a few nanogrammes or even picogrammes of material. (ii) Unlike other physical methods, mass spectrometry does not deal with a well-defined property of a molecule. The appearance of a mass spectrum depends not only on the compound itself but also upon the interval of time between ionization and detection of ions, upon the initial

Table 1.3. *Ion abundances in the mass spectrum of decahydronaphthalene.*[a]

Mass	%RA	%TIC	Mass	%RA	%TIC
41	64.02	5.25	80	5.99	0.49
42	9.79	0.80	81	86.55	7.10
43	7.19	0.59	82	78.53	6.44
50	0.67	0.06	83	15.70	1.29
51	3.02	0.25	84	17.16	1.41
52	1.82	0.15	85	0.82	0.07
53	12.99	1.07	91	2.14	0.18
54	21.23	1.74	93	2.87	0.24
55	46.84	3.84	94	3.50	0.29
56	25.85	2.12	95	60.05	4.93
57	1.94	0.16	96	99.19	8.14
63	0.46	0.04	97	11.35	0.93
65	3.73	0.31	108	0.81	0.07
66	8.24	0.68	109	30.31	2.49
67	100.0	8.20	110	9.07	0.74
68	59.97	4.92	111	0.42	0.03
69	35.17	2.89	123	0.89	0.07
70	3.19	0.26	137	0.91	0.07
71	0.45	0.04	138	99.35	8.15
77	5.02	0.41	139	10.37	0.85
78	1.08	0.09	140	0.31	0.03
79	9.43	0.77			

Notes:

[a] All peaks over m/z 40 of greater than 0.3 per cent relative abundance are included.

energy distribution in the molecular ions and on the method of ionization, and partly upon the physical characteristics of the instrument and therefore on its design and manufacture.

For these reasons it is not possible to guarantee that a mass spectrum is accurately reproducible from instrument to instrument – even for instruments of the same manufacture when operating under apparently identical conditions. Normally, variations in ion abundance of a few per cent are acceptable in comparing spectra from different laboratories, but generally, good uniformity is found and not too much difficulty is experienced from small inconsistencies. The popular method of identifying compounds by matching sample spectra against standard mass spectra stored in a computer library (section 4.4.3) is a testament to this.

1.3.   MULTIPLY CHARGED IONS

In EI mass spectrometry, ions are generally produced with single positive charges ($z = 1$) but may sometimes have two, three or more charges so that the corresponding m/z ratios correspond to half, one-third, or lower fractional masses. A molecular ion with a single positive charge corresponds to a mass m, but with two positive charges it corresponds to $m/2$. Doubly charged ions are fairly common in the electron ionization mode, especially in the mass spectra of aromatic compounds, but more highly charged ions are rarer. Figure 1.4 shows part of a mass spectrum containing some prominent doubly charged ions amongst the singly charged ones. Assignment of doubly charged ions is obvious when peaks occur at half-integer m/z values but doubly charged ions occur at integer values if their mass, m, is even. In the latter case, some integer peaks may consist of superimposed peaks from singly and doubly charged ions. This should be borne in mind particularly when examining spectra of aromatic compounds and especially if a mass

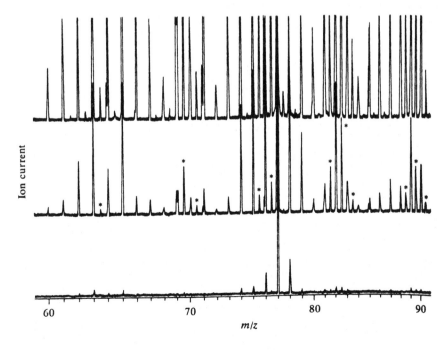

Figure 1.4. Part of a mass spectrum showing peaks due to multiply charged ions marked with an asterisk on the middle trace.

peak occurs at an m/z value corresponding to half the molecular mass of the compound under investigation. The doubly charged species may fragment either into two singly charged ions or into another doubly charged ion with ejection of a neutral particle.

There is one method of ionization, called electrospray (section 3.9), which relies on the formation of multiply charged ions. Large molecules like proteins are protonated at several (z) sites, giving $[M + zH]^{z+}$ ions. Even when the mass, m, is very large, the m/z values are not excessive for a large value of z. This makes ions of very large relative mass, and hence large molecules like proteins, compatible with simple mass spectrometers with upper m/z limits of just a few thousand.

1.4.   ISOTOPES

Many elements in their natural states contain isotopes and, because a mass spectrometer measures mass-to-charge ratios, these isotopes appear in the mass spectrum. The most abundant isotope of carbon is ^{12}C, but natural carbon contains also ^{13}C and ^{14}C. Although as a beta-ray emitter the latter is extremely valuable for radio-tracer work, its natural abundance is so low as to make it almost inconsequential to mass spectrometry. Such is not the case for ^{13}C which occurs in a natural abundance of approximately 1.08 per cent in carbon. Hence, the mass spectrum of methane shows a molecular ion at m/z 16 ($^{12}CH_4$) together with an isotopic ion at m/z 17 ($^{13}CH_4$), the two ions having relative abundances of about 99 : 1. As the number of carbon atoms in a compound increases, so also do the chances of incorporating one ^{13}C atom into the molecule rather than a ^{12}C atom. A compound with ten carbon atoms would yield a molecular ion $M^{+\bullet}$, and an isotopic ion one mass unit greater $[M + 1]^{+\bullet}$, which would be about $10 \times 1.08 = 10.8$ per cent of the abundance of $M^{+\bullet}$ (see, for example, the case of decahydronaphthalene, $C_{10}H_{18}$, for which the ^{13}C isotope peak at m/z 139 is nearly 11 per cent of the abundance of the ^{12}C molecular ion peak at m/z 138; the peak at m/z 140 represents some molecular ions containing two ^{13}C atoms (table 1.3). The chance of finding two ^{13}C atoms in the same molecule increases with increasing numbers of carbon atoms so that $[M + 2]^{+\bullet}$ ions then start to become more prominent. The chances of incorporating more than two ^{13}C atoms in a molecule are very small except with large numbers of carbon atoms and $[M + 3]^{+\bullet}$ ions from this source may be safely ignored. For a compound with ten

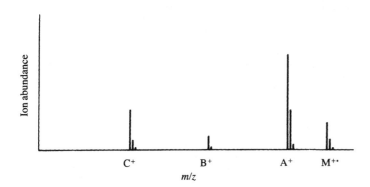

Figure 1.5. The simple mass spectrum of figure 1.1 repeated but with the addition of ^{13}C-isotopic masses for the ions $M^{+\cdot}$, A^+, B^+ and C^+.

carbon atoms, the approximate relative heights of the $M^{+\cdot}$, $[M + 1]^{+\cdot}$, $[M + 2]^{+\cdot}$ and $[M + 3]^{+\cdot}$ peaks are $100 : 10 : 0.45 : 0.01$ from which it can be seen how unimportant are carbon isotope peaks greater than $[M + 2]^{+\cdot}$. Tables are available giving the relative heights of $M^{+\cdot}$, $[M + 1]^{+\cdot}$ and $[M + 2]^{+\cdot}$ peaks for many elemental compositions (Beynon and Williams, 1963) or are incorporated as options in computer software packages for mass spectrometry. All carbon compounds yield molecular and fragment ions accompanied by isotopic ions 1 and 2 mass units greater, making the appearance of the spectrum more complex. Figure 1.5 shows figure 1.1 modified so as to illustrate the appearance of the mass spectrum with ^{13}C isotopes included. Conversely, an estimate of the number of carbon atoms in an ion may be made by measuring the relative heights of the $M^{+\cdot}$ and $[M + 1]^{+\cdot}$ peaks, but the method does not yield good results when there are more than about ten or twelve carbon atoms. Note that, in this discussion, the molecular ion ($M^{+\cdot}$) is considered to contain the isotopes of lowest mass. Then, the notation $[M + x]^{+\cdot}$ can be used for ions containing the isotopes of higher mass. Strictly, $M^{+\cdot}$ and $[M + x]^{+\cdot}$ ions are all molecular ions.

Table 1.4 gives, for some elements commonly met in mass spectrometry, the approximate natural abundances of the more significant isotopes. It should be noted that sulphur, chlorine and bromine have particularly abundant isotopes separated by 2 mass units. For this reason, chlorine- and bromine-containing compounds especially are readily recognized in mass spectrometry and, by examining the isotope pattern in the molecular ion region, the numbers of chlorine and bromine atoms in the original molecule may be determined (Beynon, 1960; Biemann, 1962). Figure 1.6 shows the appearance of the molecular ion region for a dichlorobenzene; note the major peaks spaced 2 mass units apart due to the ^{35}Cl and ^{37}Cl isotopes of the

Table 1.4. *The more important natural isotope abundances for elements commonly occurring in mass spectrometry.*[a]

Element	Isotope (percentage of natural abundance)[b]		
Hydrogen	^{1}H (99.99)		
Boron	^{10}B (19.8)	^{11}B (80.2)	
Carbon	^{12}C (98.9)	^{13}C (1.1)	
Nitrogen	^{14}N (99.6)	^{15}N (0.4)	
Oxygen	^{16}O (99.8)	^{18}O (0.2)	
Fluorine	^{19}F (100.0)		
Silicon	^{28}Si (92.2)	^{29}Si (4.7)	^{30}Si (3.1)
Phosphorus	^{31}P (100.0)		
Sulphur	^{32}S (95.0)	^{33}S (0.7)	^{34}S (4.2)
Chlorine	^{35}Cl (75.5)	^{37}Cl (24.5)	
Bromine	^{79}Br (50.5)	^{81}Br (49.5)	
Iodine	^{127}I (100.0)		

Notes:

[a] Metals frequently possess many abundant isotopes and, because of the diversity of organometallic compounds, a listing of metal isotopes has not been included here. Many reference texts contain listings of isotope abundances; see also Beynon (1960) and Kiser (1965).

[b] With the exception of hydrogen, percentages are given correct to the first decimal place. Trace isotopes are not included because they have little consequence in mass spectrometry at their natural abundance levels.

chlorine atoms and the minor peaks caused by contributions from the ^{13}C isotope. A method of calculating simple isotopic patterns is presented in section 12.2.

Normalized mass spectra are sometimes drawn with the isotopic contributions simplified. For example, for two peaks at m/z 253 and 254 of relative heights as shown in figure 1.7(*a*), the ^{13}C isotope contribution of m/z 253 contributes to the height of the peak at m/z 254; the ^{13}C isotope of m/z 254 occurs at m/z 255. If there were eighteen carbon atoms in the elemental composition of the ion at m/z 253, then approximately $18 \times 1.1 = 19.8$ per cent of its peak height is the ^{13}C contribution in m/z 254. By subtracting this ^{13}C contribution from m/z 254, the relative abundances of the ions appear as in figure 1.7(*b*). Similarly, the ^{13}C isotope peak of m/z 254 at m/z 255 disappears. This device of removing isotope contributions is

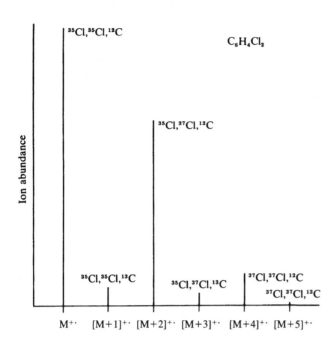

Figure 1.6. Contributions of carbon and chlorine isotopes to the pattern of peaks in the molecular ion region of a dichlorobenzene. The molecular ion (M$^{+\cdot}$) is usually considered to be the one having contributions from the isotopes of lowest mass but, strictly, all six ions are molecular ions.

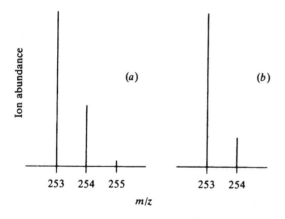

Figure 1.7. The appearance of part of the mass spectrum (m/z 253–255) of a compound having 18 carbon atoms. (*a*) With ^{12}C- and ^{13}C-isotope contributions and (*b*) with the ^{13}C-isotope contribution removed to leave only the ^{12}C-isotopes.

extremely helpful for many organometallic compounds since metals frequently have several isotopes and so yield complex-looking spectra; for example, mercury has seven isotopes spread over 9 mass units. Ions containing metal atoms can therefore afford many isotope peaks and removing their separate contributions simplifies the interpretation of such a spectrum. Figure 1.8 shows the mass spectrum of diethylmercury both with and without the isotope contributions; the spectrum without isotope peaks is much easier to interpret. For many elements, the isotope pattern serves as a

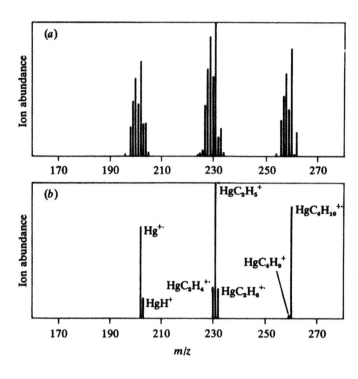

Figure 1.8. A partial mass spectrum of diethyl mercury, (*a*) including all isotope peaks, and (*b*) showing a mono-isotopic spectrum based on ^{202}Hg and ^{12}C.

fingerprint to identify the element and to define how many atoms of it are present in the molecule (section 12.2).

1.5. METASTABLE IONS

The sharp peaks of the normal ions in a mass spectrum are usually accompanied by some much broader, smaller peaks caused by decomposition of *metastable ions*. The maxima of these peaks arising from metastable ion decomposition frequently occur at non-integral m/z values and are easily distinguished from those due to normal ions (figure 1.9). The peak shapes are approximately Gaussian but occasionally flat-topped or dish-shaped ones are found extending over several mass units. The origin and fragmentation of metastable ions are discussed later (chapter 8). These ions are very useful for determining fragmentation pathways. For example, in magnetic sector mass spectrometers, if a metastable ion of mass m_1 decomposes to give an ion of mass m_2, then a peak will be found at an apparent mass, $m^* = m_2^2/m_1$. Therefore, observation of a peak at mass m^* confirms the fragmentation of a *precursor ion* (m_1) to a *product ion* (m_2). In the mass spectrum of

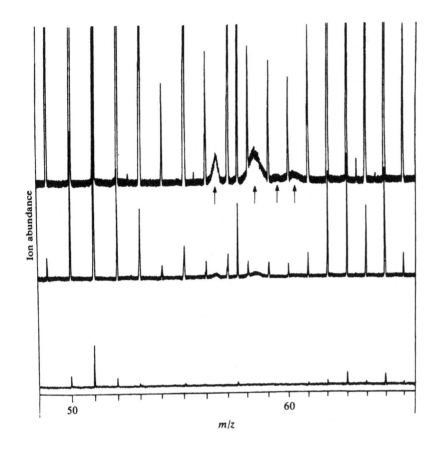

Figure 1.9. Peaks in a mass spectrum (indicated by arrows) representing product ions arising from fragmentation of metastable precursor ions. Normal product ions (produced in the ion source) give rise to narrow peaks at integral mass but product ions from metastable decomposition outside the source are usually seen as broad peaks at non-integral mass. Note also the fairly abundant doubly charged ions at m/z 57.5.

toluene, abundant ions occur at m/z 91 ($C_7H_7^+$) and m/z 65 ($C_5H_5^+$) and the appearance of a broad peak at m/z 46.4 ($= 65^2/91$) indicates that at least some of the ions at m/z 65 arise through ejection of C_2H_2 (26 mass units) from the ion at m/z 91. Computerized acquisition of data has resulted in the deletion of peaks for metastable ions from the routine mass spectrum and the output of spectra from such systems shows only narrow peaks (lines) representing normal ions. However, metastable ion decomposition can be examined conveniently by 'linked scanning' (see later – section 8.3.1).

1.6. ELEMENTAL COMPOSITIONS OF IONS

On the atomic scale, ^{12}C is given a mass of 12.0000 but other elements have fractional masses, e.g. ^1H has mass 1.0078 and ^{16}O has mass 15.9949 (table 1.5). Because of this, the masses of most ions in a mass spectrum fall near but not at integer values. For instance, the molecular ion of acetone at m/z 58

Table 1.5. *Accurate masses of some elements commonly occurring in spectrometry.*[a]

Isotope	Atomic mass	Isotope	Atomic mass
^{1}H	1.007 83	^{29}Si	28.976 49
^{2}H	2.014 10	^{30}Si	29.973 76
^{10}B	10.012 94	^{31}P	30.973 76
^{11}B	11.009 31	^{32}S	31.972 07
^{12}C	12.000 00	^{34}S	33.967 86
^{13}C	13.003 35	^{35}Cl	34.968 85
^{14}N	14.003 07	^{37}Cl	36.965 90
^{15}N	15.000 11	^{79}Br	78.918 39
^{16}O	15.994 91	^{81}Br	80.916 42
^{19}F	18.998 40	^{127}I	126.904 48
^{28}Si	27.976 93		

Notes:

[a] Given correct to five decimal places. Notice how on this atomic scale, with ^{12}C taken as standard, the masses fall either just above or just below integral values.

actually has a mass of 58.0418. The resolving power of a mass spectrometer is a measure of its ability to separate two ions of any defined mass difference. Basically, for two overlapping peaks M_1 and M_2 (figure 1.10), the resolution may be defined in terms of the mass difference (ΔM) between them such that the peaks are said to be resolved if $(h/H) \times 100 \leq 10$, where H is the height of the peaks and h measures the depth of the 'valley' between them. The resolution is then the value of $M_1/\Delta M$ when $(h/H) \times 100$ is equal to 10. For example, when two masses (100.000 and 100.005) are separated by a 10 per cent valley, the resolution of the instrument is 100.000/0.005, i.e. 20 000. There are other, less common, definitions of resolution. An instrument of even modest resolving power can distinguish readily between adjacent integral masses; to separate m/z 100 from m/z 101 a resolution of only 100 is required. A high-resolution instrument, however, can separate an ion at m/z 100.000 from one at m/z 100.005. At low resolution, a mass spectrum consists of a series of peaks at integer m/z values but at medium resolution any of these peaks may be split due to the presence of ions of different elemental compositions. As an example, an ion at m/z 28 might have the composition N_2 or CO or C_2H_4; if all three ions were present, a low-resolution spectrum would show only one peak at m/z 28 but a medium-to-high-resolution instrument would reveal the three ion types at m/z 28.0061, 27.9949 and

28.0313 respectively (figure 1.11). Thus, at higher resolution, small mass differences can be detected and this property is utilized for accurate mass measurement of ions, as described in section 4.4.1.

1.7. APPEARANCE OF THE MASS SPECTRUM

Beginning with the simple picture of ion decomposition, and introducing the complications of isotopes, multiply charged ions, metastable ions, and peak

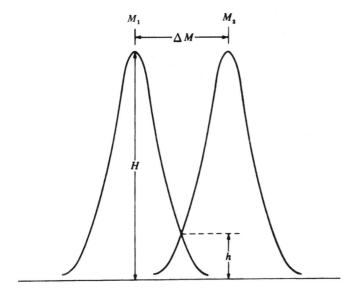

Figure 1.10. Two overlapping ion peaks M_1 and M_2 of height H and overlap h. If $(h/H) \times 100 \leq$ 10, these peaks are considered to be resolved.

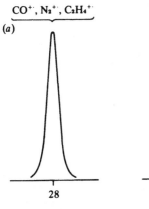

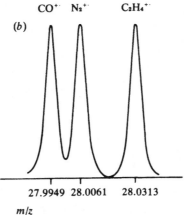

Figure 1.11. The appearance of the molecular ion region of a mass spectrum of a mixture of CO, N_2 and C_2H_4 at m/z 28 with (a) low resolving power showing no separation of masses and (b) medium resolving power (3000) showing good separation.

splitting due to differing elemental compositions of ions, an overall impression of what to expect and what to look for in a mass spectrum is gained. The mass spectrum of aniline has been chosen to illustrate these features (figure 1.12). There is a further, practical complication evident in figure 1.12: because of the great range in heights of peaks from the small ones arising from metastable ion fragmentation to the large base peak, it is necessary to be able to record the spectrum simultaneously at different sensitivities. Figure 1.12 shows three traces increasing in sensitivity from the bottom of the chart upwards. The lowest trace contains all the features of the *routine* or *low-resolution* spectrum. Figure 1.12 illustrates the mass spectrum of aniline *directly recorded*, that is, the amplified electrical signals due to ions arriving at a detector were immediately recorded via galvanometers as traces on fast-response photographic paper. With data systems and work stations, these electrical signals are digitized and processed before being displayed on a visual display unit or on paper via a printer. The processing of data is so fast that there is no perceptible delay between scanning the spectrum and its display on a screen. Figure 1.13 repeats the mass spectrum of aniline (figure 1.12) in typical computerized form. Comparison of figures 1.12 and 1.13 reveals that the ion peaks have become straight lines (sticks), masses of all ions are printed or can be read off immediately, the mass scale is linear rather than exponential and peaks arising from metastable ions have been erased. Apart from this last drawback, the advantages of computerized data processing and recording make this a vastly preferred technique over the old recording methods. Not only is the recording of a mass spectrum faster with the aid of a computerized system but also it is much simpler to compile hundreds or thousands of spectra in a computer store than it would be to save the same number of rolls of photographic paper. Nevertheless, it is worthwhile remembering what the original raw form of a mass spectrum looked like before it was computerized. For example, gaps where no ion peaks appear in a computerized mass spectrum may be due to suppression of information by the data system, as has occurred between m/z 80 and 90 in figure 1.13 compared with figure 1.12.

1.8.   FORMATION OF THE MASS SPECTRUM FOLLOWING IONIZATION BY OTHER METHODS

The description of ion formation and fragmentation appearing above has concerned itself almost entirely with processes resulting from electron

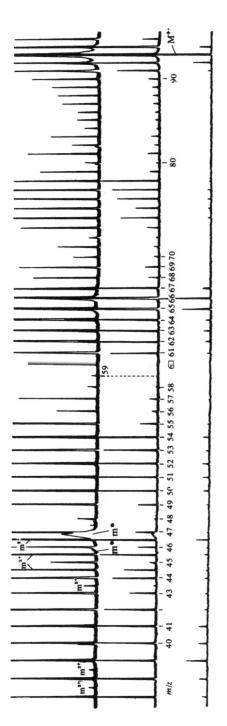

Figure 1.12. The mass spectrum of aniline. The molecular ion ($M^{+\cdot}$) and the ^{13}C-isotope ion $[M + 1]^{+\cdot}$ are shown. Products of metastable ions (m^*) and doubly charged ions (m^{2+}) are marked. Integral mass-to-charge ratios are given below the middle trace where the spectrum may be counted; where a peak does not appear, as at m/z 59, reference to the top trace will often reveal ions of low abundance.

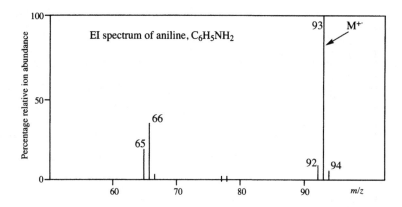

Figure 1.13. This shows the same spectrum of aniline as in figure 1.12 but now in computer-processed form. Note the change in appearance and the loss of some information. However, there is a gain in convenience with a linear mass scale, the masses already counted and marked and the whole spectrum normalized.

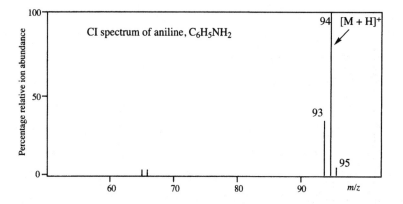

Figure 1.14. A computer output of the mass spectrum of the same substance, aniline, as in figure 1.13 but this time produced by chemical ionization. Note the molecular ion region with abundant $[M + H]^+$ ions and the relative absence of fragment ions compared with the electron ionization spectrum of figure 1.13.

ionization. Other less energetic methods of ionization are often referred to as 'mild' or 'soft'. With these mild techniques, molecular ions are formed which do not possess much, if any, excess of internal energy so that they do not fragment extensively or even do not fragment at all. For example, one of these processes is 'chemical' ionization by which a quasi-molecular ion, called a protonated molecule, $[M + H]^+$, is frequently produced. Often, these ions do not fragment in the mass spectrometer and the mass spectrum is very simple, consisting as it does only of peaks representing the protonated molecule plus any isotopic peaks. An example is illustrated in figure 1.14.

Whilst a clearly defined peak for the protonated molecule is of considerable importance in giving molecular mass information or for showing how many components of a mixture there are, the absence of significant fragmentation proscribes information which would help with structure determination for the original molecule. However, there are methods available for adding excess of internal energy to such quasi-molecular ions so that fragmentation *does* occur (mass spectrometry/mass spectrometry; see later – section 8.7) and structural information can be obtained. The methods of mild ionization are considered in greater detail in chapter 3.

2 Instrument design

2.1. INTRODUCTION

Mass spectrometers are used for a wide variety of studies, from investigations of gas-phase ion chemistry to the authentication of works of art. Mass spectrometers may be used to measure the ages of rocks or skeletons, identify illicit drugs taken by athletes or given to racehorses, probe the chemical topography of new materials, quantify trace elements in semiconductors, measure the masses of proteins, or look for signs of life on other planets. Such a diverse set of applications requires several different types of mass spectrometer and new designs are being developed all the time. Some mass spectrometers easily fit in the boot of a car, others fill an entire room. Therefore, the discussion presented here must be restricted. The types most usually encountered in chemical and biochemical applications will be featured. Instruments for measuring mass spectra are extensively described in the literature and references to such information are included where appropriate. Manufacturers of mass spectrometers frequently publish news releases which provide much detail of specific instruments and can keep the mass spectrometrist abreast of commercial developments.

Whilst diverse designs do exist, the essential features of a mass spectrometer are common and are presented in block form in figure 2.1. Each of these features is clearly seen in the photograph (figure 2.2) of a modern mass spectrometer capable of high resolving powers and used mainly for organic compounds. A brief discussion of each feature in turn is necessary for a proper understanding of how a mass spectrum is produced.

Figure 2.1. Essential features of a mass spectrometer in diagrammatic form. In all such systems, the analyser and detector regions are maintained under vacuum. Conventionally, the ion source is also under vacuum but there are several instruments that utilize atmospheric pressure ionization.

Figure 2.2. A front view of a modern, double-focussing magnetic/electric-sector mass spectrometer (the VG Autospec). 1, A linked gas chromatograph; 2, ion source; 3, direct inlet line from chromatograph into ion source; 4, septum inlet into ion source; 5, direct insertion probe into ion source; 6, magnetic sector; 7, electric sectors; 8, ion detector. Photograph by kind permission of Micromass UK Ltd.

2.2. INLET SYSTEMS

There are several means of introducing samples into the mass spectrometer and the inlet system used will normally depend on the volatility and nature of the sample, the task in hand, and the method of ionization (in particular, the gas pressure in the ion source). The introduction of samples into a mass spectrometer is not a trivial task because, in the most common instruments, the whole system must be maintained at very low pressure (high vacuum) to allow unrestricted movement of ions.

2.2.1. *Cold inlets*

Gases or compounds that are very volatile at room temperature and a pressure of about 10^{-2} Torr are allowed to 'leak' into the mass spectrometer through a glass sinter and led along a glass tube to the ion source. The pressure inside the mass spectrometer is about 10^{-6} Torr.

2.2.2. *Hot inlet systems*

The first of these is very similar to the cold inlet except that it may be heated to about 300°C to volatilize compounds, which are then led along a heated line to the ion source. To prevent metal-catalysed decomposition or rearrangement of a sample, it is preferable to fabricate the whole system from glass, in which case the unit is commonly known as an AGHIS (all-glass heated inlet system). An alternative is the septum inlet, which comprises a heated stainless-steel reservoir into which liquid samples are injected via a septum. At the low pressures and high temperatures employed, the vaporized sample diffuses through a system of valves into the ion source. A septum inlet is shown on the mass spectrometer in figure 2.2. Sophisticated septum inlets exist in which the sample vapour does not come into contact with any metal surfaces or the septum. Both of the hot inlet systems described here and the cold inlet may require up to a milligramme of sample. The hot inlets provide the most convenient methods of 'leaking' into the ion source small quantities of a reference compound, such as perfluorokerosene, required for mass calibration (see section 4.4.1).

2.2.3. *Direct probe inlets*

Compounds that are not sufficiently volatile to be introduced through the cold or hot inlets may be inserted directly into the ion source by means of a probe passing through a vacuum lock. At the low pressures of about 10^{-7}–10^{-5} Torr inside a conventional electron ionization source and with heating, very many compounds are sufficiently volatile to yield good mass spectra. The system is so easy to use and requires so little material (a few picogrammes are quite sufficient) that it is frequently the method of choice even for relatively volatile substances. These direct insertion probes are often heated by radiation from or contact with the ion source itself, but probes are available that may be heated or cooled independently of the source. Compounds with relative molecular masses up to about 2000 may be measured with a direct insertion probe using the common methods of electron and chemical ionization. Direct insertion probes are also used with milder methods of ionization, like fast atom bombardment (see section 2.3.5), allowing much larger molecules to be examined. Relative molecular masses of about 25 000 are readily manageable assuming, of course, that the analyser part of the instrument is commensurate with such a mass.

As a modification, if the probe is designed to project not just into the source, but right up to or slightly into the electron beam used for ionization, the technique is known as *in-beam* or *direct ionization*. This topic will be discussed in greater detail in chapter 3 but it may be noted here that the method is particularly useful for some large, thermally unstable molecules of low volatility such as carbohydrates.

2.2.4. *Gas chromatographic inlets*

Generally, the inlets described above are used when pure, or nearly pure, compounds are available for analysis. The mass spectra of mixtures are usually too complex to be interpreted unambiguously, thus favouring the separation of the components of mixtures *before* examination by mass spectrometry. There are several exceptions to this rule-of-thumb and they are dealt with at the appropriate points in the following chapters. Here, prior separation by gas chromatography (GC) is considered. The effluents from gas chromatographic columns consist of carrier gas, usually helium or hydrogen, admixed with the compounds being investigated. Carrier gas flow rates in modern capillary columns are in the range 0.5–2.0 ml min^{-1}. Remembering that ordinary mass spectrometers operate at very low pressures, this gas flow might appear to be a serious problem for the GC/MS combination. However, good pumping systems in mass spectrometers will maintain an adequately low pressure when the whole of such gas chromatographic effluent is transferred to it, even if an ion source requiring high electric potentials is used. Therefore, the best GC/MS interface is hardly an interface at all! The GC column is simply fed right into the ion source through a gas-tight tube. The GC/MS combination is described in detail in chapter 5.

2.2.5. *Inlets for solutions*

Gas chromatography and traditional mass spectrometry are both gas-phase techniques and, hence, reasonably compatible. The coupling of mass spectrometry and a flowing liquid stream (as with the effluent from liquid chromatography, capillary electrophoresis or flow-injection analysis) involves a different dimension of technical challenge because the liquid effluent must be vaporized and then stripped of solvent before introduction of the solute into an ion source for gas-phase ionization. If the majority of this solvent were not removed, unacceptably high pressures would develop in the ion source. It is mainly because of this difficulty that development of combined liquid chromatography/mass spectrometry (HPLC/MS, or just LC/MS) has lagged far behind that of combined gas chromatography/mass spectrometry. One widespread LC/MS interface is the thermospray device in which the liquid is heated and passed though an orifice to form an aerosol or 'spray' (nebulization). Rapid pumping removes most of the solvent and ionization is effected by chemical reactions. An adaptation of thermospray, the inaptly named plasmaspray, works on similar principles except that ionization efficiency is enhanced through generation of a plasma in the vaporizing solvent from the spray.

There are at least four other major methods for introducing liquids into mass spectrometers and, unfortunately, none is ideal for every problem that an LC/MS laboratory might wish to undertake. For example, it is possible to sweep a liquid into an ion source designed for ionization direct from the liquid phase. One such method is called *continuous-flow* or *dynamic fast atom bombardment*. Alternatively, the traditional low-pressure ionization techniques might be abandoned in favour of atmospheric pressure ionization (Bruins, 1991). Technical difficulties are much reduced when the ion source is tolerant of higher pressures. The *electrospray* is one device that comes into this category and it has been shown to produce mass spectra from low-molecular-mass materials up to solution-phase proteins with relative-molecular-masses of 300 000 and more. There is also a form of chemical ionization that can be conducted at atmospheric pressure; this too is used for combined liquid chromatography/mass spectrometry. All of these, and related, ionization techniques are described in detail in the next chapter, and their application to chromatography/mass spectrometry is discussed in chapter 5.

There is a separate approach that seeks to eliminate the liquid solvent almost entirely before the dried analyte is physically transported into the ion source. The solution is placed on a moving belt system, which differentiates between solvent and solute on the basis of their relative volatilities (section 5.3.1). The continuous moving belt is applicable to combined liquid chromatography/mass spectrometry and to rapid analysis of pure, solid samples. By depositing solutions of each sample successively onto the belt, they can be analysed at the approximate rate of one every 20 s. The fastest turnover time possible with the direct insertion probe is of the order of 5 min. The improved rate of analysis is used to best advantage with a computer that can easily process and output mass spectra at this rate.

Despite the technical difficulties, the goal of using a mass spectrometer as a *universal* detector for compounds in flowing liquid streams is worth pursuing (Yergey, Edmonds, Lewis and Vestal, 1990). It would enable analysis of mixtures without the constraints of volatility imposed by GC/MS, but with the routine reliability already realized with the latter. Such a universal liquid-phase coupling would suffice for LC/MS, capillary electrophoresis/mass spectrometry (CE/MS) and rapid automated analysis by injecting samples successively into the flowing stream, as in flow injection analysis (FIA).

Between the gaseous and liquid phases, there are supercritical fluids (Smith, 1988*a*). Supercritical fluid chromatography has also been linked to mass spectrometry, and this less common combined technique is discussed briefly in chapter 5.

2.2.6. *Total inlet systems*

A mass spectrometer may have more than one inlet. The mass spectrometer in figure 2.2 is typical in having four sample inlets: a direct probe, a septum inlet, a direct line for capillary gas chromatography and an LC/MS interface. Additionally, there is an inlet line for the reactant gas when the instrument is operating in the chemical ionization mode. In some situations it is desirable to use more than one inlet simultaneously; this is possible as long as the source pressure remains within limits suitable for the type of spectrometer being used. For instance, the instrument in figure 2.2 will perform adequately when coupled to the gas chromatograph, with a reactant gas for chemical ionization in the source and a reference compound 'leaking' into it from the septum inlet. Even at the relatively high pressure resulting from these simultaneous procedures, there is no serious arcing of the high (4 kV) electric potential in the source. For accurate mass measurement, it is important to be able to ionize both the sample under investigation and a reference compound at the same time, necessitating at least two inlets on a high-resolution mass spectrometer.

2.3. ION SOURCES

An ion source may be defined simply as the region in which ionization occurs. As shown in figure 2.3, the region is usually enclosed in a small ion chamber in which the sample is ionized. The ions produced are propelled out of the chamber towards an exit slit by a low positive potential applied to a 'repeller' plate. On leaving the ion chamber of a magnetic sector mass spectrometer, the ions are accelerated through a high potential of 2–8 kV and passed into the analyser for separation according to mass-to-charge

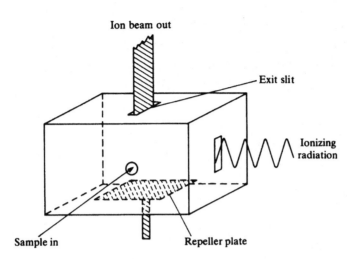

Ion beam out

Exit slit

Ionizing radiation

Sample in

Repeller plate

Figure 2.3. The essential features of an ion source for electron ionization or photo-ionization, in schematic form.

ratio. Different types of analyser have somewhat different arrangements. For example, the ion source of a quadrupole mass spectrometer is essentially similar, but the ions produced are allowed to drift out of the source under the influence of a small electrostatic field so that they have little kinetic energy on entering the quadrupolar field. In the ion trap mass spectrometer, ions are formed by pulses of electrons and then analysed in the same region. The lack of high potentials in the sources of these last two instruments makes them particularly suitable for chemical ionization and combined gas chromatography/mass spectrometry when higher pressures are required. They also have an economy of space such that they can be designed as bench-top instruments. These, and other, analysers are considered in section 2.4.

The neutral species (N) produced by fragmentation following initial ionization are, to all intents and purposes, unaffected by the fields inside the source and are not recorded by the mass spectrometer except in certain specialized instruments, which are outside the scope of this book. There are many means of ionizing compounds and the commonest of these are shown in table 2.1. These methods are dealt with briefly below and a selection of them is discussed in more detail in the next chapter.

2.3.1. *Electrons*

The commonest method of ionization is by means of electron ionization (previously and still sometimes called electron impact, but abbreviated to EI in either case). Energetic electrons from a heated filament are accelerated by

Table 2.1 *Agents bringing about ionization. The last four entries are used for elemental analyses. The acronyms are explained in the main text.*

Agent	Example
Electrons	EI
Reactions of ions	CI, APCI
Electric fields	FI, FD, EHI
Electric fields with spraying	Thermospray, electrospray, etc.
Rapidly moving atoms	FAB
Rapidly moving ions	SIMS
Photons	Laser desorption, MPI, etc.
Nuclear fission products	Plasma desorption
Inductively coupled plasma	ICP/MS
Glow discharge	GDMS
Heat	Thermal ionization
Electrical spark	SSMS

an electric field through the small volume (the ion chamber) that contains the gaseous sample. Such an electron and a volatilized sample molecule (M) will react if they pass sufficiently close for the electron to impart enough of its energy to the neutral molecule. Note that an electron is too small for this interaction to be viewed as a real 'impact' on a molecule. Not all of the energy of the electron is transferred but, if it has sufficient energy, it can cause an electron to be ejected from the molecule, forming a molecular ion.

$$M + e^- \longrightarrow M^{+\bullet} + 2e^-$$

This is not the only process that occurs in the ion source, as chapter 3 will show. In addition, some of the molecular ions ($M^{+\bullet}$) will have gained such an excess of energy that, being in a vacuum and hence not able to remove this energy by collisions, they fragment. This behaviour is a good thing for identification of compounds by mass spectrometry. If ionization gave just molecular ions, only the relative molecular masses of samples could be measured directly. Of course, one can no more identify a molecule by knowing just its mass than one could identify a human on the sole basis that he/she weighs 70 kg. So, some fragmentation is helpful because it is diagnostic of structure, but many compounds, particularly the more fragile ones, fragment too much, so that their mass spectra comprise peaks for fragment ions only, all of the molecular ions having fragmented before they reach the detector.

Before the problem of excessive fragmentation is addressed through use of milder methods of ionization like chemical ionization, it is worth emphasizing the need for keeping the analysers of mass spectrometers under high vacuum. It should now be clear that EI mass spectrometry investigates energetic charged species (ions), which would react rapidly with air and be quenched, or at least be scattered at high pressures, and therefore never reach the detector. Therefore, in the EI method, ions need to be generated in a high vacuum. However, in the following sections it will become apparent that not all ion sources are maintained at very low pressures.

2.3.2. *Chemical ionization*

The collision of an ion and a molecule can lead to a reaction giving a new charged species (Munson and Field, 1966; Mather and Todd, 1979; Harrison, 1989). For example, the reaction between a methane ion and a methane molecule gives rise to the unusual but fairly stable CH_5^+ species.

$$CH_4^{+\bullet} + CH_4 \longrightarrow CH_5^+ + CH_3^{\bullet}$$

To effect reactions like this, termed *ion/molecule reactions*, in a conventional mass spectrometer, it is necessary to operate at source pressures of about 0.1–1.0 Torr. That is, the instrument is still under vacuum but at much greater pressure than for EI sources. It is also necessary to construct the ion chamber in such a way that a pressure of 0.1–1.0 Torr may be maintained within it whilst the pressure immediately outside it is 10^{-5}–10^{-4} Torr. At a pressure of 0.1–1.0 Torr, there are many ion/molecule collisions before the ions leave the ion chamber and collisional equilibration of energies may be expected. Because of this, the ions probably possess a distribution of internal energies similar to the thermal distributions found in solution chemistry. These so-called reactant gas ions ($[A + H]^+$, such as CH_5^+ above, or NH_4^+ from ammonia) can ionize other neutral molecules (M), usually by proton transfer, giving protonated molecules:

$$[A + H]^+ + M \longrightarrow A + [M + H]^+$$

By far the most popular reactant gases for chemical ionization (CI) are ammonia, isobutane and methane. Other gases and conditions can be used to make CI a very versatile method of ionization. For instance, chemical ionization can be brought about at lower pressures in mass spectrometers that trap ions and at atmospheric pressure in 'atmospheric pressure ionization mass spectrometers'. The topic is covered in greater depth in the next chapter.

At the higher pressures conventionally employed for CI mass spectrometry, three-body collisions are more frequent and electron capture becomes fairly efficient so that production of negative ions is just as likely as production of positive ions. Hence, negative-ion CI is also a popular technique in which there are two basic ionization processes: electron capture,

$$M + e^- + B \longrightarrow M^- + B$$

$$A^- + M \longrightarrow [M - H]^- + AH$$

and reactant ion ionization,

In the electron capture process, a neutral molecule (B) is the third body needed to remove excess of energy from the collision complex of a neutral molecule (M) and an electron so as to give a stable negative ion (M^-). In the reactant ion process a negative ion (A^-), such as OH^- (which may be produced by electron bombardment of a mixture of methane and nitrous oxide) is used to abstract a proton from a molecule (M) to give an $[M - H]^-$ ion. The sensitivity of positive-ion CI is similar to that of positive-ion EI mass spectrometry, whilst negative-ion CI mass spectrometry has been claimed to be many times more sensitive than either of them, with femtogramme quantities of material giving good spectra. However, such sensitivities are recorded with atypical, highly electrophilic molecules, which favour negative-ion over positive-ion production. A more objective comparison suggests that negative-ion mass spectrometry is two or three times more sensitive than positive-ion mass spectrometry in the CI mode, but these figures are strongly dependent on the structure of the sample. Both negative- and positive-ion CI techniques, and instruments capable of performing both more or less simultaneously, will be discussed in chapter 3.

2.3.3. *Electric fields*

By applying a large positive electric potential to an electrode in the shape of a point, sharp edge or fine wire, a very high potential gradient is produced around the regions of high curvature. The electric field (molecular orbitals) of a molecule experiencing such large potential gradients is distorted, and quantum tunnelling of an electron can occur from the molecule to the positively charged electrode; this is field ionization (Beckey, 1977). The positive ion so formed is immediately repelled by the positive electrode into the analyser of a mass spectrometer. Although this is an effective way of producing ions, there are difficulties in focussing them into a beam. Because the initially produced ions are repelled by the high positive potential on the electrode immediately after formation, their residence time in the source region is short (10^{-12} s) compared with the residence time of ions in an electron ionization source. There are two different techniques employing this mode of ionization, termed field ionization (FI) and field desorption (FD). In the former method (section 3.7), the sample in the gas phase is passed through the electric field to cause ionization. It is, like EI and CI techniques, suitable for combined gas chromatography/mass spectrometry although it is not so sensitive. In the field desorption technique, the sample is first coated onto the electrode (called the emitter) so that ionization occurs in the condensed phase on or near the surface of the emitter. The precise mechanism of ionization by FD is complex and has given rise to controversy (Holland, Soltmann and Sweeley, 1976; Beckey and Röllgen, 1979; Holland, 1979; Beckey, 1979). The FD mass spectrometric method is particularly suitable for analysing thermally unstable compounds of large relative molecular mass (section 3.8).

In a third use of electric fields, electrohydrodynamic ionization (EHI), the sample is contained in a relatively involatile solvent such as glycerol and introduced to the ion chamber through a capillary tube to which is applied a large voltage. Compounds are ionized in the intense electric field without the application of heat, making it attractive for heat-sensitive materials (Simmons, Colby and Evans, 1974; Stimpson and Evans, 1978). The technique can be used for combined LC/MS.

2.3.4. *Electric fields with spraying*

Another process, particularly suited to liquid chromatography/mass spectrometry and popularly called *thermospray*, has been discovered (Vestal, 1983). A solution of the sample is vaporized rapidly by the application of heat, forming solvent vapour and a 'mist' of droplets, *viz.*, the solution is nebulized. Most of the volatilized solvent is removed by rapid pumping. The droplets, consisting of the solvent, a volatile electrolyte and the involatile solute, are charged. The electrolyte is a source of reactant ions that ionize the sample molecules. As solvent vaporizes from these charged droplets under the influence of the vacuum, the droplets decrease in size and electrical charge density increases. This electrical field, as in FI above, may facilitate the evaporation of the sample ions from the aerosol droplet. This process of generating gaseous ions from charged droplets is referred to as 'ion evaporation'. The exact mechanism of ionization in thermospray is still a matter of debate and the method could equally have been included in the section on chemical ionization because the nature of the ions observed suggests that gas-phase reactions of the molecules with ions of the electrolyte may play the dominant role. For example, in the presence of ammonium ethanoate as electrolyte, $[M + H]^+$ or $[M + NH_4]^+$ ions of polar molecules dominate the spectra. Thermospray should be considered as a descriptive name for the way in which the charged droplets are formed. Two major mechanisms of ionization for polar molecules appear to operate in the thermospray device: (i) a form of chemical ionization, and (ii) ion evaporation, which is caused by a chemically generated electrical field.

An extension to the thermospray method uses an external agent as a means of producing additional ionization, i.e., of increasing ion yield and therefore the sensitivity of the method. Solvent-induced chemical ionization can be promoted by bombarding the spray with electrons from a filament or by passing the aerosol spray through a glow discharge. The former is usually given the matter-of-fact name of *filament-on thermospray* and the latter is often called *plasmaspray*. The filament method suffers in comparison with plasma discharge because of the short lifetime of the filament itself. Both are utilized for LC/MS studies.

Electrospray is an atmospheric-pressure relation of electrohydrodynamic ionization or thermospray. The analyte solution is passed through a capillary tube, which has a potential difference of a few thousand volts between its tip and a nearby counter-electrode. The emerging liquid is charged by the field and hence dispersed into a mist of droplets. The solvent is removed in a flow of dry gas and, as in thermospray, charge accumulates on the surface of a droplet as it decreases in size. Again, the precise mechanism of ion formation is a matter of debate (Mann, 1990; Smith, Loo, Edmonds, Barinaga and Udseth, 1990; Mann and Fenn, 1992), but the electric field aids the evaporation of ions from the small droplets. This should now be familiar as 'ion evaporation'. The resulting gas-phase ions are sampled through nozzles into the high vacuum part of the mass spectrometer. The analyser could be a magnetic sector instrument but is more likely to be a quadrupole, ion trap or Fourier-transform mass spectrometer. Electrospray has the remarkable property of being applicable to small polar molecules *and* making very large molecules, like proteins with relative molecular masses over 200 000, amenable to study with a quadrupole mass spectrometer that has an upper limit of about m/z 4000! The trick will be explained in the next chapter.

2.3.5. *Rapidly moving particles*

Solid surfaces, solid samples coated on a surface and samples in involatile solvents can be bombarded with atoms (fast atom bombardment, FAB) or ions (secondary-ion mass spectrometry, SIMS) to cause ionization. The case of solution-phase analytes is particularly important for organic chemists and biochemists. Enough of the kinetic energy of rapidly moving atoms or ions is deposited in the solution to promote sputtering of ions (called 'secondary ions') into the mass spectrometer, as well as some evaporation of the solvent. Note that ionization occurs directly from the solution, so the sample does *not* have to be vaporized first. Therefore, these bombardment methods are applicable to involatile polar and ionic compounds such as peptides, oligosaccharides and phosphates. Additionally, if the solution being bombarded is continuously flowing and has come from a liquid chromatograph, the potential for combined LC/MS becomes clear. Actually, it makes no difference whether the bombarding particles are atoms (usually xenon) or ions (usually Cs^+) and this commonality of mechanism has resulted in attempts to match the names of the two key methods. Now, fast atom bombardment is sometimes referred to as 'liquid SIMS' or LSIMS ('liquid' referring to the analyte phase) and SIMS is sometimes called fast ion bombardment (FIB)! The end result is four names in widespread use instead of two. In this text, the traditional nomenclature of FAB and SIMS is used.

2.3.6. *Photons*

The ionization energies of most substances lie below 13 eV so that high-intensity, short-wavelength radiation can be used for gas-phase ionization. At energies greater than about 11 eV there are no materials with which to make windows to transmit the radiation into the source. Therefore, it is necessary to arrange for the photon source to emit radiation directly into the ion source. The helium discharge at 21.21 eV provides a convenient source of photons able to ionize all compounds. Unless a high intensity of radiation is attained inside the ion chamber, low ion yields are obtained with consequent low sensitivity of the mass spectrometer. Ions formed in a photo-ionization source reside there for about 10^{-6} s before acceleration and analysis. The utility of such sources for general mass spectrometry is limited for they offer little or no advantage over EI sources.

Lasers have been used to desorb involatile and thermally unstable molecules and ions, and to ionize neutral molecules once these have been desorbed from surfaces (Lubman, 1990). Photon-induced desorption is typically effected with a short pulse from a laser providing ultraviolet or far-infrared radiation. The ionization step may be an integral part of the desorption process (*laser desorption/ionization*) or a separate process. Neutral compounds lifted off (ablated from) the surface can be ionized with a pulse of electrons (EI), by reaction with ions (CI) or by a second laser pulse. It is possible to ionize and fragment selectively by irradiating with wavelengths specific to the desorbed species (*multiphoton ionization*, MPI, and *resonance-enhanced multiphoton ionization*, REMPI). Biomolecules with relative molecular masses of about 500 000 are amenable to a method termed *matrix-assisted laser desorption*, which is described in the next chapter. The ions, generated in pulses of a few nanoseconds, are ideal for analysis according to their mass in time-of-flight, ion trap or Fourier-transform mass spectrometers (section 2.4).

2.3.7. *Nuclear fission products*

Thermally unstable compounds, often biomolecules, which do not give good mass spectra even with field desorption ionization, may yield to the unique method of plasma desorption (Macfarlane and Torgerson, 1976; Cotter, 1988). Spectra of proteins with relative molecular masses up to about 45000 have been recorded by this means. Fission fragments of the radionuclide ^{252}Cf interact with the sample, usually coated on a nitrocellulose support, causing ionization (section 3.12). The method employs a time-of-flight analyser (see the following section). The technique is commercially available and the whole process of sample preparation and operation of the instrument is very simple.

2.3.8. *Combined sources*

It is often advantageous to be able to generate mass spectra of compounds by two or even three ionization methods to yield complementary information.

Combined electron ionization/chemical ionization sources are most commonly used. It must be stressed that, because different source designs suit different ionization techniques, these combined sources are compromises. The EI/CI source will give adequate sensitivity in both EI and CI modes, but will give ultimate sensitivity in neither. The larger mass spectrometers are expensive instruments, so there are advantages in possessing one unit that can perform in different modes. Some other reported source combinations are electron ionization/field ionization, electron ionization/chemical ionization/field desorption with or without the capability for field ionization, field desorption/electron ionization, field desorption/field ionization, thermospray/electron ionization and positive-ion chemical ionization/negative-ion chemical ionization. Undoubtedly the most popular and most advanced combined source is that of EI/CI. This has reached a point at which the switch from EI to CI mode (and *vice versa*) can take less than a second in some designs, permitting alternate scans of the mass spectra in different ionization modes. This technique is put to best use in combined gas chromatography/mass spectrometry in which it is possible to obtain several electron ionization and chemical ionization mass spectra in the time that it takes for a single compound to elute from the chromatographic column. With many classes of compound, complementary information can be gleaned from the two types of mass spectrum, leading to a surer and far more thorough analysis than when using either alone.

The more specialized ion sources, like plasma desorption, cannot be combined with other methods of ionization and require a dedicated instrument. Dedicated systems, frequently bench-top devices, for specialized uses like matrix-assisted laser desorption or electrospray are becoming increasing popular.

2.3.9.   *Other methods*

An introductory text cannot cover in any detail the many less common sources in current use, but some of these are mentioned briefly here.

For negative-ion formation, a low-pressure argon gas discharge will cause electron capture by a sample on which the discharge is focussed (von Ardenne, Steinfelder and Tümmler, 1971).

Quantitative, and some qualitative, analysis of inorganic materials is effected mainly by the last four techniques shown in table 2.1, although secondary-ion mass spectrometry, laser methods and fast atom bombardment also have a role to play (Pratt, Eagles and Self, 1987; Adams, Gijbels and Van Grieken, 1988).

In inductively coupled plasma mass spectrometry (Beauchemin, 1991; Sargent and Webb, 1993), solutions of analyte are introduced into the ion source as a gas-borne aerosol at atmospheric pressure. The aerosol punches a hole through the plasma flame and the high temperatures dry the aerosol, volatilize and atomize the particles, then excite and ionize atoms with ionization energies below about 16 eV. That includes nearly all of the elements, excepting fluorine, helium and neon. It is used for trace analysis, for example, to detect and quantify toxic elements in water.

A conducting solid sample can be made into an electrode, or placed in a hollow graphite electrode, then placed in a vacuum glow discharge. Application of a potential difference in argon causes argon ions from the resulting plasma to strike the analyte, sputtering neutral species. These particles, which are mainly atomic, are ionized in the discharge, then analysed according to their masses. The method is called glow discharge mass spectrometry (GDMS) and is particularly valuable for studies of alloys and semi-conductors (Harrison, Barshick, Klingler, Ratliff and Mei, 1990).

Conceptually, the simplest ionization method of all is thermal. It is based on the likelihood of ions sputtering when inorganic ionic compounds are heated on a filament. Modern instruments are sophisticated, highly precise and accurate, but sample preparation tends to be lengthy. It is used in geochemical, food, nuclear studies and archeological dating, amongst others (Pratt, Eagles and Self, 1987).

Spark source mass spectrometry (SSMS) is a long-established method for inorganic analysis (Harrison and Donohue, 1989). Solid samples are compressed into rods and a large pulsed radiofrequency voltage applied. The resulting electrical spark causes evaporation and ionization of some of the sample. The ions are separated in a magnetic sector mass spectrometer to provide a sensitive, but not quantitatively accurate, elemental analysis.

2.4. ANALYSERS (SEPARATION OF IONS)

As the previous section showed, for substances ranging from the elements themselves to large biomolecules, there is a wide variety of methods for converting neutral species into ions. It is probably not surprising that a single type of analyser cannot cope with this wide range of ions and, in fact, analysers come in several different guises. All ions are analysed for their mass-to-charge ratio (m/z) values.

The separation of ions according to their m/z ratios can be achieved in a number of ways, most notably with magnetic and electric fields alone or combined. Most of the basic differences between the various common types of mass spectrometer lie in the manner in which such fields are used to effect separation. It is not possible here to describe fully the many different ways in which the fields are utilized and only commercially available systems are discussed. First, a description follows for a system typically used in mass spectrometry, the double-focussing magnetic-sector instrument and then some related systems based on magnetic sectors are described briefly as variants of the basic design. The essential properties of quadrupole, ion trap, Fourier-transform and time-of-flight analysers are given. There is an additional section on ion cyclotron resonance methods in section 12.3.

2.4.1. *Magnetic sector mass spectrometers*

Double-focussing analysers. After a beam of ions has been accelerated through a potential of 2000–8000 V away from the ion source, it is divergent and metal slits are used to reduce the spread (figure 2.4). The ion beam then passes between a pair of smooth, cylindrically curved metal plates called an electric sector, where the electric potential maintained across the plates causes the ions to follow an arc and to separate according to their translational energies. Ions of the same energy are brought to a line focus at a second slit, the monitor (figure 2.4). Therefore, the electric sector disperses ions with different translational energy whilst focussing ions of like energies. If V is the potential through which the ions are initially accelerated and E is the field in the electric sector, then equation (2.1) is obtained,

$$R = 2V/E \qquad\qquad (2.1)$$

where R is the radius of curvature of the ion path. The two equations governing the motion of the ions through this energy analyser are those relating their kinetic energy, $zeV = mv^2/2$, and centrifugal force, $zeE = mv^2/R$, where v is the velocity of an ion of mass m and number of charges z, and e is the charge on an electron; simple combination of these two equations governing ion motion gives equation (2.1). Ions accelerated through a potential (V) and passing through a uniform field (E) have the same radius of curvature (R)

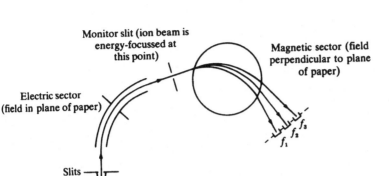

Monitor slit (ion beam is energy-focussed at this point)

Magnetic sector (field perpendicular to plane of paper)

Electric sector (field in plane of paper)

Slits

Ion source

Figure 2.4. Flight paths of ions in a double-focussing mass spectrometer of Nier–Johnson geometry; f_1, f_2 and f_3 are foci 1, 2 and 3.

irrespective of mass-to-charge ratio. Therefore, if the field E is kept constant, the electric sector sorts ions according to their translational energies.

When the narrow energy-focussed ion beam from the electric sector is passed through a magnetic field, the combination of fields can be made to converge ion beams with different kinetic energy and angular divergence at the detector. Hence, the combination of electric and magnetic sectors is called a *double-focussing* device because it provides both energy and angular focussing.

It is in the magnetic sector that mass separation is effected. As described above, after acceleration from the ion source, the ions possess a translational energy, $zeV (= mv^2/2)$. On passing through the magnetic field, the ions experience a centrifugal force (mv^2/r, where r is the radius of curvature) due to the constraint ($Bzev$) imparted by the magnetic field (B). From these quantities, equation (2.2) readily follows:

$$zeV = mv^2/2 \qquad Bzev = mv^2/r$$

and therefore

$$m/z = B^2r^2e/(2V) \qquad (2.2)$$

At constant values of the magnetic field (B) and the accelerating potential (V), each value of m/z will correspond to a radius of curvature r. Thus, after separation in the magnetic field, singly charged ions of masses m_1, m_2, $m_3 \ldots m_n$ will be brought to foci $f_1, f_2, f_3 \ldots f_n$ (see figure 2.4) corresponding to radii of curvature r_1, r_2, $r_3 \ldots r_n$. If an electrical detector, for example an electron multiplier (see section 2.5), is placed at f_1, then arrival of ions (m_1/z) will be recorded and, if other detectors were placed in an array at f_2,

$f_3...f_n$, the ions m_2/z, $m_3/z...m_n/z$ would also be recorded. Although array detectors are used for some purposes, it is simpler to vary the magnetic field (B) continuously so that each ion of different m/z value is brought to a focus at f_1 in turn, i.e., the radius of curvature (r) is kept constant; this is magnetic scanning of the spectrum. It is possible also to vary the accelerating voltage (V), with B constant, to bring each ion with a particular m/z value to the same focus (see equation (2.2)). This method of voltage or electrical scanning is less desirable than magnetic scanning because the sensitivity of the instrument decreases with any decrease in the accelerating voltage. Note that, strictly, the magnetic sector focusses ions according to their momentum, mv, as is discerned from the equation rearranged from that given above: $Bzer = mv$.

The combined use of electric and magnetic sectors described here is just one example of a double-focussing mass spectrometer. This particular layout of electric and magnetic fields (figure 2.4) is known as a *Nier–Johnson* geometry. A mass spectrometer employing this geometry is shown in figure 2.2. The energy-focussing electric sector allows this type of instrument to operate at high resolving powers and most accurate mass measurements are carried out on double-focussing spectrometers. According to equation (2.2), the larger the mass m, the greater must be the magnetic field strength to bring it to focus. This places an upper limit on the mass which can be examined, defined by electromagnet technology. Many modern magnetic sector analysers can scan up to mass 16 000 without difficulty, and some can handle even larger m/z values.

Other layouts of electrical and magnetic fields are met in other types of mass spectrometer, but the basic aim remains the separation of ions according to m/z value with as great a resolution as is needed or is possible. Less common instruments using *Mattauch–Herzog* geometry have their fields arranged as in figure 2.5. To detect the ions, a photographic plate or electron multiplier array may be placed in the plane of the ion foci $f_1, f_2, f_3...f_n$ (figure 2.5) and the magnetic field kept constant. The ions then give rise to narrow bands on the photographic plate and the density of these bands is a measure of the number of ions arriving. The sensitivity range of a photographic plate is limited and it is usually necessary to make several exposures for increasing periods of time in order to record adequately all ions from the most to the least abundant. The photographic plate method of detection is clumsier than an electrical method since after each spectrum the plate must be developed. Also, the density of the bands must be found with a second instrument, a microdensitometer, which

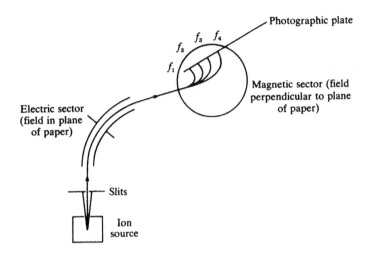

Figure 2.5. Flight paths of ions in a double-focussing mass spectrometer of Mattauch–Herzog geometry.

scans the bands and gives out optical density readings as electrical signals (section 2.5). One big advantage of the photographic plate is that fluctuations in ion abundances during the recording of a spectrum are immaterial since all ions are recorded simultaneously whereas, with a single electrical detector and magnetic scanning of the spectrum, fluctuations in ion abundances during the scan result in an alteration in the appearance of the spectrum. Electron multiplier arrays do not present the same difficulties as those of the photographic plate but resolution into m/z values is limited by how small and close together the elements of the array can be manufactured.

Single-focussing analysers. If the electric sector is omitted from the instrument it becomes a *single-focussing* mass spectrometer. Since the divergent ion beam leaving the source is not focussed for kinetic energy, this type of instrument produces only low-resolution spectra because ions of the same mass but different translational energy are not brought to a point focus. Such an instrument can usually distinguish integer m/z values up to masses of about 5000 without difficulty.

Reverse-geometry instruments. If the magnetic sector precedes the electric sector (the reverse of the situation in figure 2.4), the resulting mass spectrometer is still capable of performing in the same way as with conventional geometry, but it has additional advantages. These are concerned with the utilization of metastable ions and so discussion of this reverse-geometry is postponed until the origin of such ions has been detailed (see chapter 8).

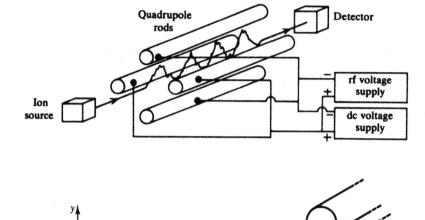

Figure 2.6. The arrangement of the quadrupole mass filter, showing schematically the complex flight path of a focussed ion beam. Ions that are not focussed collide with the rods.

2.4.2. *Quadrupole mass spectrometers*

All of the analysers discussed above are magnetic sector instruments. The quadrupole, on the other hand, functions in an entirely different way and has a different set of advantages.

Four precisely parallel rods are arranged as in figure 2.6. Between each pair of opposite and electrically connected rods, separated by a distance $2r_o$, is applied a DC voltage (U) and a superimposed radiofrequency (RF) potential ($V\cos(\Omega t)$), where V is the maximum amplitude and Ω the angular frequency of the RF voltage and t is time). Ions are propelled from the ion source into the quadrupole analyser by a small accelerating voltage (about 5 V, as opposed to several thousand volts in a magnetic sector instrument) and, under the influence of the combination of electric fields, the ions follow complex trajectories (Todd and Lawson, 1975; Campana, 1980; Todd, 1984; Dawson, 1986). If the subscript u represents motion in either the radial (x–y) plane or the axial (z) direction, then the parameters a_u and q_u are related to the amplitudes of the potentials by the following two equations:

$$a_u = 8eU/(mr_o^2\Omega^2)$$

$$q_u = 4eV/(mr_o^2\Omega^2)$$

The oscillations of ions in the quadrupole analyser may have finite amplitude, in which case the ion trajectory is said to be 'stable' and the ions are

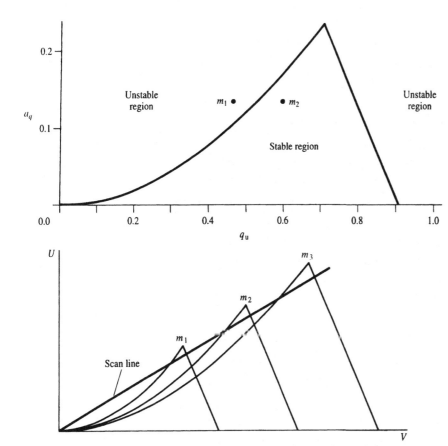

Figure 2.7. A part of a stability diagram for a quadrupole analyser.

Figure 2.8. Stability diagrams for ions of different masses, transformed onto a (U, V) plot.

transmitted through to the detector. On the other hand, the oscillations may be unstable, that is, the amplitude becomes infinite and the ions oscillate wildly. In this case, the ions collide with the rods and hence do not reach the detector. The values of a_u and q_u determine whether ions of mass m will lie in the stable region or not. A plot of values of a_u and q_u is called a stability diagram and a portion of one of these is given in figure 2.7. When the two equations are solved for the ion of mass m_1 shown in the stability diagram, it is seen to lie outside of the stable region and it will not be transmitted through the quadrupole. An ion of different mass, m_2, would execute stable oscillations: its trajectory remains within the bounds of the quadrupole and it passes through and on to the detector.

It is helpful to plot a different type of stability diagram in which the a_u and q_u values are replaced by U and V according to the equations above. The shape of the diagram is the same as in figure 2.7 but each ion of different mass-to-charge ratio will possess a different stable region. In figure 2.8, this is indicated by the three regions for ions of masses, m_1, m_2 and m_3. In the same diagram, a scan line is shown, which corresponds to the way in which the analyser is operated: with the DC/RF voltage ratio (U/V) fixed, the voltages are varied. Hence, the scan line has a fixed slope (i.e. constant a_u/q_u ratio). As the values of U and V are changed, with a constant ratio, the scan line in figure 2.8 is followed, bringing ions of different masses, m_1, m_2 and m_3 into their stable regions sequentially. Therefore, this type of fixed U/V scan effects separation of ions according to mass. At any point in the scan, only ions of one mass can pass through the system, all other ions being excluded. Because of this filtering action, the system is often called a quadrupole mass filter (Dawson, 1976) but note that it still depends on an m/z value, not just m.

The capabilities of a quadrupole instrument are about the same as for a single-focussing magnetic sector instrument, with highest detectable m/z ratio of about 4000 and maximum resolution near 4000. This practical limit on resolution can be understood by reference to the stability diagram of figure 2.8. Using the fixed U/V ratio that gives the scan line shown, it is clear that ions of mass m_1 will have a stable path at a point at which those of masses m_2 and m_3 will not. That is, ions of mass m_1 will be well resolved from their neighbours. If the slope of the scan line were reduced, it would cut through larger sections of the stable regions and hence more ions would reach the detector (i.e. the sensitivity would be greater), but ions of mass m_1, for example, would be brought into their stable region at the same time as those of mass m_2 (and possibly m_3 as well, if the slope of the scan line were reduced markedly) so resolution would be lost. Clearly, the best resolution would be achieved with a scan line of large slope (i.e. with high U/V ratio) that cuts each ion's stable region at its apex. Then, the conditions for ion transmission to the detector are severe, resulting in poor sensitivity and better resolution. This loss in sensitivity, coupled with the fact that ions enter the quadrupole filter at different positions and that ions of different mass are accelerated into the filter with different velocities (heavier ions achieving lower velocities), imposes a practical limit to the resolution such that the device is best considered as a low-resolution mass spectrometer.

The quadrupole mass filter does have several advantages over the magnetic sector instrument in that it is more robust, cheaper, physically smaller and more readily interfaced with a wide variety of inlet systems. It can be easily constructed as a bench-top instrument (often, but not always, dedicated to combined gas chromatography/mass spectrometry, GC/MS). A

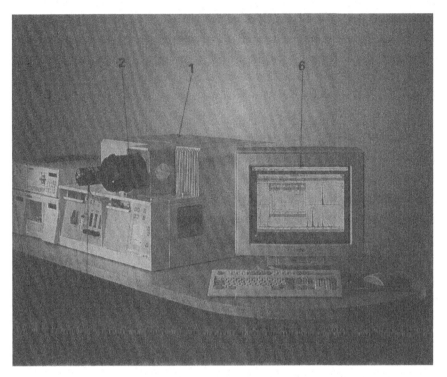

Figure 2.9. A 'bench top' quadrupole mass spectrometer (VG Platform). 1, Quadrupole housing; 2, interface from HPLC instrument to quadrupole; 3, HPLC instrument; 4, carousel for successive automated multiple sample introduction; 5, APCI probe for either HPLC or solids. Photograph by kind permission of Micromass UK Ltd.

typical commercial quadrupole mass spectrometer is shown in figure 2.9; note its relative simplicity compared with the magnetic sector instrument illustrated in figure 2.2. Like the magnetic sector instruments, the quadrupole analyser can scan the entire mass range very rapidly, acquiring many spectra in less than a second. Fast scanning is very important if compounds elute in narrow bands from chromatographic columns. Computer control of the rod voltages of a quadrupole can be effected more accurately than can the corresponding control of the magnetic field of a magnetic sector instrument. In particular, the rod voltages can be switched very rapidly from one value to another in order to focus in turn a few selected ions, an advantage that is desirable for sensitive quantitative mass spectrometry (chapter 7). Generally, the quadrupole mass filter is automated more easily than a magnetic sector instrument.

The term quadrupole refers to the four poles of the mass filter. There are other mass spectrometers that operate on the same principle but which have different numbers of poles or different analyser design. Irrespective of this, these less common instruments have loosely been called quadrupole mass spectrometers. These analysers are referred to more correctly as monopoles, dipoles, hexapoles, dodecapoles and so on (Todd and Lawson, 1975). An extension to quadrupole technology is the ion trap and this is a sufficiently important instrument to be considered separately – after a brief look at some multiple analysers.

2.4.3. *Multiple analysers*

The power of a given analyser can be increased by coupling to it another analyser of the same or a different type. Such coupled systems have special advantages for studies called *tandem mass spectrometry* or *MS/MS* in recognition of the fact that two mass spectrometric analysers have been joined together in one instrument (chapter 8). The possible combinations and permutations of analysers are enormous. Two double-focussing magnetic sector mass spectrometers can be joined to give a so-called four-sector instrument (magnetic/electric/electric/magnetic sector), or a quadrupole analyser might be positioned after a double-focussing magnetic sector system to give a *hybrid* (electric sector/magnetic sector/quadrupole). Finally here, three quadrupole mass filters can be linked to give the 'triple quadrupole' design.

The real value in joining instruments together like this will become apparent in chapter 8 on MS/MS, but an indication is given here. For the analysis of a mixture, the traditional approach is on-line chromatography/mass spectrometry in which the components of a mixture are separated before obtaining their individual mass spectra. In MS/MS, which is an alternative, complementary strategy, no initial separation occurs. First, the unseparated mixture is ionized to give an assembly of ions. Then, the first mass spectrometric analyser is used to select ions of any chosen m/z value from the assembly. The value chosen would be characteristic of, and it is hoped unique to, a component anticipated to be present in the mixture. Note that this separation corresponds to the chromatographic step of the traditional approach, but relies on the components producing ions of different mass. Each selected ion is allowed, or induced, to fragment and the resulting fragment ions are analysed according to m/z value in the second analyser. The result is the mass spectrum of the selected ion, which is characteristic of the original parent component in the mixture. Figure 2.10 compares the chromatography/mass spectrometry and MS/MS strategies. The two approaches are described in greater detail in chapters 5 and 8, respectively. Just as some components of mixtures have similar characteristics and cannot

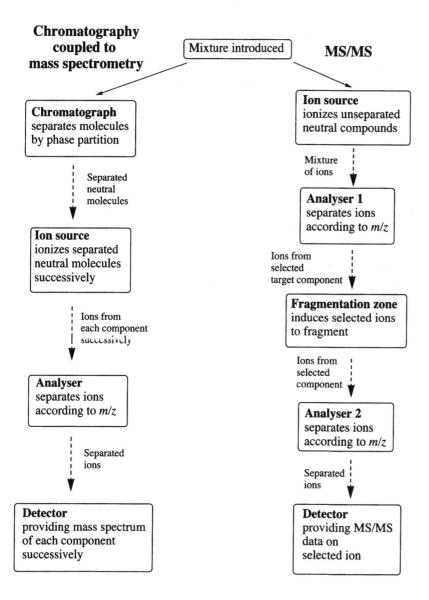

Figure 2.10. A schematic comparison of the strategies for chromatography/mass spectrometry and tandem mass spectrometry (MS/MS). Full arrows indicate transfer of compounds whereas dotted arrows signify transfer of ions.

be separated by chromatography (but do have different masses) so, in MS/MS, some components have different volatilities (and could be separated by GC) but have the same mass and cannot be differentiated.

2.4.4. *Ion traps*

An ion trap is a mass spectrometer that stores ions in an evacuated cavity by applying appropriate electric fields and can then be made to eject them selectively according to their masses. The instruments do not look like

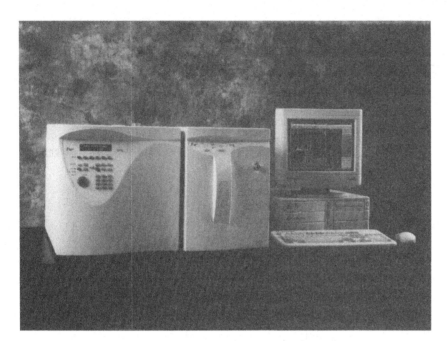

Figure 2.11. A typical ion trap mass spectrometer as an integral part of a GC/MS instrument From left to right, the components are a gas chromatograph, the ion trap mass spectrometer, and the data system. Photograph by kind permission of Finnigan MAT Ltd.

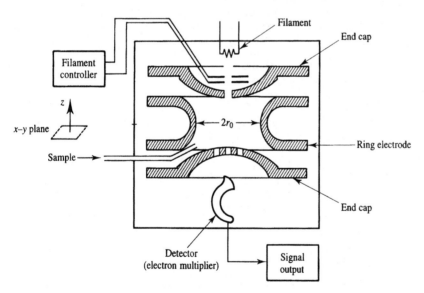

Figure 2.12. A diagrammatic representation of an ion trap.

quadrupole-based systems (figure 2.11) but are in fact an extension of the same technology (March and Hughes, 1989; March, 1991; Todd, 1991a; Todd and Penman 1991). Schematically, the layout of an ion trap resembles a two-dimensional slice through a quadrupole. This similarity can be seen in figure 2.12, in which the shaded areas correspond to the four rods in a quadrupole instrument.

Two of the electrodes, the so-called end caps, are either earthed or biased with DC or AC potentials. In reality, the other two shaded areas are joined to make a doughnut-shaped 'ring electrode' and a sinusoidal radiofrequency potential is applied to it. Therefore the ion trap is actually a three-electrode device. A batch of ions in the cavity is constrained by this oscillating electric field and some ions will have stable trajectories and remain trapped, whereas others will have unstable trajectories and will be ejected. The stability or otherwise of the ion motion is determined by the parameters a_u and q_u that were also encountered in section 2.4.2 for the quadrupole device. Letting the subscripts represent ion motion in the x, y and z directions (see figure 2.12), the stability parameters are very similar to those for the quadrupole system:

$$a_z = -2a_x = -2a_y = -8eU/(mr_o^2\Omega^2)$$

$$q_z = -2q_x = -2q_y = -4eV/(mr_o^2\Omega^2)$$

where r_o is the radius of the ring electrode. The corresponding stability diagram is shown in figure 2.13. Ions with values of a_u and q_u that fall within the stable region, like ions of mass m_1, have 'stable' trajectories and so are trapped within the device. An ion like that with mass m_3, which lies outside the stable region, will have been ejected from the ion trap and detected, whereas ions of mass m_2 at the boundary of the stable region are on the point of being ejected and detected. Note that in the quadrupole mass filter, ions with unstable trajectories are lost (i.e. not detected) whereas, in the ion trap, ions need to acquire unstable motion before they can escape the trap to be detected.

There are several modes in which the ion trap can be operated, in particular, by applying different potentials or frequently earthing the end caps. These different modes are outside the scope of this book but are discussed thoroughly elsewhere (March and Hughes, 1989; Todd, 1991a). The objective of the conventional mode of operation is (i) to create a bunch of ions by passing a pulse of electrons into the ion trap, (ii) to trap all of the ions initially and then (iii) to change the RF potential so as to eject ions to the detector sequentially, according to their m/z value, by making their motions unstable. Essentially, as the RF potential increases, ions are ejected from the cavity in turn from small to large m/z values. Many (not all) of the ejected ions leave the cavity through the hole in the lower end cap, impinge on the detector and are recorded. This scan of the RF potential is very rapid indeed. Once it is complete, the cycle is repeated: another batch of ions is

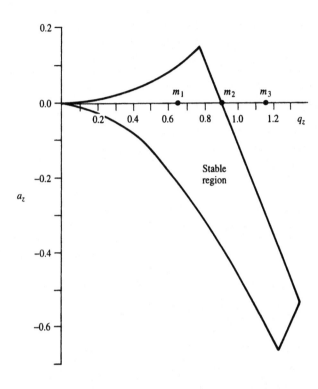

Figure 2.13. A portion of a stability diagram for an ion trap, plotted in (a_z, q_z) space.

created by another pulse of electrons and the rapid RF scanning procedure effected again under computer control. Mass spectra can be built up by this repeated routine, giving a very fast scanning device.

Ion traps are operated with a small residual pressure of helium (a *bath* gas) in the trapping cell (about 10^{-3} Torr). This increases markedly both the resolution and the sensitivity of the device. Ions in the trap collide with helium, which damps the amplitude and velocity of their motion as the RF potential increases. Therefore, they remain closer to the centre of the ion trap. When the RF potential brings ions of a particular mass to the point of being ejected, those ions then gain energy rapidly and develop unstable trajectories in a 'tighter' bunch. That is, at each m/z value, the resolution is improved, giving narrower and larger peaks.

There are scanning routines that allow chemical ionization as well as electron ionization to be performed and, in the more sophisticated ion traps, allow the chemistry of trapped ions to be explored before detection. This corresponds to MS/MS *without* the need for two linked analysers. Rather, the ions from the sample are created in the ion trap, or injected into it, and kept stable within the trap whilst all of the others are ejected. This is effected by operating with RF and DC potentials that provide a and q parameters such that only ions of a single mass have stable trajectories. The behaviour of the chosen precursor ions can then be observed in isolation. It is possible to apply a supplementary RF potential at an appropriate frequency to excite ('tickle') those ions. Without the pressure of helium, the ions would gain kinetic energy, develop an unstable trajectory and be lost from the trap. However, with helium, collisions between the selected ions and the bath gas occur, converting some of the kinetic energy of the ions into internal energy, hence exciting them internally and facilitating dissociation. A conventional scan reveals the resulting fragment ions. This collision-induced dissociation is very efficient if the 'tickling voltage' is carefully optimized. Overall, the two mass spectrometric analyses required for MS/MS are separated in time, not space. That is, isolation of the ions of interest and their subsequent dissociation are examined in the same chamber. Interestingly, one of the fragment ions can be retained in the trap and the collisional activation process repeated to explore the fragmentation of this new precursor ion. The experiment would be termed MS/MS/MS (or MS^3). In this way (trapping fragment ions and using them as new precursors), a fragmentation pathway can be followed through $(n-1)$ stages by MS^n. For example, sequential fragmentation of ions from cholestane has been studied by MS^7. The method is discussed again in chapter 8.

Note the position of the filament in figure 2.12: molecules are ionized in pulses and the resulting ions stored *and* analysed in the same region. In the other instruments discussed so far, a continuous stream of ions, produced in one device, is passed to another, the analyser, for mass separation. The arrangement in an ion trap results in an economy of space and simplicity of operation and maintenance. The system can be made readily into a bench-top device (figure 2.11). Ion traps are small, versatile and robust mass spectrometers that are gaining in popularity (March and Hughes, 1989; Nourse and Cooks, 1990; Dorey, 1992).

2.4.5. *Fourier-transform (ion cyclotron resonance) mass spectrometers*
In a sense, the Fourier-transform mass spectrometer is the 'big brother' of the ion trap. It too stores ions in a cavity, but by a different mechanism. In Fourier-transform instruments, ions are trapped in a high-vacuum chamber by crossed electric and magnetic fields. Figure 2.14 shows the arrangement of the analyser, the cubic device being in the field of a superconducting

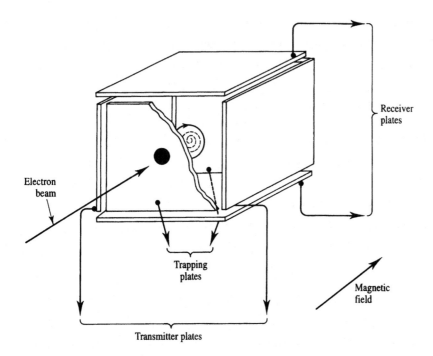

Receiver plates

Electron beam

Trapping plates

Magnetic field

Transmitter plates

Figure 2.14. A diagrammatic representation of a cubic Fourier-transform ion cyclotron resonance analyser.

magnet and the trapping potential applied to the 'trapping plates'. Ions in the cavity travel in a circular path perpendicular to the applied magnetic field and are excited to larger orbits by applying a radiofrequency pulse to the transmitter plates. This cycloidal motion of the ions generates image currents in the circuit connecting the receiver plates. The decay of the image currents is monitored as the coherent motion of the ions is destroyed by collisions. This electrical signal is converted into a mass spectrum by the mathematical process of Fourier-transformation (Nibbering, 1985; Hanson, Kerley and Russell, 1989; Marshall and Grosshans, 1991; Jacoby, Holliman and Gross, 1992).

Ionization can be effected within the cavity (as indicated in figure 2.14) or ions from an external ion source can be drawn into the cavity. The latter has the advantage of maintaining low pressure in the cavity, as is required to achieve the best performance with this system. Whilst the ions are trapped in the cavity, their chemistry can be investigated. In common with the ion trap, MS/MS in which the mass spectrometric stages are separated in time can be performed. Here, it requires an appropriate pulse programme of the RF potential. Thus, the instrument is very versatile and powerful. As long as a very good vacuum is maintained, extremely high resolutions can be attained (for example 3×10^8 for the O^+ ion). However, resolution decreases rapidly with increasing m/z value. The upper mass limit of the FT mass spectrometer depends on several parameters such as the applied magnetic field, the trapping voltage and the length of the cell. Calculations for typical instruments reveal that the ion of highest mass that can be trapped is 950 000. In practice, only a proportion of this mass limit may be achieved. Even so, the instrument has great potential for examining large molecules, probably biomolecules, by plasma desorption, electrospray, laser desorption and particle bombardment, but for truly massive ions the time-of-flight system may be the analyser of choice.

2.4.6. *Time-of-flight mass spectrometers*

These instruments do not have magnetic sectors, quadrupoles or ion trapping analysers. Rather, they separate ions of different masses by making use of their different velocities after acceleration through a potential (V). Since $zeV = mv^2/2$, the velocity (v) of an ion of mass (m) is $v = (2zeV/m)^{1/2}$. Thus, the velocity of an ion is mass dependent so that, if a bunch of ions (m_1, m_2, $m_3 \ldots m_n$) is accelerated and then allowed to pass along a field-free region, the ions will arrive at a detector at different time intervals depending on their velocities (v_1, v_2, $v_3 \ldots v_n$). If the field-free region has length l, then, for a velocity, v,

$$t = l/v$$

$$v = (2zeV/m)^{1/2} \quad \text{or,} \quad t = (m/z)^{1/2}l/(2eV)^{1/2}$$

Thus, the time taken for an ion to reach the detector is proportional to the square root of its m/z value. The larger the mass of a singly charged ion, the longer it takes to traverse the field-free region.

The resolving power of simple (linear) time-of-flight analysers is low. This drawback results chiefly from the spread of energies originally imparted to the ions. Not all ions will experience exactly the same accelerating potential, depending, for example, on precisely where they were formed. Hence, ions of the same mass will have a slight spread of kinetic energy and hence a spread of velocity. In addition, ions can be formed at different times and those formed at different locations will travel somewhat different paths. All of these effects contribute to spread in the resulting signal, that is, to poor resolution. An ion 'reflectron' device has been developed to compensate for such effects, ensuring that ions of the same mass

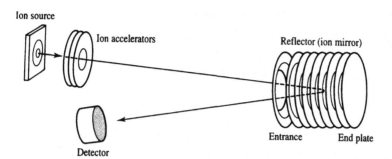

Figure 2.15. A schematic representation of a time-of-flight mass spectrometric analyser, incorporating a reflectron with an electrostatic retarding field.

reach the detector at the same time, and thus to provide a significant improvement in resolution. As its name implies, the reflectron is an 'ion mirror' and it uses an electrostatic principle (figure 2.15). Inside the mirror a homogeneous electric field is applied between the entrance and the end plate. This field retards and then reflects the ions. Ions of the same mass but with slightly greater kinetic energy penetrate further into the reflecting electric field than do their less energetic counterparts. Therefore, the faster moving ions have further to travel, spending more time in the mirror. This has the effect of bunching ions of the same mass in space so they arrive at the detector in a much shorter span of time. Enhancement of the resolution by an order of magnitude is possible and resolutions over 10 000 have been reported for time-of-flight systems incorporating ion reflectrons. Unfortunately, the improved resolution is gained at the expense of sensitivity. With the reflectron, limits of detection may be worsened by a factor of about ten, compared with a linear time-of-flight analyser. This trade-off is particularly severe with very large analyte ions and in practice these are best analysed by a low-resolution, linear time-of-flight analyser.

An advantage of time-of-flight analysers is their fast response times, so they find considerable use in instruments that generate ions in pulses (Le Beyec, 1989; Price, 1990). Such an analyser is ideal for laser desorption (section 2.3.6) and for the pulsed Cf-radionuclide source (plasma desorption, see section 2.3.7). A considerable additional advantage of the time-of-flight analyser is that, in principle, it has no upper mass range limitation, thus allowing detection of ions of very large mass, as are often generated by laser or plasma desorption.

2.5.   DETECTION AND RECORDING OF SPECTRA

There are four main ways of detecting ions and generating from them an electric current that is proportional to their abundance: by electron multiplier, scintillation counting and photomultiplier, Faraday cup or various focal plane detectors (Geno, 1992). As discussed in sections 2.4.5 and 12.3, detection of ions in Fourier-transform ion cyclotron resonance systems is characteristically different from these.

Ions are either directed sequentially by the mass analyser to a point where there is a single-channel detector (electron multiplier, scintillator or Faraday cup), as in the case of a quadrupole mass filter or time-of-flight analyser, or they are dispersed simultaneously to a plane and detected by a focal-plane (array) detector, as with a magnetic sector instrument of Mattauch–Herzog geometry. In this short section, single-channel detectors are discussed first.

The commonest detector is the electron multiplier, which consists of a series of about ten to twenty electrodes (dynodes). When a rapidly moving ion impinges on the first dynode, it causes the release of a shower of electrons, which is accelerated by an electric potential to the second dynode. Each of these electrons causes a further shower of electrons, which strike the third dynode. This cascading effect continues through the whole series of electrodes and provides gains in electric current of the order of 10^4–10^8. There is an alternative design of electron multiplier in which the discrete dynodes are replaced by one continuous dynode. This *channel electron multiplier* (CEM) is a coiled, tapered tube (not unlike a snail shell). Ions striking the inner surface near the mouth of the tube emit showers of electrons, which are drawn towards the earthed narrow end of the CEM. As they travel inwards, they collide repeatedly with the walls and thus cause more and more electrons to be emitted. The gain is similar to that of the traditional electron multiplier. The signal obtained from such detectors is then subjected to conventional electrical amplification.

The second method is the Daly detector in which the ion beam strikes a plate, again releasing electrons. The electrons are accelerated onto a phosphor screen (scintillator), which emits photons as a result. The photons are detected and the signal transformed into an electric current by a photomultiplier. This detector is known particularly for its ability to detect the ionic products of metastable ion fragmentation whilst, in one mode of operation, ignoring normal ions (see section 8.3.3).

In the third and simplest detector, ions impinge on a metal plate, which is connected to earth through a resistor. The resulting neutralization of the charge on the ions as they hit the plate leads to a flow of current through the resistor. The current can be detected and affords a measure of ion abundance. The plate is not infrequently replaced by a metal (Faraday) cup, which is a more efficient detector because it also captures the electrons that are emitted when an ion strikes the metal of the detector. These electrons amplify the signal. The only real disadvantages of this simple robust detector are its lack of sensitivity compared with the detectors described above and slow response time.

At a given time in a scan, a mass spectrometer having only a single-channel detector, such as the electron multiplier, collects only ions of one m/z value and discards the rest (unless they are stored in an ion trap). In a complete scan, the detector also spends a lot of time 'detecting' noise in the gaps between the ion arrivals that comprise mass spectral peaks. Hence, a single-channel detector inevitably involves a loss of potential sensitivity when the instrument is operating in the scan mode. This shortcoming is addressed to some extent by focal-plane detectors of which there are several designs. The photographic plate is not very common but can be useful with mass spectrometers of Mattauch–Herzog geometry for reasons described in the previous section. Where ions fall onto the plate, the resulting bands are revealed by photographic development. The method is sensitive because all of the ions are detected all of the time. As an additional advantage, fluctuations in the ion beam do not distort the ion abundances in the resulting mass spectrum. The photographic plate method does provide its own record of the spectrum, but it is not directly amenable to computer processing. To achieve the latter, the bands on the photographic plate must be converted into an analogue electrical signal by using a microdensitometer. A narrow beam of light is passed through the photographic plate (figure 2.16) and is collected at a detector, which emits an electrical signal proportional to the amount of light falling on it. The photographic plate is passed at constant speed through the light beam and the images on the plate cause fluctuations in the detected light level; the resulting electrical signal from the detector may be passed to a computer in the same way as that from an electrical or Daly detector. The method inevitably involves off-line working of the computer. This lack of

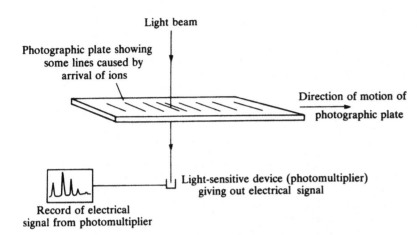

Figure 2.16. A diagrammatic representation of the action of an optical microdensitometer.

immediacy of results is inconvenient, and is addressed by more modern designs of focal-plane array detector.

The CEM version of the electron multiplier can be made so small that many of them (10^4–10^7 with diameters of 10–100 μm) can be arranged side-by-side as a channel electron multiplier array. Each CEM component of this microchannel plate acts independently and emits electrons when ions impinge on it, forming a spatially resolved array detector. Alternatively, the microchannel plate can be made to emit electrons towards a phosphor screen. Photons released from the phosphor are conducted via fibre optics to a photodiode array detector, which provides electrical signals. Both of these focal plane detectors monitor a small window of m/z values instead of a single value at a given point in a scan. The arrays are an important development because simultaneous detection of ions implies that more are collected and this greater efficiency leads to lower limits of detection than for single-channel detectors. Also, the electrical or electro-optical nature of the arrays avoids the inconvenience of the photographic plate method. Relative advantages and disadvantages of the main detectors (Faraday cage, electron multiplier, photographic plate and array) have been extensively discussed (Boerboom, 1991).

The amplified signals from a single-channel detector can be passed to a recorder. A pen recorder does not respond sufficiently rapidly to be useful unless the scan speed is very low and instead, the signals are used to deflect mirror galvanometers of different sensitivities. Ultraviolet light reflected from the mirrors is focussed onto UV-sensitive paper. The resulting trace is illustrated in figure 1.2. In more sophisticated and more widespread systems,

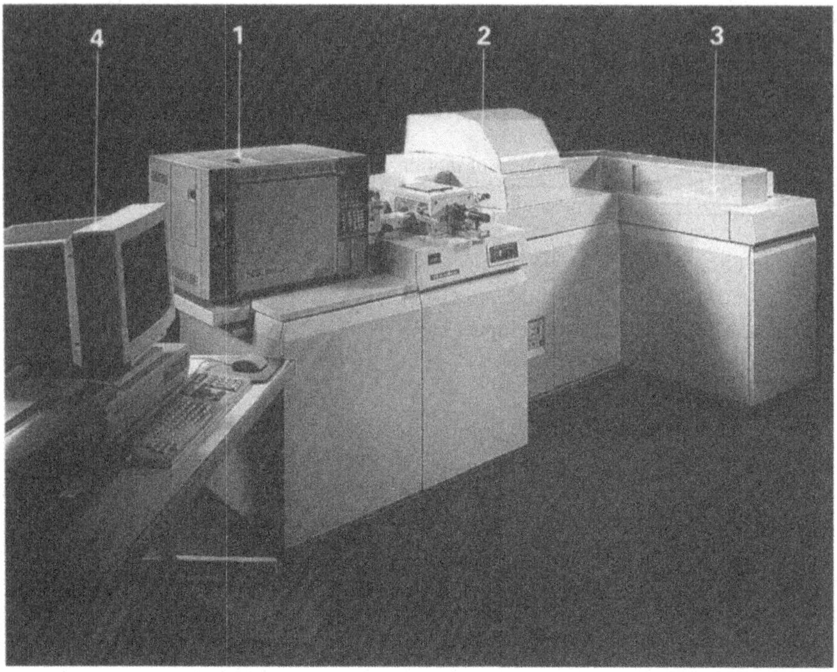

Figure 2.17. A typical general purpose mass spectrometer capable of high resolution and MS/MS studies (the VG Autospec-Q). 1, Gas chromatograph; 2, first (sector) mass spectrometer; 3, second (quadrupole) mass spectrometer; 4, workstation, including visual display unit, keyboard and computer. Pumps, electrical supplies and valves are protected by removable metal panels. Photograph by kind permission of Micromass UK Ltd.

the amplified electrical signals from the detector are passed, via a digitization device, to a computer, which evaluates the incoming data and shows the required information on a computer screen or prints it out as detailed in chapter 4.

2.6. A TOTAL SYSTEM

A modern mass spectrometer may have one or several inlet systems and methods of ionization, one or more analysers and usually one ion detector. The number and arrangement of these depend on ingenuity, commercial availability and individual needs. One popular variant uses a gas chromatograph as one of its several inlet methods, a double-focussing magnetic sector analyser followed by a linked quadrupole mass filter to enable both high-resolution studies and MS/MS work, an electron multiplier or photomultiplier as detector and a computer for instrument control, data acquisition and processing. Such a mass spectrometer/computer system is illustrated in figure 2.17. The linking of mass spectrometers to computers is described in chapter 4.

It is also true that mass spectrometers, often bench-top systems, dedicated to one specialized task, like matrix-assisted laser desorption or electrospray, are becoming increasingly popular. Such systems can be built free of the constraints and compromises that pertain to a sophisticated instrument that is designed to function in a number of different modes. The individual needs of different mass spectrometric methods will become clearer as the different methods of ionization are considered in the next chapter.

3 Methods of ionization

3.1. INTRODUCTION

The method used to ionize a substance influences markedly the appearance of its mass spectrum. This effect is illustrated in figure 3.1, which shows mass spectra of the amino acid arginine (1) obtained by electron ionization, chemical ionization, in-beam chemical ionization, field desorption, fast

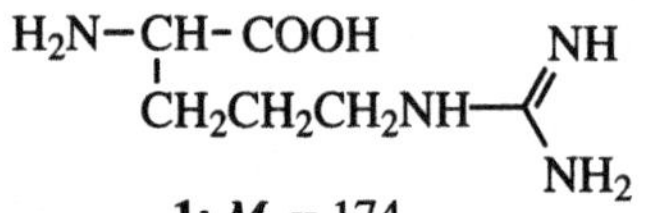

$$H_2N-CH-COOH \qquad NH$$
$$CH_2CH_2CH_2NH-\!\!\!\!\Big\langle$$
$$NH_2$$

1; $M_r = 174$

atom bombardment, thermospray, electrospray and plasma desorption. In figure 3.1, and throughout this chapter, mass spectra are of positively charged ions unless otherwise stated, simply because most mass spectrometry to date has concerned them. Using the gas-phase EI method, arginine gives some mass peaks useful for characterization either by manual interpretation (chapter 10) or by library searching (chapter 4). However, the diagnostically important molecular ions (m/z 174) are absent because, when heated, arginine decomposes before it evaporates (figure 3.1(a)). Given that chemical ionization also requires volatilization of sample molecules, it is not surprising that the CI spectrum does not show protonated molecules [M + H]$^+$ at m/z 175. On the other hand, techniques like field desorption, fast atom bombardment and electrospray that do not require the sample to attain the gas phase prior to ionization, do give [M + H]$^+$ ions (figure 3.1 (d), (e) and (g)). Also, the amount of fragmentation is much reduced with these last methods of ionization. This situation is typical for many compounds and indicates that no one ionization technique is always superior to the others. If the compound (1) had been unknown, it would not have been

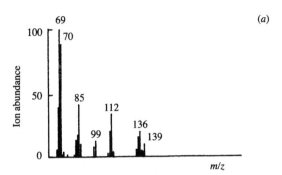

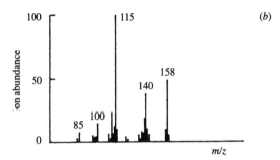

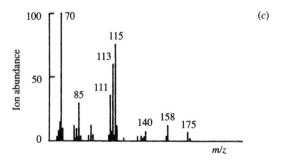

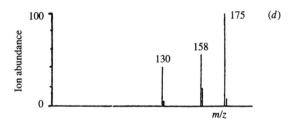

Figure 3.1. Mass spectra of arginine (1) obtained by (*a*) electron ionization, (*b*) chemical ionization with isobutane, (*c*) in-beam chemical ionization with isobutane, (*d*) field desorption, (*e*) fast atom bombardment, (*f*) thermospray, (*g*) electrospray and (*h*) plasma desorption. Note the $[M + H]^+$ ions at m/z 175, $[M + Na]^+$ ions at m/z 197, and $[M - H + 2Na]^+$ ions at m/z 219. Spectra are reproduced by kind permission from (*c*) Cotter (1979); (*f*) Blackley, Carmody and Vestal (1980); (*g*) Micromass UK Ltd; and (*h*) Torgerson, Skowronski and Macfarlane (1974).

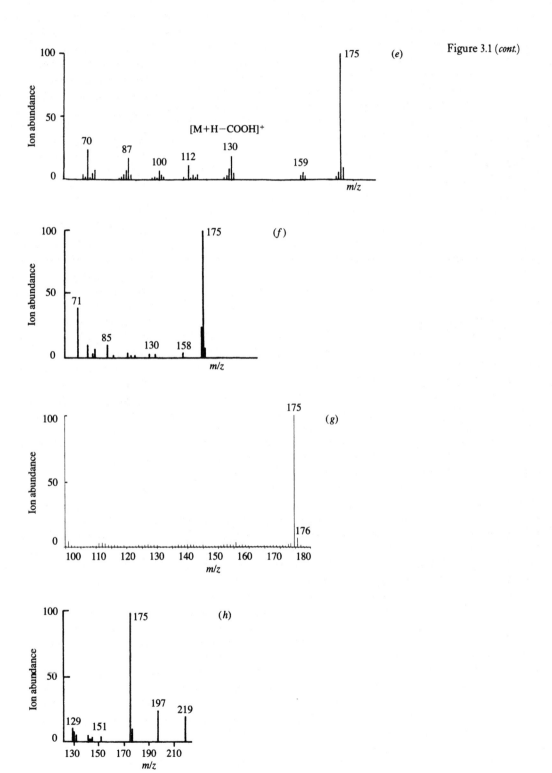

Figure 3.1 (*cont.*)

possible to assign its relative molecular mass from the EI spectrum. This information would be readily available from, say, an electrospray or a field desorption (FD) mass spectrum, but the smaller number of fragment ions produced by such methods of ionization may have made structural elucidation impossible. If a very small quantity of arginine were required to be detected and quantified by monitoring ions of just one selected m/z value (see chapter 7), then FAB, thermospray or electrospray mass spectrometry would appear to be the most suitable methods of those shown: all three afford spectra containing a mass peak accounting for a large proportion of the total ion current, thus facilitating detection of arginine at trace level. Many organic and organometallic compounds exhibit only small peaks for molecular ions in their EI mass spectra and such small responses are easily missed. In these cases, it is good practice to confirm the assignment of molecular mass by additional use of chemical ionization or another method, some others of which are represented in figure 3.1(c)–(h). Overall, the method of choice for ionization depends on the objective of the analysis and on the type of compound. A most important and general point to be gleaned is that, for structural elucidation, examination of an EI mass spectrum in conjunction with one of a number of other types of mass spectrum (CI, FAB, etc.) makes for a much easier interpretation than when only one is considered in isolation.

Many of the methods of ionization discussed in this chapter are complementary to and not alternatives for electron ionization. It is advisable to record the EI mass spectrum of an unknown substance, even when new and sophisticated techniques are available, because the behaviour of compounds under EI conditions is far better understood than it is under the conditions of any other ionization method. It is possible to draw on a vast store of experience and data on EI mass spectrometry (e.g. mass spectral libraries). Such a store is not yet available for other methods of ionization. For instance, assignment of relative molecular mass by CI, FAB and FD mass spectrometry is not always as clear as it is in figure 3.1. Chemical ionization may give $[M + H]^+$, $[M - H]^+$, or $[M + X]^+$ adduct ions, depending on the structure of M and the reactant gas used (section 3.3). Such ions, which contain the intact molecule but are not molecular ions ($M^{+\bullet}$), are collectively called *quasi-molecular ions*. Note, though, that this collective expression is best replaced by a specific term whenever possible. For instance, $[M + H]^+$ ions should be termed *protonated molecules*. Field ionization affords $M^{+\cdot}$ or $[M + H]^+$ ions, field desorption may give $M^{+\cdot}$, $[M + H]^+$, $[M + Na]^+$, $[M + K]^+$ or other adduct ions as well as cluster ions like $[M_n + H]^+$ ($n = 2, 3,\ldots$), and thermospray can give $[M + H]^+$ or $[M + NH_4]^+$ ions. It may not always be obvious which of these ions is being observed under a given set of conditions, making the assignment of relative molecular mass uncertain. However, very large or ionic molecules do not yield at all to the gas-phase EI or CI methods so that application of a method of ionization direct from the solid or solution state would be the only sensible approach in such cases.

As already seen in chapter 2, there are many methods for ionizing compounds (Milne and Lacey, 1974; Hunt, 1982) of which electron ionization is the commonest. Additional methods, such as CI, FAB, electrospray, FI and FD, are frequently termed 'soft' or 'mild' ionization techniques because the molecular, or quasi-molecular, ions are formed with less excess of internal energy than is the case in normal EI. As seen above, 'mild' ionization leads to less fragmentation and greater abundances of molecular or quasi-molecular ions. In this chapter the most popular ionization methods are described, with examples of their use, in separate sections.

3.2. ELECTRON IONIZATION

Electrons, obtained from a heated filament *in vacuo*, are accelerated by a voltage (V) and directed across an *ion chamber*; thus, the electrons have energy eV, where e is the electronic charge. The voltage (V) is continuously variable between about 5 and 100 V and, by convention, standard mass spectra are obtained at 70 V because maximum ion yield and good reproducibility are obtained near this value. In practice, mass spectra may be obtained at any voltage down to the ionization energy of the compound under investigation, and it is often advantageous to reduce the energy of the electrons much below 70 V. A volatilized sample molecule (M) and an electron of energy greater than the ionization energy of M will interact if they pass close enough for the electron to impart sufficient of its energy to the neutral species.

A moving electron can be considered as a packet of energy having a waveform somewhat analogous to that of a photon. During the approach of the electron to the molecule, the electron waves and the electric field of the molecule mutually distort one another. The distorted electron wave can be considered to be composed of many different sine waves and some of these component waves will have the correct frequency (energy) to interact with molecular electrons. However, the reaction of an electron with a molecule is rather non-specific. For example, the interaction may lead to electronic excitation in the molecule by promotion of one of its electrons from a lower to a higher orbital (as happens in ultraviolet spectroscopy):

$$M + e^- \longrightarrow M^* + e^-$$

A molecular electron may be ejected from the molecule altogether to leave a radical cation, as in equation (3.1). This is the desired outcome for mass spectrometry.

$$M + e^- \longrightarrow M^{+\cdot} + 2e^- \tag{3.1}$$

Direct capture of an electron by the molecule, to give a stable radical anion, may appear logical, but is of low probability. The translational energy

$$M + e^- \longrightarrow M^{-\cdot}$$

of an electron attaching itself to a molecule must be taken up as excess of internal energy in the new radical anion. Usually, this excess of energy leads either to the electron being 'shaken off' again or to fragmentation of the radical anion. To form stable radical anions by direct electron capture, it is necessary to have sufficient gas pressure for a 'third-body collision' to occur and remove enough excess of energy so that the radical anion does not fragment. Such higher pressure ionization conditions fall in the domain of chemical ionization and so the method is considered again in section 3.4. Other electronic effects in the molecule do occur during interaction with electrons (Cottrell, 1965) but they fall outside the scope of this book.

When ionizing transitions occur according to equation (3.1), a relatively large number of energy states in the resulting ion can be reached. Therefore, molecular ions are formed with a wide range of excesses of internal energy. Generally, the greater the electron beam energy, the greater the excess of energy appearing in the $M^{+\cdot}$ ions and hence the greater the amount of fragmentation. By reducing the electron energy well below the conventional 70 eV value, much less excess of internal energy is left in the ion following ionization so that fragmentation is much reduced.

This can be very helpful when examining fragile molecules that yield few, if any, molecular ions that are stable enough to reach the detector intact. This theme is considered again in section 9.2 and is illustrated in section 10.3.

Assuming that the mass spectrometer is tuned to analyse positive ions, any negative ions produced will be discharged at the positively charged repeller plate and, together with neutral species, will be pumped away. Since not all molecules (M) react with the electron beam and because those that do react can give a variety of products, these ion sources have low efficiencies. Probably less than 1 per cent of the sample molecules are converted into positively charged ions. To monitor negative ions, the repeller plate and acceleration potentials must be reversed in sign. Electron capture is less probable than electron removal by a factor of about 100, although this figure varies widely depending on the structure of the sample. Negative-ion mass spectrometry by EI is inherently less sensitive than positive-ion mass spectrometry for most types of compound. In electron ionization sources, the ions spend approximately 10^{-6} s after formation in the ion chamber before being accelerated into the analyser of the mass spectrometer.

3.3. CHEMICAL IONIZATION

Ion sources for chemical ionization are very similar to those for EI but operate at pressures of 0.1–2 Torr compared with about 10^{-6} Torr for electron ionization sources. A pressure of about 1 Torr is achieved by 'leaking' a reactant gas (A) into the ion source. Reactant gas ions are ionized by an electron beam to give *primary ions* ($A^{+\cdot}$), which, with mean free paths of only about 2×10^{-4} mm, collide with neutral molecules (A) many times before leaving the source. Through collision of $A^{+\cdot}$ ions with neutral A molecules, translational, vibrational and rotational energy is equilibrated between the ions and molecules so that, in chemical ionization, the primary ions are generally considered to possess thermal energies, corresponding to equilibrated ground-state species at the temperature of the ion source. Apart from equilibration, a number of ion/molecule reactions may occur during the collisions. For structural mass spectrometry, the most useful reaction is the formation of stable *secondary* (or *reactant gas*) *ions*, $[A + H]^+$ or $[A - H]^+$. Generally, ions with an even number of electrons (cations) are more stable than ions with an odd number of electrons (radical cations) like the $A^{+\cdot}$

primary ions; reaction of the primary $A^{+\bullet}$ with the excess of neutral A molecules gives more stable secondary species. Thus, methane affords principally CH_5^+, isobutane $C_4H_9^+$, ammonia NH_4^+ and hydrogen H_3^+ reactant gas ions. It is these thermally equilibrated, stable secondary ions which are the reactant species and which effect ionization of other entrained sample molecules (M) by ion/molecule reactions (scheme (3.2)). Because the

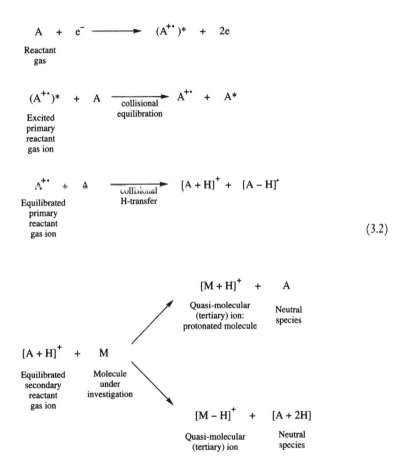

(3.2)

number of reactant gas molecules far exceeds those of sample molecules (M), little direct ionization to give $M^{+\bullet}$ occurs. Instead, collision of a reactant gas ion with a neutral molecule, both of which are massive compared with an electron, leads to a slow interaction which allows equilibration of energy between the colliding species.

To ensure that a statistically *in*significant number of sample molecules (M) is ionized directly by electron ionization, the ratio of reactant gas to sample should be at least 10^3 at a pressure of about 1 Torr. Under these circumstances, the stable secondary reactant gas ions can collide with molecules (M) of the sample to produce *tertiary ions,* which are the ions of interest to the mass spectrometrist (scheme (3.2)). The secondary ions are strong Lewis acids and, during the formation of the tertiary ions, a proton may be donated by the reactant gas ion or, less usually, a hydride ion abstracted from M (scheme (3.2)). Chemical ionization of acetone in methane is represented by equation (3.3), in which a secondary CH_5^+ ion donates a proton to acetone:

$$CH_5^+ \;+\; H_3C-\overset{\overset{\displaystyle O}{\|}}{C}-CH_3 \quad\longrightarrow\quad CH_4 \;+\; H_3C-\overset{\overset{\displaystyle \overset{+}{O}H}{\|}}{C}-CH_3 \qquad (3.3)$$

Quasi-molecular ion;
specifically,
protonated molecule

The proton is considered to be attached to the most basic site in the molecule: the oxygen atom. Therefore, the quasi-molecular ion is drawn with a protonated carbonyl group. Note that this is a situation in which the resulting $[M + H]^+$ quasi-molecular ion is best called a protonated molecule because this term describes it precisely. Almost all polar molecules contain a basic site (a heteroatom) at which protonation can occur and hence exhibit $[M + H]^+$ ions. Saturated non-polar compounds, especially alkanes, do not possess such sites and tend to display $[M - H]^+$ ions instead. Depending on the acidity of the reactant gas ion and the basicity of the sample molecule, hydrogen transfer may not occur and instead the secondary ions react with molecules (M) to give adduct ions, rather than protonated molecules, as shown in equations (3.4) and (3.5):

$$M + NH_4^+ \rightarrow [M + NH_4]^+ \qquad (3.4)$$
$$M + C_2H_5^+ \rightarrow [M + C_2H_5]^+$$
$$M + C_3H_7^+ \rightarrow [M + C_3H_7]^+ \qquad (3.5)$$

For instance, chemical ionization of an amine in ammonia can afford both $[M + H]^+$ and $[M + NH_4]^+$ tertiary ions with, respectively, masses of 1 and 18 mass units greater than the relative molecular mass of the amine M. When using methane or isobutane as reactant gas, in addition to the abundant $[M + H]^+$ or $[M - H]^+$ ions, there are often adduct ions of higher mass and lower abundance. Ion/molecule reactions of methane at 1 Torr produce some ethyl and propyl secondary ions which may undergo addition reactions (equation (3.5)). Many of these adduct ions are predictable and hence useful for confirming the assignment of relative molecular mass, but unexpected adduct or cluster ions such as $[2M + H]^+$ may be confusing. The abundance of adduct ions depends on the structure of M, the choice of reactant gas and the mass spectrometric conditions, in particular the pressure in the ion source.

Chemical ionization mass spectra, unlike EI spectra, may reflect fine differences between structural isomers. The loss of water from ketones is found commonly in mass spectrometry, but CI mass spectra of steroidal ketones exhibit marked selectivity in this reaction. The loss of water from the $[M + H]^+$ ions of 5α,3-ketosteroids is not prominent but it is very noticeable in the corresponding epimeric 5β,3-ketosteroids.

In general, quasi-molecular ions obtained by chemical ionization have greater relative abundances than molecular ions obtained by electron ionization. For example, protonated molecules $[M + H]^+$ are more stable than the equivalent molecular ions $M^{+\cdot}$ because the former are produced with little excess of internal energy and because they are even-electron species, unlike the radical cations ($M^{+\cdot}$) formed by electron ionization. Chemical ionization may afford a few fragment ions, or none at all, in which case the mass spectrum on its own is of little use for structural elucidation. If the quasi-molecular ions do have sufficient energy to fragment, they dissociate almost exclusively by loss of neutral molecules to give even-electron fragment ions. Normally, chemical ionization provides fewer but diagnostically more significant fragment ions than does electron ionization. One reason for this is that, after chemical ionization, carbon–carbon bonds tend to cleave only if the products of the dissociation are particularly stable. Frequently, the carbon skeleton remains intact and cleavage is restricted to the bonds of functional groups such as C–O, C–S and C–N bonds. This is certainly not the case with electron ionization mass spectrometry (fragmentation in both EI and CI modes is described in chapter 10). These contrasts between EI and CI mass spectra are clearly seen in figure 3.2. During analysis by EI, the molecular ions of the lactone (2) give a peak at m/z 228 of only 0.2% relative

2; $M_r = 228$

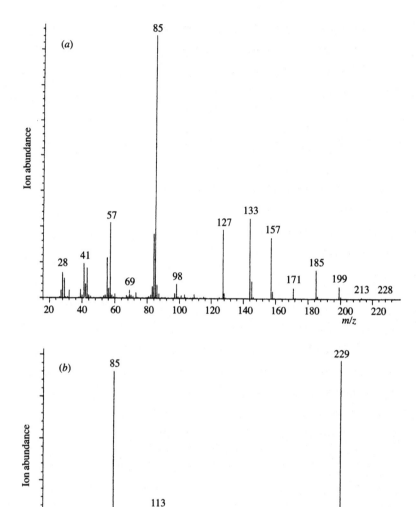

Figure 3.2. (*a*) Electron ionization, (*b*) methane chemical ionization and (*c*) isobutane chemical ionization mass spectra of the octanoyloxylactone (**2**). Note how much the amount of fragmentation decreases, and the prominence of the molecular ion or protonated molecule increases, from spectrum (*a*) to (*c*).

abundance, and the many fragment ions involve cleavages at various different sites in the molecule, including C–C bond cleavages. In contrast, its CI spectra (with methane or isobutane as reactant gas) exhibit base peaks for the protonated molecules at *m/z* 229, and far less fragmentation.

Chemical ionization allows the actual degree of fragmentation to be controlled by changing the reactant gas. Chemical ionization of acetophenone, $C_6H_5COCH_3$, gives $[M + H]^+$ ions using methane, propane or isobutane as

Figure 3.2 (*cont.*)

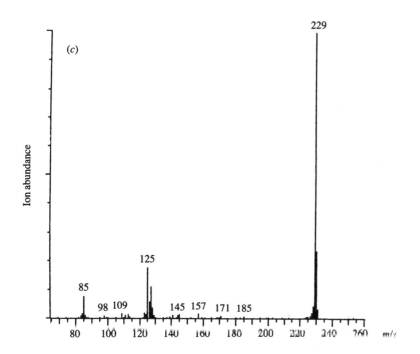

reactant gas, but the degree of fragmentation is different in each case. With isobutane, only $[M + H]^+$ ions are observed; with propane $[M + H]^+$ and $C_6H_5CO^+$ ions are found and with methane $[M + H]^+, C_6H_5CO^+, C_6H_5^+$ and CH_3CO^+ ions are formed. A similar effect is noted with lactone (**2**) in figure 3.2(*b*) and (*c*), where the use of isobutane causes less fragmentation than does methane. A CI mass spectrum can be changed by using different reactant gases for a number of reasons. Principally, the amount of excess of energy imparted to an $[M + H]^+$ ion on its formation depends on the relative proton affinities of the conjugate base of the reactant ion (CH_4, NH_3 and so on) and the compound (M). As the proton affinity of the conjugate base is decreased (i.e. the acidity of the reactant gas ion is increased), the amount of fragmentation of the protonated molecules increases because more energy is transferred to them during their formation. Since acidity increases in the order $NH_4^+ < C_4H_9^+ < C_2H_5^+ < CH_3^+ < H_3^+$, the degree of fragmentation caused by these common reactant gas ions increases in the same order as described for acetophenone. Hydrogen and methane are particularly useful reactant gases for structure elucidation because the $[M + H]^+$ ions are frequently accompanied by structurally informative fragment ions. For confirmation of the relative molecular mass of a more fragile molecule, isobutane or ammonia would be the reactant gases of choice. If the reactant gas has a proton affinity much higher than that of a compound (M), proton transfer from the reactant gas ions to M will be energetically too unfavourable. Thus, ammonia selectively protonates compounds of high proton affinity (basicity) such as amines and amides. Hydrocarbons and ethers are not ionized by CI with ammonia, whilst aldehydes, ketones, acids and carbohydrates form only the adduct ions, $[M + NH_4]^+$.

Some infrequently used reactant gases have considerable benefits under special circumstances. The reactant ions of methyl vinyl ether react with alkenes in a manner dependent on the location of double bonds (scheme (3.6)). The masses of ions (3) and (4) indicate the position of the double bond in the original structure.

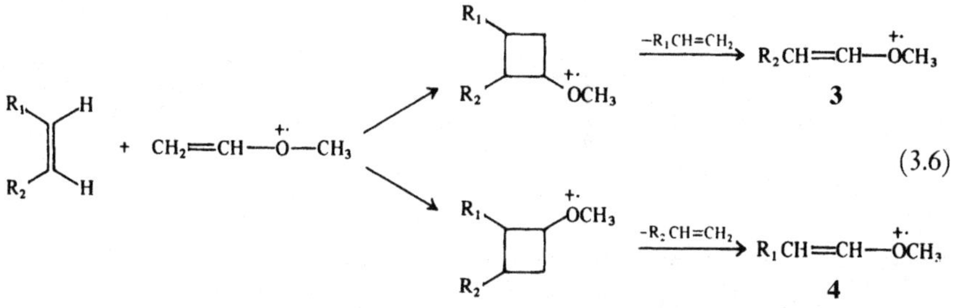

(3.6)

Chemical ionization of dipeptides with methanol affords mass spectra that are more easily interpreted than electron ionization spectra or CI mass spectra obtained with ammonia or isobutane as reactant gas. To identify a dipeptide from its mass spectrum, there should be a readily recognizable, though not necessarily particularly large, mass peak for the quasi-molecular ions and prominent peaks defining the amino-acid sequence (sections 8.3 and 10.6). Figure 3.3, comparing EI and methanol CI mass spectra of the N-acetylated, O-methylated derivative of phenylalanylglycine (5),

$$CH_3CONH-CH-CONHCH_2COOCH_3$$
$$\qquad\qquad\quad |$$
$$\qquad\qquad CH_2C_6H_5$$

5; $M_r = 278$

shows that the latter technique meets these requirements. As with all dipeptides examined to date, reasonably abundant protonated molecules (m/z 279) are accompanied by fragment ions resulting from loss of water (m/z 261) and methanol (m/z 247). These latter are used to confirm the

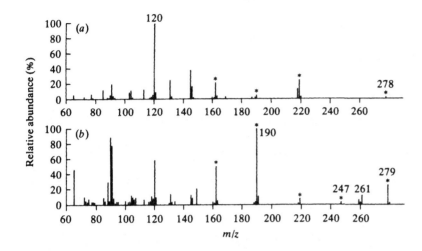

Figure 3.3. (*a*) Electron ionization and (*b*) methanol chemical ionization mass spectra of Ac-Phe-Gly-OMe (5). The mass peaks marked with an asterisk are due to ions that characterize the sequence of amino acids.

assignment of relative molecular mass. Mass peaks indicative of amino-acid sequence (marked with an asterisk) are clearer in the CI mass spectrum.

Nitric oxide as a reactant gas provides abundant $[M + NO]^+$ adduct ions with ketones, acids and esters, but $[M - H]^+$ ions with ethers and aldehydes, thereby distinguishing between these two groups of compounds. The reactant gas tetramethylsilane is useful for analysis of substances that do not give quasi-molecular ions with isobutane CI. Essentially, $(CH_3)_3Si^+$ ions are the only secondary ions and these add to a wide variety of compounds to give $[M + (CH_3)_3Si]^+$ ions, the observation of which leads to the assignment of the relative molecular mass of M.

Many other reactant gases have been used, including binary mixtures such as argon and water. The aim of some of these binary mixtures is to produce mass spectra with abundant quasi-molecular ions (typical of CI) and many fragment ions (characteristic of EI or charge exchange; see below), making the elucidation of structures easier than when using CI or EI mass spectra alone. For these, more specialized texts should be consulted (Wilson, 1971,1973; Milne and Lacey, 1974, pp. 77–85; Wilson, 1975,1977; Harrison, 1989).

Finally, it should be noted that, even in EI spectra, $[M + H]^+$ ions can be formed if the sample pressure in the ion source is allowed to become too high:

$$M^{+\bullet} + M \longrightarrow [M + H]^+ + [M - H]^\bullet$$

This process, in which sample molecules act as their own reactant gas and chemically ionize themselves, is most frequently observed when the normal abundance of molecular ions, $M^{+\cdot}$, in EI spectra is low and with molecules that contain a site of high proton affinity (like an amino acid). In these circumstances, abundances of $M^{+\cdot}$ and $[M + H]^+$ ions can become comparable. If this 'self-CI' process causes confusion in the assigment of relative molecular mass, the sample should be analysed again using a smaller amount. Because of the ensuing lower pressure, the abundance of $[M + H]^+$ ions will decrease.

3.3.1.   *Charge exchange*

If reactant gas ions are formed from monatomic species like argon, there are no vibrational degrees of freedom and the ions carry specific amounts of energy. The lowest amount of energy required to ionize an atom or molecule is the ionization energy. Argon, with an ionization energy of 15.755 eV, has just that excess of energy after ionization and there is no possibility of its internal conversion to vibrational energy. In a charge exchange reaction between reactant argon ions and molecules, M (equation (3.7)), the whole of the excess of energy (15.755 eV) is transferred to the molecule, M, which is therefore ionized and given a large excess of energy, equivalent to (15.755 − *I*) eV, where *I* is the ionization energy of the molecule, M.

$$Ar^{+\cdot} + M \rightarrow Ar + (M^{+\cdot})^* \tag{3.7}$$

The molecular ion, $M^{+\cdot}$, does have vibrational degrees of freedom and the acquired excess of energy causes rapid fragmentation. The behaviour is then very similar to the fragmentation of odd-electron molecular ions produced by electron ionization, and is unlike the chemical ionization discussed above, in which thermally equilibrated reactant gas ions transfer smaller amounts of energy to form even-electron quasi-molecular ions exhibiting greatly reduced fragmentation. The interest in charge exchange reactions with monatomic reactant gas ions lies in the knowledge of the precise amount of excess of energy given to the molecular ion. If the ionization energy of the molecule is 10.000 eV and argon is used as the reactant gas, then 15.755 − 10.000 = 5.755 eV excess of internal energy is imparted to the molecular ion. Such an amount of energy (approximately 500 kJ mol^{-1}) appearing in most of the commoner types of molecular ion, with average bond strengths of 250 kJ mol^{-1}, leads to rapid fragmentation.

Occasionally, charge-exchange mass spectra of isomeric compounds reveal fine structural differences when the corresponding EI spectra are indistinguishable. In another use of charge exchange, a carefully chosen reactant gas can sometimes ionize selectively some components of a mixture (those with lower ionization energy) and disregard those with substantially higher ionization energies. An example is the selective ionization of hazardous aromatic hydrocarbons in petroleum products; the alkane 'background' is unaffected and so does not interfere with the analysis.

3.3.2. *Chemical ionization at atmospheric pressure*

It is perfectly feasible to effect chemical ionization at atmospheric pressure. The major difficulty lies in transferring the ions into the high-vacuum analyser, and not in the creation of the ions. Atmospheric pressure chemical ionization (APCI) sources can sample ambient air (for environmental monitoring) or entire liquid chromatographic effluents (up to 2 ml min^{-1}). It is applied most to polar and ionic compounds with moderate molecular masses (up to about $M_r = 1500$). Of course, ionic compounds in solution already comprise ions so the conditions in an APCI source simply prepare the existing ions for the analyser part of the instrument. Polar compounds are ionized by corona discharge (Carroll *et al.*, 1975) or by the radioactive β-particle emitter ^{63}Ni (Horning *et al.*, 1974). A heated filament giving electrons is not used at atmospheric pressure because most filament materials would soon burn out. The corona discharge source is very sensitive owing to high ionization efficiency. It is usually used with quadrupole mass spectrometers.

When the analyte is in the liquid phase, the solution is sprayed from a capillary tube and converted into a fine mist by a heated nebulizer. Following desolvation, the gas is carried by a flow of nitrogen past a corona discharge electrode where ionization occurs. Typically, primary ions such as $N_2^{+\cdot}$, $O_2^{+\cdot}$ or $H_2O^{+\cdot}$, formed by electron ionization via electrons from the corona, collide with vaporized water molecules (or other solvent molecules) many times at atmospheric pressure, forming secondary reactant gas ions, which are clusters such as $H_3O^+(H_2O)_n$. These can be viewed as $[A + H]^+$ ions, where A is a cluster of water molecules. Polar molecules are given a charge according to the scheme

$$M + H_3O^+(H_2O)_n \longrightarrow [M + H + mH_2O]^+ + (n - m + 1)H_2O$$

The gas flow and ions are expanded through a sampling orifice into a region of intermediate pressure. Solvent molecules attached to the tertiary $[M + H]^+$ ions are stripped off as they pass into the vacuum region. The resulting $[M + H]^+$ ions enter the analyser and are separated in the usual way. Under suitable conditions, charge exchange or negative-ion chemical ionization can be made predominant.

The APCI method has become a popular ionization method for applications of coupled high-performance liquid chromatography/mass spectrometry (chapter 5). Bench-top devices for LC/MS incorporating APCI are commercially available and are particularly useful in environmental and pharmaceutical analysis.

3.3.3. *Conclusion*

Chemical ionization mass spectrometry is a versatile technique, compatible with GC/MS and LC/MS, and particularly suited to samples such as hydrocarbons, alcohols, amines, esters, amino acids, small peptides and nucleosides, which are susceptible to excessive fragmentation by the electron ionization method. When electron ionization gives molecular ions with too great an excess of internal energy to be stable, peaks corresponding to $M^{+\cdot}$ ions are not seen. In such cases, chemical ionization with a carefully selected reactant gas frequently identifies the relative molecular mass through formation of quasi-molecular ions, particularly protonated molecules. With methane and isobutane, $[M + H]^+$ ions are usually abundant unless the compound has very low proton affinity, as occurs with an alkane, in which case $[M - H]^+$ ions are prevalent. If molecular ions are absent from the EI spectrum because the neutral molecule decomposes thermally at the temperature required to volatilize it, chemical ionization also is unlikely to resolve the problem since it too requires volatilization of the sample. Then the only advantage of chemical ionization over electron ionization is that compounds may be volatilized at a somewhat lower temperature because passage of reactant gas over the sample aids evaporation. Substances that are scarcely volatile or are thermally unstable at the temperatures required for vaporization are best examined by *in-beam chemical ionization* (section 3.6), or by a method that does not require initial volatilization, such as fast atom bombardment, field desorption or electrospray mass spectrometry.

In EI studies, molecular ions often rearrange prior to fragmentation so that unexpected or misleading neutral particles may be ejected (section

10.9). This effect is much less marked with chemical ionization spectra, so the neutral particle losses tend to be more diagnostic of structure. Many of these points and aspects of structural elucidation with CI mass spectra are presented in chapter 10 and exemplified in chapter 11.

Preferably, reactant gases do not damage the ion source and filament, are readily and cheaply available, are highly volatile to allow rapid pumping out of the ion source when required and have low mass. Isobutane CI mass spectra are not useful below m/z 60 because of the presence of highly abundant ions from isobutane itself, whereas methane and ammonia CI spectra are largely free of reactant gas ions above m/z 20. An ideal reactant gas would afford only one type of reactant ion rather than a mixture of several different ions.

The large number of ion/molecule collisions occurring in an ion source during chemical ionization causes thermal equilibration to take place. As long as equilibration of vibrational and rotational energy is complete, it is permissible to use the same equations of reaction kinetics that govern the chemistry of ions in solution. For the formation of ions in the gaseous phase, rate constants are seen to obey the Arrhénius equation and activation energies and frequency factors are calculable. Proton affinities are also measurable by observing reaction (3.8) at equilibrium in the ion source at relatively high pressure.

$$[A+H]^+ + B \;\rightleftharpoons\; A + [B+H]^+ \tag{3.8}$$

The equilibrium constant (K) for the reaction is calculated through the abundances of $[A + H]^+$ and $[B + H]^+$ at equilibrium and the partial pressures of the two substances (A and B). The value of K is related to the total free energy change (ΔG°) by equation (3.9), where R is the gas constant and T, the temperature.

$$-\Delta G^\circ = RT \ln K \tag{3.9}$$

Since transfer of a proton between relatively large molecules, A and B, involves little entropy change, the total free energy change is virtually equal to the heat of reaction, ΔH°. For reaction (3.8), ΔH° equals the difference in proton affinities of A and B (equation (3.10)). If the proton affinity (PA) of A is known, that of B may be calculated:

$$\Delta H^\circ = PA(B) - PA(A) \approx RT \ln K \tag{3.10}$$

3.4.   NEGATIVE-ION CHEMICAL IONIZATION

At the low pressures obtaining in ion sources operating under electron ionization conditions, formation of negative ions is inefficient relative to that of positive ions. At pressures of about 1 Torr, there is a large increase in the number of negative ions produced. There are two mechanisms whereby this occurs, *electron capture* (equation (3.11)) and *reactant-ion chemical ionization* (equation (3.12)).

$$M + e^- + B \rightarrow (B.M^{-\bullet})^* \rightarrow M^{-\bullet} + B \qquad (3.11)$$

To form a negative ion by electron capture, a neutral molecule (M) must have a vacant, low-energy orbital in which to accommodate the extra electron. Compounds with conjugated double bonds, sulphur, phosphorus, nitro groups, or halogens will capture electrons, whereas alkanes will not. The latter compounds, methane in particular, may be used as *buffer* (*moderating*) *gases* (B) to remove the excess of energy from the negative ion as it is formed (equation (3.11)). Since electron capture is a three-body process, the pressure in the ion source should be as high as possible, at least 1 Torr. Although the process is not strictly chemical ionization because there is no *reactant* gas, the resultant mass spectra appear like CI spectra with abundant molecular anions and little fragmentation. Also, the pressure regime is akin to that in chemical ionization processes.

Electron capture can be a very sensitive method of analysis because the high mobility of the electron ensures that the rate of most electron capture processes is greater than those of typical ion/molecule reactions involving transfer of a much larger particle such as a proton. Strongly electrophilic compounds may be analysed at the femtogramme level by high-pressure electron capture. However, electron-capture mass spectra are critically dependent on the experimental conditions under which they are acquired. The temperature and pressure in the ion source, the purity of the buffer gas, the instrument used and its tuning conditions and the concentration of the analyte all affect the spectrum obtained. Therefore, electron capture ionization has a reputation for being irreproducible.

True negative-ion chemical ionization occurs by reaction of a compound (M) with thermally equilibrated, negatively charged reactant ions ($A^{-\bullet}$ or A^-). Several different types of ion/molecule reaction are observed during negative-ion chemical ionization, depending on the structure of the compound, the reactant gas and the pressure, but proton abstraction is one of the commonest and is depicted in equation (3.12).

$$M + A^- \rightarrow [M - H]^- + AH \qquad (3.12)$$

The $[M - H]^-$ ion, or *deprotonated molecule*, is frequently classed as a quasi-molecular ion. This is because the loss of a proton barely disrupts the structure of M; an $[M - H]^-$ ion can be considered to contain a 'nearly intact' original molecule. In the process (3.12) that generates an $[M - H]^-$ ion, much of the exothermicity of the reaction is carried away as vibrational energy in the new A–H bond of the neutral species, leaving a relatively stable $[M - H]^-$ ion. In positive-ion chemical ionization, much of the excess of energy in the formation of an $[M + H]^+$ ion resides initially in the newly formed bond of the ion. Therefore, negative-ion CI mass spectra show less fragmentation than many positive-ion CI spectra.

The more the basicity (proton affinity) of reactant ion A^- increases, the more likely is proton abstraction (equation (3.12)). Basicity increases in the order $Cl^- < F^- < CH_3O^- < O^{-\bullet} < HO^-$. The chloride ion, the principal product of ionization of dichloromethane at 1 Torr, is a very weak base and frequently reacts with hydrogen-bonding compounds such as acids, amides, amines and phenols to give $[M + Cl]^-$ adduct ions rather than $[M - H]^-$ ions. The reactivity of reactant ions increases with the amount of energy available in the ions, that is, in the order of their heats of formation: $F^- < Cl^- < HO^- \ll O^{-\bullet}$. The highly reactive $O^{-\bullet}$ ion, conveniently produced by ionization of nitrous oxide, causes many types of reaction, some of which are shown and exemplified in scheme (3.13). A more useful reactant ion for the elucidation of organic structures is HO^-, generated by ionization of a mixture of nitrous oxide and methane. Its high proton affinity promotes proton abstraction from virtually all compounds apart from alkanes. Being much less reactive than $O^{-\bullet}$, it does not display a wide variety of reaction types and generally it produces little fragmentation. Stearic acid (octadecanoic acid), 3-ethylpentan-3-ol and 2,4-pentanedione give only $[M - H]^-$ ions whilst ethyl ethanoate affords $C_2H_5OOCCH_2^-$ (60% TIC) and CH_3COO^- (40% TIC).

3.4.1. *Conclusion*

Negative-ion chemical ionization has many basic similarities to its positive-ion counterpart. Both are mild ionization techniques because molecules are

H· abstraction $\quad$ $O^{-\cdot} + CH_3CN \longrightarrow {}^{\cdot}CH_2CN + HO^-$

H$^+$ abstraction $\quad$ $O^{-\cdot} + C_6H_5CH_3 \longrightarrow C_6H_5CH_2^- + HO^{\cdot}$

H$_2$ abstraction $\quad$ $O^{-\cdot} + C_6H_6 \longrightarrow C_6H_4^{-\cdot} + H_2O$

$$(3.13)$$

H· displacement $\quad$ $O^{-\cdot} +$ $\longrightarrow$ $+ H^{\cdot}$

Alkyl displacement $\quad$ $O^{-\cdot} + C_2H_5COC_2H_5 \longrightarrow C_2H_5COO^- + C_2H_5{}^{\cdot}$

ionized by interaction with charged particles, the excess of energy of which has been dissipated by many collisions. Except for electron capture, the ionization step is a reaction of which there are several types. Under a given set of conditions, more than one *reaction* may be possible. Both chemical ionization modes allow the amount of fragmentation to be controlled to a degree by use of selected reactant gases. Like positive-ion CI, the negative-ion method is suited to combined chromatography/mass spectrometry. Frequently, structural elucidation is facilitated by an examination of both negative-ion and positive-ion chemical ionization mass spectra for they provide complementary information. Consider the chemical ionization mass spectra of L-phenylalanine shown in figure 3.4. Reactant gas ions, OH$^-$, are known to abstract protons from sample molecules and thus, the base peak at m/z 164 can be assigned to $[M - H]^-$ ions, revealing a relative molecular mass of 165. In contrast to the negative-ion spectrum, positive-ion chemical ionization with methane affords a base peak at m/z 166. Since $[M + H]^+$ ions are expected in this mode, the assignment of relative molecular mass is confirmed. Fragmentation in the two spectra is quite different and, as indicated in the foregoing discussion, is less extensive in the negative-ion spectrum. The only significant fragment ions in the latter arise from ejection of H$_2$ and of C$_7$H$_8$ from the $[M - H]^-$ ions. Partly because positive-ion and negative-ion CI spectra complement each other, instruments capable of measuring both more or less simultaneously have been developed (see next section).

Negative-ion chemical ionization is a useful and sensitive technique for compounds having high electron affinity, such as halides, polycyclic aromatic

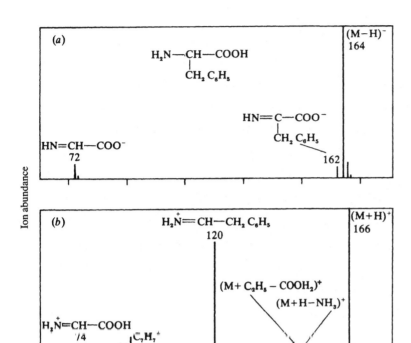

Figure 3.4. Chemical ionization mass spectra of L-phenylalanine. (*a*) Negative-ion chemical ionization with OH⁻ and (*b*) positive-ion chemical ionization with CH_5^+.

hydrocarbons, aromatic nitro-compounds and quinones. It is used to advantage particularly in the analysis of low levels of toxic polycyclic aromatic hydrocarbons, and chlorinated pesticides and herbicides. Compounds with lesser electron affinity are ionized less efficiently or not at all. The opposite situation obtains for positive-ion chemical ionization, so that, for example, hexafluorobenzene is scarcely ionized in the positive-ion CI mode. This difference in applicability makes imprudent direct comparisons of the sensitivity of the two techniques with a single compound. For analysis of small amounts of a substance, derivatization with an electronegative group (section 6.3.5) followed by negative-ion chemical ionization is likely to be two to three times more sensitive than derivatization with a relatively electropositive group followed by positive-ion chemical ionization.

Negative-ion chemical ionization (Jennings, 1977,1979; Budzikiewicz, 1986) and reactions and fragmentations of negative ions of organic, organometallic and coordination compounds have been reviewed (Bowie, 1975, 1977, 1979, 1984, 1985, 1987; Bowie, Trenerry and Klass, 1981; O'Hair and Bowie, 1989).

3.5. PULSED POSITIVE-ION/NEGATIVE-ION CHEMICAL IONIZATION

Usually referred to as PPINICI, this technique takes advantage of the fact that the analyser of a quadrupole mass filter passes ions of a given mass regardless of the sign of their charge. The method is not feasible with a magnetic sector mass spectrometer that is set up to record either positive or negative ions, not both simultaneously. The ion extraction potential of a conventional quadrupole mass filter is switched rapidly in polarity 10 000 times per second. This has the effect of ejecting from the ion source pulses (or 'packets') of positive and negative ions. After traversing the analyser, the positive ions are attracted to one detector and negative ions to another. The amplified signals from the two detectors are positive-ion and negative-ion CI mass spectra recorded essentially simultaneously (Hunt, Stafford, Crow and Russell, 1976). Of course, the reactant gas must be suitable for both positive-ion and negative-ion chemical ionization. The use of argon affords charge-exchange mass spectra (rather like EI spectra) in the positive-ion mode together with electron-capture mass spectra in the negative-ion mode. When ionized at a pressure of 1 Torr, methane gives reactant ions CH_5^+ (conventional positive-ion CI mass spectra) and electrons (electron-capture spectra). A mixture of methane and nitrous oxide provides CH_5^+ and HO^- reactant ions for simultaneous chemical ionization.

3.6. MODIFIED ELECTRON AND CHEMICAL IONIZATION

The effect of temperature on a mass spectrum is discussed in section 10.3. Often, especially for thermally labile substances, the rate of heating rather than the absolute temperature seems to be the deciding factor in the competition between evaporation (breaking of intermolecular bonds) and decomposition (breaking of intramolecular bonds). Very rapid heating favours evaporation because there is less time for the molecules to transfer energy into dissociative vibrational states and because the molecules spend less time in the region of relatively high gas density around the sample probe where bimolecular interactions may lead to decomposition. Rapid heating, sometimes called *flash desorption*, can be used in conjunction with electron ionization or chemical ionization. Rapid rates of heating vary between 10°C s^{-1} and 5000°C s^{-1} and mass spectra of involatile organic salts such as acetylcholine chloride (**6**) and thermally unstable compounds like steroid glycosides can be recorded. When electron ionization is employed, the typical fragmentations of normal EI mass spectra are observed as well as enhanced abundances of molecular ions (Daves, 1979).

$$CH_3COOCH_2CH_2-\overset{+}{N}(CH_3)_3\,Cl^-$$

6

In many mass spectrometers with direct insertion probes, once molecules of the sample are volatilized from the probe, they must travel some distance before they enter the beam of electrons where ionization occurs. During this flight time, unimolecular decomposition of thermally labile compounds is possible. In an ion source used for CI, the volatilized molecules travel to the region rich in reactant gas ions (the *ion plasma*), which is equivalent to the electron beam in EI ion sources. Because of the higher pressures used in chemical ionization, unimolecular or bimolecular decomposition may occur before ionization. Decomposition can be minimized by using long direct probes such that the molecules of sample are volatilized directly into the electron beam or ion plasma. Several terms have been coined for this technique, the common ones being *in-beam ionization*, *direct EI* or *direct CI*, *desorption EI* and *desorption CI*, but in fact the method is much older than these names. For many years, users of older mass spectrometers with direct probes of adjustable length have found that the best mass spectra are obtained with the longest probes and with the sample coated on the outside rather than the inside of the cup at the tip of the probe. This arrangement ensures greater exposure of the sample to the electron beam. The 'rediscovery' of the method has brought some refinements (Horning *et al.*, 1979). For example, instead of using simple probe tips of glass, samples have been coated on gold, Teflon, quartz or other inert surfaces in an effort to reduce irreversible adsorption of the sample on the probe. The rapid heating technique can be combined with the in-beam method when the sample is loaded onto the wire of a heating element that can be placed in close proximity to the zone of ionization. The success of the technique is due to three factors: minimal distance between sample and electron beam or ion plasma, evaporation from an inert surface and rapid heating. Examples of the in-beam CI technique are shown in figure 3.1(*c*) for arginine and figure 3.5, which compares the conventional CI and in-beam CI mass spectra of cholesterol glucoside (**7**) with ammonia as reactant gas. Note that quasi-molecular ions $[M + NH_4]^+$ at *m/z* 566 are observed

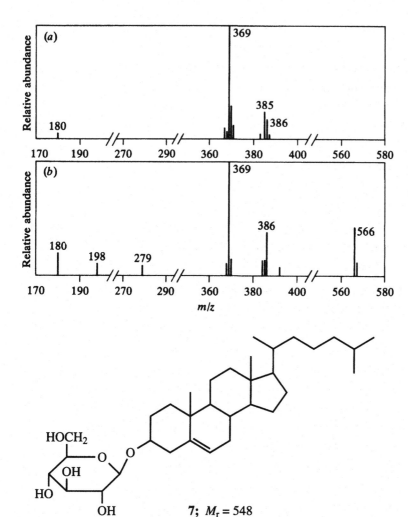

Figure 3.5. Partial mass spectra of cholesterol glucoside (7) obtained by chemical ionization with ammonia, (a) using a conventional direct insertion probe, and (b) using a longer, rapidly heated, direct insertion probe (in-beam CI spectrum).

7; $M_r = 548$

only with the in-beam method. Ions at m/z 385 and 369 are due to the steroid moiety with or without the glucosidic oxygen atom, respectively. The other ions near these two values arise from hydrogen transfers. The glucose unit is also represented at m/z 198 $[C_6H_{12}O_6 + NH_4]^+$ and m/z 180 $[C_6H_{10}O_5 + NH_4]^+$, these peaks being clearer in the in-beam spectrum.

The in-beam technique may also be used in the negative-ion mode. In-beam negative-ion CI yields useful mass spectra for polyhydroxylated compounds like (7) with OH^- as reactant gas ions, and underivatized di-, tri- and tetra-saccharides with Cl^- as reactant gas ions give abundant $[M + Cl]^-$ ions and some diagnostic fragment ions due to elimination of one or two sugar units.

These techniques are still used by mass spectrometrists but, for many applications, they have been superseded by other methods like fast atom bombardment and electrospray mass spectrometry.

3.7. FIELD IONIZATION

Molecules approaching or adsorbed on a surface of high curvature maintained at a high positive potential are subject to potential gradients of about 10^7–10^8 V cm^{-1}. Under the influence of these fields, quantum tunnelling of a valence electron from the molecule to the anode takes place in about 10^{-12} s to give a radical cation. Being positively charged, the ion is accelerated from the anode and is analysed and detected in the same way as for electron ionization. The probability of a molecule being ionized increases exponentially with the field strength. The field strength (F) may be increased by increasing the voltage (V) applied to the anode or by decreasing the radius (r) of the microscopic tips or fine points on the surface of the anode (equation (3.14)):

$$F = V/kr \qquad (3.14)$$

The value of k, a geometry factor, varies. For a sphere $k = 1$ and for other shapes $k > 1$. A wire or blade is microscopically uneven, but usually produces fields that are too low for efficient ionization. Anodes are 'activated' by growing on them microneedles ('whiskers'), usually of carbon and about 10^{-5} m in length, which provide the surface with many fine points with radii far smaller than on the original surface. The result, a *multi-point array*, increases the efficiency of ionization by one to three orders of magnitude.

For field ionization, the sample is allowed to impinge into the region of high field strength, and this arrangement is suitable for GC/MS operation. Field ionization mass spectra are generally characterized by low degrees of fragmentation compared with electron ionization (figure 3.6). The formation of the molecular ion by tunnelling is thought not to produce ions in highly excited states and therefore with not enough excess of energy for extensive fragmentation. For example, electron ionization of heptane at 70 eV, the normal operating voltage, produces molecular ions with average excess of internal energy of about 6 eV whilst the same ions formed by field ionization have only 0.5 eV excess of internal energy. Sometimes [M + H]$^+$ ions are formed, rather than M$^{+\bullet}$ ions, because proton transfer may occur by ion/molecule reactions in the ionizing regions close to the anode which are

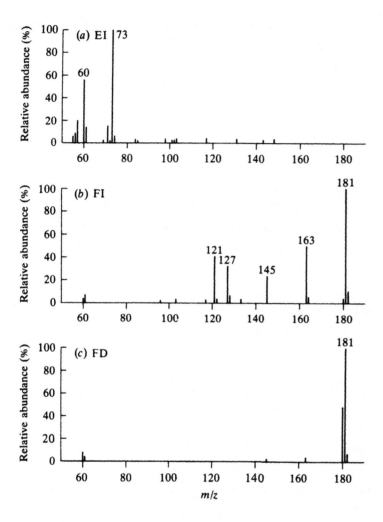

Figure 3.6. (*a*) Electron ionization, (*b*) field ionization and (*c*) field desorption mass spectra of D-glucose, $M_r = 180$.

relatively dense with molecules (M). The $[M + H]^+$ ions are distinguishable from $M^{+\cdot}$ ions because the relative abundance of the protonated molecules increases with pressure and decreases with the temperature of the anode. In field ionization, fragmentation can often be explained as cleavage of a molecular ion that is strongly polarized in the electric field, although some ions may not be true fragment ions (derived from $M^{+\cdot}$ or $[M + H]^+$ ions), but rather the molecular ions of thermal decomposition products of neutral molecules. Fragmentation may also occur through ion/molecule reactions in the condensed phase around the electrode.

Because of steric constraints (entropic factors), rearrangement processes have lower reaction frequencies than do simple bond-cleavage reactions (section 9.7). It has been suggested therefore that, because ions spend a shorter time in an ion source for field ionization than in one for electron ionization, there is little time for rearrangement during field ionization and few such processes should be observed. However, rearrangements have been observed in strong electric fields, partly at least, *before* ionization. For example, $CH_3COCH_2CH_2CH_3$ yields abundant ions at m/z 58 resulting from loss of C_2H_4 prior to ionization.

Field ionization mass spectra are frequently different for isomers and are therefore useful in structural organic chemistry. This is often not the case for electron ionization spectra.

3.7.1. *Negative-ion field ionization*

By applying a high negative voltage to the electrode in a field ionization source (FI), electrons may tunnel from the cathode to a compound having a high electron affinity. Negative-ion FI spectra are quite simple. For example, molecular ions, $M^{-\cdot}$, of halogenated quinones form virtually all of the total ion current. These ions do not fragment, but ion/molecule reactions may result in $(2M)^{-\cdot}$ ions in low abundance.

3.7.2. *Field ionization kinetics*

When ionization occurs on the surface of the electrode, maintained at a potential V_0, ions are subject to the full electric field and are drawn into the analyser with a translational energy of zeV_0 to be detected as a normal narrow peak in the mass spectrum (ze is the total charge on the ion). Ionization or fragmentation close to, but not at, the surface gives ions with less than the full kinetic energy. Such ions give rise to 'tails' on the low-mass side of normal ion peaks and are caused by decomposition of the so-called, *'fast' metastable ions*. In the case of a molecular ion of mass m_1 decomposing to an ion of mass m_2 some distance from the electrode, its translational energy zeV_i is given by equation (3.15) in which V_x is the potential at the position of decomposition:

$$zeV_i = (m_2/m_1)ze(V_0 - V_x) + zeV_x \qquad (3.15)$$

Values of zeV_i fall between zeV_0 for $V_x = V_0$ and $(m_2/m_1)zeV_0$ for $V_x = 0$. The translational energy zeV_i may be measured accurately with a double-focussing, magnetic sector mass spectrometer (chapter 8) so the value of V_x is calculable. If the distribution of potential around the electrode and the ion trajectory are calculated, it is possible to determine the time between ioniza-tion and decomposition at potential V_x. This value is called the *ion lifetime* and is of the order of $10^{-12}–10^{-9}$ s. At very short ion lifetimes, the method is prone to error because of uncertainty in the potential and exact position of ionization on the microneedles grown on the electrode. To reduce this error, field ionization may be carried out with unactivated electrodes. In practice, the kinetics of fragmentation of molecular ions at specific ion lifetimes between 10^{-11} and 10^{-9} s can be studied. Within these short lifetimes, the molecular ions are assumed not to have undergone any structural changes from the neutral molecule. The importance of field ionization kinetics lies in the ability to determine rates of decomposition of molecular ions of known structure (Derrick, 1977; Nibbering, 1984). If an ion fragments after leaving the ion source and before detection, it behaves like a normal ('*slow*') meta-stable ion as in electron ionization studies (section 1.5 and chapter 8).

Elucidation of fragmentation pathways by field ionization kinetics is aided considerably by isotopic labelling. Fragmentation of 3-phenyl-propanal in the FI mode includes loss of C_2H_2O to yield a mass peak at m/z 92. In the case of the labelled analogue (**8**), the deuterium atom is trans-ferred to the aromatic ring during fragmentation within $10^{-10.2}$ s (scheme (3.16)).

$$ \tag{3.16} $$

At longer lifetimes, the deuterium atom may interchange with the hydro-gen atoms in the *ortho*-position of the ring or in the benzylic position (Wolkoff, van der Greef and Nibbering, 1978). In electron ionization studies, which give an unresolved view of all the processes occurring in the first 10^{-6} s after ionization, the interchange processes leading to hydrogen randomiza-tion do not allow observation of the specific transfer shown in scheme (3.16).

3.7.3. *Conclusion*

For structure determination, one of the most important facets of field ion-ization is its ability to give abundant molecular ions as illustrated in figure 3.6 for D-glucose. The EI mass spectrum for this sugar contains little diag-nostic information for the purposes of structural elucidation because of thermal decomposition and excessive fragmentation, but the FI spectrum has a base peak of $[M + H]^+$ ions (m/z 181) and shows highly diagnostic, consecutive losses of water (m/z 163, 145 and 127). Field desorption, a modified field ionization technique to be described in the next section, occasions even less fragmentation. For relative molecular mass assignment, FI has the advantage over CI and FD that only $M^{+\bullet}$ or $[M + H]^+$ ions are formed; adduct ions, $[M + X]^+$, in which the nature of X may not be known with certainty, do not usually occur. For the physical chemist, field ionization is a convenient source of kinetic data of fragmentation processes occurring at specified times in a wide time scale. Normal peaks reflect reactions occurring within 10^{-12} s of ionization, fast metastable ions cover the range 10^{-11}–10^{-9} s and slow metastable ions, the range 10^{-8}–10^{-6} s.

Field ionization has been described in a book (Beckey, 1977) and reviewed (Milne and Lacey, 1974; Derrick, 1977; Schulten, 1979; Lattimer and Schulten, 1989).

3.8.   FIELD DESORPTION

The technique of field desorption has been reviewed with emphasis on applications in the life sciences (Milne and Lacey, 1974; Beckey and Schulten, 1975; Derrick, 1977; Schulten, 1977, 1979; Lattimer and Schulten, 1989). It is also described in specialized books (Beckey, 1977; Prokai, 1990). Unlike the other techniques described so far in this chapter, it is not suitable for combined chromatography/mass spectrometry. For field ionization, a compound in the gas phase is ionized near or at the surface of the field electrode. This can result in at least partial decomposition of thermally labile compounds. In the field desorption method, the ionization step occurs by the same mechanism as for field ionization but the substance is first deposited onto the activated electrode *(emitter)* as a thin layer by dipping the emitter in a dilute solution of the compound or by applying the solution to the electrode with a syringe. Once the solvent has evaporated, the emitter is inserted into the ion source and samples are ionized in the condensed phase without heating them prior to ionization, thereby reducing greatly any possible thermal decomposition. For samples of very low vapour pressure, mild heating of the electrode may be required to desorb them. As with field ionization, $M^{+\bullet}$ or $[M + H]^+$ ions may be formed; both arginine (1) and D-glucose (figures 3.1 and 3.6) yield the latter. Field desorption may also give rise to adduct ions such as $[M + Li]^+$, $[M + Na]^+$ and $[M + K]^+$ when traces of metal salts are present in the sample. This *cationization* may be induced purposely by adding to the sample a salt such as LiCl or, if the compound is sufficiently acidic, by making a salt of the sample. The uncertainty in the assignment of relative molecular mass is thereby reduced since adduct ions of known constitution are anticipated.

When heating is required to cause desorption, it is brought about directly by passing a current through the emitter, or indirectly by radiation from a separate heater or with the aid of a laser. The last of these methods has allowed observation of the molecular ion of the complex organometallic porphyrin, vitamin B_{12} (m/z 1354) together with several fragment ions. With direct heating of the emitter, the *best anode temperature* for a particular compound is the temperature at which molecular or quasi-molecular ions are most abundant. Above this temperature, thermal degradation may occur giving rise to 'fragment ions', which are, in fact, molecular ions of pyrolysis products. For many compounds, there is a very narrow range of temperature over which a reasonable FD mass spectrum is obtained. This can give rise to difficulties of reproducibility.

Arginine (1), D-glucose and compounds (9), (10) and (11) are examples of compounds that have EI mass spectra with molecular ions of low or insignificant abundance. By field desorption, all five compounds afford molecular and/or quasi-molecular ions with little, if any, fragmentation. Whilst the assignment of relative molecular mass is very important, structural identification necessitates some fragmentation. This information may be obtained in any of three ways: (*a*) by raising the temperature to observe thermal fragmentation, (*b*) by inducing fragmentation of desorbed molecular ions by collision with a neutral gas, or (*c*) by measuring the EI mass spectrum. In method (*b*), *collisional activation*, a region of the mass spectrometer is held at relatively high pressure by 'leaking' into it a gas such as argon. As molecular ions pass through this region of relatively high gas pressure, ion/molecule reactions give rise to fragmentation from which the product ions are observed. Collisional activation is described more fully in chapter 8.

Sample preparation for field desorption is often tedious, and finding and maintaining optimal experimental conditions requires both skill and luck! For many, but certainly not all, applications, FD has been superseded by the experimentally simpler method of fast atom bombardment, which is particularly favoured for polar and ionic compounds. However, there are

9 Mycotoxin (mol. wt = 466)

10 Glutamic acid
(mol. wt = 147)

11 Coproporphyrin-III
(mol. wt = 654)

situations in which only field desorption provides a worthwhile mass spectrum, particularly in the case of large non-polar analytes. For instance, ions up to m/z 11 000 have been observed in the FD mass spectrum of polystyrene. Since compounds are not volatilized prior to ionization, organic salts and inorganic compounds as well as involatile, thermally labile, nonionic compounds may be examined.

3.9.   SPRAYING AND ION EVAPORATION

Techniques like thermospray and electrospray consist of two basic processes: (i) the generation of a fine mist of charged droplets from a solution containing the analyte, and (ii) vaporization of the solvent to give ions of the analyte. Strictly, terms like thermospray and electrospray refer to the former process, not to the ionization step (ii). Ion evaporation is the generic term applied to the process whereby ions are emitted into the gas phase from the charged droplets in the spray. The concept of ion evaporation is still vigor-ously contested (Iribarne and Thompson, 1976; Iribarne, Dziedzic and Thompson, 1983; Roellgen, Bramer-Weger and Buetfering, 1987; Schmelzeisen-Redeker, Buetfering and Roellgen, 1989) but, to date, it is generally regarded as a contributing ionization mechanism in thermospray and as the major ionization process in electrospray. Despite the misgivings about ion evaporation, it will be described briefly here after considering the conditions necessary for creating charged droplets.

When a liquid is forced through a nozzle and pneumatic, thermal or electrostatic forces (or some combination of those forces) are applied, a fine spray of charged droplets can result. Several of the conceivable combina-tions have already been investigated (Fenn, Mann, Meng, Wong and Whitehouse, 1989, 1990; Smith, Loo, Edmonds, Barinaga and Udseth, 1990; Bruins, 1991). Here, the main two methods of thermospray and electrospray will be compared and contrasted, with brief reference to other devices as variants on the same theme.

For thermospray, heat is applied to the tip of a capillary tube, through which a solution containing the analyte and an electrolyte is passed. The shear and accelerating forces resulting from the sudden expansion into a vacuum produce droplets. In each droplet, there may be an excess of either anions or cations of the electrolyte. This charging is determined by statistics, rather than by physics or chemistry. Therefore, individual droplets carry a net charge but, overall, there are equal numbers of each polarity. Typically, aqueous solutions containing ammonium ethanoate are used, so some drop-lets have a net excess of NH_4^+ ions and some of CH_3COO^- ions. In contrast, electrospray is normally effected at ambient temperature and atmospheric pressure, and a typical solvent for positive-ion analysis would be water/methanol or water/acetonitrile (50:50), perhaps with a small per-centage of ethanoic or methanoic acid to act as a source of protons. The solution passes through a capillary tube to the ion source where application of a large electric field to the end of the tube (several kilovolts relative to a nearby counter-electrode) disrupts the emerging liquid surface and pro-vides a spray of highly charged droplets. The net charge on each droplet has the same polarity, either positive or negative, dependent on the polarity of the applied field. Droplet generation in electrospray can be assisted by a pneumatic sprayer in which an annular flow of gas helps to shear droplets from the liquid stream (Bruins, Covey and Henion, 1987). This is sometimes called *ionspray* and allows larger flow rates than electrospray alone (up to about 400 µl min⁻¹, as opposed to about 20 µl min⁻¹ for unassisted electrospray). Modern source designs for electrospray, incorporating devices to aid the evaporation of the solvent, permit flows of up to about 5 ml min⁻¹ but optimum sensitivity is usually achieved at much lower flow rates than the permitted maximum.

To some extent, the ionization process occurring in these devices involves the evaporation of ions from the droplets as the solvent is removed. In thermospray, rapid pumping removes the solvent whereas, in electrospray, the solvent is removed in a flow of warm, dry gas. The latter process can be aided by passing the sample through a heated capillary tube that acts as a desolvation chamber. In both thermospray and electrospray, the charged droplets shrink, thus concentrating the charge density. When the repulsive force between the charges becomes similar to the surface tension, a droplet 'explodes', producing smaller droplets that also decrease in size. Through this cascading effect, eventually the charge density on the minute droplets is great enough to desorb ions (quasi-molecular ions) suitable for analysis according to their *m/z* values. This final stage of the mechanism is akin to field desorption. However, the whole ion evaporation process, depicted highly schematically in figure 3.7, is uncertain and should not be regarded as beyond criticism (Roellgen, Bramer-Weger and Buetfering, 1987; Schmelzeisen-Redeker, Buetfering and Roellgen, 1989; Smith, Loo, Edmonds, Barinaga and Udseth, 1990). Involatile particles form a residue in the ion evaporation process so the ion source needs periodic cleaning.

When the droplets are strongly heated (thermospray), neutral molecules

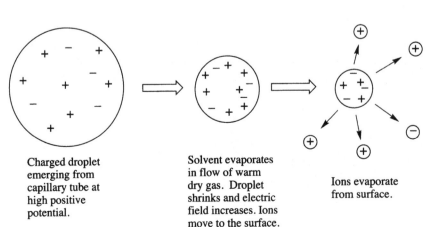

Figure 3.7. A schematic representation of the ion evaporation process.

Charged droplet emerging from capillary tube at high positive potential.

Solvent evaporates in flow of warm dry gas. Droplet shrinks and electric field increases. Ions move to the surface.

Ions evaporate from surface.

can also be evaporated. These undergo ion/molecule reactions with vaporized electrolyte ions (usually NH_4^+ ions). In other words, chemical ionization also occurs (like that observed with ammonia as reactant gas). This may be the major ionization mechanism in thermospray systems. In ambient temperature devices (such as electrospray and ionspray), chemical ionization is not a significant contributor because most analyte molecules are not sufficiently volatile to desorb as neutral molecules. Only ions are evaporated from the aerosol droplets.

In another contrast between thermospray and electrospray, multiply charged ions hardly occur in conventional thermospray devices. The chief ions observed are $[M + H]^+$, $[M + NH_4]^+$, $[M - H]^-$, $M^{-\bullet}$ or $[M + Cl]^-$. On the other hand, the cold, atmospheric-pressure conditions of electrospray encourage multiple charging of the ions. Molecules that have many sites capable of being protonated or deprotonated (e.g. basic -NH_2 groups in proteins, or acidic phosphate groups in polynucleotides) can evaporate as multiply charged quasi-molecular ions, $[M + zH]^{z+}$ or $[M - zH]^{z-}$, respectively. The choice between positive-ion and negative-ion operation would depend on the nature of the sample. The better polarity for a given sample is the one in which the analyte molecules are more able to accommodate the charge. Basic compounds are best examined as their conjugate acids in the positive-ion mode and acids are best analysed as their conjugate bases in the negative-ion mode. For instance, a protein with a relative molecular mass of 19 980 and with 20 protons attached will produce $[M + 20H]^{20+}$ ions at m/z (19 980 + 20)/20 = 1000. The effect of increasing the charge, z, well above unity is to bring molecules with large masses, m, into the m/z range of a simple mass spectrometer like a quadrupole system (although electrospray can also be effected with other instruments such as magnetic sector, ion trap and Fourier-transform systems). Molecules with relative molecular masses over 300 000 have been examined with a quadrupole system when they have been capable of carrying about 100 charges. With a Fourier-transform ion cyclotron resonance spectrometer, the relative molecular mass of a polymer has been measured as 6 500 000. Fragmentation of large molecules is avoided in the electrospray method, unlike the high-temperature thermospray method, in which some fragment ions are generally observed (figures 3.1(f) and 3.8). Figure 3.9 shows the electrospray mass spectrum of a protein, horse heart myoglobin.

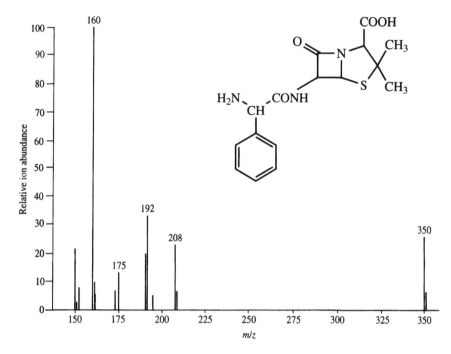

Figure 3.8. The thermospray mass spectrum of ampicillin, $M_r = 349$.

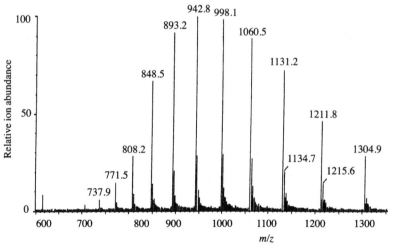

Figure 3.9. The positive-ion electrospray mass spectrum of about 1 pmol of horse heart myoglobin in water/methanol (50:50) with 1 per cent ethanoic acid. The flow rate was about 3 μl min^{-1}.

As shown in figure 3.9, the electrospray mass spectrum of a large basic compound (horse heart myoglobin) in a solution containing a source of protons (often ethanoic acid or trifluoroethanoic acid) consists of a series of $[M + zH]^{z+}$ ions, each peak differing from its neighbour by one electronic charge. That is, $[M + zH]^{z+}$ ions will be adjacent to $[M + (z + 1)H]^{(z+1)+}$ ions. The objective of the analysis is to assign the relative molecular mass, M_r, but the number of charges on any given ion is not immediately obvious from the resulting peak. Thus, simple observation of one peak at one m/z ratio is not enough because the value of z is not known. However, M_r is easily calculated if *two* peaks are examined. Take the example of two adjacent peaks, differing by one charge. It is assumed that the charge arises solely from protonation and not from addition of alkali metals, say, $[M + (z - 1)H + Na]^{z+}$ ions. It is also safely assumed that only intact molecules were ionized, that is, there are no interfering fragment ions. *Two* mass-to-charge ratios would be measured and there would be two unknown quantities (the charge, z, and the relative molecular mass, M_r). Mathematically, this can be solved for both z and M_r by setting up simultaneous equations. For example, in figure 3.9, peaks for $[M + zH]^{z+}$ and $[M + (z + 1)H]^{(z + 1)+}$ ions occur on the mass-to-charge scale at 1131.2 and 1060.5, respectively. Then:

$$1131.2 = \text{mass-to-charge ratio} = (M_r + z)/z$$

and

$$1060.5 = \text{mass-to-charge ratio} = (M_r + z + 1)/(z + 1)$$

Solving these simultaneous equations for z gives 14.99. As the charge must be an integer, this figure is refined to the nearest whole number. Substitution of $z = 15$ into the equations gives $M_r = 16\,953.0$ and $16\,952.0$. Usually, there is a series of peaks differing by one mass unit from each other and so several pairs of m/z values can be used in this computation. Therefore, this mathematical process is repeated for all of the peaks observed and an average value for relative molecular mass computed. In fact, it turns out that the measured relative molecular mass for horse heart myoglobin is 16 951.5. An accuracy of about 0.01% can be expected for such a measurement of molecular mass. Many data systems transform the initially observed series of peaks into one peak representing the hypothetical singly charged molecular ion. Such a mathematical treatment simplifies the interpretation of the spectrum. This treatment of the results is particularly helpful for interpretation of the electrospray spectra of mixtures because each series of peaks is replaced by one peak. Application of statistical computing packages can enhance significantly the interpretation of complex data from mixtures: frequently, 'hidden' (unresolved) peaks can be found and deconvoluted.

With smaller molecules, the results from electrospray mass spectrometry are just as spectacular. Again taking the positive-ion mode as an example, electrospray produces quasi-molecular ions $[M + zH]^{z+}$. As already noted, for large molecules, z can be large and values over 100 have been reported, depending mainly on the number of sites capable of being protonated, the pK_a of the compound and the pH of the medium. Small molecules are likely to have only one basic site and so z will be unity. That is, conventional protonated molecules $[M + H]^+$ will be produced and these reveal the value of M_r simply by subtracting one from the observed m/z value. The example of arginine was given in figure 3.1(g) at the start of this chapter. Likewise, molecules with one acidic site will produce $[M - H]^-$ ions at an m/z value of one less than M_r. Organic salts subjected to negative-ion electrospray will exhibit a spectrum of the anion. This effect is seen in figure 3.10 and example F in chapter 11. If fragmentation of such relatively small molecules is required for identification purposes, it can be enhanced by adjusting the voltage applied to a nozzle (cone voltage) in the ion source, that is, by appropriate tuning of the ion source.

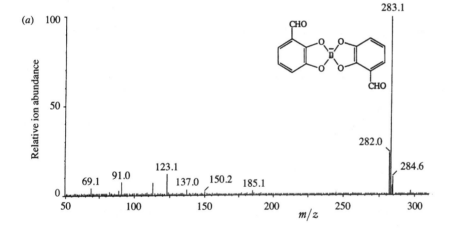

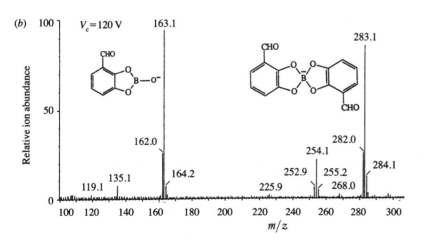

Figure 3.10. Negative-ion electrospray mass spectra of the spiroborate shown ($M_r = 283$) in water/acetonitrile (50:50) at a flow rate of 5 μl min^{-1}. Normally, the voltage on the nozzle is kept low and little or no fragmentation occurs (a). The peaks at m/z 69–185 are associated with 'chemical background' from the solvent impurities. Increasing the nozzle voltage causes some fragmentation (b).

A typical application for electrospray would be the confirmation of successful chemical synthesis of a peptide or protein. Activation of the AIDS virus relies on an enzyme called HIV-1 protease, which it would be useful to synthesize for inhibition studies. The chemical synthesis of the sequence of residues 59–99 of HIV-1 protease was followed by mass spectrometric checking for the correct structure. The crude product gave a complex ion-spray spectrum (figure 3.11(*a*)), which, nonetheless, did contain some of the expected peaks for the required product. These are marked with an asterisk. Following chromatographic purification, a far 'cleaner' ionspray spectrum was observed, as shown in figure 3.11(*b*). Thus, an indication of purity and a confirmation of the synthesis were obtained (Bayer, 1991).

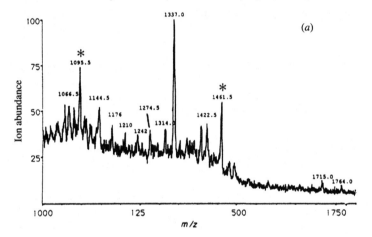

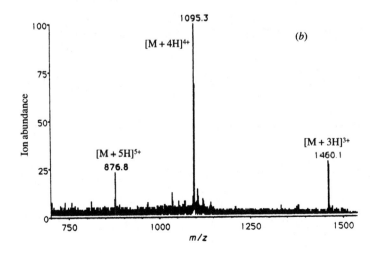

Figure 3.11. The positive-ion ionspray mass spectrum of synthetic HIV-1 protease sequence 59–99 (*a*) before purification (the asterisks indicate the peaks expected for the required product) and (*b*) after purification. The M_r was found to be 4377.8. (Taken with kind permission from Bayer (1991).)

Table 3.1. *The characteristics of thermospray and electrospray mass spectrometry.*

	Method	
	Thermospray	Electrospray
Agent producing spray of charged droplets	Heat	Electric field
Charge on droplets	Both polarities	One net polarity
Temperature	Hot	Ambient
Thermal decomposition	Thermally labile compounds are susceptible to decomposition	None
Pressure	Usually a vacuum	Atmospheric
Ionization mechanism	Chemical ionization and ion evaporation	Ion evaporation
Number of multiply charged ions	Few	Many (**if** molecules have several sites capable of carrying charge)
Fragmentation	Some	None (but can be initiated by appropriate tuning of ion source)
Type of solution	Aqueous ammonium ethanoate or similar	Water/polar organic solvent/pH adjuster
Variants	Electrons from a filament or glow discharge to aid ionization (latter is *plasmaspray*).	Warm nebulization gas to to aid solvent evaporation allows larger liquid flows (*ionspray*).
	Atmospheric pressure thermospray and electrically assisted thermospray increase multiple charging	Both pneumatic and thermal assistance aid evaporation but decrease charging

The main characteristics of thermospray (Blakley and Vestal, 1983; Vestal, 1983) and electrospray (Fenn, Mann, Meng, Wong and Whitehouse, 1990; Mann, 1990; Smith, Loo, Edmonds, Barinaga and Udseth, 1990; Mann and Fenn, 1992) are summarized in table 3.1. Their utilization as sophisticated detectors for liquid chromatography and capillary electrophoresis is discussed and illustrated in chapter 5.

3.10. HIGH-SPEED PARTICLES

Thermospray was developed in the early 1980s. The same period also saw the popularization of one of the most significant mass spectrometric methods for handling involatile samples: fast atom bombardment (Barber, Bordoli, Sedgwick and Tyler, 1981; Barber, Bordoli, Elliott, Sedgwick and Tyler, 1982; Rinehart, 1982). Like field desorption, fast atom bombardment (FAB) does not require volatilization of neutral molecules by heat prior to ionization and hence it found most application to large, polar, ionic and/or thermally unstable molecules, frequently biomolecules like peptides. The FAB method was accepted rapidly, often displacing FD, because of its operational simplicity and its ability to handle a wide range of analytes that, before the advent of modern FAB, had been difficult or impossible to examine by mass spectrometry. It transpires that it makes little difference to the mass spectrum obtained whether the sample is bombarded with atoms such as Xe or ions such as Cs^+. What is more significant is the way in which the sample is presented to the rapidly moving particles. For examination of organic compounds, the discovery that the best spectra are obtained when the sample is dissolved in a solvent (or 'liquid matrix') was of paramount importance.

3.10.1. *Fast atom bombardment*
A beam of ions such as $Xe^{+\cdot}$ can be produced by ionizing xenon atoms and accelerating the resulting ions through an electric field of about 6–8 kV. The resulting (fast) ions are directed through a xenon gas chamber when charge exchange occurs (equation (3.17)) to give fast atoms (most of the original direction and kinetic energy of the fast ions is maintained in the resulting fast atoms):

$$Xe^{+\cdot}(fast) + Xe(thermal) \rightarrow Xe(fast) + Xe^{+\cdot}(thermal) \qquad (3.17)$$

The excess of fast xenon ions can be deflected, leaving a beam of fast atoms. Other gases can be used to produce a beam of fast atoms, as for example with argon. When such beams of fast atoms impinge onto (bombard) a metal plate coated with the substance being investigated, the large amount of kinetic energy in the atoms is dissipated in various ways, some of which lead to volatilization by momentum transfer and ionization of the sample. The best results are obtained by coating the plate with a relatively involatile liquid matrix such as glycerol and mixing the substance under investigation into the liquid matrix (Gower, 1985; De Pauw, 1986, 1989). Ionization of polar molecules occurs at or just above the surface of the solution by ion/molecule reactions. Strictly, FAB does not act as the sole ionization method for compounds that are already charged in solution (so-called *pre-ionized* species). Ionic compounds, amines in acidic solution and acids in basic solution are examples of analytes that exist as ions before atom bombardment begins. In such cases, FAB provides the energy to separate the ions from their counter-ions in solution and to desorb those ions from the solution. By maintaining a suitable electric gradient in the region of the plate, either positive or negative ions can be directed into the analyser of the mass spectrometer.

In a simple modification to the approach using a direct insertion probe with a platform for a static solution, the probe can be made hollow and a continuous flow of the solution can be pumped through it to a hole or frit at its tip where the solution is bombarded by fast atoms (FAB) or fast ions (SIMS). Typically, the solution, perhaps the effluent from a liquid chromatograph or capillary electrophoresis system, would contain 1–10% of glycerol or some other solvent of relatively low vapour pressure to aid the FAB ionization process. The modification has been termed *continuous-flow FAB* or *dynamic FAB* (Caprioli, 1990*a*, 1990*b*). As well as being a simple means of effecting LC/MS (chapter 5), it is a type of 'flow injection analysis' whereby sample throughput is much greater than with a conventional direct insertion probe. Samples can be injected at rapid intervals into the solvent flowing towards the probe tip where FAB mass spectra are obtained in rapid succession. An additional advantage of the flow method of FAB lies in a decrease in the abundances of background ions ('grass') resulting from general decomposition produced by the bombardment. In the static mode of FAB, these fragments concentrate in the matrix and are ionized together with sample molecules; the ions appear as background. In dynamic FAB, the bombarded solution is continuously renewed so there is much less time for background to build up and the ions from sample molecules show up more clearly.

The presence of a liquid matrix is important to the success of the FAB method. Once the top few monolayers have been exposed to atom bombardment and the analyte molecules sputtered into the analyser in the conventional static FAB method, further molecules can diffuse in from the bulk of the solution. Thus, the solution acts as a reservoir of sample molecules continually replenishing the surface. It is this effect that causes FAB mass spectra from one solution to persist for up to 20 min. In a solid or glassy phase, diffusion is far slower and generally FAB spectra of organic solids do not persist for a convenient length of time. A particular advantage of the solution phase is that its chemistry (often the pH) can be manipulated to optimize the required analysis. For example, the yield of ions from ionic compounds is greater than that for non-ionic materials. Therefore, it is often advantageous to add to the solution a reagent that will form ions from the analyte molecules (chapter 6). A particular disadvantage of the liquid matrix is a background of peaks resulting solely from its presence (e.g. m/z 93, 185, 277, and so on for $[(glycerol)_n + H]^+$ ions). Even this disadvantage can be put to good use because the background peaks can be used as markers of accurately known mass against which to measure the accurate mass of the analyte ions (see section 4.4.1).

Several types of ions can be observed in FAB and the solution chemistry can be exploited to enhance one type of ionic species over another. True molecular ions, $M^{+\cdot}$, can be formed but, more commonly, quasi-molecular ions are produced, often with a small degree of fragmentation. Salts in solution give major peaks for the intact cation or anion, depending on which polarity is being examined:

$$A^+B^- \text{ (solution; FAB)} \longrightarrow A^+(g) + B^-(g)$$

where g denotes that the ions have been sputtered into the gas phase. Commonly, a mass spectrometer will be set up to acquire mass spectra repetitively, with every alternate acquisition being of different polarity. In this way, one analysis provides both positive-ion and negative-ion data, thus giving information on both cations A^+ and anions B^-. An example is shown in figure 3.12 for the salt 12, for which FAB provides clear evidence for the presence of a spiroborate anion in the negative-ion mode and for the pyridinium counter-cation in the positive-ion mode.

12

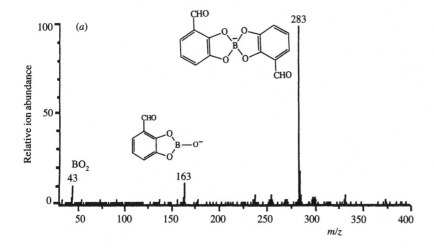

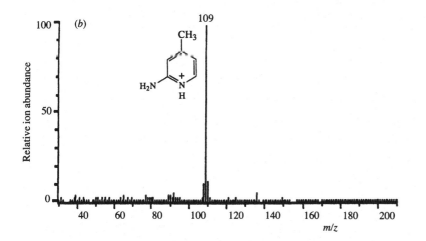

Figure 3.12. (*a*) The negative-ion and (*b*) positive-ion FAB mass spectra of the pyridinium spiroborate salt (12) in a liquid matrix of 3-nitrobenzyl alcohol. The mass of the spiroborate is 283 and that of 2-amino-4-methylpyridinium cation is 109; these two ions provide the base peaks in the negative-ion and positive-ion modes respectively. Compare the FAB spectrum of the spiroborate (*a*) with its electrospray mass spectrum, which does not contain significant fragment ions (figure 3.10(*a*)) unless the source conditions are adjusted to promote fragmentation (figure 3.10(*b*)).

Polar compounds generally exhibit abundant positive or negative quasi-molecular ions, $[M + X]^+$ or $[M - X]^-$ where X is usually H, but can be Na, K etc., and sufficient fragment ions to give valuable structural information. It is usually best to record both negative- and positive-ion mass spectra of an unknown compound because the information that they provide is complementary. However, some molecules are predisposed to one charge mode or the other because they are better able to accommodate one charge or the other. For example, an amine RNH_2 would be best examined in the positive-ion mode, especially if the liquid matrix were doped with a small amount of acid, forming RNH_3^+ ions even before atom bombardment commenced.

Peptides with relative molecular masses of up to about 25 000 have been examined in a liquid matrix of glycerol or thioglycerol with a small amount of acid, such as trifluoroethanoic acid. In a similar exploitation of solution chemistry, carboxylic acids RCOOH can be encouraged to desorb as the relatively stable carboxylate anion RCOO⁻ by adding a base to the analytical solution:

$$\text{RCOOH} + \text{NaOH} \longrightarrow \text{RCOO}^-(g) + \text{Na}^+ + \text{H}_2\text{O}$$

Analytes that are usually best examined in the negative-ion mode include oligosaccharides, glucosides, phosphates, borates and sulphates and nucleotides. Typical positive-ion and negative-ion FAB mass spectra are illustrated in figures 3.13 and 3.14.

Adding a small amount of an inorganic salt to the liquid matrix encourages the formation of adduct ions (*cationization*):

$$\text{M} + \text{X}^+ \longrightarrow [\text{M} + \text{X}]^+$$

This can be helpful in confirming the relative molecular mass and the type of a quasi-molecular ion if its nature is uncertain. It should be noted, though, that inorganic ions added purposely or simply present as contamination (which is frequently the case with biological samples) can themselves dominate the FAB spectrum. This undesirable effect is called *suppression* because ionization and/or desorption of the organic ions of interest can be completely obliterated. Suppression often occurs if mixtures are analysed by FAB. Components that are easily ionized and desorbed suppress the ionization and desorption of other components, so the array of gas-phase ions produced may not reflect all of the components in solution. For example, because ionization occurs at or near the surface, surface-active molecules (surfactants) that tend to form a

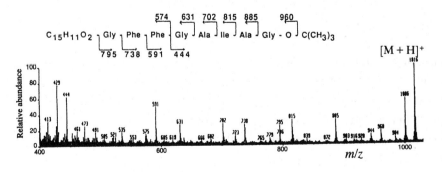

Figure 3.13. The positive-ion FAB mass spectrum of the synthetic peptide shown ($M_r = 1015$). The amino acid sequence of the peptide can be elucidated because the fragment ions arise from the cleavages at each peptide bond. However, it is unusual for peptides with M_r values over 2000 to show such fragmentation. If required, information from fragment ions is best achieved in an MS/MS experiment in which fragmentation is induced (chapter 8). (Taken with kind permission from Larsen (1990).)

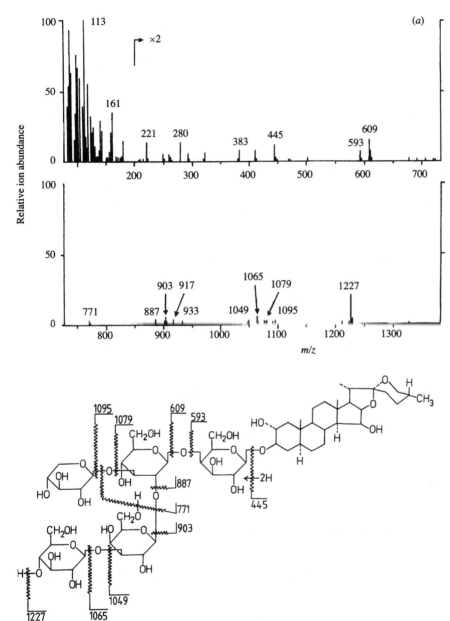

Figure 3.14. (*a*) The negative-ion FAB mass spectrum of digitonin (M_r = 1228) in a liquid matrix of poly(ethylene glycol). The accompanying structure of digitonin also shows the cleavages that give rise to the key fragment ions. Interpretation of such peaks would help to elucidate the structure. (*b*; next page)

monolayer at that surface will be more strongly represented in the resulting spectrum than a surface-inactive component, regardless of their concentrations. This type of discrimination is less serious with continuous-flow FAB than with the static probe method because in the former the liquid surface is continually being disrupted and renewed.

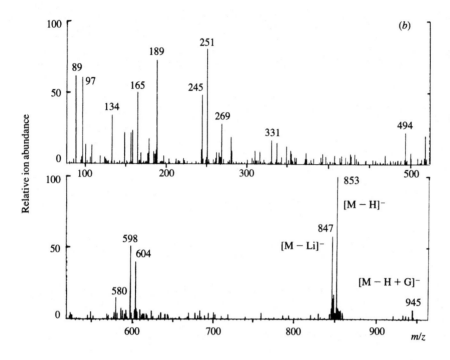

Figure 3.14 (*b*) The negative-ion FAB mass spectrum of the nucleotide shown ($M_r = 854$) in a liquid matrix of glycerol. The ion at m/z 945 is an adduct with glycerol (denoted by G).

When the fast atom gun is first turned on, there is often a burst of abundant sample ions. It is presumed that these ions are a result of sample molecules aggregating at the surface. Once these molecules have been used up, the ion abundance depends on diffusion of fresh molecules to the surface. Mass spectrometers with focal plane or array detectors (which monitor a window of m/z values simultaneously, as described in chapter 2) have been used to 'capture' this rapid initial burst of ions, resulting in quite large enhancements in sensitivity.

Industrial uses of fast atom bombardment mass spectrometry (Cochran, 1986) and applications to biopolymers (Dell and Rogers, 1989) and metal-containing compounds (Miller, 1990) have been reviewed.

3.10.2. *Secondary-ion mass spectrometry*

A substance to be analysed by *secondary-ion mass spectrometry* (*ion bombardment*) is coated on a metallic surface and bombarded with *primary ions* (e.g. $Ar^{+\cdot}$) of high kinetic energy. At or just above the surface of the metal, complex charge-exchange processes and ion/molecule reactions occur, yielding *secondary ions*. The secondary ions of interest to the mass spectrometrist are molecular ($M^{+\cdot}$ or $M^{-\cdot}$) or quasi-molecular ions (mainly $[M + H]^+$, $[M - H]^-$ or $[M + Met]^+$, where Met is a metallic element) generated from molecules (M), together with any fragment ions. When the dipeptide, phenylalanylglycine (**13**), is deposited on silver and bombarded

$$H_2N-CH-CONHCH_2COOH$$
$$|$$
$$CH_2C_6H_5$$

13

with argon ions, $C_6H_5^+$, $C_6H_5CH_2^+$, $C_6H_5CH_2CH=NH_2^+$ and $[M + H]^+$ ions, together with Ag^+ ions, are observed in the positive-ion mode whilst, in the negative-ion mode, $[M - H]^-$ ions are accompanied by $AgCl_2^-$ and low-mass ions such as H^-, C^- and Cl^- (Benninghoven and Sichtermann, 1978). The surface on which the sample is adsorbed is usually, but not always, metallic. For example, the primary-ion beam can be scanned along a developed paper chromatogram and, wherever a 'spot' of substance occurs, the component is identified by its secondary-ion mass spectrum. Natural products in some biological tissues (e.g. a fungus) may be determined directly by focussing the primary-ion beam on the tissue sample.

Many reviews of secondary-ion mass spectrometry are to be found (Benninghoven and Bispinck, 1979; Benninghoven, Rudenauer and Werner 1987; Benninghoven, Evans, McKeegan, Storms and Werner, 1990).

It is now common for SIMS to be used in exactly the same way as for FAB, that is, with a liquid matrix, but with ions replacing atoms as the bombarding species (Lyon, 1985). The mechanism of ionization is the same, irrespective of the bombarding particle and the ensuing mass spectra are very similar indeed. The technique is sometimes called *liquid SIMS* (LSIMS) to acknowledge the fact that the sample is dissolved, or *fast ion bombardment* (FIB) to stress the similarity with fast atom bombardment. Most usually, glycerol is the solvent and Cs^+, accelerated through 30–35 kV, is the primary ion. With large compounds, there is evidence that bombardment with these rapidly moving ions provides greater sensitivity than does bombardment with fast atoms.

3.11.   LASERS

Lasers play a key role in physical organic chemistry because they are used in the spectroscopy of gas-phase ions, but this discussion is restricted to the ionization method of *laser desorption* (Hillenkamp, 1989; Hillenkamp and Karas, 1990; Hillenkamp and Ehring, 1992; Overberg, Hassenburger and Hillenkamp, 1992). Popularly used lasers deliver short intense pulses of radiation in the far ultraviolet or far infrared region to polar non-volatile molecules that are coated in a thin layer on a surface, or dissolved in a solvent with finely dispersed metal powder. It is not clear whether desorption of intact molecules from the surface occurs under the influence of rapid laser heating, or requires photon absorption (electronic or rovibrational absorption). However, the more recent trend is to co-crystallize the sample with a solid organic matrix, like sinapinic acid (3,5-dimethoxy-4-hydroxycinnamic acid), various hydroxybenzoic acid isomers or nicotinic acid, that strongly absorb the wavelengths emitted by the laser. This absorption by the matrix controls the energy subsequently deposited in the sample molecules which are desorbed as protonated molecules, $[M + H]^+$, or as cationized species, $[M + Na]^+$, $[M + K]^+$, and so on, depending on the conditions and the type of laser. Sometimes, the analyte may form adduct ions with molecules of the matrix. This effect, if neglected, could lead to erroneously high values for the mass of the analyte. Fragment ions are sparse or absent but some metastable ions are formed from large molecules. It is important to select the matrix compound carefully for a given compound because the size of the resulting signal varies greatly with the matrix. It is likely that pK_a, melting temperature and the ability to sublime as well as absorption behaviour have an effect on the suitability, or otherwise, of a matrix. Also, certain impurities in the sample may adversely affect the data obtained. For example, sodium dodecyl sulphate (a common anionic surfactant) acts as a suppressant, possibly by inhibiting co-crystallization. Hence, sample preparation is crucial to the success of the method.

   This *matrix-assisted laser desorption ionization* (MALDI) is used mainly for rapid measurement of M_r values of proteins and other large biomolecules like glycosides, to within an accuracy of about 0.1%. For this purpose, useful mass spectral data can be obtained from femtomolar amounts of large biomolecules like proteins with relative molecular masses up to and over 300 000. The technique produces an uncluttered mass spectrum because it generates few interfering background ions in the region of interest for large peptides and proteins. Ions due to the matrix are confined to m/z values below 1000 (Karas, Bahr, Ingendoh and Hillenkamp, 1989).

Laser desorption offers some flexibility of operation. The laser wavelength, power or spot size can be varied to suit the analytical requirements. Changing the energy (wavelength) of the laser beam changes the selectivity towards substrates as different bonds are excited. Diminished fragmentation can be achieved by setting the laser power to produce ions near the threshold value for ionization. In matrix-assisted laser desorption, the best mass accuracy is achieved when the peaks are small, so again low laser powers are advantageous for determination of M_r values. Large spot sizes are helpful when examining a surface for contaminants that sparsely cover that surface. Desorption from a large surface area ensures a good yield of ions from any such trace (sub-monolayer) contaminant.

Typically, a low picomolar amount of the sample is added to a large excess of matrix molecules in solution. A few microlitres of that solution are carefully dried on a metal target. The target is loaded into the ion source and irradiated with pulses from the laser. Wavelengths of 266 and 355 nm are common. Figure 3.15 shows the matrix-assisted laser desorption mass spectrum of 500 fmol of the enzyme, β-D-galactosidase mixed with nicotinic acid and placed on a silver surface, which was bombarded by a Nd–YAG laser at a wavelength of 266 nm. The relative molecular mass was measured to be 117130. It is a typical laser desorption mass spectrum of a protein in that both cluster ions, $[nM + H]^+$, and multiply charged ions, $[nM + zH]^{z+}$, were detected. In general, the number of cluster ions depends on the structure of the analyte and its concentration in the matrix. The greater the concentration, the greater the extent of clustering. The presence of multiply charged ions is critically dependent on the nature of the matrix and the laser wavelength.

Several analysers are not compatible with ions as massive as those that can be generated by matrix-assisted laser desorption. Given that the ionization method intrinsically generates short pulses of ions at a well-defined point in space, the time-of-flight analyser is its ideal complement because this analyser also relies on rapidly formed bunches of ions at a precisely defined

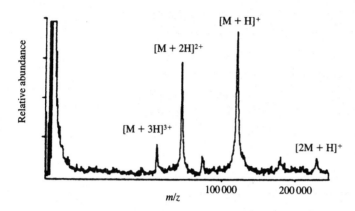

Figure 3.15. The matrix-assisted laser desorption mass spectrum of 500 fmol of the enzyme, β-D-galactosidase, mixed with nicotinic acid and placed on a silver surface. A Nd–YAG laser at a wavelength of 266 nm was used. (Taken with permission from Karas, Bahr, Ingendoh and Hillencamp (1989).)

location. In addition, the TOF analyser lacks an upper mass limit. Other likely analysers are ion trapping systems, like the Fourier-transform ion cyclotron resonance system (Speir, Gorman and Amster, 1992), which also requires pulses of ions.

Applications of lasers in pyrolysis/mass spectrometry are described in section 12.4 and substantial reviews of the uses of lasers in mass spectrometry have been published (Conzemius and Capellen, 1980; Lubman, 1990).

3.12. PLASMA DESORPTION

Another approach for determination of relative molecular mass of proteins is radionuclide ionization: ^{252}Cf *plasma desorption*. Routinely, it handles molecules up to a relative molecular mass of 20 000, but has been used to obtain spectra of complex metal coordination compounds with relative molecular masses in excess of 30 000 and proteins with relative molecular masses up to 45 000. It is yet another technique that matured into a commercially available system in the 1980s (Sundqvist and Macfarlane, 1985; Macfarlane, Hill, Jacobs and Geno, 1989).

When the radionuclide ^{252}Cf decays, it produces two fission products (e.g. $^{142}Ba^{18+}$ and $^{106}Tc^{22+}$), which move apart in opposite directions. One of the nuclear fission fragments travels towards a sample, which is coated on a very thin foil, made of aluminium or aluminized polyester. Each highly energetic ion passes through the foil and, on impact with the sample, causes a localized 'hot spot' (about 10 000 K). The sudden deposition of such a large amount of energy in the sample results in rapid volatilization before decomposition can occur. Alternatively, desorption might be a result of interactions between secondary electrons and the vibrational states of a molecule, causing a rapid expansion of the molecule, which pushes it away from the surface. It is likely that clusters of molecules are ejected from the surface and the internal energy of the cluster may be arranged such that one molecule is ejected as a positive ion, for example, leaving a negative charge to be delocalized over the remainder of the cluster. Whether molecules are protonated or deprotonated (that is, whether positive ions or negative ions, respectively, are formed) depends on the acid/base properties of the molecules in the cluster (Macfarlane, Hill, Jacobs and Geno, 1989).

In the positive-ion mode, most substances give protonated molecules, $[M + zH]^{z+}$, where z is a small integer and frequently unity. Little, if any, fragmentation is observed. The ions are drawn from the region of ionization and into the analyser, a time-of-flight mass spectrometer. The nuclear fission fragment which moves off in the opposite direction to that which causes ionization is detected and serves as a time marker for the fission event and hence the onset of ionization. Measurement of the difference in time between ionization and detection of an ion leads to mass assignment, the time taken to traverse the flight tube being proportional to the square root of the mass of the ion (section 2.4.6). A mass spectrum is developed over a period of minutes by monitoring the ions formed from each individual fission event. As with laser desorption, the time-of-flight spectrometer is used because it has no mass range limitations and because the ionization events (here, fission events) are naturally pulsed.

A problem with this method is the quenching of the spectrum by involatile compounds present on the foil as impurities. This can be overcome by first coating the foil with a thin layer of an adsorbent that is selective to the given class of analyte. For example, nitrocellulose, sprayed onto the thin aluminium or polyester foil, selectively binds peptides and proteins (Jonsson *et al.*, 1986; Macfarlane, 1988). When a sample containing a peptide is applied to the nitrocellulose surface, only the peptide is bound strongly to it. Loosely held impurities are simply washed off the surface. The pure analyte on the foil is then subjected to the nuclear fission products as before. In other words, the nitrocellulose surface acts as both a purification and preparation device. This strategy has been adopted almost universally for plasma desorption for its reliability, simplicity and the sensitivity of the ensuing analysis.

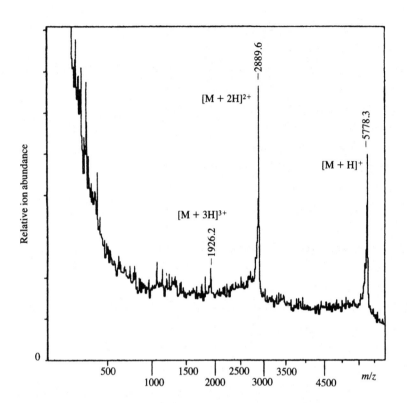

Figure 3.16. Positive-ion plasma desorption mass spectrum of porcine insulin adsorbed onto nitrocellulose. (Taken with permission from Roepstorff (1987).)

The ^{252}Cf plasma desorption method (Cotter, 1988; Macfarlane, 1988; Macfarlane, Hill, Jacobs and Geno, 1989; Roepstorff, 1989; Roepstorff, 1992) requires sample sizes in the low picomole range, but relatively few molecules are used during the analysis. In fact, the amount used during the acquisition of a spectrum is practically undetectable so the sample is available for further studies. It can be recovered from the foil afterwards or it can be chemically derivatized *in situ* followed by the recording of more mass spectra. Plasma desorption is particularly useful for natural products such as peptides, amino acids, toxins, antibiotics and oligonucleotides. As shown in figure 3.16, the protein insulin affords protonated molecules at m/z 5778.3, suggesting a relative molecular mass of 5777.3 (the true value being 5777.6).

4   Computers in mass spectrometry: data systems

4.1.   INTRODUCTION

The full potential of any mass spectrometer cannot be realized without a
data system. Whilst computers were introduced initially into mass
spectrometry to aid accurate mass measurement on double-focussing mass
spectrometers, they have proved useful also for less sophisticated instruments. If the quantity of a sample to be examined is very small, the human
operator may not be able to react sufficiently rapidly to obtain a representative mass spectrum during the brief period in which the compound
resides in the ion source. A computer can obviate this difficulty by controlling continuous repetitive scanning of the mass spectrometer from the time
that the sample is introduced into the source. The spectrum or spectra of
interest can be recalled from the data stored in the computer. If a sample is
contaminated with impurity it is often possible, but tedious and time-consuming, to deconvolute the impurity peaks from the sample peaks manually.
A computer performs this operation in seconds. It is due to the advent of
computers that the technique of combined gas chromatography/mass
spectrometry (GC/MS) has proliferated. A data system built around a computer is now a prerequisite for high-quality GC/MS and LC/MS.

Mass spectrometers have been used in conjunction with a wide range of
data systems, from microcomputers and programmable desk calculators to
large, powerful computers. The larger and more complex the data system,
the less likely it was to be dedicated to the mass spectrometer. For purely
economic reasons, such a system usually served several instruments. With
the advent of relatively cheap computers, having fast preprocessors and
microprocessors and large data storage facilities, this situation has changed
such that virtually all commercial mass spectrometers have their own dedicated computerized systems for control of instrument settings and data

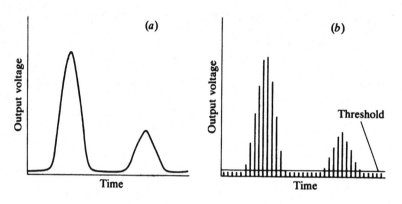

Figure 4.1. The analogue signal (*a*) of two mass peaks and their computer-compatible digitized form (*b*).

handling. The computer system on a mass spectrometer is usually capable of four basic operations: (i) control of the scanning of the mass spectrometer, (ii) data acquisition from the mass spectrometer, (iii) data processing and (iv) interpretation of data. Ideally, it should be capable of performing all four modes simultaneously. On the one hand (ground), it may be collecting data and, at the same time on the other hand (ground), it could be printing interpreted data obtained previously; this 'foreground/background' working mode is highly efficient and desirable. The subject of computers in mass spectrometry is reviewed biennially (Ward, 1971, 1973; Mellon, 1975, 1977, 1979; Sedgwick, 1981; Chapman, 1985, 1987, 1989) and has been described in detail in books (Chapman, 1978, 1993).

4.2.   DATA ACQUISITION AND MASS CALIBRATION

4.2.1.   *On-line operation*

The detection of ions has been discussed in previous chapters. It was stressed that the output from ion detectors is, or can be converted to, an electrical current which is amplified as a continuously varying voltage (an analogue signal; figure 4.1(*a*)). This signal could be acquired directly by an analogue computer but, as most computers are digital devices, the signal must be converted into digital form. This section will be concerned exclusively with digital data systems. The digitization step is accomplished by an analogue-to-digital converter (ADC) for which the mode of operation is represented in figure 4.1. The data system is equipped with a crystal oscillator (clock), which defines precise, regular time intervals at which the analogue signal voltage is to be sampled by the ADC. This results in a regular series of voltage pulses (figure 4.1(*b*)), each of which is compared with a threshold value and is accepted for transmission to the data system only if it exceeds this pre-set threshold. This limitation is applied to remove unwanted electrical (baseline) noise between mass peaks. The data system, freed from acquisition of unnecessary data, can process previously acquired data in the time intervals between the mass peaks. The duration of these intervals is often greater than the time width of the mass peaks (see figure 1.12), particularly with high-resolution spectra in which the peaks are very narrow.

It is usual for there to be a further noise-rejection facility prior to transmission of the signal to the computer. Electronic noise spikes or peaks are usually narrower than true mass peaks, so that rejection of spurious signals can be made on the basis of peak width.

A magnetic sector mass spectrometer is most often scanned from high to low mass exponentially with time. This gives rise to mass peaks of constant width throughout the mass range whilst the gaps between the peaks increase exponentially with decreasing mass (figure 1.12). For such a scan mode, the mass peak width in terms of time (t) is defined by equation (4.1),

$$t = 0.43t_{10}/R \qquad (4.1)$$

where t_{10} is the time to scan a decade in mass and R is the resolution. For example, at a resolving power of 10 000, when scanning masses from m/z 400 to 40 accurately, the ADC must be capable of taking about fifteen samples per peak in this time, that is, a sampling rate of about 35 kHz (35 000 samples per second) is required. The same rate would be required by a 0.5–1 s per decade scan at low resolution. Typical ADC devices are capable of digitization at rates of 40–200 kHz. Conversely, the maximum scan speed, t_{max} (in seconds per decade), for a given sampling rate, S (in kilohertz), is given by equation (4.2),

$$t_{max} = 2.3 \times 10^{-3} NR/S \qquad (4.2)$$

which is derived simply from equation (4.1). The number of samples (N) required across the peak is usually about 10–25.

Figure 4.2(a) shows the effect of insufficient sampling across a mass peak. Neither the position of the maximum nor its height are faithfully recorded at the lower sampling rate. With a peak that is a partially resolved multiplet, the situation is even more critical, for the multiplicity may not even be recognized at low sampling rates (figure 4.2(b)).

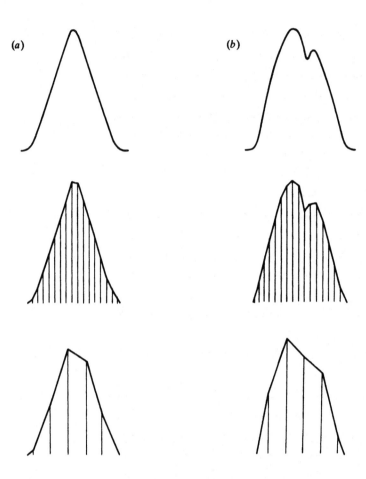

(a)

(b)

Figure 4.2. The effect of sampling rate on peak definition. The analogue signals of (*a*) a normal peak and (*b*) a doublet peak from a mass spectrometer (top row) are defined adequately by sampling fifteen times (middle row), but not by sampling five times across the peak (bottom row).

Equations (4.1) and (4.2) are not relevant for quadrupole mass spectrometers that are scanned linearly with time, producing a linear mass scale, but simple calculations based on the same principles reveal that they have similar digitization requirements. Alternatively, with data-system control, it is possible to 'scan' a quadrupole mass filter in steps by rapidly switching the rod voltages on and off. If the output of such a system is sampled at each step, the data are then already in digital form and directly amenable to acquisition by computer.

Once a mass peak has been adequately defined by sampling, it is passed to a processor where its area (abundance of ions) and its centroid (time of arrival of ions) are estimated. Ion abundance is calculated from the area or height of the peak (this area or height is often inappropriately termed 'intensity'). The time centroid (C) can be taken as the position of the largest digital sample or calculated numerically (figure 4.3) using equation (4.3),

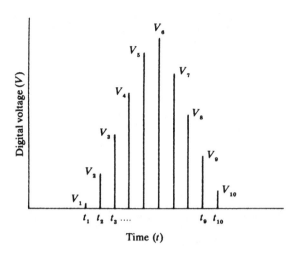

Figure 4.3. The calculation of a peak centroid.

$$C = \sum V_i t_i / \sum V_i \qquad (4.3)$$

where V_i is a digital voltage and t_i the time at which it is acquired. The latter method is more demanding computationally but more reliable, particularly for high-resolution studies. The information stored by the computer may be in either of two forms. In the first, the *profile* method, all of the digital samples are retained, the calculation of the centroids and sizes of the peaks being performed any time during or after acquisition. For the second form, to acquire *centroid* data, the calculations are carried out in real time, that is, as the data are acquired. Then, only the centroids and sizes of the peaks are stored by the data system and the original peak shapes are lost. Compared with the centroid method, the profile method requires less computer time but more storage space. It has the advantage that the shapes of the acquired peaks may be examined later for multiplicity, to determine resolution and for checking the operation of the mass spectrometer. A good data system may allow either method to be selected.

The information acquired by the computer, which is usually stored ('written') on magnetic disc, is a time/'intensity' file. The process of converting a time/intensity file to a mass/intensity file or, more properly, a mass/abundance file (a mass spectrum) is known as mass calibration and relies upon the acquisition of data from a reference (calibration) compound, the mass spectrum of which is stored in the memory of the computer as a mass/abundance file called a reference table. The commonest calibration compounds are perfluorokerosene (PFK), which is most useful for magnetic

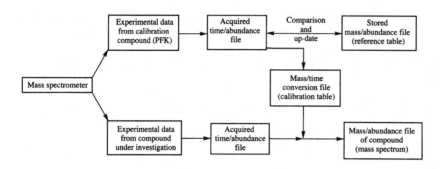

Figure 4.4. Flow chart for mass calibration.

sector instruments in the mass range m/z 30–800, and heptacosafluorotributylamine which is suitable for the mass range m/z 40–600 and is used most often with quadrupole and ion trap mass spectrometers. By matching and equating the acquired time/abundance file and the reference table, a mass/time conversion file, called a calibration table, is produced. It can be used to convert any subsequently acquired time/abundance files to mass/abundance files as long as mass spectrometric conditions are stable, that is, as long as the scan is reproducible. In practice, calibration for low-resolution mass spectra will usually suffice for at least one day. Whilst the mechanism of mass calibration is the same for both high- and low-resolution mass spectrometric data, there are special requirements for accurate mass measurement, which will be detailed in section 4.4. The process of mass calibration is represented as a flow chart in figure 4.4.

The ideal calibration compound would be readily introduced into the ion source and give a mass spectrum of closely spaced, large peaks covering a wide mass range; the m/z values would be quite distinct from those of the sample. The compound closest to this ideal for electron ionization is PFK. For calibration to high mass, fomblin oil (a polyfluoroether) gives reference ions up to m/z 3000. Substituted triazines cover a mass range up to m/z 1600 and substituted phosphazenes are useful for calibration to a mass of about 2000 for EI and FD mass spectrometry. The use of PFK for calibration in CI mass spectrometry is limited. Instead, the mass scale can be calibrated in the EI mode prior to conversion to CI conditions if a combined EI/CI source is used. Alternatively, mixtures of hydrocarbons or of halogenated aromatic hydrocarbons give useful reference peaks under CI conditions. For negative-ion mass spectrometry, PFK can be used up to a mass of about 600. For instruments using mild ionization methods that allow masses of tens or hundreds of thousands to be measured, as with electrospray and laser desorption, biomolecules such as proteins having known large relative molecular mass are used for reference purposes.

The process of mass calibration allows for any reproducible deviation of the magnetic scan from the theoretical scan law. Because a data system can control precisely and reproducibly the scanning of quadrupole mass filters, some of these instruments do not require conventional calibration compounds. Instead, with any compound in the source, the data system searches for a known mass peak onto which it can 'lock' and from which it can predict the rest of the linear mass scale from the theoretical scan law. Alternatively, the mass scale of quadrupole mass spectrometers or ion trap detectors can be calibrated as described above but using the values of the voltages applied to the rods or electrodes rather than the time variable For time-of-flight instruments, the time taken by an ion travelling from the ion source to the detector is proportional to the square root of m/z and, although only one or two reference peaks are needed to predict the remainder of a scan, it is often better to use a range of reference materials, particularly at high mass.

In chapter 1 it was seen that, except for carbon, elements have masses that are not exact integer values, that is, they have mass defects (table 1.5). Data acquisition and mass calibration of low-resolution data usually result in a mass scale accurate to four figures and rounding off to integer mass is necessary. This is not a simple process since some elements (e.g. halogens) give ions with masses just less than and others (e.g. hydrogen) just greater than the appropriate integer value. The situation is most critical at higher masses because in a large molecule the individual mass defects of the elements often add to give masses significantly different from the integer mass. For instance, with simple rounding off, ions detected at m/z 492.6 and 493.7 would appear to be separated by one mass unit, but if the former were the molecular ion of the hydrocarbon $C_{35}H_{72}$, the mass should be rounded off to a nominal mass of 492, whereas if the latter were one of the molecular ions of fully chlorinated biphenyl ($C_{12}{}^{35}Cl_{10}$), its integer mass would be 494. It is improbable that such dissimilar types of sample would be present in the source at the same time and, as long as this situation does not arise, rounding off to nominal masses can be brought about successfully in a variety of ways by assuming that there will be no sudden large changes in the mass defect throughout the mass range. The predominance of carbon and hydrogen in most organic compounds ensures that samples usually have masses slightly greater than their respective integer values and this is the reason for preferring mass-deficient, polyfluorinated calibration compounds in which mass peaks are well separated, and easily distinguished, from sample peaks.

A normal mass spectrum will contain peaks due to product ions from the decomposition of precursor metastable ions (chapter 8). These ions are of low abundance (usually less than 1 per cent of the base peak) and many will be disregarded by the data system because they fall below the threshold value of the ADC. In addition, a computer which has a maximum peak width parameter as a device for rejecting low-frequency noise will discard many peaks arising from metastable ion fragmentation because they are broader than the peaks from normal ions. Any remaining peaks caused by metastable ion decomposition are 'lost' in the procedure for rounding off to nominal mass. There are techniques for acquisition of data from metastable ions that rely on different methods of scanning the mass spectrometer and these will be detailed in chapter 8. For ion cyclotron resonance spectrometers and ion traps, details of ion detection and mass analysis are different and are discussed separately in sections 2.4 and 12.3.

4.2.2. *Transputers*

Essentially, a transputer is a microprocessor (a 'computer on a chip') which can handle data at very high speeds by having only a limited number of built-in functions but including its own random access memory and by making use of parallel rather than serial processing. Because the processor has a reduced set of instructions (functions) to consider (it is a reduced instruction set computer or RISC) and various of its operations can be carried out simultaneously (in parallel) instead of one after the other (serially), processing speeds of ten million instructions per second (10 MIPS) can be achieved. Transputers can be connected with others to give arrays of any size in which the total processing speed is effectively the number of transputers multiplied by 10 MIPS. Thus, with just two transputers, twenty million instructions per second can be handled. In newer mass spectrometer data systems, advantage is taken of this processing speed to control those sections in which data must be handled at a very high speed. Instrument control and initial (acquired) data processing require such high speeds but, when the data have been processed and compressed, they can be passed directly to a hard disc from which they can be retrieved and manipulated in a more leisurely fashion by the computer itself (sometimes called a workstation). Figure 4.5 indicates a typical layout.

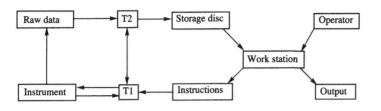

4.3. CONTROL OF THE MASS SPECTROMETER

Figure 4.5. In this simplified illustration, there are two transputers (T1, T2), the first controlling the settings on the instrument on instruction from the workstation and the second processing raw data, which are passed to a storage disc, ready for examination by the workstation. The transputers work in parallel and communicate. Since the transputers have fast processors and in-built memory, they can remove much of the routine computation from the workstation, making foreground/background working much easier.

Most data systems are not passive; they not only acquire and process data but are used interactively to control several functions of the mass spectrometer. For instance, microprocessor-controlled units are available that automatically optimize the ion source conditions for best sensitivity, and most data systems control the rate, range and type of scanning of the mass spectrometer. The latter requires a digital-to-analogue converter (DAC) to provide the mass spectrometer with the necessary analogue voltages from the digital scan instructions generated by the computer. This use of computers is particularly applicable for controlling ion processes in FTICR spectrometers and ion traps, especially for investigation of ion/molecule collision and MS/MS (section 12.3).

The DAC output can be applied directly to the rods of a quadrupole mass spectrometer. At any one value of this voltage only ions of one specific mass can pass through the filter to the detector. The relationship of mass-to-rod voltage is computed from a calibration spectrum after which the operator needs only to specify the masses at which the scan is to start and end, the type of scan and the scan rate. Then the computer and DAC generate the voltages required to 'drive' the instrument as desired. For a quadrupole mass filter there are two types of scan. The mass scale may be scanned linearly and continuously by application of a voltage varied continuously and regularly, or the applied voltage may be stepped, providing a discontinuous scan. The latter method is used for monitoring ions of a few selected masses for sensitive, quantitative mass spectrometry (chapter 7) or for acquiring full mass spectra by 'jumping' from one peak to the next and not scanning the gaps between the peaks. Since quadrupole mass filters are scanned electronically and the settling time of the low voltages involved is very short, extremely fast sweeping and switching over the entire mass range is possible.

The control of the scan of magnetic sector mass spectrometers is more complex and requires a mass marker, usually a Hall-effect sensor, which is placed between the poles of the electromagnet. The sensor measures the magnetic flux and its amplified output is calibrated against mass by use of a calibration compound. The sensor relates mass-to-charge ratios of ions to the strength of the magnetic field that deflects them onto the detector. The mass count that it provides may be displayed directly or marked on the mass spectrum or used, like the calibration of the quadrupole rod voltage against mass, to allow computer control of the scan. Once the operator has specified the mass range, the scan rate and the type of scan (most frequently exponential), the computer uses feedback from the magnetic field sensor to programme the DAC to produce the voltages required to change the magnetic field strength. Since most magnets in current use are prone to some hysteresis effects and require greater settling time than the rod voltages of a quadrupole mass filter, mass 'jumping' by switching the magnetic field discontinuously is *relatively* slow compared with a quadrupole instrument, but still fast in real time. As an alternative, for monitoring selected ions with a magnetic sector instrument, the ion source accelerating voltage (V) is switched rapidly, with the magnetic field strength constant, to focus ions of one mass after those of another according to equation (2.2). The operation is readily controlled by a data system as long as the accelerating voltage has been mass calibrated with a reference compound such as PFK and there is a second DAC to control the voltage, V. The same provisions allow for voltage scanning, which can be useful when a narrow mass range is required to be scanned at a rate faster than that possible for magnetic scanning. However, technological advances mean that newer magnets with virtually insignificant hysteresis and much faster settling times are capable of faster continuous scanning and rapid switching throughout the whole mass range for selected ion monitoring. The advantages and disadvantages of the various techniques for selected ion monitoring and their applications are discussed in chapter 7.

An advantage of a simple magnetic sector instrument over a single quadrupole mass filter is that metastable ion decomposition can be studied. Since metastable ions are of relatively low abundance, the product ions from their decomposition are best observed in the absence of the much more abundant normal ions that would otherwise wholly or partially overlap and obscure them. Normal ions can be eliminated specifically by one of several techniques involving scanning of the potential (E) of the energy-focussing electric sector. Such scanning methods can be brought under the control of a data system; these techniques and their applications are discussed in chapter 8. This lack of ability of quadrupoles to monitor metastable ions directly has been overcome by tandem mass spectrometry (MS/MS), which is applicable both to quadrupole and to sector instruments. Indeed, both can be utilized together for tandem MS, as discussed in chapter 8. In FTICR instruments and ion traps, collisional activation of selected ions can be controlled and the mass range of resulting product ions examined by special electronic 'gating' procedures in which a sequence of events is effected by a computer. Thus, initial production of ions, selection of ions of one specified mass, collisional activation of the selected ions and acquisition of the resulting fragmentation spectrum of product ions (MS/MS; chapter 8) can be carried out in about 50–200 ms. Repeated application of this routine under computer control and summation of the resulting spectra increases the sensitivity of the method.

In the foregoing discussion, several applications of computer control have been mentioned. In nearly all cases, at the heart of the process is the principle of repetitive scanning, a technique that is realistic only when a computer with a large data storage capacity is available. Repetitive scanning is most useful for combined chromatography/mass spectrometry but is also employed with simpler methods of examining samples such as with the direct insertion probe. In this last mode, as soon as the sample is placed in the inlet of the mass spectrometer, the data system is instructed to start the scan programme for full spectra, selected ion monitoring or metastable ion analysis. When the first scan ends, the programme returns the analyser to its starting point and a second scan is initiated, and so on. In this way the scanning is cycled, every 0.1–10 s in many applications, until the analysis is complete. Compared with non-repetitive, operator-initiated scans, the procedure allows for the unexpected appearance of components in the sample, that is, materials present in the source for short periods only. The technique provides a surer analysis of small amounts of sample and requires a less skilled operator or indeed no operator at all except for arranging for sample insertion. It also permits access to a vast data processing capability, which is detailed below.

4.4.   DATA PROCESSING

The better data systems are capable of data processing whenever the signals that they are acquiring concurrently fall below threshold. This facility, *foreground/background* operation, permits data of the current analysis or of a previously acquired analysis to be processed almost concurrently with data acquisition. In a correctly set system, acquisition occupies only about 1–10 per cent of total computer time. The data system must not lose any data by operating in the processing mode when it should be in the acquisition mode, so there is a strict order of priorities in which timing pulses from the clock, data acquisition and control of scanning all take precedence over data processing.

Different commercial data systems offer different processing programmes and some allow operators to write their own programmes. Therefore, a large array of data processing routines is available. The most useful programmes, forming the common core of many modern data systems, are discussed in the following subsections. Where appropriate, there is a brief discussion of some more specialized, less common routines, which, with further development, may attain wider application in the future.

4.4.1.   *Accurate mass measurement*

To calculate the elemental composition of an ion, its mass must be measured correct to three or four decimal places. This requirement usually necessitates a double-focussing mass spectrometer (but see later discussion). In a sector instrument, in the absence of a computer, accurate mass measurement can be performed manually by a process called *peak matching*, in which a reference compound such as PFK and the compound under investigation are introduced into the ion source at the same time. To measure the mass of a sample ion, M^+ (often but not always the molecular ion), an ion (R^+) of known mass, close to that of M^+, is selected from the reference spectrum. The peak from the ion M^+ is displayed on a cathode ray tube (video monitor) alternately with R^+. The ion source accelerating voltage (V) is changed until R^+ and M^+ coincide on the screen. At constant magnetic field, the mass (m) of singly charged ions is inversely related to the accelerating voltage according to equation (2.2). Therefore, the accurate mass of M^+ is calculated simply from equation (4.4), where the subscript r refers to the reference compound and s to the sample:

$$m_s = (V_r/V_s)m_r \tag{4.4}$$

When the two peaks on the screen are matched, the voltage ratio V_r/V_s is obtained from the control panel of the mass spectrometer.

The process of peak matching has been automated. In some cases the operator judges when the peaks are matched, in others the computer ascertains the matching condition.

The peak-matching technique is accurate but time-consuming, especially if several ions of the sample need to be examined. Measurements cannot be performed on compounds that are present in the ion source for short periods of time. Hence, chromatographic effluents and materials of which only a small quantity is available are excluded, a situation often obtaining in the study of natural products. For these reasons, data systems that offer alternative faster procedures for accurate mass measurement have been developed. The mass spectrometer is operated at moderately high resolution (often 10 000) with magnetic scanning over the relevant mass range. The scan rate has to be relatively slow, usually about 10 s per decade in mass, although this figure depends on the performance characteristics of the particular instrument and the available sampling rate of the ADC. A few scans of the spectrum of a calibration compound are acquired and used for mass calibration. The sample is admitted into the ion source so that, for at least one scan, both it and the calibration compound are ionized simultaneously. The data system is programmed to locate the reference peaks in the mass spectrum of the mixture using the first calibration table and to recalibrate to take into account any variation in the scanning. The accurate masses of all reasonably abundant ions of the sample are computed by interpolation between adjacent reference ions. Computer subtraction from the mixture of those ions due to the calibration compound leaves the accurate mass spectrum of the sample. Unlike in the manual peak-matching method, in one brief experiment the accurate masses of all major ions of the compound are measured. The technique is suitable for GC/MS, although if the chromatographic peaks are narrow (widths of 3–8 s are common with the better gas chromatographic columns), some compromises must be made (Meili, Walls, McPherron and Burlingame, 1979). The data system may be on-line or off-line. The accuracy of the method is generally not as high as that of peak matching, but its greater applicability and convenience render it more popular.

The accurate mass information is frequently given as a mass list, as illustrated in table 4.1 for methyl octadecanoate (methyl stearate). Typically, the list of accurate masses is accompanied by the ion abundance, a possible elemental composition for each ion and the error between the measured mass and that calculated for the given elemental composition. The data system

Table 4.1. *A computer printout of measured masses and elemental compositions of some ions in the spectrum of methyl octadecanoate.*

Measured mass	Peak height	Elemental composition					Error in mass (mmu)[a]
		$^{12}C/^{13}C$	H	N	O	Others	
299.2919	835	18/1	38	0	2	0	1.4
298.2883[b]	4919	19/0	38	0	2	0	1.1
271.2610	290	17/0	35	0	2		−2.7
		16/1	34	0	2	0	1.7
270.2558	1819	17/0	34	0	2	0	−0.1
269.2474	1222	17/0	33	0	2	0	−0.6
268.2835[c]	59	No composition calculated					
268.2716	377	17/1	35	0	1	0	−0.6
267.2687[d]	1745	18/0	35	0	1	0	−0.1
256.2358	1070	15/1	31	0	2	0	0.0
255.2306	4734	16/0	31	0	2	0	−1.8
		15/1	30	0	2	0	2.6

Notes:

[a] mmu, millimass units

[b] This is the molecular ion, $^{12}C_{19}H_{38}O_2^{+\cdot}$; the measured mass at 298.2883 differs from the mass calculated for this composition by 1.1 millimass units. Note the [M + 1]$^{+\cdot}$ ion at 299.2919 of composition, $^{13}C_1{}^{12}C_{18}H_{38}O_2$.

[c] This is a small peak in the mass spectrum for which no composition could be calculated from the given parameters. It probably represents a small amount of impurity.

[d] This ion, $^{12}C_{18}H_{38}O^+$, arises by loss of $CH_3O^{\cdot}$ from the molecular ion.

computes the elemental composition of an ion by checking, against its experimentally determined mass, the calculated accurate masses of all elemental possibilities for the particular nominal mass. Such calculation would be prohibitively time-consuming unless the operator limits the range and number of elements by stating the type and maximum number of atoms of each element to be expected as well as a maximum value for the error beyond which the potential match is considered incorrect (Dromey and Foyster, 1980). Some data systems are programmed to reject unlikely elemental compositions, such as $C_{20}H$ for an ion at m/z 241, by setting a limit on the number of double-bond equivalents.

For FTICR spectrometers, accurate mass measurement is effected by specialized mathematical procedures described in section 12.3. With these instruments, excellent resolution can be attained at low mass but, as mass increases, resolution decreases and becomes only similar to or worse than that available with a double focussing sector instrument.

It should be appreciated that the accuracy of mass measurement is a function of the product of resolution and sensitivity and does not depend solely on the attainment of high resolving powers (Smith, Olsen, Walls and Burlingame, 1971). For statistical reasons, the greater the number of ions making up the approximately Gaussian-shaped mass peak, the more accurately can its centroid be measured. However, the accuracy of mass measurement is limited by ion statistics. Consider a divergent beam of ions from the source passed through adjustable slits to reduce its spread. For a resolution of 1000, the slits are wide and transmission of ions high; to achieve a resolution of 10 000, the slits must be narrowed so that the mass peaks too are much narrower but with the penalty of reducing transmission to about one-tenth of its value at a resolution of 1000. Therefore, this increase in resolving power is achieved at the expense of a drop in sensitivity of about 90 per cent and there may be so few ions in the peak that, statistically, the measured centroid mass may not coincide with the true mass. It is a fallacy to suppose that increased resolving power necessarily leads to increased accuracy of mass measurement. Once the resolution is sufficiently high to separate ions of the sample from ions of the calibration compound at the same integer mass, it is pointless to increase the resolution further since both the accuracy and sensitivity of the analysis will deteriorate. With FTICR spectrometers, this problem of sensitivity is not so serious because, even at the highest resolution, sensitivity can be maintained by repetitive computerized summation of as many mass spectral scans as are needed.

Accurate masses *can* be measured at the lower resolving powers that preserve good sensitivity, although the technique is more cumbersome. Mass calibration with a compound such as PFK is carried out at the required resolution. Since ions of the sample under investigation and PFK would be unresolved if analysed simultaneously, the calibration compound is pumped away. Then the sample is ionized together with a *secondary reference compound*, which has in its mass spectrum only a very few ions, each of which is well removed from any sample ions (tetraiodoethylene, or a mixture of triiodomethane and phenanthrene are frequently used for this purpose). The data system is programmed to locate these secondary reference ions of accurately known mass and align their arrival times with the primary calibration table. In effect, an accurate mass scale is superimposed on the sample spectrum, allowing accurate mass measurement by interpolation. The same technique can be used for determination of elemental compositions with quadrupole mass filters, which are restricted to low resolution because of a drastic loss of sensitivity with increasing resolution. The method is equally applicable to EI, CI or FD mass spectrometry and has the advantage that, at lower resolving power but greater sensitivity, the scan rate can be sufficiently fast to allow ready determination of the accurate masses of ions of eluates from gas chromatographic capillary columns.

4.4.2. *Combined chromatography/mass spectrometric data processing (multiscan data processing)*

The following programmes are typical for the processing of GC/MS and LC/MS data and will be illustrated with reference to the former. It should be remembered that they are equally applicable to any repetitively scanned (*multiscan*) analyses. A sample that contains several components of significantly different volatilities can be analysed on the direct insertion probe by temperature programming the probe heater (Franzen, Küper and Riepe, 1974). In practice, the data processing of this fractionally evaporated mixture is the same as that of GC/MS. Similarly, programmes developed for GC/MS can often be used to remove the peaks due to impurities from the mass spectrum of a directly inserted contaminated compound. Data processing specific to selected ion monitoring and studies of metastable ions will be described in chapters 7 and 8 respectively.

Figure 4.6 shows an analysis by GC/MS of the trimethylsilyl (TMS) ether derivatives of five sterols, the last two of which co-elute. The trace is called the total ion current chromatogram and is equivalent to the chromatogram obtained from a gas chromatograph with a simple detector. It comprises the summed ion abundances of each spectrum scanned, normalized to the largest of these values and plotted against time and scan number. The computer usually has a programme for quantification of the chromatographic peaks based on the heights or areas of the peaks.

Any spectrum of the chromatogram can be identified by its scan number and recalled immediately for examination. The mass spectrum of the first peak (scan number 127), which is due to 3-trimethylsilyloxycholest-5-ene,

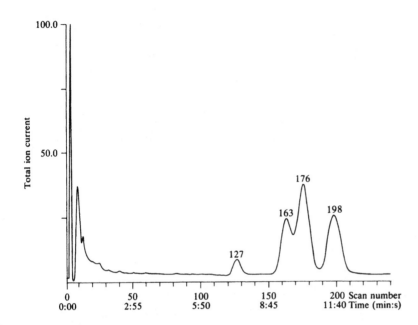

Figure 4.6. A total ion current chromatogram for the analysis of sterol derivatives by GC/MS. Mass spectra (scans) were recorded every 3.5 s; a peak is identified by the number of the scan in which it maximizes.

is shown in figure 4.7(*a*). The spectrum also contains ions of *column bleed* (decomposition products of the liquid phase of the chromatographic column), for example, m/z 75, 197, 315, 393, 451 and 529. This problem can be overcome by a numerically simple programme allowing the operator to select a spectrum (or a number of spectra), which contains only background ions and to subtract this background from the spectrum of interest. When the spectrum of figure 4.7(*a*) is processed in this way, the result is a 'clean' spectrum of the steroid derivative (figure 4.7(*b*)). Note in particular that the peaks at m/z 315, 451 and 529, due solely to ions from column bleed, disappear and that the molecular ion at m/z 458 becomes clear. The same programme can be used to remove peaks from a calibration compound if one were 'bled' into the source throughout the analysis for the purposes of accurate mass measurement.

A *mass chromatogram* is a plot of the abundance of ions of a *specified* mass against time or scan number. The technique is often termed *selected ion retrieval* but since the whole mass range is scanned, it should not be confused with selected ion monitoring. A mass chromatogram is used to advantage to test the homogeneity, or otherwise, of chromatographic peaks. When an effluent enters the source, the ions that comprise its mass spectrum will all increase in abundance to a maximum simultaneously, at the retention time. If a chromatographic peak contains more than one compound, then,

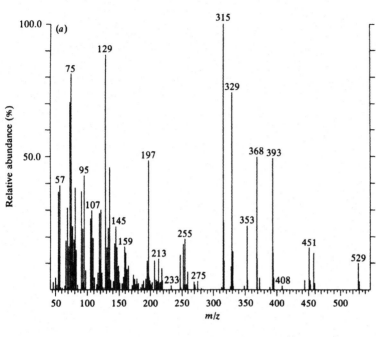

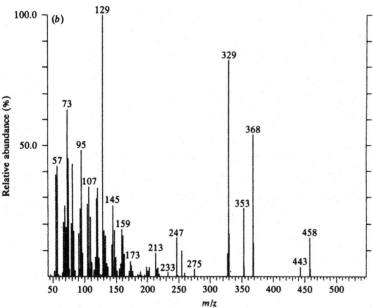

Figure 4.7. A comparison of the mass spectrum of 3-trimethylsilyloxycholest-5-ene (*a*) before and (*b*) after removal of 'background' ion peaks by computer. The spectra were obtained from the analysis by GC/MS shown in figure 4.6.

provided that each component has at least one distinguishing mass peak in its mass spectrum, mass chromatograms will reveal the true peak profiles of the components despite them being unresolved chromatographically. Such a situation is illustrated in figure 4.8, in which a chromatographic peak,

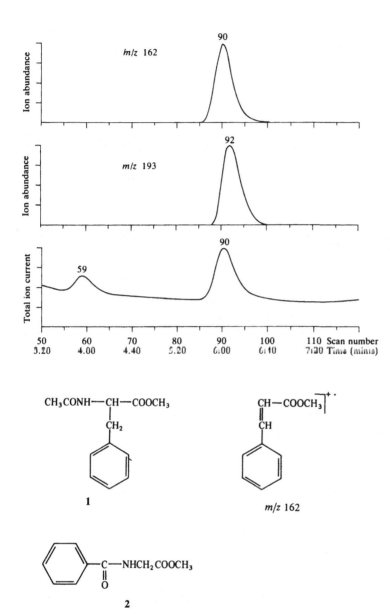

Figure 4.8. Part of the total ion current chromatogram of an analysis of a urine sample by GC/MS (bottom trace). Mass chromatograms of ions at m/z 162 (top trace) and 193 (middle trace) are also shown. Note how the mass chromatograms maximize in different scans.

apparently homogenous by inspection of the total ion current chromatogram, is shown by selected ion retrieval to be composite. Figure 4.8 shows part of an analysis of a urine sample after chemical derivatization. The spectrum of scan number 90 contained some ions that were inexplicable in terms of the assigned structure, the methyl ester of acetylphenylalanine (1). One of the contaminating ions (m/z 193) and an ion of the amino-acid derivative (m/z 162) were chosen for selected ion retrieval with the result shown in

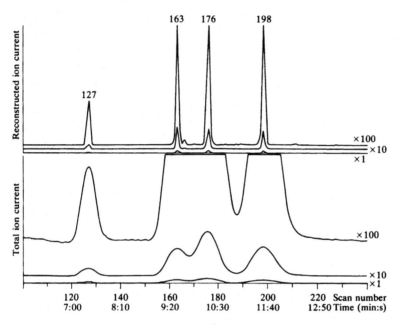

Figure 4.9. Computerized enhancement of GC/MS data. The original total ion current chromatogram (figure 4.6) is shown (bottom trace) together with the mass-resolved gas chromatogram (top trace). Note the enhanced resolution of GC eluted components.

figure 4.8. Since the two peaks maximize two scans apart, they could not originate from the same compound. The second component was identified as the methyl ester of benzoylglycine (2), the molecular ion of which has a mass of 193. It is possible to specify for this procedure not just single masses but mass ranges too. Its applications are discussed more fully in the next chapter.

In the Biller–Biemann enhancement technique, the data system examines the mass chromatogram for every integer mass in the range scanned (Biller and Biemann, 1974). When several rise to a maximum coincidentally, a chromatographic peak is defined irrespective of the total ion content. This peak definition is carried out by ignoring all ions that do not maximize collectively; a typical result is seen in figure 4.9. The lower trace shows a portion of the total ion current chromatogram from figure 4.6 and, immediately above it, the result of the Biller–Biemann enhancement. The upper trace is called the *mass-resolved gas chromatogram*, consisting of the summed ion current of all maximizing ions. The artificial increase in chromatographic resolution is striking; partially resolved components of scan numbers 163 and 176 are completely resolved. Before enhancement, the mass spectrum of the compound eluting at scan number 163 was contaminated with ions from the succeeding eluate and from column bleed (figure 4.10(*a*)) but, since these do not maximize in the same scan, they are

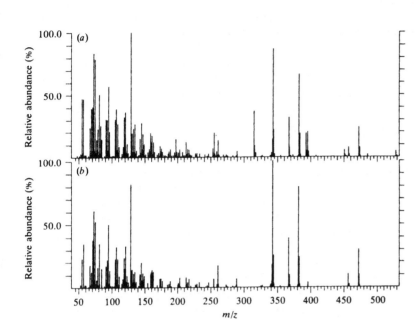

Figure 4.10. A comparison of mass spectra of 24-methyl-3-trimethylsilyloxycholest-5-ene (*a*) before and (*b*) after enhancement. The spectra were obtained from the GC/MS analysis shown in figure 4.9.

computationally erased (figure 4.10(*b*)). The enhanced spectrum is called a *reconstructed* mass spectrum in which the ordinate can be percentage of relative abundance (as shown in figure 4.10) or a percentage of reconstructed ion current (%RIC). Since the constant background ions of column bleed or calibration compound never maximize during the analysis, the enhancement eliminates these peaks altogether. However, deconvolution of the mass spectra of exactly co-eluting components cannot be effected because all component ions will maximize in the same scan. This outcome is seen in figure 4.9, in which the last peak (number 198) is due to the TMS ethers of both lanosterol and sitosterol, which co-elute under the chromatographic conditions employed. To be resolved by this technique, chromatographic peaks must maximize at least three scans apart. Hence, the faster the scanning of the mass spectrometer, the more likely it is that the method will succeed for closely eluting compounds.

The method is computationally fast and works particularly well for locating unresolved components of chromatographic peaks, even if a component is minor and masked by a more abundant one. It performs markedly less well when the unresolved components are of similar structure (e.g. isomers) and have similar mass spectra. For these cases, more sophisticated enhancement programmes are available, which require powerful computing facilities and more processing time (Chapman, 1978). In one method, the peak position of every maximizing mass chromatogram is estimated to one-third of the scan time by a least-squares interpolation about several ion current readings either side of the maximum. This procedure resolves components that are separated by $1^1/_2$ scans or more (Dromey, Stefik, Rindfleisch and Duffield, 1976).

The techniques discussed above reduce multiscan data to a single mass spectrum or a series of mass spectra, each of a pure compound and suitable for further processing, if required, by single-scan programmes.

4.4.3. *Single-scan data processing*

Determination of elemental composition by accurate mass measurement has already been detailed. The number and type of atoms present in some compounds may be estimated by examination of the natural isotope pattern in the molecular ion regions of their low- resolution mass spectra (sections 1.4 and 12.2). The technique requires that the sample give a molecular ion and contain elements such as chlorine, bromine, boron or silicon, which have abundant isotopes. The computer compares the experimentally determined isotope abundances with those computed for various elemental combinations; the goodness of comparative fit is the criterion by which it is decided whether or not the correct elemental composition has been found. The method is limited, but may also be used to calculate mono-isotopic mass spectra from isotopically complex spectra (section 1.4).

The most popular computerized method of determining the structure of an unidentified, but not novel, compound is to search through a library of low-resolution mass spectra for a match between the spectrum of the unknown and that of a known compound in the 'library'. The technique of *library searching* has been reviewed in considerable detail (Chapman, 1978). Most commercial data systems have such a library, usually stored on a magnetic disc, which will commonly contain as many as 50 000 electron ionization mass spectra. Alternatively, remote large libraries may be searched over telephone lines and sometimes via satellite. Libraries of mass spectra obtained by other methods of ionization are not so advanced, the main reason being that electron ionization has been by far the most popular method of ionization and, with the fragmentation that accompanies EI, there is more significance to a match between the mass spectra of an unknown and a known library compound. Chemical ionization and field desorption give spectra that are prone to considerable variation with experimental conditions and are less suited to matching because they usually contain few fragment ions. This problem is not now so acute, owing to the advent of MS/MS by which ions obtained by 'soft' ionization (CI, FD, FAB, electrospray, etc.) in a first mass spectrometer can be induced to fragment to give a spectrum of ions scanned by a second mass spectrometer. There are so many different approaches to library searching that only a limited description can be given here. A typical library does not contain complete mass spectra since too much storage space would be needed and excessively long search times would result. Instead, the spectra are abbreviated (*encoded*). A simple way of doing this is to store, with the name of each compound, the masses and abundances of only the 5–10 largest peaks in the spectrum although, in this and most other encoding procedures, the important molecular ion is included even if it is of low abundance. Of this type of encoding procedure, the most well known is the 'eight-peak index' which, in book form, is also used by many chemists for manual assignment of structure. A particular disadvantage of the method is that, with compounds of higher molecular masses, the most abundant ions tend to be of low mass and of low diagnostic value because they are common to many different structures. There are several ways of circumventing this problem. Firstly, there can be 'weighting' towards the more diagnostic high-mass ions by defining a new parameter, the product of the original peak height and the mass of the ion. Secondly, starting at low and progressing to high mass, the n largest peaks every m mass units are stored. The best known of these procedures is the encoding of the two ($n = 2$) largest peaks every fourteen ($m = 14$) mass units (Hertz, Hites and Biemann, 1971), but other variants are used ($n = 1$, $m = 7$ or 14; $n = 2$, $m = 25$; $n = 3$, $m = 14$). Lastly, attempts may be made to assign and store automatically the peaks of the greatest structural significance (Grönneberg, Gray and Eglinton, 1975; Pesyna, Venkataraghavan, Dayringer and McLafferty, 1976) or information content (Wangen, Woodward and Isenhour, 1971; van Marlen and Dijkstra, 1976). In more rigorous encoding methods, elements from all three approaches are evident. In another type of encoding, some other characteristics of the mass spectrum are stored together with the largest peaks (Kwok, Venkataraghavan and McLafferty, 1973). These characteristics are chosen to be highly diagnostic of structure, as with: (a) losses of small neutral molecules, such as H_2O, HCN and H_2CO, from the molecular ion, (b) series of ions occurring every fourteen mass units (e.g. m/z 43, 57, 71, ... for $C_nH_{2n+1}{}^+$ or $C_nH_{2n-1}O^+$) and (c) some low-mass ions characteristic of functional groups, e.g. m/z 30 ($CH_2{=}NH_2^+$), indicating a primary amine. With this method some interpretative aspects are brought into play as well as simple 'fingerprinting'.

It is not possible to reproduce mass spectra exactly from instrument to instrument, but good methods of encoding minimize anomalies. Quadrupole and magnetic sector instruments give somewhat different spectra for the same compound and, even with the same mass spectrometer, the spectrum of a sample can vary with the method of introduction into the ion source, the condition of the source and so on. To some extent a library search is adversely affected by peaks due to impurities in the spectrum of the unknown compound, and enhancement by one of the techniques described in the previous section is imperative. This situation applies particularly to spectra obtained by GC/MS in which contamination by column bleed might otherwise prevent a correct match. Frequently, it is useful to construct a library of one's own by encoding and storing mass spectra of standard compounds to ensure that the content of the library and the conditions under which the spectra are recorded are relevant to the work carried out.

A typical search of a large library of several thousand spectra may take 15–30 s, a time that can be reduced markedly with a data system which allows additional information to be stored with the encoded spectra. Molecular mass and elemental composition are often included and the search time can be lessened by instructing the computer to consider only certain library entries. For example, the search would only concern itself with reference spectra of nitrogen-containing compounds if the unknown compound were an amine. Other properties that may be stored and used to reduce search times are melting or boiling points, nuclear magnetic resonance spectral details or chromatographic retention times. A large library may be classified according to structure so that, in effect, it is a series of sub-libraries of steroids, ketones, amines and so on. In many studies, the class of the compound under investigation is known and then only one sublibrary need be searched. For specialized laboratories it may be unnecessary and wasteful to have a large library. In these cases, storage of a small amount of relevant spectral data can be most useful and complete mass spectra may be stored since demands on storage space are not so great. For compounds with very few differences in their mass spectra, such as isomeric terpenes, reference to complete spectra may be necessary in order to distinguish them (Adams, Granat, Hogge and von Rudloff, 1979).

The matching routine involves calculation by the data system of a *similarity* index, *match factor* or *purity* between the unknown spectrum and library (reference) spectra (McLafferty, 1977). Common scales are 0 for a complete mismatch and 1 or 1000 for a perfect match. To avoid time-consuming calculations of purity against all entries in a large library, there is usually a *pre-search* or *filter,* which rapidly eliminates grossly dissimilar spectra by requiring that a candidate reference spectrum must contain at least some of the largest peaks of the unknown spectrum. Only the entries that show the most matches in this test are selected for the more rigorous calculations of purity of match. The technique by which an unknown and a reference spectrum are compared varies amongst different data systems (Chapman, 1978). Often, at each m/z value in turn, the ratio (R) of the mass peak heights in the normalized unknown and reference spectra is calculated. The smaller height is always divided by the larger so that $0 \leq R \leq 1$ (where $R = 0$ if there is a peak in one spectrum but not in the other). The individual R values are summed and the average value is normalized to produce the final purity figure for the particular reference spectrum. The entries with the highest figures are displayed as the result of the library search.

In the search techniques described above, all of the peaks in the unknown spectrum are compared with all of the peaks in the reference spectrum. If the only peaks considered for the search are those in the reference spectra, the technique is called *reverse searching.* The so-called *fit* is analogous to and calculated in the same way as that for purity but peaks present only in the unknown spectrum are ignored. Therefore, extraneous peaks due to impurities or unresolved components will not prevent a correct match if the compound is in the library. The technique should be used with caution since a good fit will occur if the unknown and the reference compound are not identical but have a substructure in common. For instance, a reference spectrum of benzoic acid matches well with the spectrum of ethyl benzoate because the molecular ion of the ester loses ethylene to give an ion that fragments further like the molecular ion of the acid. It is often useful in analyses by GC/MS or LC/MS to perform a mass spectral search for a specific compound thought to be present in the sample being investigated. The technique, sometimes called inverse searching, involves computer comparison of a reference spectrum of the material in question with each spectrum of the repetitively scanned analysis. If a plot of the calculated similarity indices against time or scan number maximizes at the retention time of the compound, its presence in the sample is

LIBRARY SEARCH   DATA: ENHANCED SPECTRUM NUMBER 127

SAMPLE: TMS DERIVATIVES OF STEROLS

25409 SPECTRA IN LIBRARY SEARCHED FOR MAXIMUM PURITY
 174 MATCHED AT LEAST 3 OF THE 16 LARGEST PEAKS IN THE UNKNOWN SPECTRUM

RANK	NAME
1	3.BETA-TRIMETHYLSILYLOXYCHOLEST-5-ENE
2	3.BETA-TRIMETHYLSILYLOXYCHOLEST-4-ENE
3	3.BETA, 5.ALPHA-BIS-TRIMETHYLSILYLOXYCHOLESTANE
4	3.BETA, 6.BETA-BIS-TRIMETHYLSILYLOXYCHOLESTANE
5	3.BETA-ETHOXYCHOLEST-5-ENE

RANK	FORMULA	MOL. WT	PURITY	FIT
1	C30.H54.O.SI	458	780	966
2	C30.H54.O.SI	458	658	934
3	C33.H64.O2.SI2	548	649	899
4	C33.H64.O2.SI2	548	495	721
5	C29.H50.O	414	470	722

Figure 4.11. A computer printout of the result of a library search for mass spectral matches to the first-eluting component in figure 4.6.

indicated. Environmental monitoring benefits by searching for target pollutants in this fashion (see section 5.2.2)

The enhanced spectrum of the first-eluting component in figure 4.6, when subjected to a library search, gave the result tabulated in figure 4.11. The report shows that the library contains 25 409 reference spectra and that, of these, only 174 pass the pre-search requirement. For each of these candidate spectra, the similarity to the unknown spectrum is evaluated by the data system on the scale of 0 for a complete mismatch and 1000 for a perfect match. The search is conventional in that the entries are ranked in order of highest purity, but the fit values for these matches are also calculated. In figure 4.11, both values are high for the entry ranked first and the assignment of 3-trimethylsilyloxycholest-5-ene is correct. Had the compound not been in the library, a result affording some structural information about the unknown compound would still have been gained since compounds of similar structure would have been retrieved. This gain in structural information can be seen in figure 4.11, in which the entry ranked second is an isomer of the correct compound and the three entries ranked third to fifth are all structurally related to cholest-5-en-3-ol. Whatever the fit or purity values, the result of a library search should be assessed by careful visual examination of the matching spectra when presented as shown in figure 4.12 for the highest ranked library entry of figure 4.11. A match against an uncontaminated sample spectrum should be judged correct only if the differences in the two spectra can be attributed to peaks eliminated in the encoding of the library spectrum. This is the case in figure 4.12; notice that virtually all the peaks in the reference spectrum are present also in the spectrum of the

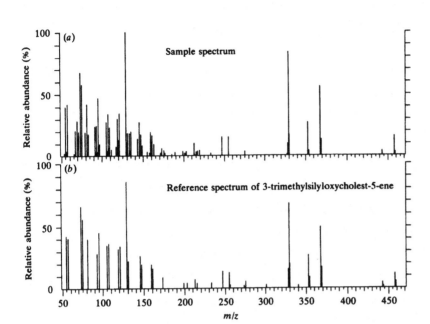

Figure 4.12. Comparison of (*a*) the sample spectrum and (*b*) the abbreviated mass spectrum of 3-trimethylsilyloxycholest-5-ene stored in a mass spectral library.

sample, and that the relative heights of the matching peaks are approximately the same.

When an unknown compound is novel or not present in the library, the methods discussed above are of limited use. There are two approaches to computerized structural analysis that do not require there to be a standard spectrum of the unknown compound. Both methods are far less popular than library searching and discussion of them will be correspondingly brief. Several groups of workers have attempted to condense into computer-compatible rules the approaches that a mass spectrometrist would use for structure elucidation (chapter 10). A data system programmed with these rules has interpretative ability and performs as well as, but faster than, a trained mass spectrometrist in some instances (Slagle, 1971; Chapman, 1978). The technique (*automatic spectral interpretation*) is said to use *artificial intelligence* or the *heuristic method*, but the latter name is confusing since it implies self-learning, which is used only in *learning machine* methods, to be described later. Artificial intelligence requires powerful computational facilities, large storage areas and high-level programming language with consequent slow processing. Both metastable ion and accurate mass measurement data can be used if these have been acquired (see earlier). The best known automatic spectral interpretation is DENDRAL, in which the functional groups present in an unknown compound are ascertained from its mass spectrum by

way of a qualitative theory of fragmentation processes, any available metastable ion data and the empirical or molecular formula. Potential structures
are generated, taking into account any other information supplied, such as
the nuclear magnetic resonance spectrum. For each candidate structure
generated, a mass spectrum is predicted, using known fragmentation mechanisms previously obtained by examining standard spectra of similar, known
structures. These theoretically predicted spectra are checked against the
experimentally determined spectrum of the unknown compound and those
which are consistent are ranked in order of best matching and are printed
out. A fairly successful programme designed to predict the mass of a molecular ion, whether it is present in the spectrum or not, is also available. As
before, this programme operates in much the same way as a mass spectrometrist does, only more rapidly (Dromey, Buchanan, Smith, Lederberg
and Djerassi, 1975).

The last method of automated structure elucidation discussed here is that
of *pattern recognition* or *machine learning* (Meisel, 1972; Isenhour, Kowalski and
Jurs, 1974; Chapman, 1978, pp. 150–186). The mass spectra of known compounds (a *training set*) are represented as points in a multidimensional space
(an *n*-fold vector), the coordinates of each point being the mass and abundance of each ion. It is reasonably argued that since a mass spectrum is
characteristic of structure, the points of compounds with similar structure
will tend to cluster in hyperspace. When the spectrum of an unknown
material is similarly treated, it is hoped that it will fall within one of the clusters, thereby categorizing its structural type. There are many ways in which
this cluster analysis has been performed (Justice and Isenhour, 1974; Mellon,
1975, pp. 127–132). Structural features that have been identified readily by
the method are the presence of nitrogen or oxygen, the ratio of carbon to
hydrogen, the phenyl group (although an experienced human interpreter
would usually recognize this feature from the mass spectrum at a glance)
and the phosphonate and carbonyl groups. Interesting studies include the
analysis of isomeric compounds (e.g. sugar derivatives, alkylbenzenes), the
spectra of which are too similar to be distinguished by other methods, the
classification of bacteria by subjecting the mass spectra of their pyrolysis
products to cluster analysis (section 12.4) and determinations of
pharmacological activity of a limited range of drugs. In this last case, it is
assumed that, just as a mass spectrum is related to structure by an obscure
function, so structure is related to pharmacological activity by a complex

function. Thus, the application of pattern recognition has been validated to some extent, and some successful predictive ability exhibited.

4.5.   DATA DISPLAY UNITS

Interaction between the data system and operator is through a *visual display unit* (VDU; workstation monitor) and a keyboard or 'mouse' or both. Often, operation of the display unit or workstation monitor is controlled by simply selecting 'icons' from displayed 'menus' of options. In the best systems, this simple selection procedure can be enhanced by the operator selecting a string of the options and assembling them into a routine (a simple programme) by using extended programmable automation language (EPAL). Also, the monitor screen can be split into sections so that several sets of results can be displayed simultaneously (e.g. the various factors controlling the operation of the instrument together with peak shape). All commands to and communications from the computer can pass through these units. A mass spectrometer produces and a computer processes data at high rates. For example, in an analysis by GC/MS, the printing out of the results is likely to be the most time-consuming part of the analysis so that a fast VDU for almost immediate display of results and a fast output device for hard copying of those results needing to be printed is recommended. Ideally, the data system is programmed to allow unsupervised data processing and hard copying so that these steps can be carried out at times when the mass spectrometer is not in use for data acquisition. In such circumstances, the speed of the hard copy unit is of lesser importance.

5 Combined chromatography and mass spectrometry

5.1. ## INTRODUCTION

It is possible to collect (trap) compounds as they elute from gas or liquid chromatographic columns and afterwards to obtain their mass spectra. Such gathering of mass spectrometric data is cumbersome, especially for complex mixtures, and has been superseded by coupling the chromatograph directly to a mass spectrometer. The present chapter is concerned with the unified, continuous processes of gas chromatography/mass spectrometry (GC/MS) and liquid chromatography/mass spectrometry (LC/MS), and there will be no discussion of the trapping of chromatographic effluents prior to mass spectrometric analysis. The combinations of thin-layer chromatography, supercritical fluid chromatography and capillary electrophoresis with mass spectrometry (TLC/MS, SFC/MS and CE/MS respectively) are also considered, despite the fact that, strictly, the last of these methods is not chromatography. In purpose, though, capillary electrophoresis shares a common goal with the true chromatographic methods: the components of mixtures are separated in a flowing carrier stream prior to detection. Discussion of the chromatographic and electrophoretic processes themselves is outside the scope of this book, although general references are provided to facilitate access to this basic information (Poole and Schuette, 1984; Grob, 1985; Miller, 1988; Smith, 1988*a*, 1988*b*; Poole and Poole, 1991).

GC/MS is an established technique for the analysis of complex mixtures, holding a prime position in analytical chemistry because of its combination of sensitivity, wide range of applicability and versatility. LC/MS instruments have not yet reached the degree of reliability that is achieved daily with GC/MS but some LC/MS devices have become established for materials too involatile or thermally unstable to pass unchanged through a gas chromatograph. Both combined techniques are capable of obtaining complete mass spectra of a few picogrammes or nanogrammes of each component. Below this level, analysis is still possible but special scanning techniques, detailed in chapter 7, are necessary. A further method for the analysis of mixtures, involving separation by mass rather than by chromatography (MS/MS), is discussed in chapter 8.

Combined chromatography/mass spectrometry is important in analytical chemistry because of the complementary nature of the individual techniques. On their own, chromatographs are excellent for separating the components of a mixture but they do *not* allow the identities of those components to be assigned with certainty. The only criterion for the identity of an eluting substance is the time that it takes to pass through the chromatograph (or through the capillary tube in electrophoresis) – the retention time. Unfortunately, any one of a number of compounds could have the same retention time. The chromatograph needs to be linked to an instrument capable of giving structural information. The most common devices are the photodiode array for ultraviolet/visible spectroscopy, an infrared spectrometer or, in the most widespread combination, a mass spectrometer. Mass spectrometry is a powerful method for identifying pure substances but the mass spectra of mixtures are usually too complicated to be useful. Therefore, combined chromatography/mass spectrometry provides a very effective tool for the qualitative characterization of complex mixtures by exploiting first the resolving power of chromatography to obtain the pure components (in a flowing stream) and then the strength of mass spectrometry to identify those separated compounds.

Of the chromatography/mass spectrometry combinations, that with gas chromatography (GC/MS) is the most developed and routine (Message, 1984; Jaeger, 1987). Systems vary from modest commercial bench-top devices to very large instruments capable of a vast array of GC/MS and GC/MS/MS experiments. The latter are covered in chapter 8. The other combined chromatography/mass spectrometry instruments are less well established in analytical science (Catlow and Rose, 1989). For example, it is possible to purchase any one of at least five fundamentally different LC/MS designs. To date, none of these interfaces can claim to meet the demands of every liquid chromatographic problem that a laboratory might wish to tackle. A broad analytical capability in this area may require access to more than one design of interface.

References to GC/MS and LC/MS are to be found in the preceding chapters and, in particular, computer programmes to acquire and process the data of such analyses are described in sections 4.4.2 and 4.4.3. In this chapter it will be assumed that the combined chromatograph/mass spectrometer is coupled to a data system.

In chromatography, the components (i) emerge sequentially from a tube as separated 'peaks' that have Gaussian shapes, the bases of which are just a few seconds wide, and (ii) are entrained in some carrier which is a gas, liquid or supercritical fluid. This arrangement imposes some constraints on a combined chromatograph/mass spectrometer. For example, as the amount of analyte in the ion source changes constantly during the emergence of a chromatographic peak, the mass spectrometer must be capable of scanning a spectrum quickly; otherwise, ion abundances would be grossly distorted by the change in concentration of the sample during the scan. This is only a problem for older mass spectrometers that scan more slowly or for mass spectrometers operating at very high resolution. Modern mass spectrometers are adequate to meet these more demanding tasks. More seriously, the outlet of a chromatograph is at atmospheric pressure but the ion sources of conventional mass spectrometers operate at low pressure so that, in general, the chromatographic effluent must be transferred to the mass spectrometer with concomitant rejection of most of the carrier. The difficulty of removing a *liquid* solvent for LC/MS, as opposed to a carrier *gas* for GC/MS, is one reason why LC/MS technology has lagged behind that of GC/MS. For combining mass spectrometry with liquid chromatography, the current trend is towards the use of ion sources that tolerate higher pressures and/or use the solution phase directly. In the following sections, different ways of combining (*interfacing*) chromatographs and mass spectrometers are discussed, with examples of their use.

5.2.   GAS CHROMATOGRAPHY/MASS SPECTROMETRY

This combination of two gas-phase analytical techniques can be applied to a very wide range of problems. With a good interface, any compounds that elute from a gas chromatograph may be examined by GC/MS. Chemistry, medicine, biochemistry, forensic science, pharmacology, environmental science, food chemistry, geochemistry and gas analysis are some of the areas that have benefited from GC/MS. Information on the full range of applications is best obtained by reference to the biennial reviews in *Mass*

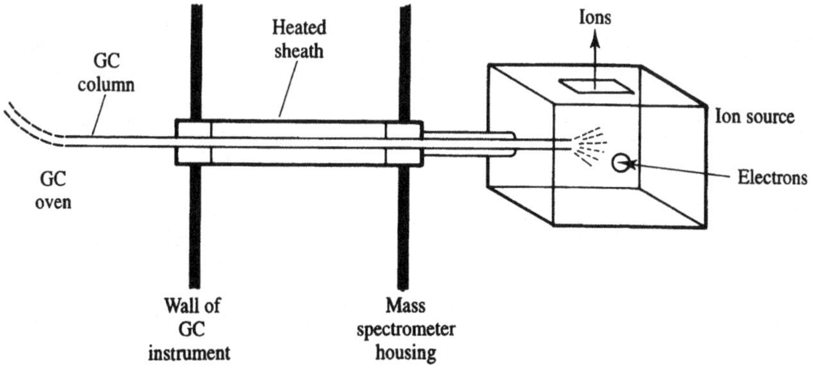

Figure 5.1. A schematic diagram of the direct coupling approach to combined gas chromatography/mass spectrometry.

Spectrometry (Brooks and Middleditch, 1975,1977,1979; Mellon, 1981; Rose, 1984, 1985; Evershed, 1987, 1989).

The components in the hot effluent of a gas chromatograph are in the gas phase, so compatible ion source designs are electron ionization, positive-ion and negative-ion chemical ionization and field ionization. This chapter will illustrate the method with EI and CI applications.

5.2.1. *The GC/MS interface*

The simplest interface is a direct line from the end of the chromatographic column to the mass spectrometer. Since modern capillary columns are made from flexible fused-silica, the best arrangement is to feed the column through a gas-tight heated sheath – even if the path to the mass spectrometer is curved. The end of the column is then located in the ion source, just short of the region of ionization (figure 5.1). The main advantages of this *direct coupling* are: (i) all of the sample is transferred into the ion source, thus maintaining good limits of detection; (ii) there is no dead volume that would otherwise degrade the resolution; and (iii) the components do not come into contact with any surface other than the stationary liquid phase coated on the inside wall of the column itself (for instance, hot metal tends to act as a catalytic surface on which decomposition can occur).

In this direct coupling arrangement, the carrier gas is passed into the ion source itself. This is acceptable as long as the gas is highly diffusible; helium is the usual choice. A modern mass spectrometer with efficient pumps operating in the electron or field ionization mode can maintain an adequate vacuum in the ion source with a flow of helium carrier gas from the column of about 2 ml min^{-1} or less. With gas chromatographs coupled to mass spectrometers that

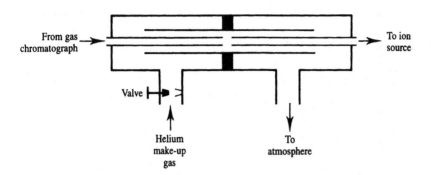

Figure 5.2. A schematic diagram of the open split approach to combined gas chromatography/mass spectrometry.

have chemical ionization sources, it is possible to use the carrier gas in the ion source as the CI reactant gas. Methane and hydrogen are most often used for such purposes, but the method is not particularly popular because it restricts the choice of CI reactant gas to those gases which are also suitable as GC carriers. It is preferable to admit the reactant gas for CI into the ion source by an independent inlet.

Note that, with direct coupling GC/MS, the end of the GC column is not at atmospheric pressure, as distinct from stand-alone gas chromatography. The vacuum outlet conditions in GC/MS do disturb somewhat the gas flow characteristics in the chromatographic column such that the resolution decreases slightly. This loss of chromatographic efficiency is rarely important to an analysis. If a critical separation is unduly deteriorated by this effect, it is possible either to compensate for the vacuum outlet conditions by increasing the flow rate of the carrier gas or to use an alternative interface design, which maintains the column outlet at atmospheric pressure. In the *open-split* interface (figure 5.2), the end of the column is open in the sense of being at atmospheric pressure, and the gas flow from it is split. One exit is the ion source, which takes a constant flow of about 1 ml min^{-1} through a restrictor, and the other is a vent to the atmosphere. If the flow from the column is less than 1 ml min^{-1}, it is supplemented by more helium ('make-up' gas). If the flow from the column is greater than 1 ml min^{-1}, the excess is vented to the atmosphere. This interface allows the same chromatographic performance in GC/MS as in GC alone under the same conditions. It does have three disadvantages: (i) at high flow rates, some sample is vented, increasing the limit of detection; (ii) contact with the hot glass walls of the device may bring about adsorption or decomposition of analyte molecules; and (iii) it is not as simple as direct coupling.

Whichever interface is used, it is worth stressing that it must be maintained at a temperature that is at least as high as the highest temperature to which the column itself is subjected. Any cold spots in the system can cause condensation of the analyte molecules and hence poor resolution and a loss – sometimes a total loss – of sensitivity.

Older GC/MS systems used packed columns rather than modern fused-silica capillary columns. These older columns could involve carrier gas flow rates up to 60 ml min^{-1}. Such flows could not be admitted directly to the ion source because the vacuum would be compromised. The problem was managed with enrichment devices called *molecular separators*, which sought to remove most of the small carrier gas molecules whilst selectively passing the larger molecules of interest into the mass spectrometer. The most common device, the *jet separator*, relied upon differential diffusion of solute and carrier gas after the effluent had passed through a small jet into a vacuum. For details of this, and other enrichment devices that have largely been made obsolete by advances in GC columns and vacuum technology, an older text should be consulted (McFadden, 1973).

5.2.2.   *Applications*

Discussion of quantitative aspects of GC/MS is presented in chapter 7. Here, some of the ways in which a mass spectrometrist can identify the components of mixtures by GC/MS are illustrated by reference to typical examples in the clinical and environmental areas.

The first example concerns a clinical application of GC/MS. In some cases of sudden unexpected infant death, victims have been shown to be deficient in medium-chain acyl-CoA dehydrogenase, a key enzyme for β-oxidation of fatty acids. This disease, and several other inborn errors of metabolism leading to organic acidurias and acidemias, are characterized by elevated concentrations of certain acylcarnitines (1) in urine. Therefore, an analytical procedure for identifying urinary acylcarnitines would facilitate diagnosis of, and monitoring of patients with, these serious and sometimes life-threatening diseases. There is one problem to solve before any GC/MS analysis for acylcarnitines can be attempted. Gas chromatography relies on analytes eluting through the column *in the gas phase*. The acylcarnitines are ionic and hence not sufficiently volatile to be vaporized intact. To convert them into volatile compounds amenable to gas chromatography, they were chemically derivatized according to scheme (5.1). This derivatization relies on heat to cause cyclization to volatile acyloxylactones (2). These lactone

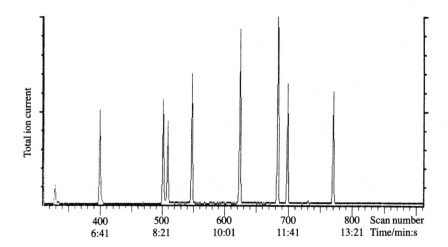

$$\text{(5.1)}$$

Figure 5.3. Part of the ion current chromatogram of a standard mixture of nine acylcarnitines after chemical derivatization according to scheme (5.1). Spectra were recorded repetitively every second.

derivatives were analysed on a conventional capillary column, with a GC/MS system employing direct coupling, repetitive scanning and, initially, electron ionization. A set of nine acylcarnitine standards gave the total ion current (TIC) chromatogram shown in figure 5.3.

If this work had been carried out with GC alone, the information in figure 5.3 would be the only data available. The presence of any of the nine derivatives in a treated urine sample could only be assessed by a coincidence in retention time with the peaks in figure 5.3. Such a comparison is not totally reliable. In GC/MS, the presence of a target molecule is determined by a coincidence of retention time with the authentic compound *and* identical mass spectra, within experimental error. This is much firmer evidence on which to base an identification. Of course, an authentic sample of a compound may not be available. When this situation pertains, GC cannot be used to confirm an identity, but GC/MS provides a mass spectrum that can be interpreted to provide structural information.

Corresponding to each peak in figure 5.3, there are mass spectra that have been stored in computer memory and can be recalled. For example, the lactone from octanoylcarnitine (2, R = $(CH_2)_6CH_3$) elutes near 13 min (scan number 770) and its EI mass spectrum was shown in figure 3.2. This

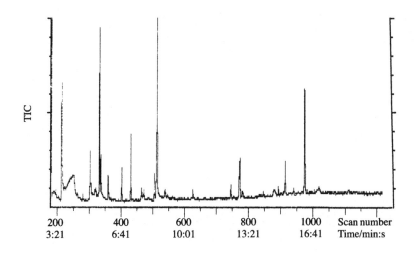

Figure 5.4. A total ion current chromatogram of a urine extract, from which acylcarnitines were isolated and subjected to the conditions of the chemical derivatization shown in scheme (5.1). The conditions are the same as those in figure 5.3.

spectrum was obtained by summing a few mass spectra acquired near the top of the GC peak, then subtracting a 'background' spectrum taken at a nearby point where there are no peaks. This background subtraction removes the contribution of constantly eluting substances like the breakdown products of the stationary phase ('column bleed'). By storing the mass spectra of the nine standards represented in figure 5.3, a small 'library' of relevant compounds is accrued. Knowing the mass spectra and retention times of the lactones (2) from some of the key acylcarnitines, a urine sample can be examined.

The urine sample of one baby, after work-up, derivatization and GC/MS, gave the TIC trace shown in figure 5.4. There are several ways of approaching the problem of identifying the constituents of this mixture. It is possible to recall mass spectra for examination whenever the total ion current chromatogram indicates that there is a chromatographic peak and interpret each spectrum. For an analysis concerned only with the acylcarnitine content, it is not appropriate to examine all components of the mixture. To reduce processing time, it is possible to focus attention on only the components of interest by the use of some form of 'target analysis'. That is, the data from the urine sample and from the authentic standards are compared and the following question is asked: are any of the reference mass spectra also present in the analytical sample at the appropriate retention times? The answer to the question can be found in a number of ways. Manual control of the computer or completely automatic computer processing can be used. Here, a manual method employing mass chromatograms is described first.

The reference mass spectra of derivatized acylcarnitines eluting after propanoylcarnitine (**2**, R = $(CH_2)_nCH_3$, where $n > 1$) all contained abundant ions at m/z 85 and 144, which can be drawn as structures (**3**) and (**4**), respectively. The mass chromatograms of these two m/z values, together with the total ion current, are shown in figure 5.5. It is seen that peaks due to the compounds of interest become apparent wherever the two mass chromatograms rise and fall in unison, such as at scan numbers 511, 626, 745 and 773. Such characteristic responses can be selected by the operator or automatically by the computer. The operator can then view a full spectrum at each of the scan numbers at which there is a peak corresponding to m/z 85 and 144. Manual interpretation or library searching enables the identity of those components to be verified. For example, the mass spectrum at scan number 511 corresponds to the position expected for the derivative of isovalerylcarnitine (R = $(CH_3)_2CHCH_2$). Figure 5.6(*a*) shows the spectrum originating from the urine sample and figure 5.6(*b*) gives the reference spectrum from figure 5.3. Within

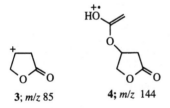

3; m/z 85   **4**; m/z 144

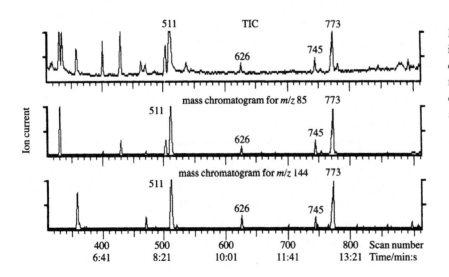

Figure 5.5. Part of the total ion current chromatogram of the urine extract as in figure 5.4 (top) and the mass chromatograms of ions at m/z 85 and 144.

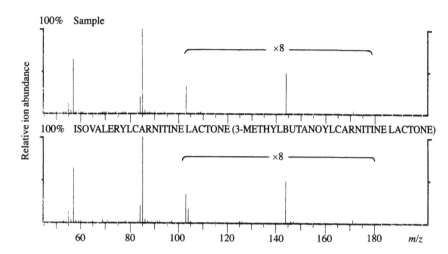

Figure 5.6. (*a*) The EI mass spectrum of the component eluting at scan number 511 in the urine extract, and (*b*) the reference mass spectrum of the lactone of isovalerylcarnitine eluting at the same retention time. In parts of the spectra, the ion abundance has been multiplied by a factor of eight for clarity.

experimental error, they are identical and, since they both appear at the same retention time, the identity of that particular component can be confidently assigned to the lactone of isovalerylcarnitine.

Discussion of mass chromatograms has concentrated so far on the simultaneous detection of several related compounds by use of common characteristic ions in their mass spectra. Mass chromatograms can also be used to search for one specific compound using ions that, ideally, are unique to that compound. This approach is useful when a whole reference mass spectrum of a target compound is not available. At least one suitable m/z value must be predicted from the structure of the compound suspected of being present. Often, its molecular ion, or quasi-molecular ion, is chosen. An example is taken from the same urine sample but subjected to GC/MS in the chemical ionization mode (GC/CIMS).

In the suspected disease, it is known that octanoylcarnitine is usually accompanied by at least one unsaturated analogue, an octenoylcarnitine. Figure 5.7(*a*) shows the spectrum of a compound eluting just before the lactone from octanoylcarnitine, which may be due to an unsaturated ana-logue, but the lack of significant molecular ions (m/z 226) makes the assign-ment tenuous. Also, the spectrum contains only low abundances of the informative high-mass ions. These uncertainties can be addressed by using chemical ionization.

Generally, when the ion source is changed from EI to isobutane chemical ionization conditions, the resulting mass spectra tend to show abundant proto-nated molecules at one atomic mass unit greater than the relative molecular

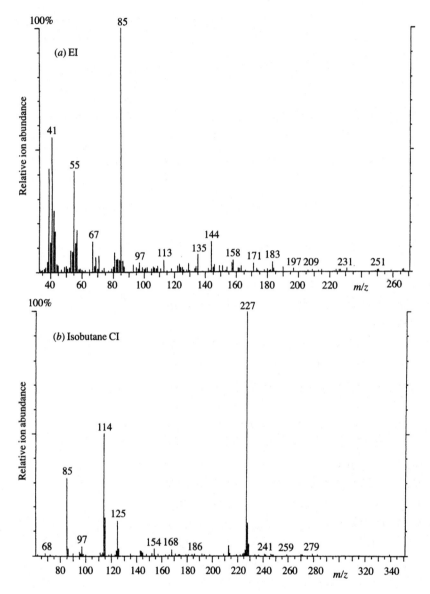

Figure 5.7. (*a*) The EI mass spectrum of the component, thought to be an octenoyloxylactone, eluting just before octanoyloxylactone (**2**, R = C_7H_{15}) in the urine extract, and (*b*) the CI mass spectrum of the same component, confirming that $M_r = 226$.

mass (section 3.3). A sample of authentic lactone from an octenoylcarnitine was not available but, in CI, a large peak at *m/z* 227 would be confidently predicted for the protonated molecule (based also on the analogous behaviour of the octanoyloxylactone whose EI and CI spectra appear in figure 3.2). The TIC trace and mass chromatogram of *m/z* 227 for a GC/CIMS analysis is shown in figure 5.8. The largest response in the mass chromatogram occurs at scan number 762, just before the octanoyloxylactone elutes at scan 766. The CI mass spectrum at scan number 762 is given in figure 5.7(*b*) and clearly confirms

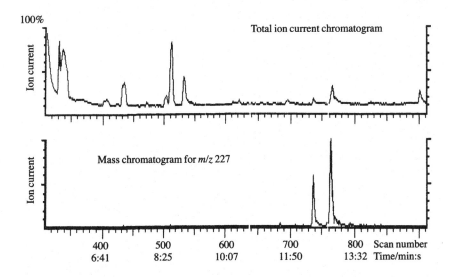

Figure 5.8. The total ion current chromatogram of the urine extract analysed by GC/CIMS (top) and the mass chromatogram of ions at m/z 227.

the relative molecular mass of the lactone from an octenoylcarnitine. It should also be noticed from figure 5.8 that this compound is a very minor component; its peak in the total ion current chromatogram is simply a shoulder on the peak due to the octanoylcarnitine lactone. Even so, the mass chromatogram approach highlights its presence so that it could not be missed. The second clear peak in the mass chromatogram (figure 5.8, scan number 738) corresponds to a substance with a CI mass spectrum that is almost identical to that in figure 5.7(b). This component would be assigned as a second, isomeric octenoylcarnitine derivative.

In the CI mode, the choice of characteristic ions used to search for specific compounds by mass chromatograms is made easy since it can be assumed that $[M + H]^+$ ions will be abundant. Several advantages accrue from this emphasis on the higher-mass ions. In particular, low-mass ions are relatively non-specific, frequently being common to many different structures.

The structures of unknown and unexpected compounds are far easier to elucidate when their relative molecular masses can be determined, so consideration of their CI mass spectra, together with their EI spectra, is beneficial. To facilitate analyses requiring both EI and CI, like the acylcarnitine study, some mass spectrometers are designed to combine the electron ionization and chemical ionization approaches by alternate EI/CI switching so that, for example, all the odd-numbered scans are EI and the even-numbered scans CI spectra.

It should be appreciated that a single mass chromatogram which rises to a maximum at the retention time of a particular compound is not sufficient evidence for its identification. Such a signal is used only as a means of locating the mass spectrum or spectra, which must be examined to ascertain the presence or absence of the compound. For a compound present in very small quantities, only the major ions in its mass spectrum may be detectable. Under such circumstances, if the mass chromatograms of two or more characteristic ions rise and fall in abundance simultaneously at the correct retention time, the presence of the compound is indicated but should not be considered as proven.

In an alternative automated approach, the computer can be programmed to search for whole mass spectra from a reference file within a known window of chromatographic retention times. This 'automated targeting' or inverse library searching is very fast and effective, typically taking less than 2 s per analyte. Thus, several hundred target analytes could be located (and quantified) in a GC/MS analysis of 1 h duration. This method is based on a type of library search (section 4.4.3): a comparison of each reference spectrum with the mass chromatograms of all of the m/z values in the appropriate retention time window. Examples of this methodology occur in environmental investigations.

Pollutants are identified unambiguously by acquiring their full mass spectra during GC/MS of environmental samples. To expedite routine monitoring of so-called priority pollutants, data processing is automated. For example, over 200 target analytes are required to be checked in one waste-water analysis of about 60 min duration. Before any such analysis is carried out, a standard mixture must be examined. This ensures that the GC/MS system is working properly. Retention times, chromatographic peak shape and mass spectra should match those known for the reference compounds in the standard mixture. The integrity of both the GC separation and the mass spectrometer tuning is thus assessed before the sample is analysed. In an additional quality control procedure, a waste-water sample is 'spiked' with compounds representative of the pollutants. Measurement of the recovery of these known amounts after the work-up tells the analyst whether the whole analytical procedure is producing valid results.

Volatile components in a waste-water sample might be isolated by bubbling an inert gas through the sample for a few minutes. The gas with entrained analytes is passed through a porous polymer (Tenax) that traps the volatile organic material. Later, heating and backflushing of the Tenax cartridge releases the pollutants, which are led into the inlet of a GC/MS system for analysis. Alternatively, activated carbon or simply a cold surface may be utilized to trap volatile components in such *purge-and-trap* devices. Analysis of any less volatile pollutants remaining in solution would be undertaken after extraction and concentration steps. In subsequent GC/MS analyses, the data system uses the automated targeting method to detect mass spectra of the priority pollutants. Basically, a programme would utilize known retention times (relative to those of internal standards) to assign narrow time windows in which to search for given pollutants, and reference mass spectra for each of those target pollutants. Within the time window set for a given pollutant, a library comparison would require the relative ion abundances of a putative pollutant to rise and fall in unison and to match those of the authentic compound, within experimental error. (If the presence of a given pollutant is confirmed the programme also quantifies it by measuring the area of its peak in a given mass chromatogram, but discussion of this topic is postponed to chapter 7.) As with normal library searching, a 'score' for the degree of matching is calculated, but here it also includes a factor for the goodness of fit with the predicted retention time. Doubtful matches are highlighted so that the expert can review them later. Figure 5.9 illustrates the complexity of the TIC trace of a waste-water analysis, and figure 5.10 shows that automated targeting locates the polycyclic aromatic hydrocarbon, phenanthrene, spiked into the water sample, even though the amount is so small that it does not elicit a peak in the TIC trace. In fact, phenanthrene elutes in a valley between two peaks: the computer found the component whereas inspection by human eye would not reveal anything. Typically, the processing needed to find this substance would take 0.2–2 s. There are two reasons why automated data processing can be more effective than human interpretation: it locates components that are hidden to the eye and it is considerably faster. Of course, a data system does not object to working overnight either!

In summary, the utility of mass chromatograms has been stressed because they have been shown to be useful not only for specific detection of a single compound but also for locating simultaneously several members of the same class of compound. In other studies, benzylic compounds could be located in chromatographic analyses by examination of the mass chromatograms of m/z 91 ($C_7H_7^+$) and 65 ($C_5H_5^+$), trimethylsilyl derivatives by looking for m/z 73 ((CH_3)$_3Si^+$) and so on. The selection of ions specific to compounds of

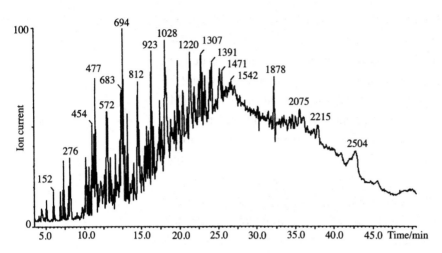

Figure 5.9. The total ion current chromatogram from an analysis of waste water. (Kindly supplied by VG MassLab Ltd.)

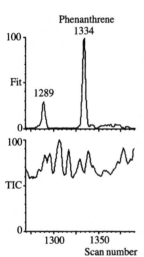

Figure 5.10. Automatic targeting of phenanthrene in waste water. An expanded part of the ion current chromatogram in figure 5.9 is shown at the bottom. At the top, the goodness of fit with the standard spectrum of phenanthrene is plotted. The predicted retention time occurs at scan number 1330 so the better fit at scan number 1334 is likely to correspond to phenanthrene. Note that phenanthrene elutes in a valley between two peaks in the TIC trace. (Kindly supplied by VG MassLab Ltd.)

interest in complex mixtures is of prime importance for selected ion monitoring and is discussed in chapter 7.

The results described so far could be obtained on a bench-top or larger GC/MS system, employing a quadrupole, magnetic-sector analyser or an ion trap. Some of the systems, most notably those based on magnetic sectors, can be operated at high resolution so that the masses of all abundant ions are accurately determined. Operating in this way is particularly helpful when extra selectivity is required because elemental formulae of the ions can be calculated. For example, an accurate mass chromatogram is, in effect, a plot of the occurrence of ions of a particular elemental composition against time or scan number. It is a more specific indicator of structure than is the simpler integer mass chromatogram. Measurement of accurate masses makes it possible to distinguish readily background ions (such as those due to column bleed) from sample ions when these occur at the same integer mass but have different accurate masses. The methodology for GC/MS with mass spectrometers of high resolving powers has been reviewed (Kimble, 1978). Perhaps the greatest use for measuring accurate masses during GC/MS involves selected ion monitoring at high resolving power and this is described in chapter 7.

The following additional examples of GC/MS are included to illustrate briefly the wide scope of the method. The Viking mission to Mars carried equipment for analysis by GC/MS of any organic material in the Martian atmosphere and soil (Biemann, 1979). Organic compounds were not detected in the samples despite the high sensitivity. The experiments were carried out with a small magnetic-sector instrument. Space research provided considerable impetus for the development of GC/MS with quadrupole mass filters, which are more robust and much more easily miniaturized than magnetic-sector instruments. For food flavour and odour research, the GC/MS interface is often a simple splitter, with the eluent that is not transferred to the mass spectrometer being passed to a 'sniffing port' at which a person evaluates the odour of eluents. The structures identified by mass spectrometry are thus related to aroma. The aroma of any food is rarely attributed to a single component and results from a complex mixture. For example, 2-isobutylthiazole, unsaturated aliphatic alcohols and aldehydes, and metabolites of carotenoid compounds have been shown to contribute to the odour of tomatoes. Toxicological aspects of food chemistry also rely on GC/MS. For instance, contamination of wine by ethylene glycol has been detected by use of GC/MS.

GC/MS is used in forensic science to detect cocaine, naturally occurring cannabinoid compounds, anabolic steroids and opiate compounds amongst many others. Such is the limit of detection of the method that amphetamine can be detected in a single hair of a drug user. In a remarkable analysis, GC/MS was used to identify bile acids (acidic steroids) in the gall bladder of a 3200-year-old Egyptian mummy. After taking account of changes induced by bacteria and the environment, the qualitative and quantitative composition of the acids was found to be the same as that in modern humans, indicating that the pathways for cholesterol metabolism in humans have not changed in that time. Analyses by GC/MS of gaseous samples are common.

Cigarette smoke, human breath and pollutants in air have all been studied. The volatile products resulting from pyrolysis of the polymeric materials of meteorites were shown by GC/MS to include alkanes, alkenes, alkylbenzenes and various thiophens. For on-line pyrolysis/GC/MS (discussed in detail in section 12.4), products of pyrolysis are passed immediately to a GC/MS instrument. Such a system is useful for analysis of paint, fibres, greases, adhesives, dye pigments, polymers and bacteria, and has been used for diagnosis of cancer.

5.3.   LIQUID CHROMATOGRAPHY/MASS SPECTROMETRY

The solute emerging from a liquid chromatographic column is not only in the condensed phase but also highly dilute. A flow of 1 ml min^{-1} of water becomes roughly 1250 ml min^{-1} of vapour (compared with perhaps 1 ml min^{-1} of gas from a GC capillary column). Often, involatile inorganic compounds may be present in the eluent, as buffers for example. Conducting solutions can lead to arcing of the high potential inside the ion source of a magnetic sector mass spectrometer. If gradient elution is used in the HPLC step, the composition of the solvent changes during the analysis. In addition, the solutes themselves are frequently involatile and thermally unstable (if they were sufficiently volatile to elute from a gas chromatograph, they would be analysed by GC/MS since the latter method is better developed and the necessary equipment is more widely available and less expensive). For all these reasons, the technical difficulties of combining (interfacing) liquid chromatography and mass spectrometry exceed those for the GC/MS combination. Potentially, though, LC/MS is a more widely applicable method than GC/MS. Liquid chromatography is suitable for large, polar, ionic, thermally unstable and involatile compounds. Some of these troublesome compounds may be made amenable to GC/MS by chemical derivatization (chapter 6) but, if there are no suitable derivatives or if the yield of the derivatization reaction is low, LC/MS can be applied with advantage to the underivatized mixture. Also, work-up of samples prior to analysis by LC/MS is often less extensive than that needed for GC/MS.

Typical flow rates for LC are 0.5–5.0 ml min^{-1}, which, for common solvents after vaporization, are equivalent to gas flows in the range 100–3000 ml min^{-1}. Conventional ion sources, employing electron, chemical or field ionization, cannot tolerate such large volumes of vaporized solvent so direct coupling of liquid chromatographic columns to these types of mass spectrometer is precluded. If a standard analytical column for HPLC (4–5 mm diameter) is replaced by a so-called microbore column (1–2 mm diameter) the typical flow rate decreases to 10–100 μl min^{-1} of solvent, which still represents over 2 ml min^{-1} of gas. It is not surprising that mass spectrometers allowing ionization at atmospheric pressure are making a particular impact in the area of LC/MS. Even when such methods are used for LC/MS, there are still restrictions on the LC solvent and difficulties with involatile buffers, but there is a greater degree of inherent compatibility between the liquid chromatograph and the mass spectrometer.

This section is concerned mostly with a discussion of the currently popular interfaces for LC/MS (Voyksner, 1989; Arpino, 1990a; Brown, 1990; Yergey, Edmonds, Lewis and Vestal, 1990; Niessen and van der Greef, 1992). There will be no need to examine in detail any one particular application because the approaches to the processing of data obtained by GC/MS, discussed in the previous section, are relevant also for LC/MS and additional computer programmes to handle chromatographic aspects are not required.

5.3.1. *The moving belt interface*

A continuously moving, endless belt provides a versatile LC/MS interface (Arpino, 1989). As illustrated in figure 5.11, the HPLC effluent is applied continuously to the moving polyimide belt, which is a few millimetres wide. Much of the solvent is evaporated by an infrared heater. Flow rates compatible with this approach depend on the volatility of the solvent: the more

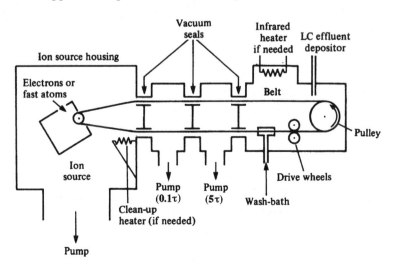

Figure 5.11. A schematic diagram of a moving belt interface for LC/MS.

volatile the solvent, the higher the acceptable flow rate. The maximum flow rate of aqueous solutions is about 0.5 ml min^{-1} whereas that of non-polar solvents is about 2.0 ml min^{-1}. The relatively less volatile solute remains on the belt and is mechanically transported through two evacuated chambers via a series of narrow gaps (vacuum locks). Here, the residual solvent is pumped away, leaving a solid on the belt to be carried into the ion source, where it is evaporated by a powerful heater. This design, carrying the solutes right into the ion source before they are vaporized, ensures best results with fragile compounds (compare with in-beam ionization in section 3.6). Any residue remaining on the belt is removed by another heater after it leaves the ion source.

The interface is suitable for operation of the mass spectrometer in the EI or positive- or negative-ion CI mode without severe restrictions on the choice of LC solvent or CI reactant gas. The use of involatile salts as buffers is permitted since they can be scrubbed off the belt during its 'return journey' to the inlet chamber, and gradient elution is not problematic. The interface does not cause significant loss of chromatographic resolution and the efficiency of transfer can be as high as 80 per cent, but 40 per cent is more typical. A few nanogrammes of sample provide complete mass spectra.

Typical successful applications of this interface have been to porphyrins, coumarins, aflatoxins, nucleosides, antibiotics and drug metabolism studies. The moving belt system is suitable for quadrupole or magnetic sector mass spectrometers, was commercially available, and, as long as the belt protrudes into the ion source, allows ionization by bombarding the solutes with ions (SIMS), atoms (FAB) or photons (laser desorption). Hence, the interface is strong on versatility (Rose, 1987). However, it also has several weaknesses. The more volatile components of the sample tend to be vaporized and pumped away with the solvent. The system is rather expensive, mechanically complex and cumbersome to use. It occupied a key position amongst LC/MS interfaces in the 1970s and 1980s but has not stood the test of time very well. It has largely been eclipsed by the 'particle beam interface', which is compatible with a broadly similar range of analytes.

5.3.2. *The particle beam interface*

Compared with the moving belt, the particle beam interface has the advantage of robustness and ease of operation: it has no moving parts. It is a commercial development of the 'monodisperse aerosol generator for interfacing

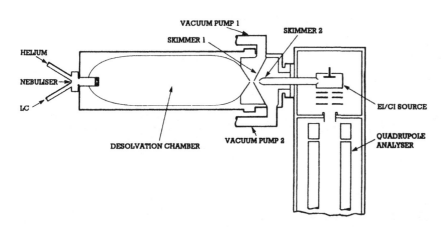

Figure 5.12. A schematic diagram of a particle beam interface for LC/MS. (Taken with kind permission from Mellon (1991).)

chromatography', called MAGIC by its proponents (Willoughby and Browner, 1984; Winkler, Perkins, Williams and Browner, 1988). A schematic diagram of the interface is shown in figure 5.12. The HPLC eluent, at a flow rate of 0.1–1.0 ml min^{-1}, is passed into the device together with a nebulizing flow of helium. This generates an aerosol of droplets, which traverse the heated 'desolvation chamber' where the solvent vaporizes to give a mixture of helium, vapour and particles of analyte. The mixture enters a vacuum region through a small nozzle. This creates an expanding jet, which is directed towards two further nozzles (or skimmers), which form a two-stage separator relying on differential diffusion of the solute particles and the gases. Helium and the vaporized solvent molecules diffuse faster than the much heavier beam of analyte particles from the expanding jet. Hence, at this point, the majority of the unwanted gases has diffused far enough from the mainstream to be pumped away ('skimmed'), leaving the analyte particles, with their greater momentum, to travel directly into the ion source. In the source the residual particles strike a hot wall, or pass through a hot wire mesh, causing vaporization of analyte molecules. These are then ionized by electron or chemical ionization. Usually, a few nanogrammes of each component is required.

Usually, this LC/MS interface provides a coupling to a quadrupole mass spectrometer but it can also be made compatible with magnetic sector instruments. The particle beam system shares with the moving belt the advantage of generating classical EI mass spectra, which can be interpreted or compared with the vast stores of data in computerized libraries of EI spectra. That is, it provides much information for determination of unknown structures, in contrast to those methods that tend to produce quasi-molecular ions such as $[M + H]^+$ ions and few, if any, fragment ions. In this sense, the particle beam approach is often seen to be complementary to the thermospray method. Samples must be partially volatile, though, so the particle beam method is not suitable for large, involatile or thermally labile compounds. Also, some samples with low volatility are cleared from the ion source region rather slowly and so produce 'memory effects.' In addition, the interface has only a limited tolerance to involatile buffers.

5.3.3. *The thermospray interface*

This device is a widely established and reliable LC/MS method for many materials, including involatile ionic compounds (Blakley and Vestal, 1983; Vestal, 1983; Arpino, 1990*b*). The ionization mechanisms operating in the thermospray method were discussed in section 3.9. The column effluent, which contains a volatile electrolyte, is heated as it passes through a capillary and is forced through a small orifice, forming a stream of vapour and aerosol (figure 5.13). The vapour comprises about 95 per cent of the solvent whilst the aerosol droplets carry a charge and contain most of the solute. The vacuum encourages vaporization of the solvent so the droplets shrink. This is thought to cause 'ion evaporation' but chemical ionization between solute (M) and electrolyte ions also occurs. The resulting ions, usually $[M + H]^+$ and $[M - H]^-$, are repelled through an orifice into the mass spectrometer. In general, the yield of ions is not high. Figure 5.13 also shows an added discharge electrode. This, or a filament, is used to ionize the analyte if a volatile electrolyte like ammonium ethanoate cannot be added to the effluent, if the

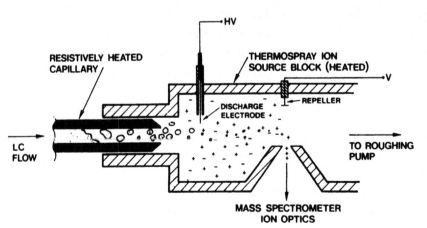

Figure 5.13. A schematic diagram of a thermospray interface for LC/MS. (Taken with kind permission from Mellon (1991).)

analytes are non-polar, or if non-aqueous solvents are employed. For compounds that produce only quasi-molecular ions and no fragmentation from which structure could be deduced, a higher than normal voltage can be applied to the repeller electrode shown in figure 5.13. This promotes fragmentation possibly by accelerating the ions into the high-pressure region of the ion source, where collisions can cause an increase in internal energy of the ions.

Liquid flow rates that are compatible with the thermospray device are roughly 0.5–2.5 ml min^{-1}, but it does not tolerate involatile buffers like potassium phosphate. The interface is applied advantageously to nucleosides, nucleotides, peptides, alkaloids, glycosides and oligosaccharides. However, optimum performance is dependent on the experimental conditions and on the type of compound. Detection limits can be as poor as 10 mg for a full spectrum but are more typically in the region of 1 ng. The more thermally sensitive compounds have the poorest detection limits. Large and/or thermally labile compounds are better examined by either of the next two interfaces to be described.

5.3.4. *Continuous-flow fast atom bombardment*

Fast atom bombardment (section 3.10) can be effected by placing a few microlitres of the analyte solution on a probe or, of more relevance here, by continually pumping the solution to a hole or frit at the end of a hollow probe located in the evacuated ion source (Ito, Takeuchi, Ishii, Goto and Mizuno, 1986; Ishii and Takeuchi, 1989; Caprioli, 1990*a*, 1990*b*). The tip is warmed somewhat to prevent freezing as the solvent evaporates in the ion source. The solution should contain 3–10% of a viscous solvent like glycerol, which can be part of the mobile phase or can be added after the HPLC separation. Unfortunately, only small flows can be accommodated by this method, up to 10 μl min^{-1} in most designs. The requirement for small flows is the most severe limitation of the LC/FABMS method. Splitting of the HPLC effluent and rejection of some of it is necessary, unless capillary columns with bores of 50 μm to 0.35 mm are used (Ishii and Takeuchi, 1989; Moseley, Deterding, Tomer and Jorgenson, 1991*b*). Bombardment can be by atoms (usually Xe for FAB) or by ions (usually Cs$^+$ for secondary ion mass spectrometry, SIMS). Given that FAB is sometimes called 'liquid SIMS', a dazzling array of names has been used in the literature of this method: continuous-flow FAB, frit FAB, dynamic FAB and the three LSIMS equivalents. They are all fundamentally the same simple device.

As stated in section 3.10, the liquid surface in continuous-flow FAB is being continually disrupted and replenished. Unlike with the static mode of FAB, there is little time for the bombardment to cause degradation at the tip and so accumulate by-products. Therefore, mass spectra obtained in the continuous-flow system have less background interference than do those from the conventional static FAB method. With favourable solutes, pico-mole amounts provide good FAB mass spectra and suppression is much reduced. The range of analytes is the same as for static FAB. That is, polar and ionic compounds are ideal, and they may have fairly large relative molecular masses.

5.3.5. *Atmospheric pressure ionization interfaces*

The ionization process underlying the electrospray device is described in section 3.9. An LC/MS interface using the electrospray principle is illustrated schematically in figure 5.14 (Hiraoka and Kudaka, 1990; Smith, Loo, Edmonds, Barinaga and Udseth, 1990; Bruins, 1991). Like the FAB device in the previous section, the liquid flow capacity of the electrospray method was severely restricted: only 1–20 μl min^{-1} were once used. However, the inherent advantages of the method outweighed this disadvantage in many applications. Addition of an annular flow of gas (a pneumatic sprayer) assists solvent evaporation and increases the flow capacity to 400 μl min^{-1}. This has been called *ionspray* (Bruins, Covey and Henion, 1987; Huang, Wachs,

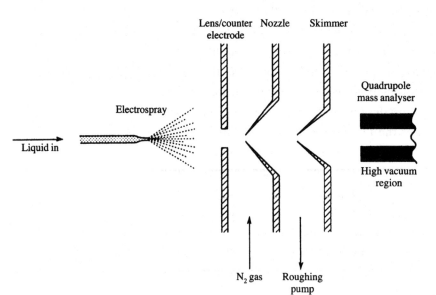

Figure 5.14. A schematic diagram of an electrospray interface for LC/MS. (Taken with kind permission from Mellon (1991).)

Conboy and Henion, 1990; Bruins, 1991). Further improvements in the design of the electrospray source, such as the inclusion of a heated capillary tube that acts as a desolvation chamber, have increased the flow capacity; flow rates of the order of 1–5 ml min^{-1} have been achieved. However, ion suppression can occur at the high flow rates that obviate the need for splitting the flow from the HPLC column. The signal maximizes at a flow rate that depends on source design, the solvent, the type of analyte and other factors. Therefore, the maximum response is likely to be obtained by splitting of the HPLC effluent.

The liquid emanating from the capillary tube is charged by application of a voltage in the low kilovolt range. Charged droplets are formed and 'ion evaporation' occurs at atmospheric pressure. A stream of warm dry gas, usually nitrogen, flowing between the counter-electrode and the sampling nozzle aids evaporation of the solvent molecules clustering around the ions and prevents the vapour from entering the evacuated part of the mass spectrometer. The ions are drawn through the nozzle and skimmer into the analyser, which is maintained at high vacuum and is frequently a quadrupole system.

The interface has several strengths. Most notably, it is compatible with a remarkable range of analytes, from small to large (polar compounds with relative molecular masses from about 60 to over 300 000 have been examined with simple quadrupole mass spectrometers, the latter being made possible by the effect of multiple charging as discussed in chapter 3). Molecules are not required to be thermally stable. It also has good limits of detection, many analytes being examined at the low femtomole level. Electrospray is probably the mildest method of ionization known at present. It generates only quasi-molecular ions in the positive- or negative-ion mode and it allows relative molecular masses to be assigned to within an accuracy of 0.01%. Fragmentation, if required, must be induced.

The related derivative method of atmospheric pressure chemical ionization (APCI), discussed in section 3.3.2, also provides a means of LC/MS coupling (Huang, Wachs, Conboy and Henion, 1990; Bruins, 1991). The APCI method readily accepts flow rates up to 2.0 ml min^{-1}, so that there is no need to split the flow from a standard HPLC column. The effluent is driven from a pneumatic or gas sprayer and the resulting droplets are swept by a gas into a hot tube (up to 150°C) in which desolvation and vaporization occur. The heated gas is led into the ion source of the APCI instrument, where exposure to a corona discharge causes chemical ionization according to the reactions described in section 3.3.2. A flow of dry gas prevents solvent, neutral sample or buffer from entering the quadrupole analyser region. This means that the system is compatible with both volatile and involatile buffers. Whereas LC/MS with electrospray allows mixtures of very large involatile molecules to be examined, the APCI approach is more suitable for moderately sized molecules (up to about $M_r = 1500$) like many pesticides and pharmaceuticals.

5.4. SOME APPLICATIONS OF LIQUID CHROMATOGRAPHY/MASS SPECTROMETRY

5.4.1. *Steroid glucuronides and thermospray LC/MS*

Intact steroid glucuronides cannot pass through gas chromatographic columns without very careful derivatization but they can be examined directly by liquid chromatography. Figure 5.15(*a*) shows a standard mixture of three underivatized steroid glucuronides analysed by reverse-phase HPLC in a solvent of 0.1 M aqueous ammonium ethanoate (70%) and acetonitrile (30%) at a flow rate of 1.25 ml min⁻¹. The column effluent was passed into a thermospray device operating in the negative-ion mode. The negative-ion mode was chosen because compounds with many OH groups (polyols) are readily able to support a negative charge as $[M - H]^-$ ions. In the positive-ion mode excessive fragmentation occurs, giving highly complex mass spectra. The chromatographic peaks in figure 5.15(*a*) are not perfect Gaussian shapes and are not as narrow as typical GC peaks, but the degree of separation achieved is good. The negative-ion thermospray mass spectra of the first and last eluting components, 11-hydroxyandrostene- and pregnanediol-glucuronide, are also shown in figure 5.15. The mass spectra are dominated by $[M - H]^-$ ions at m/z 481 and 495 respectively.

5.4.2. *Acylcarnitines and the continuous-flow fast atom bombardment approach*

The acylcarnitines 1 that are found in urine can be examined by GC/MS after derivatization (section 5.2.2, scheme 5.1). They can also be detected by LC/MS using fast atom bombardment mass spectrometry (Millington, Norwood, Kodo, Moore, Green and Berman, 1991). The key advantage of using an approach based on HPLC, as opposed to one based on gas chromatography, is the power of LC/MS to examine the zwitterionic compounds

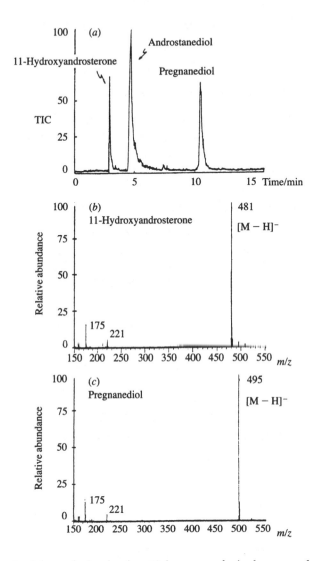

Figure 5.15. (*a*) The TIC trace obtained for reverse-phase HPLC/MS analysis of a mixture of the glucuronides of 11-hydroxyandrosterone, androstanediol and pregnanediol. (*b*) The negative-ion thermospray mass spectrum of the first eluting component, showing a dominant peak for $[M - H]^-$ ions at m/z 481. (*c*) The mass spectrum of the third component, dominated by $[M - H]^-$ ions at m/z 495. (Reproduced from Shackleton, Gaskell and Liberato (1987).)

1 without derivatization. Using an analytical, reversed-phase HPLC column (150 mm length and 3.9 mm internal diameter), the total flow emerging from the column, 1.0 ml min^{-1}, was split such that 1 per cent of the eluant (10 μl min^{-1}) was passed into a probe for continuous-flow FAB analysis. The solvent contained 2 per cent of glycerol. The need for flow splitting reduces the sensitivity of the analysis but the relatively large amounts of acylcarnitines in the urine of patients with a defect of fatty acid metabolism were still above the limit of detection. Figure 5.16 shows summed ion current traces for a set of seven acylcarnitine standards and a suitably prepared urine sample from a patient suspected of having a metabolic disease, characterized

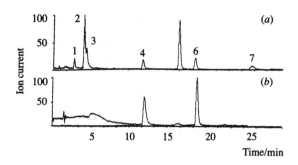

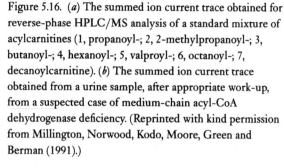

Figure 5.16. (*a*) The summed ion current trace obtained for reverse-phase HPLC/MS analysis of a standard mixture of acylcarnitines (1, propanoyl-; 2, 2-methylpropanoyl-; 3, butanoyl-; 4, hexanoyl-; 5, valproyl-; 6, octanoyl-; 7, decanoylcarnitine). (*b*) The summed ion current trace obtained from a urine sample, after appropriate work-up, from a suspected case of medium-chain acyl-CoA dehydrogenase deficiency. (Reprinted with kind permission from Millington, Norwood, Kodo, Moore, Green and Berman (1991).)

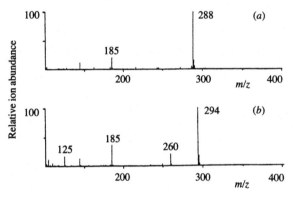

Figure 5.17. FAB mass spectra from the trace shown in figure 5.16(*b*). (*a*) The component eluting near 19 min, showing a dominant peak for [M + H]$^+$ ions at *m/z* 288. (*b*) The components eluting near 12 min, showing [M + H]$^+$ ions *m/z* 294 and 260. (Reprinted with kind permission from Millington, Norwood, Kodo, Moore, Green and Berman, 1991).)

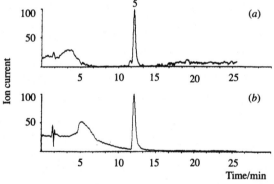

Figure 5.18. Mass chromatograms of *m/z* 260 (*a*) and *m/z* 294 (*b*) from the clinical analysis shown in figure 5.16(*b*). (Reprinted with kind permission from Millington, Norwood, Kodo, Moore, Green and Berman (1991).)

by elevated urinary levels of hexanoyl-, an octenoyl-, octanoyl- and a decenoylcarnitine. Based on retention times alone, the major peaks in the summed ion current trace corresponded to these expected components (figure 5.16(*b*)). The identities were confirmed by examining the FAB spectra at those retention times: the spectra were consistent with the expected structures. For example, the component assigned to octanoylcarnitine showed, after background subtraction, the typical features of an FAB mass spectrum: a base peak for [M + H]$^+$ ions (*m/z* 288) and a few fragment ions (such as that at *m/z* 185) as given in figure 5.17(*a*).

The peak at about 12 min in figure 5.16(b) was coincident with that for authentic hexanoylcarnitine in figure 5.16(a). The mass chromatogram for its [M + H]$^+$ ion (m/z 260) is shown in figure 5.18(a). As expected, a prominent peak appears at about 12 min. However, the mass spectrum at this point has a base peak of m/z 294 (figure 5.17(b)), and indeed the mass chromatogram of m/z 294 maximizes at the same time (figure 5.18(b)). Being greater than the mass of hexanoylcarnitine, this peak must be due to a larger, co-eluting compound. In fact, the spectrum is consistent with a mixture of hexanoylcarnitine and 3-phenylpropanoylcarnitine, the [M + H]$^+$ ion of which occurs at m/z 294. Analysis of authentic 3-phenylpropanoylcarnitine confirmed that its retention time under the conditions of the analysis was the same as that of hexanoylcarnitine. 3-Phenylpropanoic acid had been administered to the patient as a clinical test and 3-phenylpropanoylcarnitine is a metabolite of that acid.

5.4.3. *Desulphoglucosinolates and thermospray LC/MS*

An extract from rape-seed, when subjected to thermospray LC/MS, gave the TIC trace shown in figure 5.19. The peaks corresponded to desulphoglucosinolates, 5, and the peak appearing at about 11 min was deemed to be of particular interest. Its thermospray spectrum is shown in figure 5.20. This spectrum was ascribed to desulphogluconapin, which gives [M + H]$^+$ ions

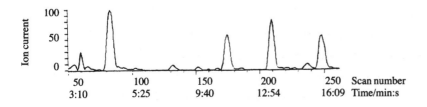

5 Desulphogluconapin R = CH$_2$CH$_2$CH=CH$_2$
Desulphoglucoalyssin R = (CH$_2$)$_5$SOCH$_3$

Figure 5.19. The TIC trace obtained for reverse-phase HPLC/MS analysis by thermospray of a rape seed sample, after appropriate work-up. The flow rate was 1.5 ml min^{-1}, and solvent programming was used such that the concentration of acetonitrile in 0.1 M ammonium acetate increased from 0.2 to 20 per cent over 20 min. (Taken with kind permission from Mellon, Chapman and Pratt (1987).)

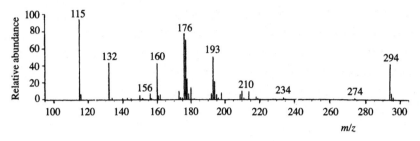

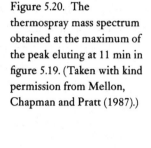

Figure 5.20. The thermospray mass spectrum obtained at the maximum of the peak eluting at 11 min in figure 5.19. (Taken with kind permission from Mellon, Chapman and Pratt (1987).)

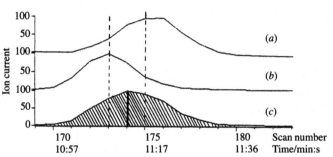

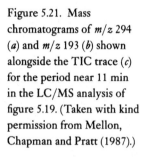

Figure 5.21. Mass chromatograms of m/z 294 (a) and m/z 193 (b) shown alongside the TIC trace (c) for the period near 11 min in the LC/MS analysis of figure 5.19. (Taken with kind permission from Mellon, Chapman and Pratt (1987).)

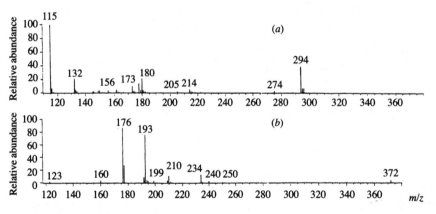

Figure 5.22. The enhanced thermospray mass spectra obtained at scan number 175 (a) and at scan number 173 (b). Note that the original 'mixed' mass spectrum (figure 5.20) has been cleanly resolved computationally into two. One is ascribed to desulphogluconapin with $M_r = 293$ (a), the other to desulphoglucoalyssin with $M_r = 371$ (b) by comparison with authentic spectra and retention times. (Taken with kind permission from Mellon, Chapman and Pratt (1987).)

at m/z 294, but several of the observed peaks, such as m/z 177 and 193, do not occur in the mass spectrum of authentic desulphogluconapin. To explain this discrepancy, the mass chromatogram for an ion that is characteristic of desulphogluconapin (m/z 294) was plotted alongside the mass chromatogram of one of the 'mystery' m/z values (m/z 193). The result is shown in figure 5.21. Since the two peaks maximize two scans apart, they could not originate from the same compound. In other words, the chromatographic peak near 11 min consists of (at least) two components and these have retention times that differ by just 8 s. Deconvolution of the overlapping peaks by the data processing method described in section 4.4.2 gave the two 'clean' mass spectra shown in figure 5.22. The spectrum in figure 5.22(a) was

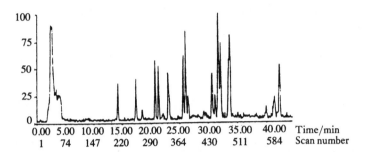

Figure 5.23. The TIC trace obtained for reverse-phase HPLC/MS analysis of the tryptic peptides representing a 70 pmol aliquot of methionyl human growth hormone. The flow rate through the 1 mm × 10 cm column was 40 µl min⁻¹, and solvent programming was used such that the concentration of acetonitrile in water increased from zero at 1% min⁻¹. Trifluoroethanoic acid was constant at 0.1 per cent. (Taken with kind permission from Covey, Huang and Henion (1991).)

clearly that of desulphogluconapin and that in figure 5.22(*b*) was assigned to desulphoglucoalyssin.

Exactly the same data enhancement can be applied to GC/MS data when two compounds elute very close to each other. However, such enhancement of gas chromatographic data is required less frequently because of the great separation efficiency of modern capillary GC columns.

5.4.4. *Peptides and electrospray LC/MS*

One strategy for determining the sequence of amino acid residues in a protein is to hydrolyse the protein with a proteolytic enzyme that cuts the backbone of the protein at specific locations, producing smaller, more manageable portions of the parent protein. For example, the peptides produced from methionyl human growth hormone by digestion with trypsin can be separated by HPLC and analysed by on-line electrospray mass spectrometry. One such approach utilizes a microbore reverse-phase column and pneumatically-assisted electrospray (ionspray), which is compatible with the flow rate of 40 µl min⁻¹ of water/acetonitrile/trifluoroethanoic acid. The experiment employed 50 nmol of methionyl human growth hormone and the tryptic peptides representing a 70 pmol aliquot were subjected to LC/MS following the enzymatic digestion. The total ion current chromatogram from this analysis is shown in figure 5.23. In fact, in this illustrative case, the amino acid sequence of the growth hormone is already known and, considering the specificity of trypsin, a number of peptides was expected in the hydrolysate. All of the predicted peptides (of greater than relative mass of 300) are accounted for in figure 5.23. For instance, a dodecapeptide was anticipated to be present:

Asp-Leu-Glu-Glu-Gly-Ile-Gln-Thr-Leu-Met-Gly-Arg

This peptide has a calculated relative molecular mass of 1360.66. Figure 5.24 shows the electrospray spectrum of the component eluting near 32 min.

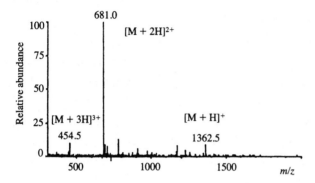

Figure 5.24. The electrospray mass spectrum obtained for the peak eluting near 32 min in figure 5.23. (Taken with kind permission from Covey, Huang and Henion (1991).)

It is consistent with the expected dodecapeptide because, by applying the calculation given in section 3.9 to determine M_r values, the peaks at m/z 454.5, 681.0 and 1362.5 can be shown to be due to $[M + 3H]^{3+}$, $[M + 2H]^{2+}$ and $[M + H]^+$, respectively and the average M_r to be 1360.67. As long as each separated peptide gives protonated molecules $[M + zH]^{z+}$ with at least two different z values, the simultaneous equations can be solved to determine each relative molecular mass. However, so far, evidence for the *structure* of each peptide is lacking.

Structural information is gained with a multiple analyser, usually a triple quadrupole system (section 2.4.3). During the elution of each of the peptides from the LC column, the first quadrupole mass filter is set to pass protonated molecules at one m/z value only; for the peptide represented in figure 5.24, the base peak at m/z 681 due to $[M + 2H]^{2+}$ ions would be the most logical choice. These ions would then be induced to fragment, and the resulting fragment ions analysed in another quadrupole mass filter. Interpretation of such MS/MS spectra for all of the tryptic peptides would allow most of their primary structures to be determined, and hence much of the sequence of the original protein is delineated after one LC/MS and one so-called LC/MS/MS experiment (Covey, Huang and Henion, 1991). Details of MS/MS are described in chapter 8.

5.5.   CAPILLARY ELECTROPHORESIS/MASS SPECTROMETRY

Capillary electrophoresis (CE) utilizes a voltage gradient through an open fused-silica tube to provide high-resolution separations of ionic compounds, or neutral compounds that can be made ionic by, for example, manipulation of pH. In practice, the method can also be made amenable to neutral compounds but this aspect falls outside the scope of this text.

Typical analytes include peptides and proteins (cations at low pH), poly-nucleotides (anions at higher pH) and acids (again, anions at high pH). Clearly, if a CE/MS combination is to be successful, the mass spectrometer must be capable of handling ionic compounds, both small and large. In addition, capillary electrophoresis requires a largely aqueous solution and operates with extremely low flow rates (less than 1 µl min^{-1}). These attributes make the method readily compatible with the electrospray method and with continuous-flow fast atom bombardment, both of which utilize liquid flow rates in the low µl min^{-1} range. Coupling to continuous-flow FAB is being used (see, for example, Caprioli, 1990(b), pp. 121–136; Moseley, Deterding, Tomer and Jorgenson, 1991a) but interfacing with electrospray would appear to be the more natural combination: both CE and electrospray are employed to greatest effect with large ionized biomolecules, often proteins or nucleic acid components, at femtomole levels and both operate at atmospheric pressure. This striking correspondence has led to the development of CE/MS systems based on the electrospray or ionspray approaches (Lee, Mueck, Henion and Covey, 1988, 1989; Smith, Olivares, Nguyen and Udseth, 1988; Smith, Loo, Edmonds, Barinaga and Udseth, 1990). The fact that electrospray operates at atmospheric pressure minimizes the perturbation of the flow through the electrophoretic column.

The capillary column can be connected to the electrospray mass spectrometer via a 'liquid junction' or through a multi-axial probe, as illustrated in figure 5.25. At the junction or probe tip, the flow is made up to a few µl min^{-1} by adding a solution to provide the optimum composition and flow for operation of the electrospray ion source. The interface is held at a certain potential to act as a counter-electrode, which is required to maintain a voltage gradient along the capillary column. For example, the injector end (anode) of the column might be held at 25 kV and the liquid junction or probe (cathode) at 3 kV, providing a net voltage gradient of 22 kV.

In an analysis of a mixture containing a few picomoles of each of five closely similar peptides, 6–10, the anode (inlet end) of the capillary electrophoretic equipment was maintained at + 30 kV and the cathode (detector end) at + 3 kV.

Tyr-Gly-Gly-Phe-Leu-Arg-Arg-Ile-Arg	6
Tyr-Gly-Gly-Phe-Leu-Arg-Arg	7
Tyr-Gly-Gly-Phe-Leu-Arg-Arg-Ile	8
Tyr-Gly-Gly-Phe-Leu-Arg	9
Tyr-Gly-Gly-Phe-Leu	10

(a)

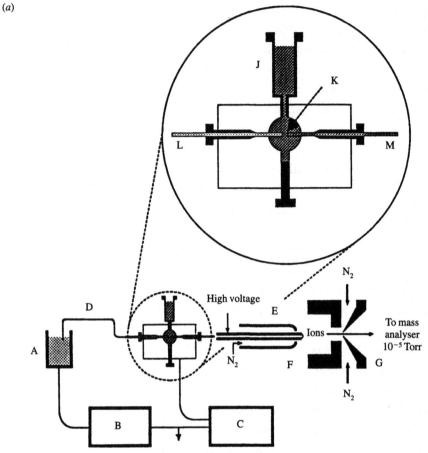

Figure 5.25. Designs for coupling a capillary electrophoresis column to an electrospray mass spectrometer (a) via a liquid junction or (b) through a direct probe. ((a) is reproduced by kind permission of PE Sciex Instruments.)

Key for (a): A, buffer reservoir; B and C, power supplies; D, CE column; E, ionspray source; F, electrode; G, skimmer; J, buffer make-up reservoir; K, liquid junction; L, end of CE column; and M, outlet to ionspray source.

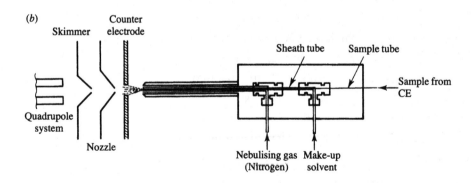

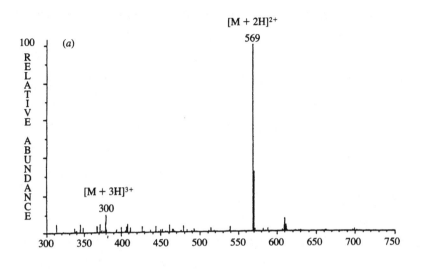

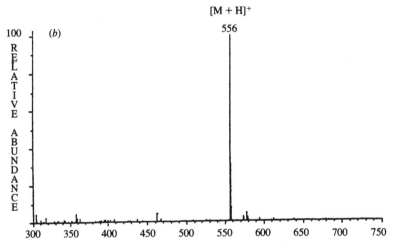

Figure 5.26. The electrospray mass spectra of peptides **6** (*a*) and **10** (*b*) obtained from a CE/MS experiment. (Taken with kind permission from Lee, Mueck, Henion and Covey (1988).)

The electrolyte was a buffer of pH 4.8, at which acidity the peptides exist as cations and hence move towards the cathode. Scanning with a quadrupole mass spectrometer from m/z 300 to 750 during electrophoresis provided electrospray mass spectra of each separated peptide (Lee, Mueck, Henion and Covey, 1988). Specimen spectra are given in figure 5.26. Alternatively, the quadrupole system could be programmed to detect just the major protonated molecules $[M + zH]^{z+}$ for each component. In the resulting summed ion current trace (figure 5.27), all five peptides are seen to be resolved. The elution order of the peptides is dependent primarily on the charge (the greater the number of basic amino groups, the greater the number of positive charges, and

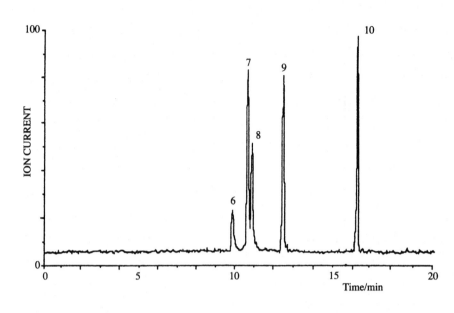

Figure 5.27. The summed ion current trace for selected ion monitoring of the major $[M + zH]^{z+}$ ions of peptides **6–10** during CE/MS. Each peak is assigned to one of the peptides, which were present at the 2 pmol level. The electrolyte was a 1:1 mixture of acetonitrile and 30 mM acetate buffer (pH = 4.8) and the net voltage gradient along the column was 27 kV. (Taken with kind permission from Lee, Mueck, Henion and Covey (1988).)

the faster the elution) and secondly on size (the smaller the molecular size, the faster the elution). Information on the structure, that is, the amino acid sequence, of each peptide was obtained by inducing fragmentation of the major protonated molecules and performing an on-line MS/MS experiment to record all of their product ions (chapter 8).

5.6. SUPERCRITICAL FLUID CHROMATOGRAPHY/MASS SPECTROMETRY

Analytes can be swept through a chromatographic column with various carriers: a gas, liquid or supercritical fluid. A supercritical fluid is a material that is held above its critical temperature, at which point it is, by definition, a gas and cannot be liquified, no matter how high a pressure is applied to it. For SFC, a large pressure is applied to the supercritical fluid to give a 'dense gas'. This fluid has properties intermediate between those of a gas and a liquid: it is a dense gas with great solvating power. When a molecule dissolves in a supercritical fluid, the process is akin to volatilization but, for the commonly used fluids, it can be brought about at relatively low temperatures (40–120°C is typical). As a mobile phase for chromatography, a supercritical fluid allows high resolution separations (like capillary-column GC) but also solvates thermally unstable, polar and/or large molecules without derivatization (as in HPLC). The method is explained in a book (Smith, 1988*a*).

The SFC/MS combination is more difficult to achieve than is GC/MS because the flow rate of mobile phase for SFC is greater than that for GC, and because conditions needed to maintain a supercritical fluid must be preserved in any SFC interface. On the other hand, flow rates in SFC are lower than those in HPLC and so interfacing to a mass spectrometer employing low-pressure ionization is easier for SFC than for HPLC. There are several different designs of SFC/MS interface and they are described in depth elsewhere (Smith, Kalinoski and Udseth, 1987; Smith, 1988a, pp. 159–174; Arpino, 1990a; Olesik, 1991; Jinno, 1992). For example, the effluent from SFC packed columns can be introduced into a mass spectrometer via a thermospray, electrospray or particle-beam device or via a moving belt, as used in LC/MS. Here, the discussion is restricted to one effective interface that is used with capillary-column supercritical fluid chromatography (figure 5.28).

The end of the capillary column for SFC (frequently of 50 μm internal diameter) is connected to a restrictor, which is a few centimetres of narrow fused-silica tubing (about 4 μm internal diameter). The restrictor, which could alternatively be a 1 μm laser-drilled orifice in an otherwise sealed capillary, a taper or a frit, maintains the pressure in the chromatographic

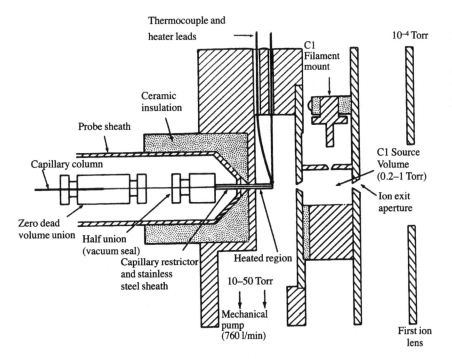

Figure 5.28. A design for coupling a capillary SFC column to a mass spectrometer via a restrictor. (Taken with kind permission from Smith and Udseth (1987).)

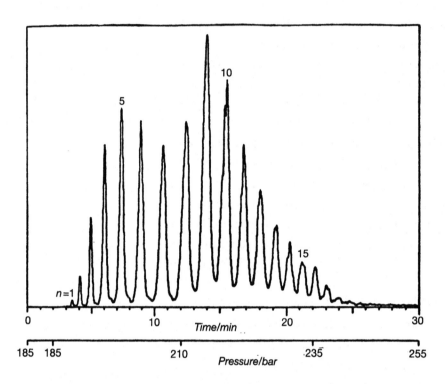

Figure 5.29. A total ion current trace for the analysis of the oligomers of Triton X-100, **11.** The mobile phase was carbon dioxide at 100°C and an initial pressure of 185 bar was used (programmed to increase at 2.5 bar min^{-1} from 2 min). The fused-silica column was 30 m long and of 100 μm internal diameter, and was coated with non-extractable SE54. (Reproduced by kind permission from Smith and Udseth (1987).)

column. The mobile phase is decompressed rapidly as it emerges from the restrictor and the whole interface volume needs to be warmed to prevent condensation during decompression. The jet caused by the rapid pressure drop is directed into a heated expansion zone, where the vacuum, shock fronts and collisional processes break up clusters prior to introduction into the ion source for chemical ionization. The most common supercritical fluid is carbon dioxide; this is not a familiar reactant gas for CI operation and so conventional CI spectra are obtained by a separate addition of a reactant gas such as methane, isobutane or ammonia to the ion source.

Small modifications to this arrangement, including the use of a more 'open' ion source to reduce the pressure of carbon dioxide, allow electron ionization mass spectra to be recorded. In fact, ionization with the more open source probably occurs by a mixture of EI and charge exchange with carbon dioxide, the two methods giving very similar results. However, the EI version is less sensitive than the SFC/MS interface for chemical ionization.

The interface in figure 5.28 was used in the analysis of the non-ionic surfactant, Triton X-100 (**11**). The resulting TIC trace is shown in figure 5.29. It

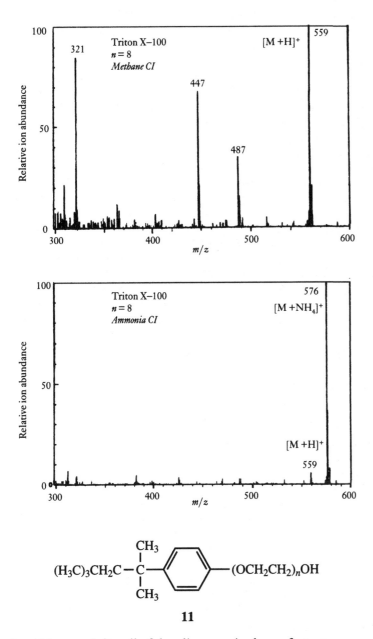

Figure 5.30. The methane and ammonia CI spectra of an oligomer of Triton X-100 (11, $n = 8$) obtained by introducing either reactant gas into the ion source during the analysis shown in figure 5.29. (Reproduced by kind permission from Smith and Udseth (1987).)

should be noted that all of the oligomers in the surfactant are separated and that the temperature was only 100°C (Smith and Udseth, 1987). Typical CI spectra, obtained with methane or ammonia as reactant gases, are illustrated in figure 5.30. Even large amounts of carbon dioxide in the ion source do not affect the CI spectra because of the very low proton affinity of CO_2 with respect to that of methane or ammonia. As would be expected (section 3.3), the methane CI spectrum shows more fragmentation than does the ammonia CI spectrum.

It may once have been hoped that SFC/MS could replace LC/MS because of the former's greater resolution or speed of analysis, sensitivity and ease of coupling. However, advances in LC/MS coupling (such as the versatile electrospray interface), doubts over the ability of SFC to handle truly involatile materials and the emergence of capillary electrophoresis/mass spectrometry have removed some of the impetus from futher development of SFC/MS. Even so, the method has a niche between GC/MS and LC/MS, particularly for analytes with insufficient stability for high-temperature GC/MS. Supercritical fluids have frequently been used to extract analytes from food, environmental samples, polymers and biological samples. The on-line combination of supercritical fluid extraction/SFC/MS is particularly attractive for its speed and convenience (Jinno, 1992).

5.7.   THIN-LAYER CHROMATOGRAPHY/MASS SPECTROMETRY

In chapter 3, several ionization methods were described that involve bombarding a substance on a surface, often dissolved in a small amount of a liquid matrix, with atoms, ions or photons. In planar chromatography, after elution, the components lie as separated spots on the chromatographic surface, so it is not surprising that methods of ionizing substances on surfaces have been applied as means of detecting the separated compounds following thin-layer chromatography and planar electrophoresis (Busch, 1987, 1990; Busch, Brown, Doherty, Dunphy and Buchanan, 1990). This section concentrates on the very widely used method of thin-layer chromatography (TLC). It does not cover off-line methods in which samples of the adsorbent are removed and, in turn, placed on the probe tip of a mass spectrometer that effects ionization direct from that surface. Such an approach can be very effective and it does not require specialized, combined chromatography/mass spectrometric equipment.

Research into interfaces for TLC/MS has not been pursued with the same vigour as with those for GC/MS and LC/MS, but commercial instruments for TLC/MS are available. If gas, supercritical fluid and liquid chromatography are termed 'dynamic' methods, in which the mass spectrometer must acquire spectra during chromatography (i.e. at a rate that is

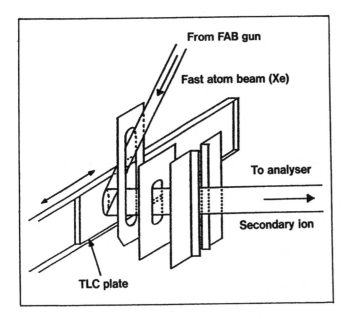

From FAB gun

Fast atom beam (Xe)

To analyser

Secondary ion

TLC plate

Figure 5.31. A design for scanning a developed TLC plate with a beam of ions or atoms and detection of the secondary ions by mass spectrometry. The beam is directed to a fixed point and the plate is moved to bring spot after spot to that point. (Reproduced by kind permission of Jeol Ltd.)

compatible with the chromatographic process), TLC is a 'static' method. The mass spectra are acquired after the chromatographic step is complete so that the time required to develop a TLC plate and that required to acquire mass spectra do not have to be matched. A developed TLC plate can be scanned again and again at any time, and the components on the plate can be examined in any order. Typically, a beam of ions, atoms or photons is produced in a fixed direction and sample spots are brought in turn to the target area by physically moving the developed plate (figure 5.31).

Before undertaking such a mass spectrometric examination of a TLC plate, a liquid matrix is applied to the plate. Solvents similar to those used in conventional FAB mass spectrometry are utilized. The smear of solvent extracts analytes from the solid chromatographic support. This effect makes the spectra independent of the nature of the underlying support and, at the same time, creates a reservoir of sample molecules in much the same fashion as does the liquid matrix in conventional FAB analyses (section 3.10).

In comparison with dynamic forms of chromatography, TLC provides poorer resolution. In a TLC/MS system, scanning with a very narrow beam of caesium ions can enhance the effective spatial resolution of the separation. The secondary ion mass spectrometer is used to scan over the plane of the composite spot whilst monitoring secondary ions that are specific to the unresolved components. In this way 'maps' or profiles are generated that

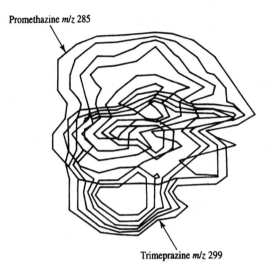

Promethazine *m/z* 285

Trimeprazine *m/z* 299

Figure 5.32. A result of profiling a composite TLC spot by secondary ion mass spectrometry. The contour lines join equal abundances of the protonated molecules of two almost co-eluting drugs, promethazine and trimethazine. The overlapping contour lines define the true spot shape for the two components and reveal those areas of the plate from which a pure sample is available. (Reproduced by kind permission from Busch (1987).)

reveal the true shape of each component's spot, hence deconvoluting one from another, as long as the individual spot shapes are not identical. Figure 5.32 shows such an analysis for overlapping spots of the phenothiazine drugs, promethazine and trimethazine. The contour lines for the protonated molecule of promethazine (*m/z* 285) show that this drug occupies just the top part of the spot. Likewise, the contour lines for the protonated molecule of trimethazine (*m/z* 299) show that it resides only at the bottom of the spot. The middle of the spot contains both drugs. Uncontaminated mass spectra for promethazine and trimethazine could be acquired by effecting complete scans near the top and bottom of the composite spot, respectively.

With respect to LC/MS and GC/MS, the TLC/MS combination is cumbersome and suffers from poorer limits of detection. Generally, on a TLC plate, the components are not sharply resolved so the molecules are more spread out than those emerging from, say, a GC column. Therefore, at any moment, the signal from a TLC spot is relatively weak. However, the situation can be improved for TLC/MS by scanning back and forth over a spot to build up a sizeable ion current. This method suffers from the inconvenience of being time-consuming but takes advantage of the independent time domains of the chromatographic and mass spectrometric partners.

6   Uses of derivatization

6.1.   INTRODUCTION

The commonest forms of mass spectrometry require the sample under investigation to be in the gaseous phase for analysis. In the case of studies by GC/MS, samples also spend considerable time in the vapour phase during their elution through the chromatographic column and transfer into the ion source of the mass spectrometer. To be suitable for examination by electron, chemical and field ionization mass spectrometry, compounds must be both volatile and stable with respect to thermal decomposition and rearrangement in the gaseous phase. Many substances do not meet these requirements and, consequently, are not directly amenable to these types of mass spectrometry. In such cases, the analysis may be approached in three distinct ways. Firstly, formation of a derivative that imparts volatility and gas-phase stability to a troublesome compound may render it suitable for traditional (EI, CI or FI) mass spectrometry. Secondly, recourse may be made to the more specialized techniques for the analysis of involatile compounds, such as laser desorption and electrospray mass spectrometry. Lastly, and usually less satisfactorily, volatile pyrolysis products of the involatile material may be analysed to gain some information about the original structure. Most of this chapter is concerned with the first of these approaches: derivatization. For this part of the discussion, it is assumed that a mass spectrometer suitable for examining involatile substances is not available and that intact molecules (as opposed to pyrolysis products) are to be analysed. Here, the term 'involatile' refers to a substance that does not have significant vapour pressure at the temperature and pressure of an ion source or, when referring to analysis by GC/MS, is not sufficiently volatile to pass through a gas chromatographic column. Later in the chapter (section 6.3.7), derivatization with the opposite aim – making analyte molecules *more* polar – is considered. This type of derivatization, often the conversion of a molecule into an ionic derivative, has become popular with the advent of methods like fast atom bombardment (FAB), which work best with polar compounds and salts. In these cases, the objective is to use a chemical derivatization to convert a neutral molecule, M, into some ionic form such as the salt $[M + H]^+Cl^-$ by adding HCl, and a technique like FAB to desorb such pre-formed ions directly from the condensed phase.

Derivatization as an aid to the analysis of organic compounds is a very old concept, possibly the best known application being the confirmation of a structural assignment by measurement of the melting temperature of a derivative of an unknown compound. For this determination, derivatization is a necessity if the unknown substance is a liquid. If it is a solid, derivatization may not be necessary but it can be advantageous. An analogous situation exists in the field of mass spectrometry. To examine involatile materials by a gas-phase ionization method, derivatization is a necessity, but for some compounds that are directly amenable to mass spectrometry, prior derivatization may offer several advantages. It is a versatile process because the type of derivative may be chosen such that its properties are more suitable for the ensuing analysis than those of the original substance. Derivatization does not just affect the nature, volatility and stability of a molecule. Having a structure different from that of the original compound, the derivative will have a mass spectrometric fragmentation pattern different from that of its parent. This change in fragmentation pattern can be advantageous or disadvantageous and is discussed more fully in later sections.

At an ion source pressure of about 10^{-6} Torr, many compounds that would be classified as non-volatile at atmospheric pressure have vapour pressures high enough to give good EI, CI or FI mass spectra from the direct insertion probe. The thermal stability required for GC/MS is more stringent than that required for vaporization directly into the ion source. During GC, samples are in the gaseous phase for a longer time and at far higher pressure. Therefore derivatives suitable for GC/MS should not only possess volatility and long-term thermal stability but also good chromatographic properties. Figure 6.1 shows the gas chromatograms obtained for 6-methylpurine and its trimethylsilylated derivative. The underivatized sample interacts strongly with the stationary phase, resulting in an unsatisfactory peak shape and poor chromatographic resolution. In comparison, the derivatized compound gives rise to a sharp peak with improved chromatographic resolution. Derivatives commonly used for gas chromatography are

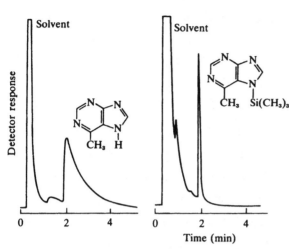

Figure 6.1. Gas chromatograms for 6-methylpurine (left) and its trimethylsilyl derivative (right).

often useful also for combined GC/MS. Trimethylsilylation (see below) is the most ubiquitous derivatization procedure for GC/MS, the only serious disadvantage being that the resulting derivatives, such as that shown in figure 6.1, are susceptible to hydrolysis by traces of water. The problem is often overestimated and is overcome by carefully avoiding contact of the sample with sources of water. Short reviews of derivatization for GC/MS have been published (Brooks, Edmonds, Gaskell and Smith, 1978; Rose, 1984, 1985; Evershed, 1987, 1989). Derivatives for all forms of chromatography, including GC, and analytical derivatization reactions have been described in detail in handbooks by Blau and Halket (1993) and Knapp (1979), respectively.

For best results, any derivatization reaction should satisfy several criteria, particularly for analyses in which the quantity of sample is very small. With amounts of sample in the picogramme range, its manipulation should be minimized to prevent accidental loss and to reduce the effect of unavoidable losses. For instance, transfer of the sample between several different vessels is best avoided because each transfer inevitably entails some loss of that sample. If possible, the reaction should be designed to require little or no work-up, which can be achieved when the reagents and/or by-products do not interfere with subsequent analysis. Trimethylsilylation is one such example since usually the reaction mixture can be injected directly onto the gas chromatographic column. Ideally, derivatization should proceed rapidly and quantitatively to one product. If each component of a complex mixture affords more than one product, the analysis is made much more difficult. The reagents for derivatization must affect molecules in a predictable and reproducible way. Reagents may be either general or selective in their interaction with functional groups. Alcohols, phenols, acids, amides and amines are all readily trimethylsilylated (Pierce, 1968), whereas diazomethane reacts selectively with phenols and carboxylic acids to give methyl ethers and esters respectively. In spite of the hazards associated with diazomethane, it is an excellent derivatizing reagent: it is not only volatile itself but the only by-product of its quantitative, almost instantaneous reactions is nitrogen. Some reagents that react with several different functional groups may be made selective by proper choice of reaction conditions. Acyl halides and anhydrides react with alcohols, phenols and amines to give, respectively, aliphatic esters, aromatic esters and amides and, under controlled conditions, amino-alcohols can be derivatized to amido-alcohols.

Reactions that may not normally be thought of as derivatizations are often useful. For instance, reduction of acids and amides with lithium aluminium hydride converts them into the more volatile alcohols and amines. In a second example, in which the aim is to convert the analyte molecules into a salt-like form by derivatization prior to FAB mass spectrometry, amines can be quaternized to impart a positive charge.

In the following sections, it is not possible to present a comprehensive survey of derivatives used in mass spectrometry because the topic is extensive. Rather, the aspects discussed and exemplified are those which determine the circumstances under which derivatization is advantageous and the criteria by which suitable derivatives are chosen.

6.2.   IMPARTING VOLATILITY AND THERMAL STABILITY

Involatility of a compound is caused by its having large relative molecular mass or strong intermolecular forces or both. For substances with large relative molecular mass, the common derivatives only exacerbate the situation inasmuch as they add to the relative molecular mass but, if at the same time they decrease intermolecular forces, the compounds are nevertheless made more volatile. Hydrogen bonding and the electrostatic associations of zwitterionic compounds are common causes of involatility. The effects of hydrogen bonding can be prevented by removing active hydrogen atoms. For example, the strongly hydrogen-bonding carboxylic acids and amides may be alkylated (scheme 6.1), involatile poly-alcohols like sugars may be acetylated or trimethylsilylated (scheme 6.2) and the zwitterionic amino acids may be acetylated and methylated (scheme 6.3):

$$RCOOH \xrightarrow{\text{diazomethane}} RCOOCH_3 \tag{6.1}$$

$$RCONH_2 \xrightarrow{\text{base/iodomethane}} RCON(CH_3)_2$$

$$ROH \xrightarrow{\text{ethanoic anhydride}} ROCOCH_3 \tag{6.2}$$

$$ROH \xrightarrow{\text{bis-trimethylsilylacetamide (BSA)}} ROSi(CH_3)_3$$

$$^+NH_3.CHR.COO^- \xrightarrow[\text{(ii) diazomethane}]{\text{(i) ethanoic anhydride}} CH_3CONH.CHR.COOCH_3 \tag{6.3}$$

Generally, derivatization should convert the polar groups of a molecule into functionalities of lesser polarity, which are unlikely to form strong intermolecular associations. By increasing their volatility, compounds of quite large relative molecular mass can be vaporized in the ion source. A decapeptide is unlikely to be sufficiently volatile to give a conventional EI or CI mass spectrum without considerable thermal decomposition. However, acetylation of the terminal amino group, methylation of the terminal carboxyl group and prevention of hydrogen bonding by methylation of all the amide (peptide) bonds affords a derivative that is readily volatile even with a relative molecular mass of about 1500 (scheme 6.4):

$$\tag{6.4}$$

With compounds of large relative molecular mass, it is important, particularly for GC/MS, that derivatization does not cause an inordinate increase in relative molecular mass. Derivatives that form cyclic products from bifunctional molecules are helpful in this respect. The increases in relative molecular mass caused by the derivatization of 1,2-diols are 144 on trimethylsilylation (scheme 6.5a), 40 on formation of the cyclic ketal (scheme

6.5b) and 24 for the cyclic boronate (scheme 6.5c), and thus the last would be favoured.

$$(6.5)$$

Because high temperatures may cause rearrangement or decomposition and add to the internal vibrational energy of a molecule, the lower the temperature required to volatilize the sample, the less likelihood there is of changes in structure or of excessive fragmentation after ionization. Therefore, it is preferable to convert compounds, especially those having thermal instability, into volatile derivatives, which require less heating for vaporization. Thermal instability, like involatility, is often associated with active hydrogen atoms, which can form reactive sites in the structure, so derivatization to confer volatility frequently also confers thermal stability on a compound. Sterols with several hydroxyl groups are prone to decomposition through dehydration on gas chromotographic columns, so conversion to a polyether by trimethylsilylation or methylation, or to a polyester by acetylation, is advisable.

6.3.   MODELLING THE MOLECULE FOR THE ANALYSIS

Judicious selection of derivatives is of inestimable value for maximizing the efficiency of many analyses. Several examples are discussed individually below to illustrate this principle. The topic has been reviewed by Anderegg (1988).

6.3.1. *Structure elucidation*

The molecular ion is one of the most informative ions for elucidating the structure of an unknown compound from its mass spectrum but, for many substances, the molecular ion fragments as soon as or soon after it is formed so that it is not seen in the EI spectrum. In such cases, once the class of compound is known, a derivative may be selected that preferentially gives abundant molecular ions. For example, alcohols, that lose water from their molecular ions very readily, can be derivatized to introduce a group associated with a low ionization energy (i.e. a group from which an electron is easily ejected). Such a step would block the loss of water and allow a larger number of molecular ions to be produced with a smaller excess of internal energy. Etherification through trimethylsilylation or esterification with an aromatic acid chloride would yield derivatives with molecular ions of greater stability and hence greater abundance. Unknown polyhydroxylic compounds should not be acetylated in general because the resulting polyethanoates afford complex electron ionization mass spectra, which rarely contain molecular ions of significant abundance (they undergo a facile fragmentation involving elimination of ethanoic acid). Typically, chemical ionization is more likely to yield relative molecular mass information than is electron ionization mass spectrometry (sections 3.2 and 3.3) so, when chemical ionization is employed for this purpose, the choice of derivative is not so critical. Other techniques that produce ions containing the intact molecule are described in chapter 3, and methods of enhancing the abundance of molecular ions by electron ionization are given in sections 10.2 and 10.3.

When derivatives are prepared specifically to induce a particular mode of ion decomposition, it is known as *directed fragmentation*. The positions of substitution in a cyclic ketone can be determined from the masses of the fragment ions derived by cleavage of ketal derivatives (scheme 6.6). The location of double bonds in underivatized alkenes and unsaturated lipids is difficult or impossible to determine because of the ease with which these bonds appear to migrate in the molecular ion before fragmentation. Many derivatives have been used to accomplish this analysis by 'fixing' the double bond so that the fragment ions in the mass spectra define the position of the double bond in the original molecule. Three examples of this strategy are shown in scheme 6.7. Unsaturated fatty acids can be converted to pyrrolidide amides or 3-picolinyl esters, which fragment in a manner indicative of the position of unsaturation. Methods for the determination of the position of the double bond have been reviewed (Schmitz and Klein, 1986). A further method for locating double bonds relies on specialized chemical ionization methodology and is more suitably described in section 3.3.

$$(6.6)$$

$$(6.7)$$

1

Methylation of peptides (scheme 6.4) not only increases volatility by preventing hydrogen bonding but also causes directed fragmentation. The breaking of the bonds shown (1) is more favoured for tertiary amides than for the original secondary amides. The resultant fragment ions define the amino-acid sequence of the peptide.

6.3.2. *Selected ion monitoring*

The technique of selected ion monitoring, which is almost always performed by combined chromatography/mass spectrometry, is described in detail in the next chapter. Here, selected ion monitoring during GC/MS is considered. The mass spectrometer is adjusted to focus ions of one mass or successively a small number of ions of different masses; the selected ions should be characteristic of the compound or compounds under investigation. Also, each m/z value monitored must be as specific as possible to one compound. In particular, ions that are present as 'background', say from column bleed, would not be a wise choice for monitoring. The analysis is often carried out on natural products or pollutants that may comprise only a small percentage of the total sample. For example, the components of interest may occur at the level of femtogrammes to nanogrammes per gramme of sample. Therefore, there is a clear requirement for the derivatization to be quantitative so as to preserve the detection limit of the analysis. Proper choice of derivatives greatly enhances the value of studies by selected ion monitoring by imparting volatility, thermal stability and good gas chromatographic

(6.8)

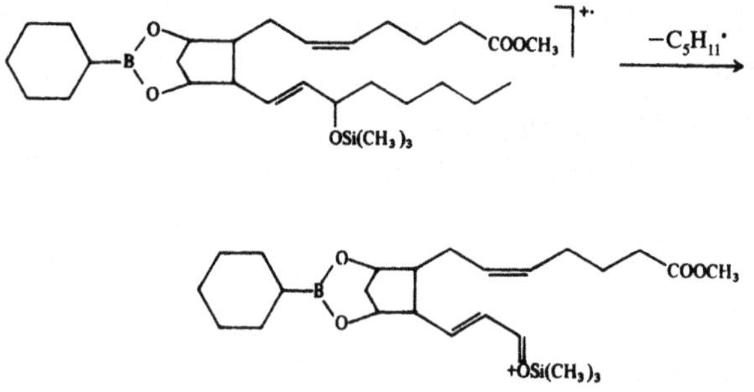

$$(6.9)$$

properties to the sample and by increasing the sensitivity and specificity of detection.

Because the presence and amount of a substance are measured by monitoring selected ions, the greater the proportion of the total ion current that resides in those ions, the smaller the amount of sample that can be detected. A derivative is required that affords particularly stable and therefore abundant ions, preferably of high mass (see below). The loss of $C_4H_9^{\cdot}$ from *tertiary*-butyldimethylsilyl ethers (scheme 6.8) and $C_5H_{11}^{\cdot}$ from the derivative of prostaglandin $F_{2\alpha}$ shown (scheme 6.9) give suitably abundant fragment ions. With many compounds, the trimethylsilyl (TMS) ether grouping yields abundant molecular or $[M - CH_3^{\cdot}]^+$ ions, which are frequently useful for selected monitoring. However, the $[M - C_4H_9^{\cdot}]^+$ ion of a *tertiary*-butyldimethylsilyl ether is usually more abundant than the $[M - CH_3^{\cdot}]^+$ ion of the corresponding TMS derivative because of the relief in crowding when the bulky *tertiary*-butyl group departs, and because of the greater amount of excess of energy that is accommodated in the departing *tertiary*-butyl radical as compared with that in the methyl radical.

Many materials in the smaller relative molecular mass range (up to about 250) pose two special problems in selected ion monitoring during GC/MS. Firstly, some of these compounds are likely to pass rapidly through gas chromatographic columns, eluting shortly after the solvent together with the many impurities that may be present. Secondly, structurally important ions in the mass spectra of compounds of small relative molecular mass are of relatively small mass and may well be common to those of the impurities. Both of these two related sources of interference are overcome by selecting a derivative that adds considerable mass to the sample, but not to the impurities, and increases the retention time on the column. *Tertiary*-butyldimethylsilyl and perfluoroacyl derivatives of amines, alcohols and phenols (see, for example, schemes 6.8 and 6.10) have been used to increase their relative molecular masses so that they are removed from the region containing interfering compounds and thereby the specificity of the analysis is improved.

$$M_r = 191 \qquad\qquad\qquad M_r = 597 \tag{6.10}$$

Another means of increasing the specificity of selected ion monitoring involves the operation of the mass spectrometer at high resolution. Accurate, rather than integer, masses of ions are monitored. The method is specific in that only one elemental composition is detected at each mass. For example, at the integer mass 282, three of many possible elemental compositions are $C_{20}H_{42}$, $C_{18}H_{34}O_2$ and $C_{22}H_{18}$, having accurate masses 282.3284, 282.2557 and 282.1408, respectively. These three mass peaks are separable with a mass spectrometric resolution of 4000 or over, so by monitoring only m/z 282.2557 at high resolution, analysis of the heteroatomic species, $C_{18}H_{34}O_2$, can be effected even in the presence of the two hydrocarbons. In conjunction with this technique, a derivative containing elements that are not common to the impurities or background is useful because the added elements shift the accurate masses of ions containing those elements to distinctive values. Frequently, a natural product is contaminated with compounds of high hydrocarbon content (lipids). Derivatization of the natural product with a polyfluorinated reagent, such as for example $(C_nF_{2n+1}CO)_2O$, to give compounds similar that shown in scheme 6.10, enables ions of the sample to be distinguished from those of the background at the same integer mass.

6.3.3. *Isotope studies*

Many selected ion monitoring methods require as a standard an isotopically labelled compound that is otherwise identical to the substance under examination (chapter 7). The label may be introduced by derivatization through the use of isotopically labelled reagents like $(CD_3CO)_2O$ for acetylation, CD_3OH for methyl esterification, CD_3I or $^{13}CH_3I$ for methylation and fully deuteriated silylating reagents for the introduction of $(CD_3)_3Si$ groups.

Isotopically labelled derivatizing agents are sometimes used to reduce ambiguity of results. Methylation of the charged molecules, acylcarnitines (scheme 6.11), prior to analysis by fast atom bombardment mass spectrometry, results in a product with the same mass and elemental composition as the free acid of an analogue with an additional CH_2 in the side chain (R). For example, the predominant M^+ ions for the methyl ester of iso-valerylcarnitine **2** occur at the same m/z value as the $[M + H]^+$ ions of hexanoylcarnitine **3** (both $m/z = 260$). Incomplete methylation of an analytical sample could therefore lead to erroneous conclusions. The problem is solved by derivatizing with CD_3OH in place of CH_3OH. Whereas methyl esterification ($COOH \rightarrow COOCH_3$) causes a net increase of 14 daltons in the relative mass of the quaternary ammonium ion, tri-deuteriomethylation ($COOH \rightarrow COOCD_3$) causes a net increase of 17 daltons in the molecular mass, forming isotopically labelled compounds such as ester **4**, which are distinct in mass from underivatized homologues.

$$\overset{+}{(CH_3)_3NCH_2CH(OCOR)CH_2COOH} \xrightarrow{CH_3OH/H^+} \overset{+}{(CH_3)_3NCH_2CH(OCOR)CH_2COOCH_3}$$

$$(6.11)$$

$$\overset{+}{(CH_3)_3NCH_2}\underset{\underset{OCOCH_2CH(CH_3)_2}{|}}{CHCH_2COOCH_3}$$

2; m/z 260

$$\overset{+}{(CH_3)_3NCH_2}\underset{\underset{OCO(CH_2)_4CH_3}{|}}{CHCH_2COOH}$$

3; m/z 260

$$\overset{+}{(CH_3)_3NCH_2}\underset{\underset{OCOCH_2CH(CH_3)_2}{|}}{CHCH_2COOCD_3}$$

4; m/z 263

For a polyfunctional natural product that has been derivatized by acetylation prior to analysis, the possibility exists that one or more of the acetyl groups was present naturally before derivatization. The use of $(CD_3CO)_2O$ would allow ready determination of any acetyl groups originally present.

Another useful technique is to add to a sample a selective derivatizing reagent containing elements with isotopes having characteristic abundance ratios. In section 1.4, elements with readily recognizable isotope abundance ratios were discussed, the most familiar ones being chlorine, bromine, sulphur and boron. Examination of the resulting mass spectrum for the abundance of isotopes in the molecular ion region reveals whether a compound has been derivatized by such a reagent. This is useful in the analysis of mixtures containing compounds of different classes because, for each component, the type of derivative formed will lead to identification of the functional group present. Distinctive isotope abundance ratios may be introduced artificially by use of a mixed reagent such as one containing equal proportions of $(CH_3CO)_2O$ and $(CD_3CO)_2O$ for selective acetylation of amines. In the ensuing analysis, observation of mass spectra containing a 'doublet' of equally abundant ions, three mass units apart, immediately confirms the presence of derivatized amines. In a related method, the precursor for a metabolic study is isotopically labelled, usually with ^2H, ^{13}C or ^{15}N, at a site that is not affected by and does not significantly affect the metabolic pathways of interest. After incubation, products derived from the precursor may be unambiguously assigned because their mass spectra will contain the same isotopes in the same ratio as for the parent compound (see also section 12.2).

Isotope labelling may be used also to investigate mechanisms of mass spectrometric fragmentations but this does not involve derivatization in the context of the present chapter. The topic is discussed in section 12.2.

6.3.4. *Metastable ion and MS/MS studies*

After ionization, some compounds tend to produce relatively abundant metastable ions. Such ions are often associated with functional groups that have been introduced by derivatization specifically for analysis of samples by examination of the decompositions of their metastable ions. The method may be considered as a specialized application of directed fragmentation and is suitable for aiding structure elucidation of unknown compounds (chapter 8) and for selected reaction monitoring (chapter 7). An example of the latter technique is the specific and sensitive analysis of *tertiary*-butyl-dimethylsilyl ethers of sterols by tuning the mass spectrometer to monitor metastable molecular ions fragmenting by loss of $C_4H_9^{\bullet}$ (scheme 6.12). In this study, there is an additional aspect of derivatization in that a keto group is converted to a methyl oxime to prevent its enolization and subsequent silylation.

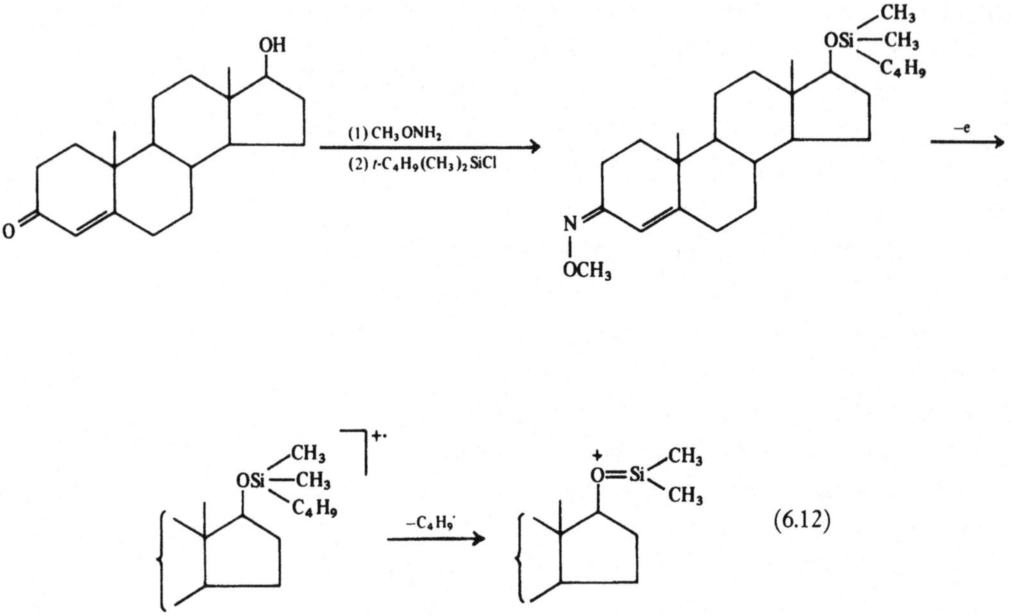

$$(6.12)$$

It has been found that addition of alkali metal salts (Met^+X^-, where Met is Li, Na, K, Rb or Cs and X is I or OH) to a solution of fatty acids prior to its analysis by fast atom bombardment mass spectrometry, causes the formation of $[M + Met]^+$ and $[M + 2Met - H]^+$ ions of the acids instead of $[M + H]^+$ ions. On subsequent fragmentation, initiated by collisional activation in an MS/MS experiment (chapter 8), the $[M + Met]^+$ ions gave many more fragment ions that were diagnostic of structure (such as double-bond position and branch points) than did the $[M + H]^+$ ions (Adams and Gross, 1987). Thus, the *in situ* derivatization by 'cationization' directed fragmentation to provide improved structural specificity in the MS/MS experiment.

6.3.5.   *Negative-ion mass spectrometry*

Typical derivatizations for the analysis of amines by positive-ion mass spectrometry are acetylation and benzoylation (scheme 6.13).

$$RNH_2 \xrightarrow{\text{ethanoic anhydride}} RNHCOCH_3$$

$$RNH_2 \xrightarrow{\text{benzoyl chloride}} RNHCOC_6H_5$$

$$(6.13)$$

The same derivatives would not be ideal for negative-ion studies since they are relatively inefficient at stabilizing a negative charge . To take best advantage of negative-ion techniques, electronegative derivatives should be employed because they increase ionization efficiency and hence sensitivity. For amines, the derivatives shown in scheme (6.14) are more satisfactory than those shown earlier for positive-ion studies (scheme 6.13):

$$RNH_2 \xrightarrow{(CF_3CO)_2O} RNHCOCF_3$$

$$RNH_2 \xrightarrow{C_6F_5COCl} RNHCOC_6F_5 \qquad (6.14)$$

$$RNH_2 \xrightarrow{O_2NC_6H_4COCl} RNHCOC_6H_4NO_2$$

6.3.6. *Examination of enantiomeric samples by chromatography/mass spectrometry*

It is increasingly important for analyses of chiral compounds to differentiate between enantiomers. This is largely because different enantiomers interact differently with living organisms. On conventional chromatographic columns, enantiomers co-elute and, to be distinguished, the enantiomers must experience an asymmetric environment. Such a condition can be provided by derivatization with a homochiral reagent. If one enantiomer of a chiral derivatizing agent is used then a chiral substrate affords diastereoisomers, which, because of their different physical properties, may be resolved by chromatography with an achiral stationary phase. The mass spectra of the separated diastereoisomers are still frequently identical within experimental error.

When the diastereoisomers produced from the derivatization of racemic octan-2-ol with homochiral reagent 5 (scheme 6.15) were analysed by GC/MS, reasonable separation was achieved (figure 6.2). Mass spectrometry confirmed the identity of the separated diastereoisomeric pair, **6a** and **6b**, inasmuch as their EI spectra were consistent with their structures but virtually identical (figure 6.3). The assignment to individual isomers relies on their chromatographic retention times: the elution order for the octan-2-ol diastereoisomeric derivatives had to be determined by separate analyses of commercially available, enantiomerically pure (*R*)- and (*S*)-octan-2-ols.

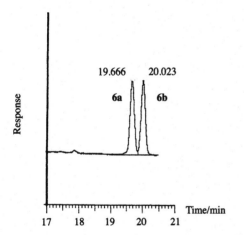

(6.15)

Figure 6.2. Part of the gas chromatogram obtained when racemic octan-2-ol is reacted with reagent **5** according to scheme (6.15). The peaks are assigned to the resulting diastereoisomers, **6**.

6.3.7. *Making the molecule more polar*

Until the emergence of those mass spectrometric techniques that desorb ions directly from the condensed phase, this section would have been unwarranted and indeed alien. The aim of derivatization for traditional, gas-phase mass spectrometry is to prepare non-polar, thermally stable and volatile products. However, derivatization of neutral molecules to produce involatile ionic species is appropriate for methods like fast atom bombardment,

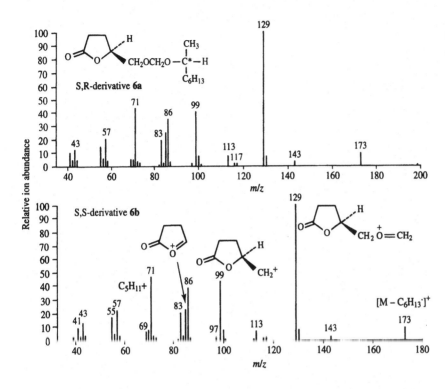

Figure 6.3. The EI mass spectra of the octan-2-ol derivatives, 6.

secondary ion mass spectrometry, electrospray and laser desorption, because they are particularly suited to the analysis of salts. Generally, the sensitivity of such techniques and the quality of the mass spectra are better for salts than for neutral compounds. Most neutral organic compounds are ionized rather inefficiently: less than one in a thousand of such molecules are converted into ions. Instead, molecules may be ionized by chemical manipulation prior to the mass spectrometric examination, to give so-called pre-ionized analytes, in yields approaching 100 per cent. In a mass spectrometer, these ionic products need only be desorbed from the condensed phase, be it solid or solution, and separated from their counter-ions before transfer into the analyser. Whilst fast atom bombardment, for example, *is* a method of ionization, it is not being used as one if derivatization has already given the molecules a charge. Then, FAB and similar methods simply provide the energy to desorb pre-formed ions. Derivatization can also improve detection limits in solution-phase methods like FAB by imparting surfactant-like properties on analyte molecules. As explained in chapter 3, surface-active molecules tend to concentrate at the surface and so will be more strongly represented in the resulting spectrum than a surface-inactive

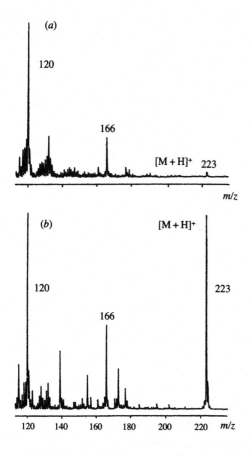

Figure 6.4. Positive-ion SIMS spectrum of Gly-Phe before (*a*) and after (*b*) the addition of *para*-toluenesulphonic acid to the analyte solution. (Reprinted by kind permission from Busch, Unger, Vincze, Cooks & Keough (1982).)

compound. The requisite properties can be introduced by chemical derivatization and usually involve the addition of a large hydrophobic alkyl group (Anderegg, 1988).

The mass spectrum obtained from the dipeptide, glycylphenylalanine, by bombarding a sample on graphite with argon ions (that is, by SIMS) contains a peak for the protonated molecule that is hardly visible above the background peaks. If the same sample is treated with an organic acid, which protonates the dipeptide, the $[M + H]^+$ ions give a very prominent peak, as is clear in figure 6.4 (Busch, Unger, Vincze, Cooks and Keough, 1982).

The negative-ion FAB mass spectra of underivatized aldopentoses are too similar to constitute a reliable means of distinguishing one from another. If the liquid matrix contains a boronic acid, three hydroxy groups in an analyte will participate in a reaction with the boronic acid if they are close in space (scheme 6.16). The product is a negatively charged cage compound, which provides excellent negative-ion FAB mass spectra. For example, figure 6.5

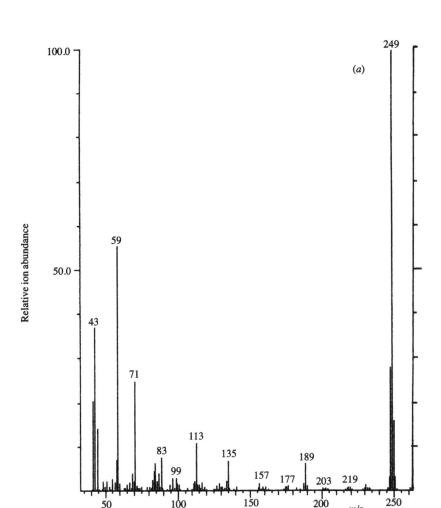

Figure 6.5. Negative-ion FAB mass spectrum of (*a*) ribose and (*b*) arabinose in the presence of 4-tolueneboronic acid. Only in the former case is there a significant peak at *m/z* 249 corresponding to the negatively charged boronate complex.

compares the spectrum obtained from D-(−)ribose, which has several strain-free conformations that can react readily with 4-tolueneboronic acid (e.g. structure 7) to give abundant ions at *m/z* 249, and that of the isomeric D-(−)arabinose, which cannot achieve such a favourable orientation of its hydroxy groups and gives few ions at the same mass. This type of derivatization on the FAB probe tip, or in solution prior to electrospray mass spectrometry, can be used to pre-ionize and analyse polyhydroxylated compounds or boronic acids, to provide some insight into the configuration and conformation of polyhydroxylated compounds that is missing from the spectra of the neutral compounds, and to obtain a measure of the affinity of substrates for boronic acids (Rose, Longstaff and Dean, 1983). In another analysis of a sugar by FAB mass spectrometry, the limit of detection of the

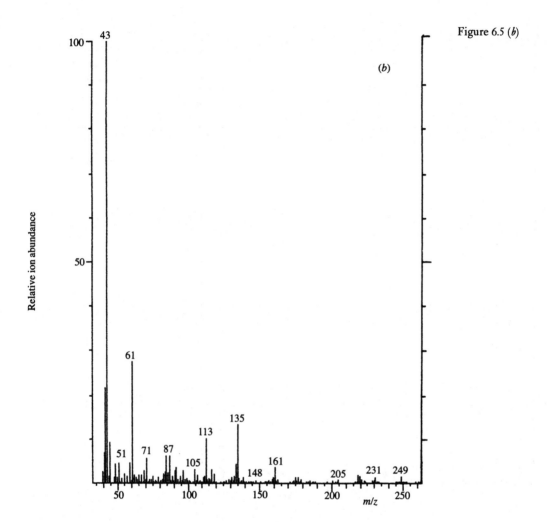

Figure 6.5 (*b*)

disaccharide, maltose, was improved by a factor of 100 when it was derivatized with Girard's reagent, which imparts a charge to aldehydes and ketones as shown in scheme 6.17.

The role of reactions, both intentional and unintentional ones, during various desorption methods has been discussed in several reviews (De Pauw,

7; *m/z* 249

$$\boxed{\text{Analyte molecule}} \quad + \quad R-B\begin{smallmatrix} OH \\ \\ OH \end{smallmatrix} \quad \longrightarrow$$

$$\boxed{\text{Analyte molecule}} \quad + \quad 2H_2O \; + \; H^+ \qquad (6.16)$$

$$(C_2H_5)_3\overset{+}{N}CH_2CONHNH_2 \; + \; O=C\begin{smallmatrix} R^1 \\ \\ R^2 \end{smallmatrix} \longrightarrow (C_2H_5)_3\overset{+}{N}CH_2CONHN=C\begin{smallmatrix} R^1 \\ \\ R^2 \end{smallmatrix} \; + \; H_2O$$

Girard's reagent

$$(6.17)$$

1986; Fenselau and Cotter, 1987; Anderegg, 1988; Detter, Hand, Cooks and Walton, 1988; Vekey and Zerilli, 1991).

6.4.   INORGANIC COMPOUNDS

Whilst it is true that much mass spectrometry is concerned with organic compounds, inorganic substances are also analysed frequently. Volatile inorganic compounds may be analysed directly, examples being rhombic sulphur (S_8 rings), the hydrides of boron and elements of Groups IV and V, and liquids such as arsenic trichloride and mercury. Involatile compounds such as metallic salts may be analysed by the condensed-phase techniques described in chapter 3 or derivatized to volatile organometallic compounds. Some diketone or ketoester chelates of metals are so volatile that they can be distilled, passed through gas chromatographic columns or introduced into an EI ion source at room temperature. The propionylacetonates, for example that of copper (**8**), and trifluoroacetonates are useful for obtaining spectra of metal ions. Generally, as the proportion of organic groups in the organometallic compound increases, so does volatility. Thus, diketones with bulky alkyl groups are useful derivatizing agents. 8-Hydroxyquinoline (oxine) is well known in solution chemistry for precipitating most metal ions. These metal oxinates give good mass spectra. Transition metals may be studied by use of such derivatives as the carbonyl (e.g. $W(CO)_6$ and $Mn_2(CO)_{10}$) and

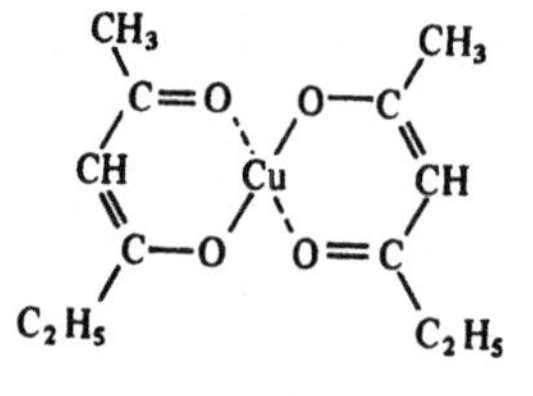

π-cyclopentadienyl complexes (e.g. ferrocene, **9**). The mass spectrum of such a compound is discussed in chapter 11 (example G).

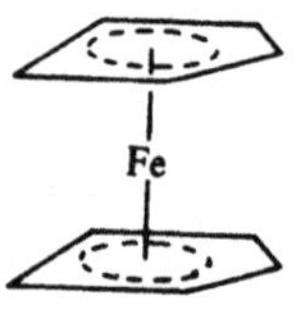

Organometallic derivatives pose some problems not normally encountered in organic chemistry. By several mechanisms, polymerization may occur, so that a metal complex, $MetL_2$, where L is a ligand and Met is the metal, may show a 'mixed' mass spectrum of $MetL_2$, Met_2L_4 and Met_3L_6. When determining the components of any mixture, it is hoped that, if several substances volatilize at the same time, the resulting mass spectrum comprises the superimposed spectra of individual components. This is not necessarily so with organometallic compounds because of reactions occurring in the ion source yielding compounds not present in the original mixture. Exchange reactions are quite common, as shown in scheme 6.18. Misleading results may also be due to associations such that two complexes, $MetL_2$ and $Met'L$, give rise to $Met\text{-}Met'L_3$. An awareness of such reactions should prevent erroneous deductions.

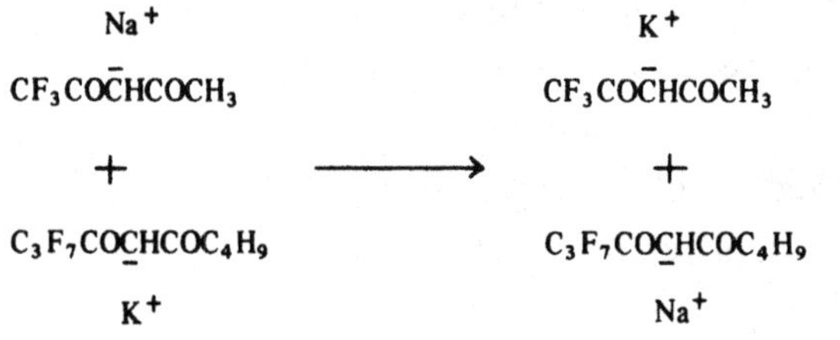

$$
\begin{array}{ccc}
Na^+ & & K^+ \\
CF_3CO\bar{C}HCOCH_3 & & CF_3CO\bar{C}HCOCH_3 \\
+ & \longrightarrow & + \\
C_3F_7CO\bar{C}HCOC_4H_9 & & C_3F_7CO\bar{C}HCOC_4H_9 \\
K^+ & & Na^+
\end{array}
\qquad (6.18)
$$

7 Quantitative mass spectrometry

INTRODUCTION AND PRINCIPLES

Frequently, analytical problems are not considered to be solved until both the identity *and* the quantity of a substance have been determined. In the areas of drugs and environmental science, quantitative analysis is particularly important. For example, the threat posed by a particular pollutant depends on the amount of it that is in the environment, and the quantity of a drug reaching the active site will determine the extent of the biological effect. Given that some substances exert an effect on their surroundings at very low concentrations, methods are often required to be highly sensitive as well as quantitative. The technique of mass spectrometry, known for its great analytical sensitivity, is clearly a candidate for quantitative analysis.

Estimation of quantities of substances by mass spectrometry is not straightforward because mass spectrometric measurements are not exactly reproducible (section 1.2). The response to a sample at the detector depends on several parameters that are difficult or impossible to control, including the condition, temperature and pressure of the ion source and the condition of the detector. It is important to recognize that equimolar amounts of different compounds do not give an equal response because only a proportion of the total number of molecules is ionized and this proportion (the *ionization efficiency*) depends partly on molecular structure. Despite these complications, quantification by mass spectrometry is possible because, when a compound is ionized, the absolute abundance of any of its ions is related to the amount of that substance, albeit by a complex function that varies during day-to-day operation of the mass spectrometer and that cannot be applied to any other compound. To achieve quantification, the mass spectrometer must be calibrated with known amounts of the compound under investigation either just before the assay is carried out on the true sample (on the assumption that instrumental conditions do not change significantly during calibration and analysis) or in a manner that makes the measurement independent of instrumental variability (section 7.2).

For quantification, the abundance of ions may be measured in three distinct ways. During the residence of the compound in the ion source, repetitive scanning with a mass spectrometer (section 4.3) can provide several complete mass spectra. Note that, when such full spectra are obtained, the instrument is devoting much time to detecting regions of the spectrum in which there are few ions. After full scan analysis, the absolute abundance of ions of a selected m/z value can be plotted against time or scan number and a *mass chromatogram* obtained (section 4.4.2). The height or area of a peak in the mass chromatogram is a measure of the quantity of substance giving rise to that peak. The method necessitates a computer for data processing. On the other hand, even the most unsophisticated mass spectrometer may be tuned to monitor continuously ions of only one mass by applying an appropriate fixed magnetic field or analyser voltage. This second method of measuring a mass peak, reported as early as 1959, is more sensitive than the first because, for the entire time that a compound resides in the ion source (typically 2–60 s), the detector records only the selected ions. Regions devoid of peaks and the gaps between peaks are ignored. For a typical complete mass spectral *scan*, in which mass spectra may be recorded every second, any one m/z value would be sampled 2–60 times, being focussed on the detector each time for a few microseconds only, depending on the resolution. Therefore, by monitoring ions of one specific mass instead of the whole spectrum, an approximate thousand-fold increase in sensitivity is attained. Single ion monitoring is often capable of measuring amounts of substances in the femtogramme range, whereas several picogrammes or nanogrammes of sample are required to obtain a complete mass spectrum.

Between the monitoring of ions of a single mass and the recording of complete mass spectra, there are several techniques of intermediate sensitivity involving measurement of a limited number of mass peaks. This third group includes repetitive, continuous scanning of a relatively small mass range and rapid, discontinuous switching to bring into focus, in turn, ions of a few different masses characteristic of the compound or compounds under investigation. Generally, the limit of detection of an analysis decreases (that is, improves) as the number of mass peaks monitored decreases. Unfortunately, a great deal of analytical information is lost at the highest sensitivities because only one m/z value or a few m/z values are observed. The consequences of this are discussed in section 7.3.

The discussion above is applicable to quadrupole mass filters and most magnetic-sector mass spectrometers. The less common magnetic-sector double-focussing instruments of Mattauch–Herzog geometry detect simultaneously all ions of a sample (section 2.4.1). Therefore, they record complete mass spectra with a sensitivity as high as that of a mass spectrometer of Nier–Johnson geometry when it is set to monitor only one mass peak. They are used to good effect in the quantification of traces of metals by spark source ionization (Chapman, 1978). The inconvenience of, and difficulty of quantification from, the photographic plates usually used as detectors for such instruments makes them relatively unpopular. Ion trap mass spectrometers (section 2.4.4) are also an exception. These instruments capture the full range of ions and store them all until they are detected in turn. Therefore, ions at a particular m/z value are not lost whilst the other ions are being detected. Because of this trapping effect, the device is very sensitive even when acquiring full mass spectra and quantitative analysis would be based on mass chromatograms.

Because of the high sensitivity available, quantification of a compound is often brought about by monitoring just one mass peak of its mass spectrum, together with a mass peak from a chemically similar reference compound called an *internal standard* (see next section). The methodology for monitoring a few selected mass peaks was reported first in 1966, when it was termed *mass fragmentography*. This is not an ideal name because mass peaks from molecular ions as well as from fragment ions may be monitored. As the technique developed and proliferated, different researchers gave it different names, with corresponding acronyms. Together with mass fragmentography, some common names to be found in the chemical literature are *multiple ion detection, single ion monitoring* (when the mass spectrometer is tuned to one m/z value only), *multiple ion monitoring, multiple peak scanning, selected ion monitoring, selected ion recording* and *selected ion detection*. The term preferred here is *selected ion monitoring* because it describes the method most aptly and because it is applicable irrespective of the number of mass peaks monitored. Its popular abbreviation SIM should not confused with SIMS (secondary-ion mass spectrometry).

Selected ion monitoring is especially useful in the fields of biochemistry, medicine and environmental science and in those cases in which the compound or compounds to be quantified are present in complex mixtures. The majority of applications utilizes GC/MS or LC/MS to fractionate samples prior to mass spectrometric analysis. In the area of SIM during GC/MS, where the volatility of the analyte(s) is important, a carefully chosen derivatization can often improve quantitative measurements considerably (section 6.3). Substances introduced directly into the ion source, such as on a direct insertion probe, may also be quantified (Millard, 1978a, pp. 91–115). A modern strategy for quantitative analysis without chromatography utilizes the specificity of MS/MS (chapter 8). Here, a component of a directly ionized mixture is separated by selecting unique ions at one specific m/z value with the first analyser; quantification is based on the monitoring of the fragmentation of those ions with a second analyser. In effect, this is *selected reaction monitoring* by MS/MS (section 7.7).

Many of the mass spectrometric techniques described in this book are amenable to quantitative analysis, but only the most useful of these are discussed in this chapter. Further information on the subject of quantitative mass spectrometry may be found in reviews (Bjoerkhem, 1979; de Leenheer and Cruyl, 1980; Garland and Barbalas, 1986; Gelpi, 1986; Gaskell and Finlay, 1988) and in a detailed book (Millard, 1978a). Sources of error and criteria for the selection of internal standards for quantitative measurements have been treated well by Millard (1978a, 1978b).

7.2. CALIBRATION AND INTERNAL STANDARDS

One method for relating the quantity of a substance to the signal it causes at the detector is to assay successively different, known amounts of the substance. A plot of signal sizes against quantities then serves as a *calibration graph* for subsequent determination of unknown amounts. The compound is said to be an *external standard* because it is not added to the sample to be measured. The method, and variants of it, tend to be used with non-chromatographic methods of introducing the sample into the mass spectrometer. Techniques based on external standards are prone to error since changes in the condition of the instrument, such as a slight variation in pressure of the ion source, affect the magnitude of the signals obtained. Also, to quantify a compound in a sample accurately and precisely, it is not just the errors associated with the mass spectrometric stage that need to be considered. The efficiency of any preparatory steps prior to the mass spectrometric analysis (the *recovery*) must also be measured. Samples for analysis are frequently worked up by one or more extractions, a purification process and possibly a derivatization reaction, all prior to mass spectral analysis. These manipulations make quantification complex and the chances of error high, because deviations from the measured yields at any of the steps affect the recovery and hence the final result. Approaches employing an external standard do not accommodate such sources of error.

A better procedure is to add to different, known amounts of the substance to be analysed a constant, known amount of a reference compound called an *internal standard*. The different mass spectrometric responses to the various amounts of substance are measured against the response to the fixed amount of internal standard. The ratio of these two responses is plotted against the amount of substance to obtain a calibration graph (see figure 7.5 below). For actual analysis, the same amount of internal standard is admixed with the crude, unknown sample and both are worked up together. Then, a ratio and not an absolute value is measured by mass spectrometry. As long as the internal standard and substance to be quantified behave identically during work up, the procedure is not affected by losses of sample, due to spillage or incomplete extraction or derivatization, for example, because the ratio of the two compounds does not change. The end result – the ratio measured – is valid irrespective of the efficiency of, and errors in, the sample preparation. Similarly, instrumental changes are made irrelevant because they affect only absolute values, not ratios. Because all these effects would introduce error or even invalidate the results of methods utilizing external standards, those based on internal standards are much preferred. A typical procedure for quantitative measurement of a sample is presented schematically in figure 7.1 and the method is exemplified in section 7.4.

The two commonest types of internal standard are homologues and isotopically labelled analogues of the compound of interest. Both have their advantages but the one using isotopic labels is more theoretically sound and popular, when available (Garland and Barbalas, 1986).

An internal standard should compensate for any losses during work-up by behaving identically towards extraction or purification measures applied to the compound to be quantified. Therefore, potential internal standards for a compound are of the same class of compound and structurally similar. Using a standard mixture of a candidate internal standard and a compound under investigation, it is a straightforward matter to determine whether they act

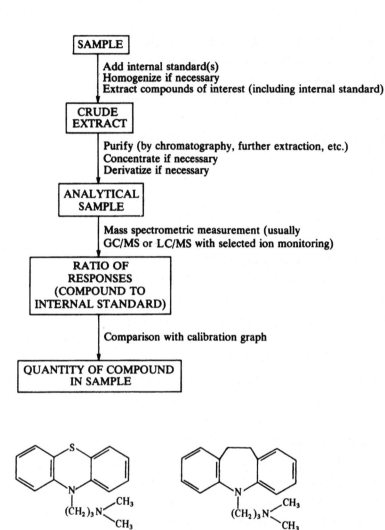

SAMPLE

Add internal standard(s)
Homogenize if necessary
Extract compounds of interest (including internal standard)

CRUDE
EXTRACT

Purify (by chromatography, further extraction, etc.)
Concentrate if necessary
Derivatize if necessary

ANALYTICAL
SAMPLE

Mass spectrometric measurement (usually
GC/MS or LC/MS with selected ion monitoring)

RATIO OF
RESPONSES
(COMPOUND TO
INTERNAL STANDARD)

Comparison with calibration graph

QUANTITY OF COMPOUND
IN SAMPLE

Figure 7.1. A schematic representation of the typical analytical procedure for quantitative mass spectrometry.

1

2

identically during work-up (the ratio of the two should be the same, within experimental error, before and after work-up). As an example, the tricyclic compound (1) was established as a suitable internal standard for quantification by selected ion monitoring of the antidepressant drug imipramine (2). There is an additional advantage of using this type of internal standard. If it is possible to discover a suitable homologue, which contains in its spectrum a mass peak in common with that of the substance to be quantified, then only this one mass peak need be monitored to detect both compounds. This procedure affords the advantage of the simplest, most precise and most sensitive method of quantification: single ion monitoring. The trifluoroethyl derivative (3) is such an internal standard for the derivatized drug (4), for both have the same side-chain, which affords a common ion (scheme 7.1). This method of quantification necessitates the use of GC/MS or LC/MS so that the substance and its internal standard are separated prior to mass spectrometric monitoring of the common ions, otherwise the responses from each compound would be superimposed. For mass spectrometric quantification of substances without fractionation, each mass peak monitored should be unique to only one of the unseparated components.

An isotopically labelled internal standard acts like a homologue that does not provide a mass peak in common with the substance to be quantified. For example, the tetradeuteriated analogue (5) is a suitable internal standard for assay of the drug, cyclophosphamide (6). Analogues containing stable isotope labels are useful as internal standards because they are practically identical in chemical and chromatographic properties to the respective unlabelled compounds whilst being readily distinguishable by mass spectrometry because of their mass difference. Since the analyte and labelled internal standard will have the same, or very nearly the same, chromatographic retention time, the ion of the internal standard that is used as a reference must retain the isotopes that distinguish it from the unlabelled compound. If the ions from both compounds were the same, a single composite peak would result. Internal standards labelled with ^{13}C, ^{15}N, ^{18}O and ^{37}Cl are even more similar chemically to the corresponding unlabelled compounds than those labelled with ^{2}H (D), but they are not as popular because they are expensive and generally less easy to synthesise. One example of the use of these less common types of standard is the quantification of polychlorinated biphenyls (PCBs) in the environment with ^{13}C analogues as internal standards. In general, the most precise *overall* assays are attained with isotopically labelled internal standards because of the great similarity in chemical behaviour of an analyte and an isotopically labelled analogue during sample preparation.

It is shown in sections 1.4 and 12.2 that, for all carbon compounds, there exists a natural isotopic pattern in their mass spectra such that ions 1 and 2 mass units greater than the molecular ion are of significant abundance. The $[M + 3]^{+\cdot}$ ions may be ignored unless the compound contains Cl, Br or other elements with extensive natural isotopes. Even when analysed by GC/MS, an unlabelled compound is hardly separated from its labelled analogue, so the latter should ideally contain sufficient isotopes to shift the relative molecular mass by at least 3 mass units, so that the mass peaks monitored are not common to both compounds. Likewise, it is disadvantageous but not intolerable for the labelled analogue to contain a small amount of the unlabelled analogue as impurity. Of course, it is possible for both types of interference to occur at the same time as, for example, with a dideuteriated internal standard that contains some unlabelled impurity. The more the number of isotopic labels incorporated into a structure increases, the greater is the likelihood that the labelled and unlabelled analogues will behave differently (including being separated by the chromatographic method employed). Therefore, the optimum number of ^{2}H atoms is usually three to five.

When a compound is quantified by GC/MS, some of it may be adsorbed irreversibly on the associated glassware or on active sites in the gas chromatographic column. This becomes a problem with small amounts of sample since much or all of it may be adsorbed. The addition of a relatively large amount of an isotopically labelled analogue of the compound is widely assumed to ameliorate this problem because statistically it is much more likely that the labelled compound will be adsorbed instead and this amount will be made insignificant by the large quantity present. This so-called 'carrier' effect has not been demonstrated universally for compounds labelled with deuterium and so the use of large amounts of such analogues as both carriers and internal standards is not recommended because (a) the most precise quantitative measurements are made when the quantities of internal standard and compound of interest are similar, and (b) the small percentage of unlabelled analyte that is present as impurity in most isotope labelled analogues gives a significant background signal when large amounts of the analogue are used. Compounds labelled with ^{13}C, ^{15}N or ^{18}O are more likely to be 'carriers' because the adsorption sites are unlikely to distinguish between ^{12}C and ^{13}C compounds, for instance. Hence, the use of relatively large quantities of these internal standards may be justified, but the disadvantages noted above, (a) and (b), still apply. For a fuller description of calibration, the reader is referred to a more specialized text (Millard, 1978a, pp. 56–90).

7.3.   SELECTED ION MONITORING

Quadrupole mass filters offer the advantage of simplicity over magnetic-sector mass spectrometers for selected ion monitoring. Under computer control, the voltage applied to the rods of the quadrupole analyser may be switched very rapidly from one value to any other to detect successively several ions of different mass. Switching rapidly between different mass peaks by changing the magnetic field of older magnetic-sector instruments is less straightforward because it takes a relatively long period for the field to settle to a constant value after switching (owing to hysteresis). Because the mass of an ion is inversely proportional to the accelerating voltage (V; equation (2.2)), the magnetic field (B) is kept constant and instead the accelerating voltage is switched rapidly. This voltage scan imposes a restriction that is not found with quadrupole mass filters, namely, that the lowest and highest m/z values monitored at any one time should be within about 10% of each other (for example, within a mass range of m/z 450 to 500) because sensitivity decreases with decreasing accelerating voltage. However, magnetic-sector instruments, unlike quadrupole mass filters, permit quantification with high mass spectrometric resolution.

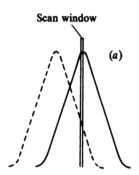

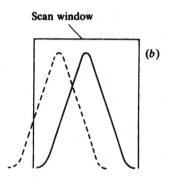

Figure 7.2. Monitoring a mass peak and the effect of drift (*a*) in the static mode and (*b*) in the scanning mode. The original peak position is drawn as a continuous line. Instrumental instability may cause a drift in the relative position of monitoring and the mass peak (dashed line).

Most mass spectrometers allow switching among up to 25 different mass peaks. Also, the computer software allows the set of monitored ions to be changed during a chromatographic analysis. This latter facility can be used to circumvent the restricted mass range during SIM with older magnetic-sector mass spectrometers. For example, in one GC/MS experiment, the computer can be programmed to monitor appropriate low-mass ions for the smaller components eluting at short retention times, then re-set to monitor higher m/z values appropriate to the larger compounds eluting later in the analysis. The m/z values monitored within each set should be selected to be within about 10% of each other, but any two or more sets of m/z values monitored at different times could be very disparate in mass.

There are two different ways of monitoring a mass peak (figure 7.2). During monitoring, the voltage may be static such that only the top of the peak is focussed on the detector (figure 7.2(*a*)), or it may be swept through a pre-set range so that some part or the whole of the peak is scanned (figure 7.2(*b*)). Whilst the former method provides the greater sensitivity, it is more prone to error because of instrumental instability. Slight changes in the volt-ages of the instrument during an analysis manifest themselves as a drift in the position of monitoring or of the mass scale (figure 7.2, dashed lines). In the static mode (figure 7.2(*a*)), the result is a large reduction in sensitivity and hence a large error because only a small proportion of the total number of ions impinge on the detector, whereas in the scan mode of figure 7.2(*b*), the number of ions reaching the detector is little changed. The problem of instrumental instability resulting in drift can be addressed by monitoring not just the analyte peaks but also a reference peak (say, from perfluo-rokerosene). The computer can be programmed to measure the mass of the reference peak periodically, check it against the known mass and, if neces-sary, compensate for any drift by changing appropriate instrumental settings. Even after this precaution has been taken, there are advantages in recording the whole peak profile rather than simply monitoring at the peak maximum (Tong, Giblin, Lapp, Monson and Gross, 1991). Examination of the exact shape and centroid of the peak can reveal the presence of any co-eluting interfering substances. As a result, the analyst is more certain that the correct compound is being measured with the peak profile approach than with peak top monitoring. The latter measures a very narrow window so that changes in peak shape would be unnoticeable.

For a chromatographic analysis by SIM, in which more than one peak is monitored, the amount of time spent monitoring each m/z value must be chosen with care. The total time taken to monitor each peak and to return the system to its starting point ready for the next cycle must allow 10–20 executions of the cycle during the residence time of the compound(s) in the ion source. If the cycle time is too long to take at least ten samples across the chromatographic peak, the shape of that peak will not be adequately defined and its height and area will be inaccurately recorded (section 4.2). Hence, the quantification will be inaccurate. Therefore, the summation of the dwell time at each m/z value plus the re-setting times must be a tenth or less of the width of the chromatographic peak. To correct a total cycle time that is too long, fewer peaks would have to be monitored or the dwell times reduced. With modern software, it is quite possible to specify a different dwell time for each m/z value. In particular, it would be advantageous to collect data for a longer time period when a given mass peak is small, and specify a shorter dwell time for a larger peak. In calculating peak sizes after completion of the analysis, the computer takes into account any differences in dwell times.

7.3.1. *Selecting ions for monitoring*

Before a substance can be analysed quantitatively by SIM, its mass spectrum must be examined for at least one mass peak suitable for selective monitoring. The spectrum is best not examined in isolation but rather with a knowledge of the masses of ions that are likely to be present as 'background'. For studies by GC/MS, it is useful to have a mass spectrum of 'bleed' from the gas chromatographic column to be used so as to avoid selecting ions common to this spectrum. Impurities present in the sample, which ionize at the same time as the substance of interest or its internal standard, such as co-eluting components in GC/MS, will not affect the assay as long as they do not have ions at the m/z values selected for monitoring. An exception to this occurs if the quantity of the impurity varies from sample to sample and is large enough to change significantly the pressure in the ion source, because this will then affect unpredictably the ionization efficiencies of the compounds to be quantified. Generally, the lower the mass of an ion, the more likely it is to be present as background. To avoid this interference, ions of large mass (over m/z 300) should, if possible, be chosen for monitoring. The use of derivatization to impart large relative molecular masses is described in section 6.3. It is also advantageous to select ions of even mass because ions of odd mass occur more frequently in mass spectrometry, so that the latter are the major contributors to background. The selected ions should be abundant since this will improve the detection limit of the assay. The size of the mass peak is best gauged as a percentage of total ion current rather than as percentage relative abundance (section 1.2). A peak that is 75 per cent of the total ion current would dominate a mass spectrum since all other peaks together would constitute only a further 25 per cent. This peak would be a highly favoured candidate for monitoring. A peak that is 75 per cent in relative abundance may be a poor candidate since there could be any number of larger, more suitable peaks between 75 and 100 per cent relative abundance.

Compounds afford more large-mass ions when analysed by positive-ion or negative-ion chemical ionization, fast atom bombardment or electrospray than by electron ionization (chapter 3). Frequently, such mild ionization mass spectra are very simple, containing only a few mass peaks (see figure 3.1). Compared with electron ionization mass spectra, the total ion current is divided amongst fewer ions, allowing a more sensitive quantitative analysis. However, it should be remembered that, if a substance is ionized less efficiently by chemical ionization, for example, than by electron ionization, then the overall sensitivity may be less for the CI method despite the fact that a peak with a larger proportion of the total ion current is monitored. It is prudent to test different ionization modes for the best response to a standard quantity of the compound of interest before making a final choice. A further, related advantage of mild methods of ionization is that any impurities in a sample will also afford mass spectra with a small number of peaks, thereby reducing the number of potentially interfering ions (i.e. increasing specificity). This effect may be exploited best in the CI mode by careful choice of reactant gas. Ammonia is an excellent reactant gas for amines and carbohydrates, but it does not ionize hydrocarbons or gas chromatographic column bleed (section 3.3). For assay by GC/MS of an alkaloid, for instance, with

contaminating hydrocarbon lipids, the use of ammonia CI mass spectrometry would not cause background interference.

7.3.2. *Sensitivity and specificity*

The discussion so far has intimated that sensitivity is of prime importance;
this is often so, but not always. When a nanogramme or more of the substance to be measured is available, the most sensitive methods are not
required. Repetitive scanning of the entire mass range is superior to selected
ion monitoring in this case because it provides more analytical information
(complete mass spectra) so there is more confidence in the *identification* of
compounds. Also, the m/z value on which quantification is to be based can be
selected from the full mass range *after* the analysis. When using selected ion
monitoring, the peak or peaks must be chosen and fixed before the analysis
proceeds. Even so, at or below the level of picogrammes, quantification
usually requires a selected ion monitoring method (unless an ion trapping
device is used), in which the compound of interest is identified on the basis
of retention time and the presence of one or more m/z values known to be in
the mass spectrum of that compound. When ions of several masses are monitored to detect a given compound in a mixture, they should have the same
relative abundances as those in the standard mass spectrum of that compound, within experimental error. If the relative abundances are at variance,
the identity of the detected substance would have to be questioned. The
most sensitive quantitative technique, monitoring of ions of a single,
selected mass, is most prone to interference since an impurity that has a mass
peak at the monitored m/z value and elutes at the same retention time as the
suspected substance will cause an erroneous identification. Ideally, before
this technique is used, the presence of the compound to be quantified should
be proved, not merely anticipated. This is not possible if there is so little of it
that there are no other techniques capable of detecting it. In such circumstances, the *specificity* of the analysis should be as high as possible.

An assay of high specificity rejects signals from chemicals that are not the
analyte, or at least differentiates them from those of the true analyte. When
GC/MS or LC/MS is used for a quantitative analysis, the retention time
and shape of the chromatographic peak are characteristic of the analyte.
Any change to the expected retention time or peak shape would suggest
interference, that is, poor specificity. When using an isotopically labelled
analogue as internal standard, it is possible to programme a computer to

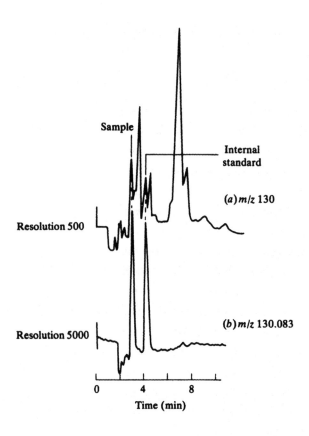

Figure 7.3. A comparison of selected ion current profiles from analysis of a food extract by GC/MS. (*a*) Selected ion monitoring of all ions with mass 130 and (*b*) selected ion monitoring at the accurate mass of ions of a flavour component and its internal standard. (Figure reproduced by kind permission of Micromass UK Ltd.)

compare the chromatographic peak profiles of the analyte and co-eluting internal standard. Any discrepancy in the two profiles alerts the user to possible interference.

If chemical interference occurs, changing chromatographic conditions can sometimes improve specificity. If this is unsuccessful, recourse may be made to high mass spectrometric resolution since it is then possible to monitor ions of selected elemental composition rather than selected integral mass. The effect is illustrated in figure 7.3, which shows the analysis of a component contributing to the flavour of a complex food extract. To detect the component, ions at m/z 130 (accurate mass 130.083) were monitored at low resolution (figure 7.3(*a*)) and at high resolution (figure 7.3(*b*)). The plots of ion current against time are called *selected current profiles*, or sometimes *mass fragmentograms*. It is clear from figure 7.3 that the specificity is better at higher resolution because ions with a nominal mass of 130, but different accurate mass (elemental composition) from the ions of interest, are not detected. The area or height of the peak profiles for the flavour component and its internal standard may be measured reliably only at high resolution. Despite the lower absolute sensitivity of high resolution compared with low resolution operation (section 4.4.1), quantification of smaller amounts of substance may be possible in the high-resolution mode in such circumstances (Hsu, 1993). In other words, the smallest quantity of a compound that can be quantified by a particular method is affected markedly by the specificity of that method. Generally, it is advisable to use mass spectrometric resolution just great enough to separate mass peaks of interest from those of interfering substances. Since methods of high specificity are less susceptible to interference by impurities, it is sometimes possible to reduce or eliminate altogether purification procedures prior to analysis. Examples illustrating many of these points are presented in the following sections.

7.4.   APPLICATIONS BASED ON GAS CHROMATOGRAPHY/MASS
        SPECTROMETRY

The concentration in blood plasma of desipramine (7), a metabolite of the drug imipramine (2), may be measured by use of GC/MS in the chemical ionization mode. For calibration, different amounts of compound (7) in the range 25–250 ng ml^{-1} of plasma and a standard amount (320 ng ml^{-1}) of a tetradeuteriated analogue of (7) were added to drug-free plasma samples. After basification, the aqueous samples were extracted with hexane and the combined organic fractions were evaporated to dryness. Trifluoroacetylation gave a mixture of compounds (8) and (9). These derivatives possess better chromatographic properties and higher mass than the parent secondary amines. They afford electron-ionization mass spectra with ions of low abundance above m/z 220, but the base peaks in

7   8; M_r = 362   9; M_r = 366

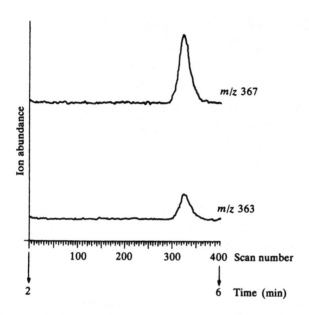

Figure 7.4. Mass chromatograms of $[M + H]^+$ ions of compounds (8) and (9). (Taken with kind permission from Claeys, Muscettola and Markey (1976).)

their chemical-ionization spectra are $[M + H]^+$ ions, formed by proton transfer in the ion source from ions of the methane reactant gas. These $[M + H]^+$ ions occur at m/z 363 and 367 for compounds (8) and (9), respectively. Since the quantity of metabolite to be extracted from the plasma of patients taking imipramine was relatively large (greater than 20 ng ml^{-1}), the relatively insensitive method of rapid, repetitive scanning was applied for quantification. A computer, coupled to a quadrupole mass filter, recorded mass spectra as the standard mixtures eluted from the chromatograph and printed out mass chromatograms of the $[M + H]^+$ ions (figure 7.4). It is noteworthy from figure 7.4 that the deuteriated analogue was not separated from the unlabelled compound by gas chromatography. The ratio of the areas of the peaks was plotted against the known concentration of compound (8) in the plasma to give the calibration graph shown (figure 7.5), a straight line in the range under observation. The same amount of internal standard was added to true samples of plasma; work-up and analysis were performed as before to determine the ratios of responses. By reference to the calibration graph, the concentration of the metabolite in the plasma was determined. The measurement provided clinically important information for adjusting the dose levels of this antidepressant drug (Claeys, Muscettola and Markey, 1976).

In an environmental study (Sen, Miles, Seaman and Lawrence, 1976), very small amounts of highly toxic nitrosamines in cured meats were

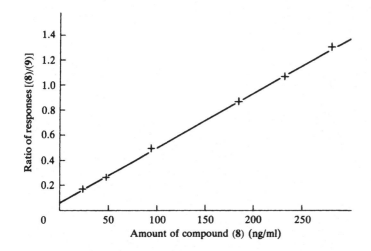

Figure 7.5. The calibration graph for quantification of desipramine. (Redrawn with permission from Claeys, Muscettola and Markey (1976).)

detected and quantified by electron ionization mass spectrometry. These carcinogenic compounds are thought to arise from the interaction of naturally occurring amines and added nitrite salts. In the previously discussed case, each mass peak was specific to one compound, but in this example an ion common to all nitrosamines (NO^+; m/z 30) enabled selected ion monitoring of that single mass peak for all compounds of interest. At low mass-spectrometric resolution (500), the technique would be highly sensitive, but non-specific since the m/z value is so low that it is common to many compounds. Some elemental compositions with nominal mass 30 are NO, ^{13}CHO, CH_2O, H_2N_2, CH_4N, $^{12}C^{13}CH_5$ and C_2H_6 with accurate masses of 29.9980, 30.0061, 30.0106, 30.0218, 30.0344, 30.0425 and 30.0470, respectively. A mass spectrometric resolution of over 3700 (30/(30.0061 − 29.9980)) is necessary to separate NO^+ from its nearest neighbour, $^{13}CHO^+$. In fact, a resolution of 5000 was used to monitor ions at m/z 29.9980, allowing specific detection of nitrosamines.

Measurement of low levels of the toxic compound 2, 3, 7, 8-tetrachlorodibenzodioxin (TCDD, or simply dioxin) in human tissue requires a specific, sensitive assay. The molecular ion region in the mass spectrum of TCDD (**10**) is shown in figure 7.6. It is seen that the four chlorine atoms give rise to a complex, characteristic isotope pattern. The eluant from a GC capillary column was monitored selectively at m/z 320, 322 and 324 at a resolution of 10 000. Figure 7.7 shows the three selected ion current profiles, offset for clarity, obtained from injecting 10 pg of TCDD. The specificity of the assay is high because correct identification could be verified by

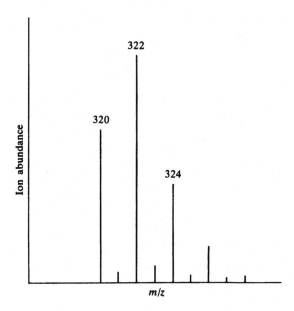

Figure 7.6. The molecular ion region of 2,3,7,8-tetrachlorodibenzodioxin.

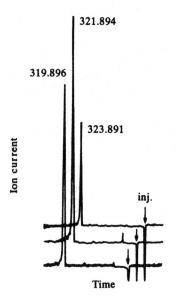

Figure 7.7. High-resolution selected ion current profiles for an analysis of 10 pg of TCDD. The three traces are offset (see injection (inj) points) for clarity; all three profiles maximize simultaneously. (Reproduced by kind permission of Micromass UK Ltd.)

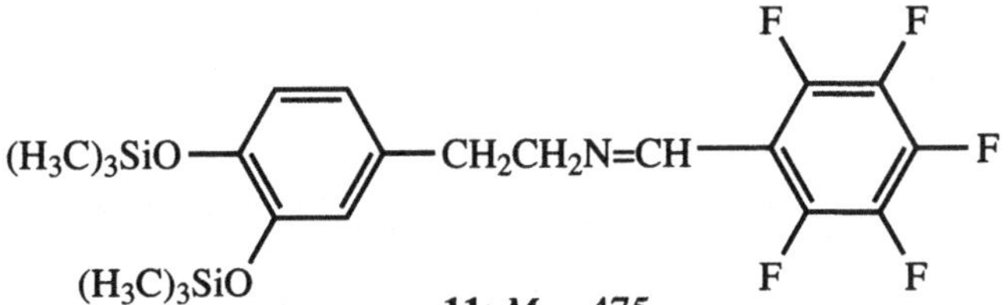

$(H_3C)_3SiO$ — ⟨⟩ — $CH_2CH_2N=CH$ — [polyfluoro ring] — F

$(H_3C)_3SiO$

11; $M_r = 475$

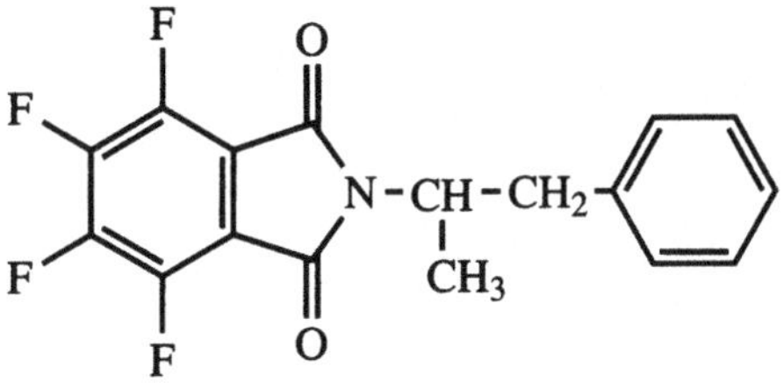

N—CH—CH$_2$— ⟨⟩

CH$_3$

12; $M_r = 337$

OCO — [polyfluoro ring] — F

C_5H_{11}

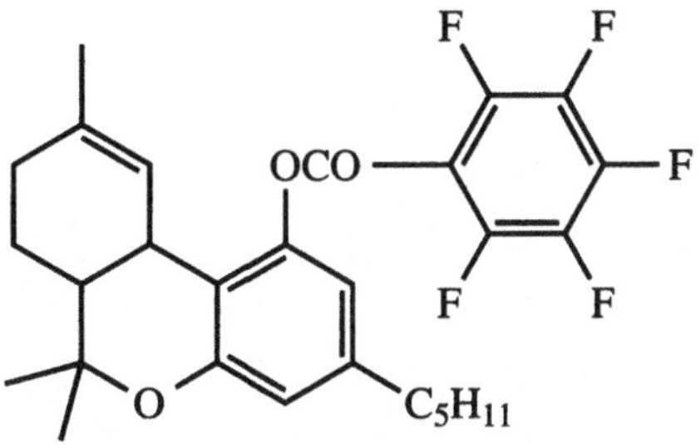

13; $M_r = 508$

several parameters: (a) the gas chromatographic retention time, (b) the presence of ions of reasonably high and even mass, (c) the presence of ions of correct elemental composition by high-resolution mass spectrometry and (d) three selected ion current profiles of the same relative heights as the corresponding mass peaks in the standard mass spectrum. Such an approach has been used for the detection and quantification of environmental contamination by TCDD down to levels of 1.25 parts in 10^{15} (corresponding to 1.25 fg of TCDD per gramme of serum, for example).

Selected ion monitoring with a quadrupole mass filter operating in the negative-ion chemical ionization mode (section 3.4) allowed detection of derivatives of dopamine (**11**), amphetamine (**12**) and Δ^9-tetrahydrocannabinol (**13**) at the level of 10–25 fg (Hunt and Crow, 1978). This study is notable for its good use of derivatives in increasing both specificity and sensitivity of analysis. The polyfluorinated groups not only impart high electronegativity, and hence high ionization efficiency in the negative-ion mode, but also add considerable mass. The derivatives stabilize the molecular ions, $M^{-\bullet}$, which then account for a large proportion of the total ion current and are suitable for selected ion monitoring. Natural products that need to be analysed are usually isolated as components of complex mixtures. Therefore, their analysis is often more difficult than that of a pure compound or of a mixture of pure compounds in which the level of other substances is artificially low. By

14   15

a similar method, Garland and Min (1979) quantified the drug clonazepam
(14) in human plasma by reference to the labelled internal standard (15).
The $[M - H]^-$ ions at m/z 314 (for 14) and 321 (for 15) were monitored.
The detection limit was estimated to be about 100 pg ml^{-1} for compound 14
in the original plasma, corresponding to an injection of an aliquot of 4 pg for
each assay. The method was about twenty times more sensitive than a similar
positive-ion chemical ionization technique developed by the same workers.

7.5.   APPLICATIONS BASED ON DIRECT INLET

Samples for quantification may be inserted directly into the ion source via
hot or cold inlets, or on a direct probe (Millard, 1978a, pp. 91–115). The last
of these methods is the commonest and forms the subject of most of this
section. The same principles and scanning methods are used for quantification by direct probe as for chromatographic methods of inlet, although the
probe method was given a different name, the *integrated ion current technique*.
It was first used to identify and quantify *para*-tyramine (16) in rat brain
(Majer and Boulton, 1970). The direct probe inlet has little potential for
fractionating samples unless the components differ widely in volatility.
During quantitative analysis, the components of interest in mixtures
remain largely unseparated from each other and from impurities, causing
many background ions. In the case of *para*-tyramine, the signal at m/z 108
was due to the selected fragment ion (17, $C_7H_8O^+$ at m/z 108.0575) and to
hydrocarbon impurity ($C_8H_{12}^{+\cdot}$ at m/z 108.0939). One way of overcoming
this problem of low specificity is to use high mass-spectrometric resolution.
A resolution of 3000 is sufficient to separate $C_7H_8O^{+\cdot}$ and $C_8H_{12}^{+\cdot}$ ions and
allow selective detection. Background interference is worst at low mass so
that derivatization provides another means of overcoming the problem.

16   **17** m/z 108

18; $M_r = 603$

The 5-dimethylaminonaphthalene-1-sulphonyl (dansyl) derivative (**18**) of *para*-tyramine is measurable in the picogramme range by monitoring molecular ions at m/z 603.1861. If the quantity of impurities is large enough to affect the pressure in the ion source, the resulting variation in ionization efficiency introduces error in the estimation no matter how specific it is. The cure to this problem is to increase the specificity of the work-up by removing as much of the contamination as possible.

With instruments that allow the direct probe to be heated independently of the ion source, some degree of fractionation is possible by *temperature programming*. Figure 7.8 shows the assay of some steroids from human ovarian tissue by monitoring molecular ions at a resolution of 10 000. Each line in figure 7.8 represents the abundance of the selected ion as the compounds evaporate from the probe. The tops of the lines form an *evaporation profile*, the area under which is related to the amount of substance; the relationship is determined by use of external standards. It can be seen that temperature programming of the direct probe, portrayed at the bottom of figure 7.8, causes some separation of the steroids. At the high resolution used, the method was so specific to the steroids that no work-up of the samples was required, a small quantity of the dried tissue being placed directly on the probe (Snedden and Parker, 1976).

Field desorption and fast atom bombardment mass spectrometry are suitable for thermally unstable and involatile compounds. Quantification of organic compounds, inorganic cations, especially those of the alkali

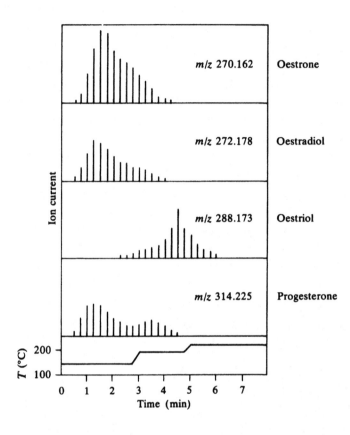

Figure 7.8. Evaporation profiles of four steroids, together with the temperature programme for the direct insertion probe. In each case, molecular ions were monitored at high resolution. (Adapted with permission from Snedden and Parker (1976).)

metals, and organic cations, such as ammonium salts, in the FD or FAB mode is technically difficult, but possible when internal standards are used (Schulten, 1979, pp. 111–112 and pp. 234–244; Gaskell and Finlay, 1988). Generally, the detection limits of cations are lower than those of neutral organic compounds. The mass spectra in these modes are characteristically simple, often consisting only of molecular ions or cationized molecules, for example, $[M + H]^+$ and $[M + Na]^+$. Since ions of high mass account for a large percentage, or all, of the total ion current, the method is both specific and sensitive, allowing analysis in the picogramme range for some salts. Assays may be carried out at low or high mass-spectrometric resolution.

In addition, atmospheric pressure ionization mass spectrometry, such as electrospray, provides very sensitive techniques, which readily allow detection of compounds in the femtogramme range by selected ion monitoring. With this approach, compounds in a suitable solvent are injected directly into the ion source.

7.6.   APPLICATIONS BASED ON LIQUID
       CHROMATOGRAPHY/MASS SPECTROMETRY

The use of LC/MS combines the advantages of the direct probe and
GC/MS methods because unstable and involatile substances may be examined but with efficient fractionation. There are both advantages and disadvantages in quantitative analysis by LC/MS compared with that by
quantitative GC/MS. It is generally true that samples subjected to LC/MS
require less preparation than those for GC/MS. There are many reasons for
this observation. For instance, LC/MS copes with involatile analytes, so
derivatization is rarely required. Liquid chromatography columns are
usually more tolerant of dirty samples than capillary columns for gas chromatography, the efficiency of which is soon undermined by contamination.
Also, the ionization methods used in LC/MS, like thermospray and electrospray, frequently produce only quasi-molecular ions and no fragmentation.
Therefore, contaminants produce fewer potentially interfering ions than
they would by electron ionization during GC/MS. Taking the example of
LC/MS by thermospray, its sensitivity towards a given substance is very
dependent on thermospray conditions (section 5.3.3). Turning this effect to
advantage, conditions can be selected carefully to optimize ionization of the
required analyte and to minimize the ionization efficiency of contaminating
components. In favourable cases, the unwanted components in a relatively
crude extract will elicit little if any response in the mass spectrometer so
that such components would not have to be removed from the sample prior
to LC/MS analysis. A comprehensive protocol for working up crude
samples is thus obviated. The same effect, of marked variation in sensitivity
with compound type and thermospray conditions, can be a disadvantage. It
is difficult to predict the relative sensitivity of thermospray mass spectrometry for different compounds, so each one must be tested for efficacy before
quantitative analysis proceeds. Some compounds cannot be detected by the
method at sufficiently low concentrations for a given quantitative analysis.
All LC/MS systems in which some solvent is carried into the ion source
suffer to a degree from 'background' interference. Ions from the solvent and
any additives will cause large signals in the low-mass range. For thermospray
and electrospray applications, such interference becomes less significant at
masses above about m/z 100 and m/z 150, respectively. There is no equivalent constraint for GC/MS in the EI mode (but reactant gases used for

CH_2OH

HO — O — OH

HO — OH

OH

19

chemical ionization during GC/MS cause similar, abundant background ions at low mass).

The selected ion monitoring method described above for GC/MS applies in exactly the same way to LC/MS in its different guises. For example, glucose (and other sugars) in blood or amniotic fluid can be quantified by thermospray LC/MS after a simple work-up procedure and without chemical derivatization. Under thermospray conditions, glucose (**19**) gives mainly $[M + NH_4]^+$ ions at m/z 198, with some consecutive losses of water molecules. To obtain reproducible full mass spectra, 1–10 nmol of glucose were required. Using $[^{13}C_6]$glucose as internal standard and monitoring $[M + NH_4]^+$ ions at m/z 198 and m/z 204, the limit of detection of glucose was about 50 pmol (9 ng) and the relative standard deviation, that is the precision, was 8% (Esteban, Liberato, Sidbury and Yergey, 1987).

7.7. SELECTED REACTION MONITORING

The topic of metastable ions was introduced in section 1.5 and is discussed in some detail in the next chapter. Compounds giving metastable ions may be detected and quantified by tuning an appropriate mass spectrometer to monitor the fragmentations of those metastable ions selectively, rather than those of normal peaks. Since metastable ions are almost always of low abundance compared with normal ions, the technique is relatively insensitive but, like selected ion monitoring at high resolving powers, selected reaction monitoring makes up in specificity for that which it lacks in sensitivity. The molecular ions of the methyl oxime, *tertiary*-butyldimethylsilyl ether derivatives of testosterone (**20**) and epitestosterone (**21**) fragment by loss of $C_4H_9^{\cdot}$ (m/z 431 → m/z 374 in both cases). Ions undergoing this reaction may be monitored by special scanning techniques (chapter 8) as the compounds elute from the chromatograph of a GC/MS system.

Figure 7.9 compares the ion current profiles obtained by selected reaction monitoring (figure 7.9(*a*)) and selected ion monitoring of molecular ions at high (figure 7.9(*b*)) and low resolution (figure 7.9(*c*)). It is clear that, at low

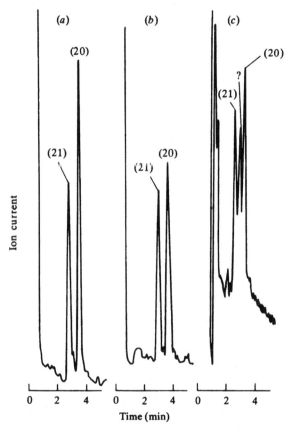

Figure 7.9. GC/MS with (*a*) selected reaction monitoring of the reaction m/z 431 → 374, (*b*) selected ion monitoring of m/z 431.3220 at a resolution of 8500 and (*c*) selected ion monitoring of m/z 431 at low resolution. Testosterone derivative (**20**), epitestosterone derivative (**21**) and an unknown interfering substance (?) are marked. (Adapted with permission from Gaskell, Finney & Harper (1979).)

resolution, selected ion monitoring does not allow accurate measurement of the size of the peaks due to compounds (**20**) and (**21**) because of an unknown, interfering component. The greater specificity of selected reaction monitoring or monitoring the molecular ions (m/z 431.3220) at a resolution of 8500 enables quantification because the impurity is not detected. The compound (**21**) may be used as an internal standard for the determination of the testosterone derivative (**20**) at the low nanogramme

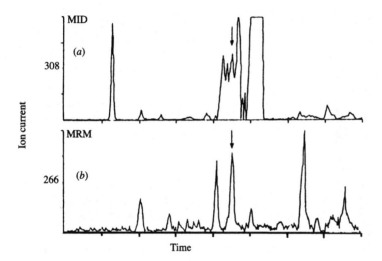

Figure 7.10. GC/MS for the detection of 1 ng ml⁻¹ of the *tris*-trifluoroacetate of deacetylmetipranolol. (*a*) Conventional selected ion monitoring of the fragment ion correspoding to the side chain (m/z 308) at low resolution. (*b*) Selected reaction monitoring of the loss of C_3H_6 from that side-chain ion (m/z 308 → 266). The arrows indicate the expected retention time of the analyte. (From *Tandem Mass Spectrometry* edited by F.W. McLafferty, © 1983. Reprinted by permission of John Wiley and Sons, Inc.)

level in hamster tissue. The detection limit for the steroid (**20**) by the method based on metastable ions was 30 pg (Gaskell, Finney and Harper, 1979). The method is very useful with derivatives, such as the *tertiary*-butyl-dimethylsilyl ethers, which favour the production of metastable ions of high mass.

The importance of specificity for attaining low limits of detection is further illustrated in the quantification of the *tris*-trifluoroacetate derivative of deacetylmetipranolol (**22**), a drug metabolite, in serum. Conventional SIM, using chemical ionization during GC/MS, provided quantification only down to the ng ml⁻¹ level because of inadequate specificity. Selected reaction monitoring with a triple quadrupole mass spectrometer linked to a gas chromatograph exhibited much improved specificity, as seen in figure 7.10, and enabled quantitative analysis down to levels of 100 pg ml⁻¹ (Richter, Blum, Schlunegger and Senn, 1983).

In this latter approach with an MS/MS instrument, fragmentation was induced to occur by collisional activation before the ions entered the final mass analyser. Reactions induced in this region can be monitored in the same way as for the fragmentation of metastable ions and can form the basis of a quantitative assay (Gaskell and Finlay, 1988). The MS/MS method is described in the next chapter.

8 Metastable ions and mass spectrometry/mass spectrometry

8.1. THE ORIGIN OF METASTABLE IONS

The term 'metastable' has been applied to those ions in a mass spectrometer that have just sufficient energy to fragment some time after leaving the ion source but before arriving at the detector. The excess of internal energy imparted to these ions during ionization is sufficient to give them a rate of decomposition such that the latter occurs during the ion *flight-time*. The product ions from such 'in-flight' fragmentation have less than the full kinetic energy originally imparted to the precursor metastable ion when it left the ion source because the initially imparted momentum must be shared between the products of decomposition. As one of these products is itself an ion that is necessarily of smaller mass than its precursor metastable species, it follows that the product ion must have less momentum than the precursor. It is this reduced momentum that leads to the products of metastable ions having an *apparent* mass different from the corresponding 'normal' product ions formed *in the ion source*. The difference is explained more fully below.

To understand the origin and decomposition of metastable ions, it is convenient to consider electron ionization in a conventional double-focussing, magnetic-sector mass spectrometer as depicted in figure 8.1. For a mass spectrometric fragmentation reaction in which an ion M^+ of mass m_1 yields a fragment ion A^+ of mass m_2 and a neutral particle N, equation (8.1) holds:

$$M^+ \rightarrow A^+ + N \qquad (8.1)$$
$$(m_1/z) \quad (m_2/z) \quad (m_1-m_2)$$

As was shown earlier in chapter 1, ions M^+ and A^+ will be present in the ion source and, after acceleration from there, each will possess translational energy equal to zeV, where V is the accelerating voltage, e the charge on an electron and z the number of electronic charges on the ion. By suitable adjustment of the magnetic field flux, ions M^+ and A^+ are collected at the

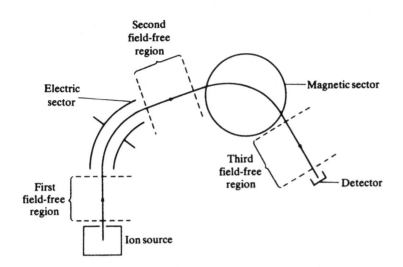

Figure 8.1 The flight path of the main ion beam from ion source to ion detector through a conventional mass spectrometer, showing the first, second and third field-free regions.

detector according to the formula, $m/z = B^2r^2e/(2V)$ (section 2.4); these may be called 'normal' ions and they are seen as 'sharp' or narrow peaks in the mass spectrum.

When a substance is ionized, ions are produced with a range of internal energies. Particularly with the common electron ionization, some M^+ ions will possess a sufficiently large amount of internal energy that the rate of decomposition is high and so, even though they reside for only a short time in the ion source, there is still sufficient time for them to fragment to A^+ ions; these last ions are extracted from the ion source and give 'normal' A^+ ions, i.e., these ions are seen as narrow peaks in the mass spectrum. The smaller the excess of internal energy of an ion, the slower is its rate of decomposition. Therefore, any ions, M^+, which have only a small excess of internal energy sufficient that their rate of decomposition is very small or even zero, reach the detector before any possible fragmentation can occur; they are seen as narrow peaks for 'normal' M^+ ions in the mass spectrum, viz., the *total* time for M^+ ions between their formation and residence in the ion source plus their flight time to the detector is too short for any decomposition to have occurred and they are observed as narrow peaks in the mass spectrum. On the other hand, there are some ions, M^+, which possess excesses of internal energy that are intermediate between those M^+ ions that have so much that they fragment in the ion source and those M^+ ions that have so little that they reach the detector unchanged. This intermediate group of M^+ ions may fragment *after* leaving the ion source and *before* reaching the detector.

The product ions, A^+, from fragmentation of M^+ in the ion source are seen as narrow peaks in the mass spectrum at an m/z value correct for the mass and charge on the ion, A^+. The product ions, A^+, from fragmentation of M^+ during flight between the ion source and detector are seen in the mass spectrum as broad peaks, centred at m/z values that are *not* correct for the mass and the charge on the A^+ ions. The appearance of these broad peaks in mass spectra obtained with the old photographic recording methods (sections 1.5 and 4.2.1) gave rise to the term 'metastable ion peaks'. However, it should be clear from the above that these broad peaks do not represent metastable M^+ ions themselves but represent *products of decomposition* of metastable ions. As is often the case with frequently used terminology, the term 'metastable ion peak', which is meant to imply a peak for ions such as A^+ arising *from* decomposition of a metastable ion, M^+, has been loosely applied in such a way that it can be associated erroneously with the mass and charge on the metastable precursor ion, M^+, itself. The cause of the A^+ ions from metastable ion decomposition being *detected* differently from 'normal' A^+ ions by the mass spectrometer is due to their different momenta. This instrumental aspect is discussed more fully in later sections in this chapter concerned with the specific detection of metastable ion decomposition in various kinds of ways with various kinds of instruments. It is perhaps worth pointing out that modern computerized data processing methods positively discriminate against seeing the 'metastable ion peaks', which are readily observable in the older photographic methods of recording data, and this is why they do not appear in the usual 'stick' spectra printed from a data system. However, many other methods for detection of metastable ion decomposition have been developed, which are computerized in such a way that automatic 'metastable ion scanning' can be instituted readily on suitable spectrometers. This computerized omission of metastable ion decomposition data is not a grievous loss because of the advent of the newer, more powerful methods for their acquisition described below.

Typically, for electron ionization, ions spend 10^{-6} s in the ion source and take about 10^{-5} s after leaving the source to reach the detector. Normal ions can have large excesses of internal energy and hence high rate constants for decomposition ($>10^6$ s^{-1}) so that they fragment in the ion source or, at the other extreme, small excesses (rate constants $<10^5$ s^{-1}) so that they are stable within the lifetime of the analysis. Ions that fragment with rate constants in the range 10^5–10^6 s^{-1} are 'metastable' in a conventional mass spectrometer in the sense that, within a particular time-frame (ion flight-time), the instrument analyses them to a different m/z value from those of the corresponding 'normal' ions. With this conventional sector mass spectrometric view of the decomposition (equation (8.1)), the observation or time window occurs so long after ionization that most ions, M^+, that are going to fragment will have done so in the ion source and relatively few ions will be classified as metastable. Therefore, the abundances of metastable ions are much less than those for normal ions. In other mass spectrometers, metastable ions may not be so easy to examine. For example, in field ionization, ion collection occurs at such a short time after formation that the time window for observation is very short; under these circumstances, ionic products from metastable ions appear as leading edges to normal ion peaks, giving them a skewed appearance (section 3.7.2). There is a similar effect in time-of-flight mass spectrometry. In quadrupole mass filters, metastable ions are not differentiated from normal ions.

For many of the 'soft' or mild ionization methods, it is frequently the case that the initially formed ion has very little excess of energy and, after formation, may lose much of that which it has through collisional transfer to neutral gas species present at the higher source pressures usually prevailing for mild ionization. This is particularly true of chemical or atmospheric pressure ionization (see chapter 3 for a fuller discussion). Therefore, these ions also have very small or zero rates of decomposition and are not observed to fragment on the time scale of the mass spectrometer. Although this effect is very good for providing abundant molecular or quasi-molecular ions and hence excellent relative molecular mass information, the lack of fragmentation removes most structural information. In these instances it becomes necessary to induce decomposition of the initially formed ions and this is usually carried out whilst the ions are in flight. This induced decomposition gives results that are comparable to those of the metastable ion decompositions observed for initially formed ions that do have enough energy to fragment in flight between the source and the detector. Induced decomposition is discussed more fully below.

When reaction (8.1) occurs outside the ion source, affording an A^+ ion and a neutral species, N, the translational energy in M^+, namely zeV, must be shared between A^+ and N in accordance with the law of conservation of momentum. Thus, A^+ from metastable ion decomposition will have only a part of the original energy zeV and will not pass through a magnetic field in the same way as 'normal' A^+ ions having the full translational energy, zeV.

Therefore, in conventional sector mass spectrometry, the A^+ ion products of metastable ion fragmentation have the same mass and charge as normal A^+ ions but are collected differently because they do not have the same translational energy. If the new translational energy is equivalent to V' (where $V' < V$), then $m/z = B^2 r^2 e/(2V')$. Because m/z is the same for normal and metastable ions and $V' < V$, a weaker magnetic field flux is required to focus the ions arising from metastable ion decomposition than for those arising from similar fragmentation inside the ion source.

8.2.   THE USEFULNESS OF METASTABLE IONS

Normal fragment ions in a routine mass spectrum are products of reactions, some of which may be complex. The precursor ions, M^+, of any given product ion, A^+, are assigned by educated conjecture since normal ions provide no information other than mass with regard to their origin. The lack of any formal connection between M^+ and A^+, except that the mass of the first is greater than that of the second, makes molecular structure determination difficult. Imagine attempting to solve a jig-saw puzzle blindfolded and with only the masses of the pieces as a guide! Observation of metastable ion decomposition removes much of the guesswork since metastable ions define reaction pathways, i.e., they show that $M^+ \rightarrow A^+$. In a complex mass spectrum, the additional information gained through metastable ions is invaluable for spectral interpretation of structure. Consider an unknown molecule, M, giving molecular ions, $M^{+\bullet}$, and fragment ions at m/z values less than that of $M^{+\bullet}$ by 15 and 42 mass units. The $[M - 15]^+$ ion will be due almost certainly to elimination of a methyl radical from $M^{+\bullet}$ but the character of the $[M - 42]^{+\bullet}$ ion is less easy to rationalize. It could be due to loss of 42 mass units (e.g. C_2H_2O or C_3H_6) from $M^{+\bullet}$ or of 27 mass units (e.g. HCN) from $[M - 15]^+$. The simple low-resolution mass spectrum cannot differentiate between these possibilities. The structural inferences in each case are rather different, a loss of the elements of ketene (CH_2CO) being characteristic of the acetyl group and of propene (C_3H_6) a propyl group, whereas elimination of hydrogen cyanide is indicative of an aromatic amine or nitrile. Therefore it is important to be able to distinguish between $M^{+\bullet} \rightarrow [M - 42]^{+\bullet}$ and $[M - 15]^+ \rightarrow [M - 42]^{+\bullet}$. If there were metastable ions undergoing the latter reaction, then this would be consistent with elimination of HCN and there would be no basis for postulating an acetyl or propyl group. If such metastable ion fragmentation were not observed, it does not *necessarily* rule out the reaction $[M - 15]^+ \to [M - 42]^{+\cdot}$ but does suggest that it is unlikely. Conversely, if a metastable ion fragmentation is observed for $M^{+\cdot} \to [M - 42]^{+\cdot}$, usually, but not always, it implies that a single reaction has occurred (rapid successive loss of $CH_3^\cdot$ and HCN could afford a metastable ion for the composite process). When metastable ion data are combined with accurate mass measurement of the normal ions, an unambiguous fragmentation scheme can often be obtained (Rose, 1981).

For compounds with subtle differences in structure, such as stereoisomers, and positional and geometric isomers, the vast majority of peaks due to normal ions in their routine mass spectra are frequently due to common ions having similar abundances and so obscure any small differences that would have distinguished them. Selective detection of metastable ion products is more sensitive to fine differences in structure because it allows the mass spectrometrist to ignore the common ions and observe in isolation the behaviour of those ions that are most likely to differentiate between similar structures, *viz.*, those ions having only small excesses of internal energy. This principle has been discussed with reference to the analysis of isomeric carotenoid compounds by linked scanning (Rose, 1982).

Care must be taken in assigning the masses of precursor and product ions from metastable ion data, especially when the method of recording mass spectra gives rise to broad peaks. When looking for fragmentations of metastable ions in the absence of normal ions by specialized techniques, there may be several complicating factors. During studies of dissociations in one particular region of a mass spectrometer, product ions of metastable ion decomposition formed in another (often within the electric sector) may be focussed accidentally, affording interfering peaks. Metastable ions decomposing in one region may yield product ions with sufficient energy to fragment in another region. Such consecutive reactions, when studied intentionally, can provide useful information on fragmentation pathways.

Metastable ions are not just useful for characterizing inorganic and organic compounds (particularly natural products). The ability to study the reactions of relatively long-lived (low-energy) ions and the energetics of such processes permits fundamental research on the structures of ions. Further information on the topic of metastable ions may be obtained from more specialized literature (Jennings, 1971; Cooks, Beynon, Caprioli and Lester, 1973; Beynon and Caprioli, 1980; Holmes and Terlouw, 1980).

An ion formed in the ion source with insufficient excess of internal energy to fragment further is transmitted to the detector as a normal ion and, as discussed above, these are of greater abundance than metastable ions. If, *during flight*, normal ions can be induced to fragment by giving them more internal energy, then the fragmentation process will appear to be exactly the same as if natural metastable ions had decomposed. There is a major difference in that the ability to impart this extra energy to all or a substantial fraction of normal ions enhances the possibilities for detecting such processes as that shown in equation (8.1). For example, whereas the abundances of the usual metastable ions are far too low for adequate trace analysis, examination of induced fragmentation of normal ions allows highly specific analyses of complex mixtures, often down to the parts per million level. The deliberate induction of decomposition of normal ions can be done by a variety of techniques and the methods for doing it and for examining decomposition pathways have led to a whole new area of mass spectrometry, *viz.*, tandem mass spectrometry or mass spectrometry/mass spectrometry (MS/MS). This chapter discusses first the conventional metastable ions and then goes on to describe some of the techniques and uses of MS/MS.

8.3.   METASTABLE IONS IN CONVENTIONAL MASS SPECTROMETERS

The actual place in the mass spectrometer where the partition of translational energy occurs, i.e. where the metastable M^+ ions decompose to give A^+ ions and neutral particles, N, determines how metastable ion fragmentation may be detected. To discuss how this is done, the layout of three distinct regions in a conventional magnetic sector mass spectrometer will be considered. These may be called the first, second and third 'field-free' regions (figure 8.1).

8.3.1.   *The first field-free region*
This lies between the ion source and the electric sector of the double-focussing spectrometer; it does not exist in the single-focussing instrument. Any A^+ ions formed in this region have the wrong energy to pass through the electric sector because the latter is an energy-focussing device designed to transmit only those ions that are of the correct translational energy. Therefore, A^+ product ions produced from metastable ions and having only

a fraction of the full translational energy of normal A^+ ions will not be focussed by the sector and will not be observed in the mass spectrum. If one wishes to investigate metastable ions decomposing in the first field-free region, it is necessary to alter the accelerating voltage at the ion source or the electric field strength of the electric sector so as to allow the passage of their ionic products. If an ion M^+, of mass m_1, decomposes to give a product ion A^+ of mass m_2 and a neutral particle N of mass $(m_1 - m_2)$, *after* accelera- tion through a potential (V), the initial translational energy (zeV) of M^+ must be shared between A^+ and N as in equation (8.2):

$$zeV = [(m_2/m_1)zeV] + \{[(m_1 - m_2)/m_1]zeV\} \qquad (8.2)$$

If the accelerating voltage is increased to V' (without changing the field at the electric sector) such that $V' = (m_1/m_2)V$, then the translational energy of the product ion of mass m_2 from metastable ion decomposition becomes $(m_2/m_1)ze \times (m_1/m_2)V = zeV$, i.e., it then has the correct translational energy to pass through the electric sector and be detected. Changing the voltage at the ion source in this way prevents passage of normal ions. Because the normal ions are defocussed, the term *defocussing technique* is used for this method of investigating metastable ion reactions (Barber and Elliott, 1964; Futrell, Ryan and Siek, 1965; Jennings, 1965). Perhaps it ought to be termed a focussing technique since the products of metastable ions that are being investigated have to be focussed and the normal ions defocussed.

In using this method, suppose that a normal A^+ is selected and that A^+ may be formed from any or all of the precursor ions M_1^+, M_2^+, M_3^+ and M_4^+ (figure 8.2). It is necessary to search for a product ion, A^+, arising from the possible *metastable* precursor ions M_1^+, M_2^+, M_3^+ and M_4^+. After adjust- ing the instrument to focus at the detector only normal ions, A^+, the ion source accelerating voltage is gradually increased whilst maintaining a con- stant voltage on the electric sector. As the acceleration potential is changed,

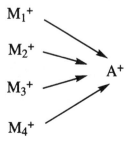

Figure 8.2. Fragment ion A^+ is formed by decomposition of any or all of the ions, M_1^+, M_2^+, M_3^+ and M_4^+. To find which fragmentation path(s) is followed, a search is made for transitions from the possible metastable precursor ions M_1^+, M_2^+, M_3^+ and M_4^+ to fragment ion A^+.

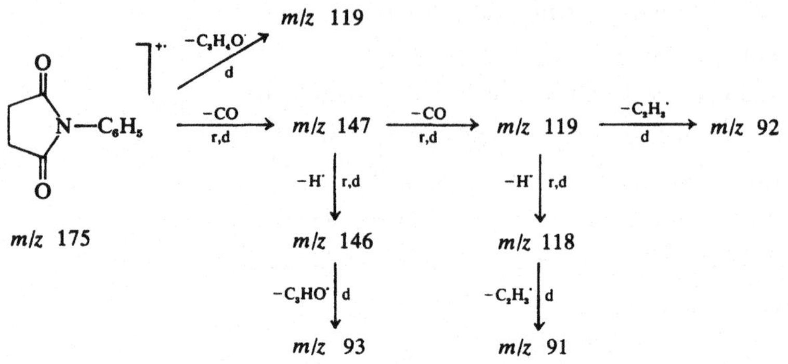

The figure shows fragmentation pathways:

$$\text{m/z } 175 \quad \xrightarrow[\text{r,d}]{-CO} \quad \text{m/z } 147 \quad \xrightarrow[\text{r,d}]{-CO} \quad \text{m/z } 119 \quad \xrightarrow[\text{d}]{-C_2H_2\cdot} \quad \text{m/z } 92$$

with $-C_2H_4O\cdot$ (d) giving m/z 119; $-H\cdot$ (r,d) giving m/z 146 and m/z 118; $-C_2HO\cdot$ (d) giving m/z 93; $-C_2H_2\cdot$ (d) giving m/z 91.

Figure 8.3. Some pathways elucidated by observation of metastable ion transitions in the mass spectrum of *N*-phenylsuccinimide by the defocussing technique (d), compared with those observed directly in the routine (r) spectrum recorded on photographic paper. More transitions are revealed by the defocussing method.

the normal A^+ ions go out of focus (are no longer detected) but, if any A^+ ions are formed from metastable precursors, then they will come into focus (they will be detected) and the corresponding fragmentation must occur. The new accelerating voltage (V') at which A^+ (mass m_2) ions come into focus is noted. If the original voltage set for normal ions, A^+, is V then the mass (m_1) of the precursor to A^+ must be given by $m_1 = m_2 V'/V$. Because all of the quantities on the right-hand side of this equation are known m_1 can be calculated easily. The process has been automated. It may be that only one or some or all of the ions, $M_1{}^+$, $M_2{}^+$, $M_3{}^+$ and $M_4{}^+$, fragment to give A^+, in which cases only one or some or all four of the fragmentation routes will be found to occur. The metastable ion fragmentations in the mass spectrum of *N*-phenylsuccinimide found by 'defocussing' are shown in figure 8.3; products from metastable ions observed in the routine spectrum recorded on photographic paper are also indicated. It can be seen that more metastable ion fragmentations are found by the defocussing method than appear in the routine spectrum. The defocussing technique gives more definitive information than the 'metastable ion peaks' appearing in a routine spectrum (see below). The positions of the latter peaks depend on the masses of the precursor (m_1) and product ions (m_2), which may not be exactly definable in a routine spectrum containing many ions, i.e., there can be ambiguity about which ions are precursors to a given product. In the defocussing method, product ions are examined individually and the precursor ions giving rise to them are found precisely.

An alternative method of focussing the products of metastable ion decomposition is to keep the ion-accelerating voltage constant and to vary the voltage on the electric sector. During routine operation of the mass spectrometer, ions are focussed in the space between the electric sector and the

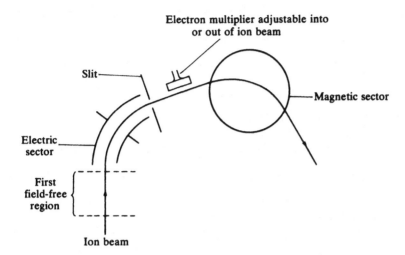

Figure 8.4. The arrangement for ion kinetic energy spectroscopy. This is like the electric and magnetic sector arrangement shown in figure 8.1 but with the addition of an extra ion detector, which can be moved into and out of the ion beam; this electron multiplier is located behind a metal slit placed after the electric sector.

magnetic field (figure 8.4) but products from metastable ions are not focussed. By altering the voltage on the electric sector, any normal ions are defocussed but product ions from metastable fragmentations can be focussed. This method of examining metastable ion dissociation has been termed ion kinetic energy spectroscopy (IKES) (Beynon, Caprioli, Baitinger and Amy, 1970; Beynon and Cooks, 1975). A second ion detector, which can be raised or lowered into place after the electric sector (figure 8.4), and an arrangement for continuously varying the voltage on the electric sector plates without affecting the ion-accelerating voltage are the extra requirements for using this method.

In general, as shown in section 2.4.1, after acceleration from the ion source through a potential of V volts, an ion with z charges will have translational energy zeV and will pass through an electric field E on a curved trajectory of radius, $R = 2V/E$. Thus, a metastable ion of mass m_1 with x positive charges fragmenting, either by unimolecular decomposition or as a result of collision with a neutral particle, to give an ion of mass m_2 with y positive charges in the first field-free region, will have a trajectory with a radius of curvature (R_1) given by equation (8.3):

$$R_1 = [(m_2x/m_1)](2V/E) \qquad (8.3)$$

The radius is not correct for focussing. To achieve the correct radius (R), the actual radius must be multiplied by m_1/m_2x, viz., $R_1m_1/(m_2x) = [m_1/(m_2x)](m_2x/m_1)(2V/E) = 2V/E$. The radius is then correct for focussing. However, the product ion of mass m_2 has y positive charges and the effect of the field E is magnified y times above that of a singly charged ion and it will have a new radius of curvature (R_2) given by equation (8.4):

$$R_2 = [(m_2/m_1)(x/y)](2V/E) \tag{8.4}$$

To focus this multiply charged product ion, the radius of curvature must be corrected again. Therefore, to constrain the ion of mass m_2 into the required radius of curvature (R) correct for focussing, the radius R_1 must be adjusted y times. The total adjustment required is $(m_1/m_2)(y/x)$ so that $R_2(m_1/m_2)(y/x)$ = $2V/E$. If V is constant, changing the electric sector field from E to $(m_2/m_1)(x/y)E$ will change the radius of curvature of the ion of mass m_2 to R and focus it at a slit placed between the electric and magnetic sectors (figure 8.4). Examination of the expression for the required voltage $(m_2/m_1)(x/y)E$ leads to a number of possible results. For $x = y = 1$, a common occurrence, the electric field need only be changed by the ratio of the masses of the product and precursor ions (m_2/m_1); because one of these masses will be known, the other can be calculated easily. If $x = 2y$ and $m_2 = m_1$, i.e. for a reaction in which such a process as $M^{2+} + N \rightarrow M^+ + N^+$ occurs (charge exchange; see section 3.3.1), where N is a neutral species, then $(m_2/m_1)(x/y)E$ = $2E$; the ions of mass m_2 appear at twice the normal electric sector field, E.

Suppose that the doubly charged ion, M^{2+}, decomposes to two singly charged ions, P^+ and Q^+, then $x = 2y$ and $(m_2/m_1)(x/y)E$ becomes $(P/M)2E$ or $(Q/M)2E$ where P, Q and M are the masses of P^+, Q^+ and M^{2+}, i.e. there is a product ion peak at each of these voltages. An example of this behaviour can be seen amongst the metastable ions in the spectrum of naphthalene. The doubly charged ion $C_{10}H_6^{2+}$ (126 mass units) fragments to yield two singly charged ions, $C_7H_3^+$ (87 mass units) and $C_3H_3^+$ (39 mass units):

$$C_{10}H_6^{2+} \longrightarrow C_7H_3^+ + C_3H_3^+$$
$$126 \qquad\qquad 87 \qquad 39 \text{ m.u.}$$

The product ions occur at electric sector voltages of $1.38E$ $(= (87/126)2E)$ and $0.6E$ $(= (39/126)2E)$, respectively; one product ion occurs at an electric field greater than the original and one at a lesser field. When $x = y$ but $m_2 <$ m_1, the required sector voltage for focussing $((m_2/m_1)E)$ is less than E. Thus, fragmentation of metastable ions may be observed at electric sector voltages below or above the normal operating voltage E without interference from normal ions. The extra ion detector placed behind the slit (figure 8.4) enables focussing of the ionic products of metastable ions to be carried out by adjusting the electric sector voltage continuously. If the ion detector is raised out of the ion beam, the ions pass on and into the magnetic sector for mass analysis. The mass analysis is necessary to discover which metastable ions are being investigated. In the example given above of the ion, $C_{10}H_6^{2+}$, decomposing to give ions at $C_7H_3^+$ and $C_3H_3^+$, mass analysis of the product ions found at a sector voltage of $1.38E$ showed that they had a mass-to-charge ratio of 87, i.e., $m_2 = 87$. Thus $1.38E = (87/m_1)2E$, from which $m_1 = 126$; mass analysis of the product ions focussed at $1.38E$ confirmed their mass and this combined with the electric-sector voltage gave the mass of the precursor ion. The doubly charged ion at mass 126 must have been the precursor of the singly charged ion at mass 87. Similarly, it was found that the ion of mass 126 was also the precursor of the singly charged ion of mass 39.

There are other variants of the above methods of scanning mass spectrometers to detect metastable ions without interference from normal ions. These techniques are collectively called *linked scanning* and have been automated so that the complex modes of scanning are brought under the control of a computer. In these variants, two of the three terms V, E and B are changed in a linked manner and 'linked scan' refers to the fact that the voltage (E) of the electric sector is varied simultaneously with either the accelerating voltage (V) or the magnetic field (B) so as to maintain a specified relationship between them throughout the scan. A treatment of combinations of B, V and E is available (Boyd and Beynon, 1977) but only the commoner ones are discussed here.

For the first linked-scan method, the magnetic field is kept static whilst the accelerating and electric sector voltages are varied such that E^2/V remains constant (Weston, Jennings, Evans and Elliott, 1976). Let the value of the accelerating voltage V be changed to $(m_1/m_2)^2V$. The kinetic energy of product ions, A^+, from metastable decomposition, is therefore changed from $(m_2/m_1)zeV$ to $(m_2/m_1)(m_1/m_2)^2zeV = (m_1/m_2)zeV$ and, for the electric sector to pass these ions, its voltage, E, must be changed to $(m_1/m_2)E$. Therefore, as the accelerating voltage is changed by a factor of $(m_1/m_2)^2$, so the voltage of the electric sector must change simultaneously by (m_1/m_2). During a scan, the values of E and V are linked in the sense that E^2/V is a constant. Precursor ions, M^+, have translational energy $zeV = (m_1v^2)/2$, from which their momentum, $m_1v = (2zem_1V)^{1/2}$. The initial momentum of A^+ ions is $(m_2/m_1)(2m_1zeV)^{1/2}$ but, since V has been changed by a factor of $(m_1/m_2)^2$, their new momentum is $(m_2/m_1)[2m_1(m_1/m_2)^2zeV]^{1/2} = (2m_1zeV)^{1/2}$, which is the same as that for M^+ ions. Therefore, during the scan, the magnetic field is kept constant at the value required to transmit M^+ ions. By focussing initially on a selected precursor ion M^+ (often but not always the molecular ion) and then performing this linked scan, ions that are products of fragmentations of M^+ in the first field-free region may be determined unambiguously. If sufficient sample is available, many different precursor ions may be examined individually, so that whole fragmentation pathways can be characterized. The method is complementary to the defocussing method, which determines precursor ions of a chosen product ion (see above). The E^2/V linked-scan spectrum of molecular ions of n-decane characterized the fragmentations shown in scheme (8.5):

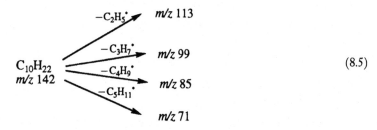

$$(8.5)$$

By performing the analysis also on the ions at m/z 99 and 57, the pathway depicted in scheme (8.6) was determined:

$$C_{10}H_{22}^{+\bullet} \xrightarrow{-C_3H_7^{\bullet}} C_7H_{15}^{+} \xrightarrow{-C_3H_6} C_4H_9^{+} \xrightarrow{-CH_4} C_3H_5^{+}$$
$$m/z\ 142 \qquad\qquad m/z\ 99 \qquad\qquad m/z\ 57 \qquad\qquad m/z\ 41$$

$$(8.6)$$

Both the E^2/V linked-scan method and the defocussing method suffer from limitations incurred by varying the accelerating voltage: the mass range is limited and the efficiency of the ion source changes during the scan. To overcome these problems, the accelerating voltage is kept constant and linked scanning of the magnetic field and electric sector voltage is performed instead. When the ratio B/E is kept constant during a scan, the metastable ion spectra obtained are analogous to the E^2/V scan in that all product ions of a chosen precursor are observed (Millington and Smith, 1977; Bruins, Jennings and Evans, 1978). The translational energy of product ions, A^+, from metastable ion decomposition is $(m_2/m_1)zeV$. These ions will be transmitted by the electric sector only if its initial voltage is changed by a factor of m_2/m_1. When the magnetic field is set to focus precursor M^+ ions, then $m_1 v_1^2/r = Bv_1 ze$ or $B = m_1 v_1/(zer)$, where v_1 is the velocity of M^+ ions.

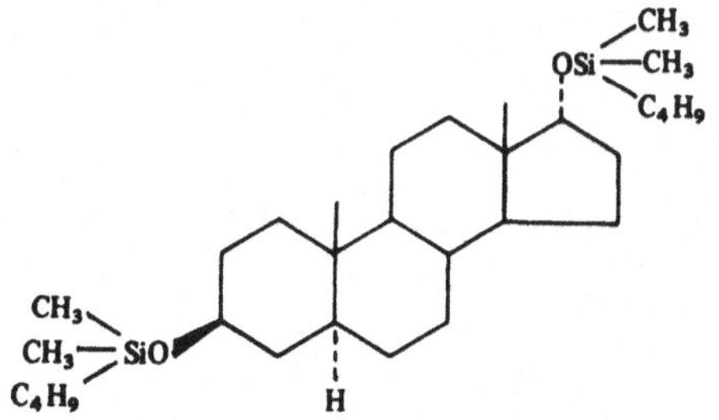

CH₃
OSi—CH₃
C₄H₉

CH₃
CH₃—SiO
C₄H₉

H

1

CH₃
OSi—CH₃
C₄H₉

CH₃
CH₃—SiO
C₄H₉

H

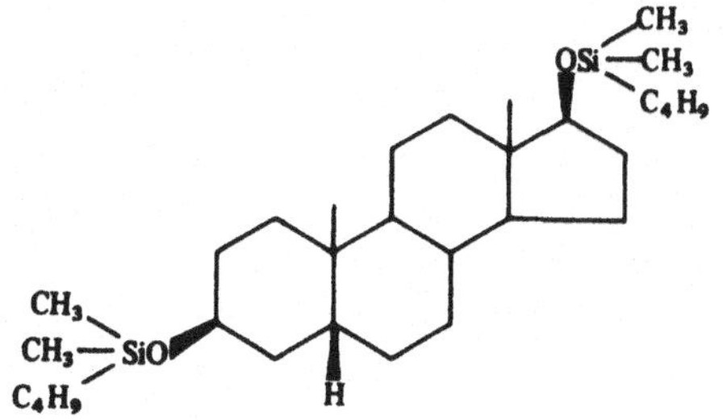

2

Ions, M^+, which decompose in the first field-free region afford product ions A^+ with the same velocity, v_1. Hence, to focus these A^+ ions, the magnetic field must be changed to $m_2 v_1/(zer)$, i.e. by a factor of m_2/m_1. Since B and E are changed by the same factor, the magnetic field and electric-sector voltage may be scanned simultaneously such that B/E remains constant and this locates all product ions resulting from dissociation of a selected metastable precursor ion.

As expected, the B/E linked-scan spectrum of n-decane is virtually identical to its E^2/V linked-scan spectrum (see above). The isomeric androstane-diol derivatives (**1**) and (**2**) give very similar conventional mass spectra (table 8.1), but the B/E linked-scan spectra of the ions at m/z 463, resulting from loss of $C_4H_9\cdot$ from the molecular ion, enable the compounds to be differentiated readily by reference to relative product ion abundances at m/z 331 and 255 as seen in table 8.1 (Gaskell, Pike and Millington, 1979).

A linked scan of the magnetic field and electric-sector voltage in which B^2/E remains constant provides an alternative to the defocussing method. At constant accelerating voltage, the magnetic field is set initially to focus normal A^+ ions with momentum $(2m_2 zeV)^{1/2}$. The electric sector voltage must be changed from E to $(m_2/m_1)E$ to transmit A^+ ions, which possess momentum

Table 8.1. *Routine and linked scan mass spectra of steroid (1) and (2).*

m/z	Percentage relative abundance			
	Routine scan		B/E linked scan	
	Compound (1)	Compound (2)	Compound (1)	Compound (2)
505	1	2	—	—
463	100	100	Precursor ion	
387	8	9	100	100
373	—	—	3	13
345	—	—	1	2
331	6	12	41	84
255	66	56	70	21

$(m_2/m_1)(2m_1 zeV)^{1/2} = (m_2/m_1)^{1/2}(2m_2 zeV)^{1/2}$. Therefore, to focus these ions on the detector, B must be changed simultaneously to $B(m_2/m_1)^{1/2}$. Observation of metastable precursor ions for a chosen product ion, A^+, requires scanning such that B^2/E is constant. The technique may be used in the same way as described above for the elucidation of fragmentation pathways of N-phenyl-succinimide. The steroids (1) and (2) both give ions at m/z 255.2118 ($C_{19}H_{27}^+$) in which two silanol groups have been lost. A linked scan such that B^2/E remains constant showed that m/z 463, 387, 373, 345 and 331 were all precursors of $C_{19}H_{27}^+$. From these data, coupled with the B/E linked-scan results shown in table 8.1, several fragmentation pathways were discovered, as shown in figure 8.5 (Gaskell, Pike and Millington, 1979). Since the ion at m/z 255 was shown to be a product of m/z 463 and that at m/z 463 a precursor of m/z 255, rapid successive loss of both silanol groups is possible. This illustrates the general principle that the observation of metastable ions does *not* necessarily imply that a single reaction has taken place. When two (or more) successive reactions occur sufficiently rapidly in comparison with the flight-time of the metastable ion in the field-free region of interest, data for the composite process may be observed. Metastable ion data for the individual reactions may also be found, depending on fragmentation rates (figure 8.5).

The final linked-scan method to be described here selectively detects all product ions that result from a fragmentation involving elimination of a neutral species with a chosen mass. To do this, the accelerating voltage is

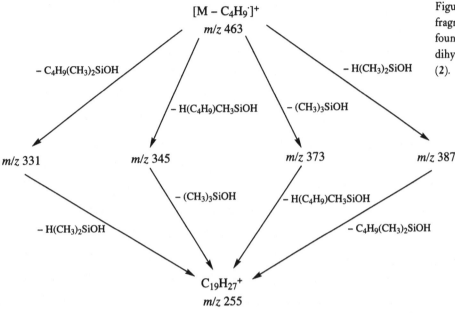

Figure 8.5. Some of the fragmentation pathways found for silylated dihydroxysteroids (1) and (2).

held constant and the voltage of the electric sector and the magnetic field are varied such that $(B/E)(1 - E)^{1/2}$ remains constant. In an alternative method, which is less satisfactory because the accelerating voltage is changed, the magnetic field is static whilst a linked scan of accelerating and electric sector voltages is performed (Lacey and Macdonald, 1979). For a single compound, several reactions may occur that give rise to a common neutral fragment. For example, several ions of a derivatized peptide eliminate carbon monoxide and the masses of ions involved in such reactions define the amino acid sequence of the peptide (scheme (8.7)). Many hydrocarbon ions lose hydrogen (H_2), as has been demonstrated through a linked scan of the ions produced from electron ionization of decane. The major fragmentations of this type are $C_4H_7^+ \rightarrow C_4H_5^+$, $C_3H_7^+ \rightarrow C_3H_5^+$ and $C_3H_5^+ \rightarrow C_3H_3^+$. Since only the difference between the masses of precursor and product ions is required, product ions resulting from loss of the common neutral species, H_2, are all detected in one scan. For the analysis of a mixture, components with common structural features may be detected selectively by the so-called *constant neutral* spectrum. Negative-ion chemical ionization of carboxylic acids affords $[M - H]^-$ ions which undergo loss of carbon dioxide. By performing a linked scan for the loss of 44 mass units, carboxylic acids are detected and identified.

$$\sim\!\!\sim\!\!\sim\!\!\sim\!\!-NHCHCONHCH-C\!\!\equiv\!\!\overset{+}{O}$$
$$\underset{R_{n-1}}{|} \quad \underset{R_n}{|}$$

$$\Big\downarrow -CO$$

$$\sim\!\!\sim\!\!\sim\!\!\sim\!\!-NHCHCO\overset{+}{N}H\!\!=\!\!CHR_n$$
$$\underset{R_{n-1}}{|}$$

$$\Big\downarrow -NHCHR_n$$

$$\sim\!\!\sim\!\!\sim\!\!\sim\!\!-NHCH-C\!\!\equiv\!\!\overset{+}{O}$$
$$\underset{R_{n-1}}{|}$$

$$\Big\downarrow -CO$$

$$\sim\!\!\sim\!\!\sim\!\!\sim\!\!-\overset{+}{N}H\!\!=\!\!CHR_{n-1}$$

(8.7)

In the discussion so far, it has been assumed that, during reaction (8.1), the internal energy of a metastable M^+ ion is partitioned between its product ion, A^+ and the accompanying neutral species, N, in proportion to their respective masses. However, some of the *internal* energy in M^+ may be released as *kinetic* energy during the decomposition and this added kinetic energy imparts an energy spread to the A^+ ions. The effect results in broad peak shapes for the products arising from metastable ions. Useful kinetic data may be gleaned from analysing the resulting peak shapes (see section 8.3.2) but, in many situations, kinetic data are not required and the breadth of the peaks prevents accurate determination of their position. In particular, when the peaks are so broad that they overlap, the masses of product or precursor ions are difficult to estimate. During linked scanning such that E^2/V or B/E is constant, the mass spectrometer discriminates against extremes in kinetic energy so that all peaks are equally narrow, irrespective of the amount of internal energy converted to kinetic energy during fragmentation. In these scanning modes, the masses of product ions are readily assigned. On the other hand, the defocussing and B^2/E linked-scanning methods preserve energy broadening of the peaks. Therefore, in these modes, kinetic data are available but accurate determination of all precursors of a given product ion is difficult. This situation is illustrated in figure 8.6, which shows for the terpene, limonene, (*a*) its routine mass spectrum, (*b*) the B/E linked-scan

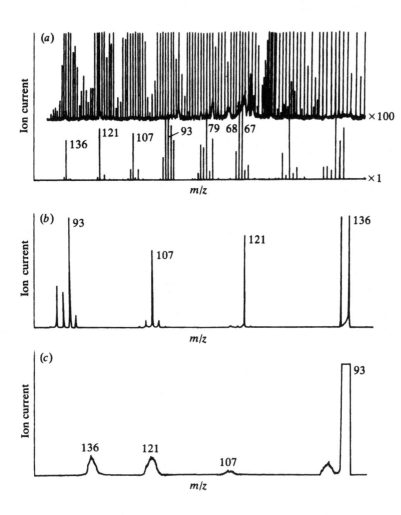

Figure 8.6. Mass spectra of limonene. (*a*) Routine mass spectrum recorded on UV-sensitive paper at two different sensitivities, (*b*) the *B/E* linked scan spectrum of molecular ions ($C_{10}H_{16}^{+\cdot}$) at *m/z* 136 and (*c*) the B^2/E linked scan spectrum of fragment ions ($C_7H_9^+$) at *m/z* 93. Note that the resolution of spectrum (*c*) is much lower than of (*b*), and that, in (*a*), broad peaks arising from metastable ions can be observed at ×100 sensitivity.

spectrum of its molecular ion at *m/z* 136 and (*c*) the B^2/E linked-scan spectrum of its fragment ion at *m/z* 93. In spectrum (*b*), metastable ion processes that lead to losses of methyl (15 mass units), ethyl (29 mass units) and propyl radicals (43 mass units) are observed with good resolution of mass peaks. On the other hand, the B^2/E linked-scan spectrum (*c*) is of much lower resolution, containing only broad peaks. It is observed that, in the first field-free region, ions at *m/z* 136, 121 and 107 are all precursors of the product ions at *m/z* 93. Combining the results of these two metastable ion analyses, the following fragmentation pathway can be constructed for limonene:

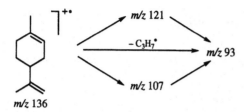

Linked scanning of the electric-sector voltage and magnetic field may be applied to mass spectrometers of conventional or reversed geometries (see below). Automation of the scanning requires a monitor for the strength of the magnetic field and this is usually a Hall-effect probe (section 4.2). This device measures magnetic flux (B) and its output is used to control the scanning of the electric sector voltage to maintain the desired relationship of B to E. Alternatively, control of scanning may be based on a calibration table, which correlates scan time with mass during mass calibration (section 4.2). This latter method defines more accurately the magnetic field and leads to more precise mass assignment. With automated systems, it is possible to obtain linked-scan mass spectra in a few seconds, making the technique compatible with GC/MS. Also, a data system may be used to control a conventional scan of a mass spectrometer until a preselected mass peak is detected. At this point, a linked-scan spectrum of that ion is initiated by the computer. When this scan ends, the instrument is programmed to continue the original scan. Hence, a routine mass spectrum and linked-scan spectra of ions of interest in that spectrum may be obtained in one experiment. The alternate use of two different linked-scan modes can achieve accurate mass assignment from the sharp peaks in one mode and analysis of kinetic energy distribution from the broad peaks in the other mode.

Selected (metastable ion) reaction monitoring in the linked-scan mode is a sensitive method for detecting and quantifying small amounts of substances (section 7.3). Characteristics of some selective methods for observing metastable ion reactions are summarized in table 8.2.

8.3.2. *The second field-free region*

This region extends from the end of the electric sector up to the beginning of the magnetic field. In a single-focussing instrument, it is the space between the ion source and the magnetic field. Because decompositions of metastable ions in the second field-free region occur after the electric sector, there is no discrimination against their product ions, which are therefore

Table 8.2. *Characteristics of scanning modes for observation of metastable ion decomposition in the first field-free region.*

Type of scan (Peak shape)	Metastable ions determined (Energy analysis)
Defocussing (Broad)	All precursors of a chosen product ion (Possible)
E^2/V linked scan (Sharp)	All products of a chosen precursor ion (Not possible)
B/E linked scan (Sharp)	All products of a chosen precursor ion (Not possible)
B^2/E linked scan (Broad)	All precursors of a chosen product ion (Possible)
$(B/E)(1-E)^{1/2}$ linked scan (Sharp)	Ejection of common neutral particles (Not possible)

mass-analysed in the magnetic field and appear in routine spectra from both single- and double focussing instruments as broader peaks than those arising from normal ions (section 1.5). Focussing in the magnetic field is affected by both the m/z value of an ion and its translational energy. Therefore, the product A^+ ions, which have less translational energy than normal A^+ ions, are deflected differently in the magnetic field. The ionic products of metastable ions have translational energy $(m_2v^2)/2 = (m_2/m_1)zeV$, so they have the same velocity, $v = (2zeV/m_1)^{1/2}$, as their metastable precursor ions. On passing through the magnetic field, the product ions experience a centrifugal force, $Bzev = m_2v^2/r$. Combination of the latter two equations gives the radius of curvature (r) of product ions of metastable M^+ ions in the magnetic field (equation (8.8)). When equation (8.8) is compared with that derived from equation (2.2) for normal M^+ ions (equation (8.9)), it is seen that the product ions of metastable decomposition will occur in routine mass spectra at an apparent mass, $m^* = m_2^2/m_1$. If a peak is observed at m^* in the routine mass spectrum, it is only necessary to apply this formula to determine the precursor and product ions from which it arises. For example, the mass spectrum of aniline contains large peaks at m/z 93 and 66, and a broad peak at m/z 46.8. Since $66^2/93 = 46.8$, it is evident that at least some of the ions at m/z 66 arise from ions at m/z 93 by ejection of 27 mass units. Perhaps it should be stressed again here that, because of the way that computerized data acquisition and processing programmes have been written, data systems operating on normal routine scans eliminate all traces of the presence of such metastable ions; these last are best observed by direct recording of a mass spectrum onto photographic paper. The behaviour of metastable ions discussed in this section was observed in just such a manner.

$$r = \{(m_2{}^2/m_1)[2V/(zeB^2)]\}^{1/2} \tag{8.8}$$

$$r = \{m_1[2V/(zeB^2)]\}^{1/2} \tag{8.9}$$

If there are many peaks in the mass spectrum, it is time consuming to check manually all of the possible precursor and product ions that could give rise to an observed broad peak at apparent mass, m^*, and so tables and nomograms (McLafferty and Turecek, 1993) are available to speed the checking. Also, it is simple to programme a computer to do this checking. It is worth remembering that, when a mass spectrum is recorded on UV-sensitive paper, which is run through the recorder at a uniform speed (as in figure 11.1, chapter 11), the m/z values are actually presented on an exponential scale. This fact, coupled with the formula, $m^* = m_2{}^2/m_1$, dictates that, on a photographic recording of the mass spectrum, the distance between the normal *precursor* and product mass peaks is the same as that between the *normal* product ion peak and the broad peak at m^*. Therefore, simply by moving a ruler along the trace, potential precursor and product ions are determined. The method is fast and convenient, but not particularly accurate, so that tentative assignments must be checked using the formula. Two factors determine that it is not always possible to assign unambiguously m_1 and m_2. Firstly, because product ion peaks are usually broad with rounded tops, the precise position of m^* may be difficult to measure. Secondly, because two unknown variables (m_1 and m_2) are needed for the single formula, $m^* = m_2{}^2/m_1$, there is no unique solution. However, the values of m_1 and m_2 are restricted to the m/z values of observed normal ions and, for there to be observable product ions from metastable ion decomposition, *normal* precursor and product ions must be reasonably abundant and therefore the values of m_1 and m_2 giving m^* are usually obvious. If this is not the case, recourse must be made to other, unambiguous techniques described in this chapter, such as linked scanning (see above).

Usually the peaks arising through metastable ion decomposition have roughly Gaussian shapes in the mass spectrum, frequently with a 'tail' on the high-mass side (figure 8.7(a)). Sometimes the peaks appear as broad and flat,

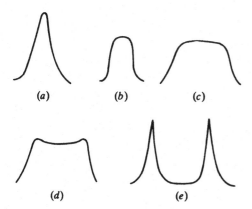

Figure 8.7. Some observed shapes of product ion peaks (a)–(d) resulting from fragmentation of metastable precursor ions. Peak (a) is the shape normally found, (b)–(d) are wide, flat peaks,. The theoretically expected shape of one of these flat-topped (broad or dish-shaped) peaks is shown in (e).

extending over several mass units (figure 8.7(b)–(d)); this arises because, during the actual fragmentation process, some of the internal energy in the decomposing ion appears along the reaction coordinate as kinetic energy. Figure 8.7(e) shows the theoretically expected peak shape which is rarely observed so clearly because energy spread in the ion beam causes filling of the region between the two sharp peaks. Theoretically expected peak shapes can be observed by using slits to cut down the spread of the ion beam. The energy (T) released in giving rise to these broad peaks may be calculated from formula (8.10):

$$T = [dm_1/4(m_2^2)]^2 (zeV/\mu) \qquad (8.10)$$

where d is the width of the peak in mass units, m_1 and m_2 are the masses of the precursor and fragment ions, and μ is a reduced term equal to $(m_1 - m_2)/m_2$ (Beynon, Saunders and Williams, 1968; Holmes and Terlouw, 1980).

There is another method for examining metastable ions dissociating in the second field-free region. Since their product ions have less translational energy than the corresponding normal ones, the two types may be differentiated by having an ion repeller voltage at the detector. Suppose that normal ions possess translational energy corresponding to a voltage V, and the product ions from metastable ones to V'. If the repeller voltage (V_r) is greater than V (and therefore V') all ions are turned back and strike a metal plate causing secondary electrons to be emitted (figure 8.8). These secondary electrons are detected by a scintillator-photomultiplier. If the repeller voltage (V_r) is made less than V but greater than V' then normal ions are not repelled but the products of metastable ion decomposition are, again causing emission of secondary electrons. Therefore, the method allows a

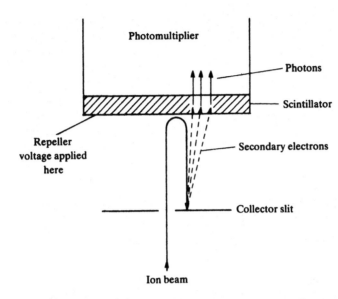

Figure 8.8. The arrangement of scintillator/photomultiplier and ion beam repeller voltage for detection of product ions from metastable precursor ions.

normal mass spectrum to be obtained at one voltage, $V_r (>V)$, and then, after changing the repeller voltage to $V>V_r>V'$, a second scan records only the product ions arising from fragmentations in the second and third field-free regions. Since the potential required just to repel these product ions may be measured, this system (the Daly detector), enables analysis of their kinetic energy. Like the defocussing and linked-scan techniques, it also reveals metastable ion data that might be obscured by normal ion peaks in a routine mass spectrum.

8.3.3.   The third field-free region

After the end of the magnetic field and before the detector there is another small field-free region (figure 8.1) where decompositions of metastable ions may take place. The region lies after all focussing has occurred so there is no spatial separation of normal ions and product ions from metastable fragmentation and both types arrive at the collector at the same m/z value. However, A^+ ions formed as products of metastable M^+ ions still possess less translational energy than normal A^+ ions formed in the ion source and a detector with a repeller voltage can discriminate between them. The Daly detector (see previous section 8.3.2) is suited to this end and can detect product ions formed in either the second or the third field-free regions. By changing the repeller voltage at the detector, those ions with less than the full translational energy of normal ions can be differentiated. Thus, if the detector is receiving normal M^+ ions, and some of these decompose in the third field-free region, the resultant product A^+ ions will continue at the same velocity (but different momentum) on the same trajectory and be recorded as M^+ ions. The detector is set at a suitable repeller voltage to cause all ions to be turned back onto the scintillator. Then, by reducing the repeller voltage at the detector, normal M^+ ions continue on and are lost but the product A^+ ions formed by fragmentation in the third field-free region are still repelled and strike the scintillator, thereby being detected. Thus, detection of product ions, A^+, originating in the third field-free region requires only the recording of both types of ion at mass m_1 by turning back all ions at a repeller voltage V and then recording again at a voltage V' ($< V$) to record only the product ions of metastable ion decomposition. These last ions appear as sharp peaks in the same (mass) positions as the precursor ions in this method since no mass separation occurs after the end of the magnetic field. The mass of the product ion must be obtained by measuring the ratio of the voltage (V) at which M^+ and A^+ ions begin to be observed (total of precursor and product ions at the mass m_1) and then reducing the voltage to V' at which all ions disappear; $m_1/m_2 = V/V'$. As m_1 is known, m_2 can be calculated.

8.3.4. *Other regions*

The other places along the flight path of the ions between the source and the detector where metastable ions may fragment lie in magnetic or electric fields and observing them is not simple. At the beginning of the magnetic field (the end of the second field-free region), the relation $m^* = m_2^2/m_1$ holds, but at the end of the magnetic field (the beginning of the third field-free region), $m^* = m_1 = m_2$. Inside the magnetic field, any ionic products of metastable ions will be detected between m_2^2/m_1 and m_1 and this leads to a long tail appearing on the high-mass side of the routinely observed product ion peak and extending up to the normal precursor ion peak (figure 8.9). Because these peaks from dissociating metastable ions are, in any case, not large, this diffusion of the peaks into a long tail is only infrequently noticeable at the usual amplification required for recording a normal mass spectrum on photographic paper and they are never observed following computerized data acquisition (section 4.2). Product ions formed within the electric sector are not readily observable, but may cause interfering peaks during linked scanning for metastable ions fragmenting in the first field-free region.

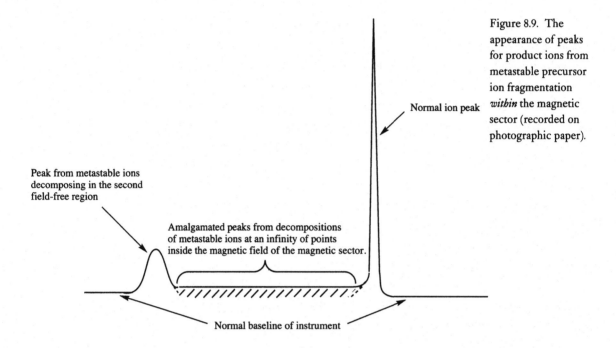

Figure 8.9. The appearance of peaks for product ions from metastable precursor ion fragmentation *within* the magnetic sector (recorded on photographic paper).

Normal ion peak

Peak from metastable ions decomposing in the second field-free region

Amalgamated peaks from decompositions of metastable ions at an infinity of points inside the magnetic field of the magnetic sector.

Normal baseline of instrument

8.4. METASTABLE IONS IN MASS SPECTROMETERS OF REVERSED GEOMETRY

Double-focussing mass spectrometers of conventional geometry are arranged so that ions pass first through the electric sector and then through the magnetic sector (figure 8.1). Low-resolution and high-resolution mass spectrometry may also be performed with mass spectrometers that have the reverse arrangement of magnetic and electric sectors. A mass spectrometer of reversed geometry is depicted in figure 8.10 and this should be compared and contrasted with figure 8.1. The altered design depicted in figure 8.10 is particularly useful for observing metastable ions dissociating in the field-free region between the magnetic and electric sectors. The major advantage of the reversed geometry is that the magnetic field may be set to focus only ions of one selected mass into the field-free region. Some of the selected ions may decompose there, giving product ions with kinetic energy $(m^*/m)zeV$, where m^* is the mass of the product ion and m that of the precursor ion. All product ions with different masses, m^*, formed from the chosen precursor, have different kinetic energies and can be separated in the energy-focussing electric sector. With the magnetic field set at a constant value so as to transmit the precursor ion of interest, a single sweep of the voltage of the electric

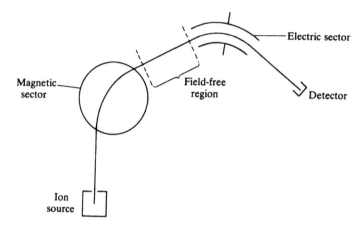

Figure 8.10. The path of the main ion beam through a double-focussing mass spectrometer of reversed geometry. The ions pass through a magnetic sector before passing through the electric sector. The field-free region indicated between the sectors is the most important one for studying metastable ion fragmentations.

sector identifies all product ions. The technique is called mass-analysed ion kinetic energy spectroscopy (MIKES) since it is in principle the same as ion kinetic energy spectroscopy but obviates the need for a separate mass analysis (section 8.3.1). The mass peaks obtained are broad due to release of internal energy as kinetic energy during decomposition. Hence, kinetic-energy data are available by analysis of peak shapes, but the attendant limited resolution makes mass measurement difficult. Resolution may be increased somewhat by increasing the accelerating voltage since this increases the kinetic energy of all ions, thereby reducing the effect of small kinetic energy changes caused by fragmentation. To this end, an additional accelerating potential may be applied between the magnetic and electric sectors.

Inasmuch as all product ions of a given precursor are examined, MIKES is equivalent to E^2/V or B/E linked scans with a mass spectrometer of conventional geometry, although the latter techniques give sharp peaks and no kinetic data. Compounds with the same basic skeleton but different side chains, like steroids or substituted aromatic compounds, are frequently difficult to differentiate from their routine mass spectra. With the MIKES technique, it is possible to ignore the ions characteristic of the basic skeleton and hence common to a whole series of compounds, by setting the magnetic field to pass selected ions containing the side chain. For instance, a series of steroids (3) with different side chains (R), when ionized, gives ions R^+ and $[R + C_3H_6]^+$. The metastable ion spectra of either of these types of ion are much more characteristic of differences in structure than are the routine mass spectra.

Like the linked-scanning methods discussed in the previous section,

$$\xrightarrow{-e} \quad R^+, (R + C_3H_6)^+$$

MIKES is applied with advantage to the direct analysis of mixtures. The MIKES technique has been suggested as an alternative to combined chromatography/mass spectrometry. Instead of physically separating components of a mixture by chromatography, the whole mixture is ionized and a single component is examined by setting the magnetic field to transmit ions characteristic of that particular component (frequently the molecular ion). Metastable ion mass spectra then characterize the substance without interference from other components. The method is faster than combined chromatography/mass spectrometry since separation occurs in about 10^{-5} s rather than in minutes or hours for a chromatographic analysis. The sensitivity is as good as that for GC/MS and selected ion monitoring for quantitative analysis is possible with either technique. The MIKES method suffers from several disadvantages. The composition of the ionized mixture may not reflect accurately the composition of the original mixture since some components may suppress the ionization of others. Also, if ionization leads to fragmentation, as with electron ionization, each component generally affords many different ions so that the mass spectrum of the mixture is very complex and several different substances may contribute to the signal at any one m/z value. 'Soft' ionization techniques such as positive-ion and negative-ion chemical ionization, field ionization and field desorption and electrospray give less fragmentation than does electron ionization or none at all. A mixture ionized by FD, for instance, is likely to give rise to molecular ions only, so the mass spectrum of the mixture will be relatively simple; to obtain metastable ion data, these molecular ions would need to be given more energy so as to induce fragmentation, leading to the same problem of overlapping fragmentation pathways as with electron ionization, unless the molecular ions are first selected and examined separately. The instrument shown in figure 8.10 is not capable of separating such mixtures of ions because they are not energy-focussed before passing through the magnetic field, but instruments that allow separation are available (section 8.6). In addition, if a mixture contains isomeric compounds, they cannot be distinguished by mass. Despite these complications, many mixtures have been analysed by MIKES. Direct ionization of plant tissues followed by MIKES has allowed identification of natural alkaloids. Barbiturates in crude drug preparations have also been examined successfully and notable results have been obtained from pyrolysing biological samples in the ion source of a mass spectrometer. For example, the pyrolysis of deoxyribonucleic acids produces a mixture of purine and pyrimidine bases, each of which may be characterized rapidly by MIKES. Any modified or unusual bases may also be identified by the same method.

Mass-analysed ion kinetic energy spectroscopy and related methods have been described extensively in many reviews (Beynon, Morgan and Brenton, 1979; McLafferty, 1980; Russell, McBay and Mueller, 1980).

8.5.   TRIPLE QUADRUPOLE ANALYSERS

In ordinary quadrupole mass filters, metastable ions cannot be studied because they are not distinguished from normal ions. However, when quadrupole mass filters are linked together, fragmentation of metastable ions occurring between analysers gives rise to observable product ions. A triple quadrupole instrument embodying this principle is shown in block form in figure 8.11, its capabilities for studying metastable ions being roughly the same as those of the reversed-geometry, magnetic-sector mass spectrometer. The first analyser is used to transmit selected ions but the second is not used for mass separation; it is a region in which dissociations may occur and from which any resulting product ions are transmitted to the next quadrupole. The final mass filter is scanned to detect in turn all product ions from decompositions occurring in the second filter. For instance, for an alcohol or mixture of alcohols, all ions eliminating water may be detected selectively, regardless of their masses, by scanning the instrument such that, when the first analyser transmits ions of mass m, the second is set to transmit ions of mass $(m - 18)$. This method is far simpler than the complex scan required to record a constant neutral mass spectrum with a magnetic sector instrument. The properties of quadrupole mass filters that make them particularly compatible with computers and useful for combined chromatography/mass spectrometry and chemical ionization (sections 2.3 and 2.4) also favour the use of triple quadrupole analysers for studying metastable ions. Further advantages over magnetic-sector instruments are outlined below.

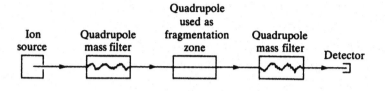

Figure 8.11. A schematic diagram of the triple quadrupole analyser showing the path of the main ion beam through two normal mass filters with a third, operated under different conditions, placed between them.

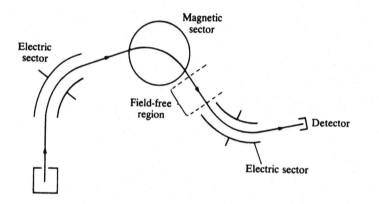

Figure 8.12. The path of the main ion beam through a triple-focussing mass spectrometer having one magnetic (B) and two electric (E) sectors. Only the most widely used field-free region is indicated.

8.6. TRIPLE-FOCUSSING MAGNETIC-SECTOR MASS SPECTROMETERS

Mass-analysed ion kinetic energy spectroscopy, as described above, is hampered by its inability to separate ions with the same integer mass but different elemental compositions. If an additional electric sector is placed between the ion source and magnetic sector of an instrument of reserved geometry, high-resolution MIKES is possible. The design of the mass spectrometer is shown in figure 8.12; it can be considered as a conventional high-resolution mass spectrometer (cf. figure 8.1) up to the field-free region of interest, after which is placed an electric sector for analysis of the ionic products of metastable ions by ion kinetic energy spectroscopy. During an analysis of a mixture of N-methylaniline and methylphenols with the instrument shown in figure 8.10, the selected precursor ions at m/z 107 (schemes (8.11) and (8.12)) fragmented in the field-free region to give a complex mixture of product ions, indicating more than one decomposing entity. Operating at a resolution of 15 000, a triple-focussing mass spectrometer separated $C_7H_7O^+$ ions with mass 107.0497 from $C_7H_9N^{+\bullet}$ ions with mass 107.0735. The different precursor ions were then characterized

$$\text{(8.11)}$$

$$\text{(8.12)}$$

independently through their metastable ion mass spectra. In the field-free region after the magnetic sector, $C_7H_7O^+$ ions were observed to dissociate by loss of CO and H_2CO (reaction (8.11)) and $C_7H_9N^{+\bullet}$ ions by loss of HCN and CH_3N (reaction (8.12)). The decompositions (8.11) and (8.12) are typical of phenols and aromatic amines, respectively (see chapter 10), and the components were readily distinguished and identified. The increase in specificity attendant upon high resolution makes feasible analysis of complex mixtures. For example, the steroid glycoside digitoxin may be determined in blood or urine by direct analysis. Separation of isomers in a mixture remains problematic, irrespective of mass resolution.

8.7. ACTIVATION OF NORMAL IONS

Normal ions formed in the ion source and which have mostly stopped fragmenting on the time scale of a mass spectrometer can be induced to fragment further by addition of more internal energy after they leave the source. This induced fragmentation can be applied to ions formed by any of the ionization methods (EI, CI, FI, FAB, electrospray, etc.). The induced decomposition gives fragment ions, similar to those produced during the 'non-induced' fragmentation of metastable ions and within the same field-free regions. The same techniques used for investigation of metastable ions can be applied to the induced fragmentation of normal ions. Since ions decomposing in a field-free region have been selected by one sector (MS) and are then examined in another (MS), the technique has been termed MS/MS or tandem MS. These tandem methods are described in greater detail in the next section. Here, some current means of inducing fragmentation are described.

8.7.1.   *Collision-induced decomposition*

Singly charged ions. Collision-induced decomposition or dissociation (CID) of normal ions, sometimes called collisional activation (CA) or collisionally activated dissociation (CAD), is achieved by making use of their kinetic energy, gained on acceleration from the ion source. Commonly, a small gas cell is placed across the beam of ions. With no gas in the cell, the ion beam enters and leaves through small holes at opposite sides of the cell. When a gas, such as argon, helium or nitrogen is allowed into the cell, collisions between the fast-moving ions and the gas molecules moving with thermal energies convert some of the joint kinetic energy into an excess of rotational, vibrational and even electronic energy in the ion. Provided the collisions are not too direct, ions that have collided with gas molecules leave the gas cell approximately along the same flight path as the one they had before entering the cell. Some scatter of ions does take place but, provided that the gas pressure in the cell is not too high, the scatter in the ion beam can be accommodated by the analyser region. Of course, if the gas pressure is too high, all ions will be scattered by collision and none will get through the cell. The art of successful CID depends on having a high enough gas pressure so that multiple collisions produce sufficient fragment ions for the following sector to detect; for reproducibility of results, it is important to be able to reproduce the chosen gas pressure exactly. Often, a pressure is chosen that attenuates the ion beam by a half but this is only a guide because sensitivity to induced fragmentation varies with the mass of an ion and the mass of the collision gas molecules. As cell gas pressure is increased, many ion/molecule collisions occur so that, although the excess of internal energy deposited in the ion also increases it is only realized at the expense of unacceptable scattering of the ion beam. Because an ion undergoes several ion/molecule collisions during flight through the gas cell, consecutive decompositions may occur and product ions will be seen for most ions processed in the mass spectrum.

Two collisional activation regimes are in use, one at low kinetic energies corresponding to acceleration of ions through about 0–100 eV and one at high kinetic energy, corresponding to acceleration through more than 1 keV. At the lower collisional energies required for quadrupole sectors, the amount of internal energy deposited in the ion is mostly rotational and

vibrational. The *abundances* of product ions produced by low-energy collisions are highly dependent on the actual collision energy, which can be calculated from equation (8.13), in which m_i and m_g are the masses of ion and
collision gas molecules respectively, E_{LAB} is the energy of collision measured
in the laboratory and E_{CM} is the actual collision energy experienced by an
ion and a gas molecule with reference to their centre of mass. For example,
with argon ($m_g = 40$) as collision gas and an ion, $C_4H_9^+$ ($m_i = 57$) accelerated
through 100 V, $E_{CM} = 100(40/97) = 41$ eV but, for a bigger ion,
$C_{12}H_{20}N_3O_5^+$ ($m_i = 287$), E_{CM} is only 12 eV. Note that only a small fraction
of the collisional energy, E_{CM}, appears as excess of internal energy in the ion,
which reaches a maximum of about 10 eV (McLuckey, Ouwerkerk,
Boerboom and Kistemaker, 1984). The greater the molecular mass, m_g, of the
collision gas, the greater is E_{CM} for any particular ion, m_i. For this reason,
argon ($m_g = 40$) is more widely used than is helium ($m_g = 4$) in these low-
energy regimes.

$$E_{CM} = E_{LAB}[m_g/(m_i + m_g)] \tag{8.13}$$

For magnetic sector instruments, the ion kinetic energy is very high,
usually of the order 4–8 keV. At these higher collision energies, some of the
collisional excitation appears as an excess of electronic energy, similar to the
excitation caused by electron ionization. The electronic excitation rapidly
converts to excess of rovibrational energy in the ion, which then fragments;
under these high-energy conditions, the abundances of fragment ions produced are not very sensitive to the actual collision energy, unlike the situation at low energies. However, charge exchange (section 3.3.1) can be a
serious problem for collision gases having relatively low ionization energies
(the primary ions are converted into neutral species). Therefore, although
helium is not the preferred target gas for low-energy collisions, it is for high-
energy ones, mainly because its high ionization energy reduces loss in the
primary ion beam through charge exchange. It also has the advantages of a
high relative cross-section for electronic excitation in the colliding ion
whilst, because of its small mass and size, reducing any loss of ions through
scattering. Depletion of the ion beam through scattering has been counteracted by slowing down the incoming ion beam before the collision cell and
then providing post-collisional acceleration of product ions. Unlike for low-
energy collisions, the features of a mass spectrum of product ions resulting
from high-energy collisions do not change much with the nature of the

target gas. Although *inherent efficiencies* of low- and high-energy collisions appear to be about the same (Bricker, Adams and Russel, 1983; McLuckey, 1984), instrumental factors intervene in such a way that the low-energy process appears to be more efficient than the high-energy one. Thus, nitrophenol can be detected at a level of 10 pg with high-energy collisions in a reversed geometry magnetic sector instrument whilst 120 fg can be detected with a quadrupole instrument utilizing low-energy collisions. For FTICR mass spectrometers and ion traps (section 8.9.2), good reproducibility of product ion spectra after collisional activation seems to be finely dependent on actual gas pressure and the time between introduction of the gas and application of the ion extraction pulse (Scrivens and Rollins, 1992). Observation of so-called charge permutation processes, such as those shown in scheme (8.14), where B is a molecule of the neutral collision gas, can lead to valuable analytical information for elucidating ion structures:

$$M^+ + B \rightarrow M^{2+} + B + e^-$$
$$M^+ + B \rightarrow M^- + B^{2+}$$

(8.14)

The power of collisional activation is such that other techniques, such as mass-analysed ion kinetic energy spectroscopy, are rarely used without it. Collisional activation is highly useful when used in conjunction with a 'soft' ionization technique (chapter 3). Field desorption, negative-ion chemical ionization and electrospray in particular are likely to yield only molecular or quasi-molecular ions. This makes it possible to assign molecular mass but little information concerning the structure of an unknown compound is gained. In such cases, once the molecular or quasi-molecular ions have been produced, fragmentation may be induced by collision. In the one experiment, data with the combined advantages of mild ionization, EI-like fragmentation and metastable ion mass spectra are available. Electron ionization of derivatized peptides yields useful ions indicative of structure but molecular ions are of low abundance or absent. If, instead, peptides are ionized by positive-ion or negative-ion chemical ionization, quasi-molecular ions, $[M + H]^+$ or $[M - H]^-$, are observed and may be induced to fragment by collision. The resulting product ions, recorded by MIKES with a mass spectrometer of reversed geometry or by B/E linked scanning with a conventional instrument, provide the information required for complete structural analysis. The same approach has been used for characterization of saccharides and polynucleotides ionized by field desorption.

$$C_2H_5-CD_2 \quad \overset{CH_2}{\underset{|}{|}} \xrightarrow{-e} \quad \overset{+}{\underset{CH_2-CH_2}{CD_2 \quad CH_2}} \xleftarrow{-e} \quad C_2H_5-CH_2 \quad CD_2$$

$$\overset{|}{CH_2}-\overset{|}{CH_2} \qquad\qquad \overset{|}{CH_2}-\overset{|}{CH_2}$$

4   5

Collisional activation is useful also in studies of ion structures. Collisional activation mass spectra of $C_4H_6D_2{}^{35}Cl^+$ ions from the two differently labelled 1-chlorohexanes (4) and (5) are identical, suggesting that both neutral compounds afford a common ion. The formation of a chloronium ion, often postulated for reactions in solution chemistry, would be consistent with these results. Unfortunately, ions can rearrange during collision with another body so that such studies cannot give unequivocal evidence of ion structure.

Multiply charged ions. Most of the above discussion in section 8.7.1 has implicitly concerned itself with singly charged ions such as protonated molecules. Part of the topic of collisional activation has been described in terms of passing an ion beam in sector or quadrupole instruments through a region containing a neutral gas held at a pressure a little above the general vacuum of the spectrometer. With these arrangements, collisions take place over a very short period of time for each ion as it passes through the neutral gas and there is a spatial separation of ionization and collision processes (figure 8.13). For other instruments such as ion traps (IT) and ion cyclotron resonance (ICR) devices, the ions are constrained within one space for varying lengths of time during which, into this same space, some neutral gas is admitted. Ion/molecule collisions take place as the ions circulate within the space over a period of time; this is a temporal separation of ionization and collision (figure 8.14). In IT and ICR experiments, the ions can be given more energy by accelerating them in their orbits or, for sector and quadrupole machines, the ions in a 'linear' beam are accelerated so as to increase the energy transferred at collision. For some atmospheric pressure ionization sources operating also as inlet systems, this extra collisional energy can be

Figure 8.13. A schematic diagram illustrating the placing of a gas collision cell between two mass spectrometric analysers. The analysers may be of sector or quadrupolar type. Note the *spatial* separation of analyser and collision zones. The times between analysis and collision regimes are normally very small.

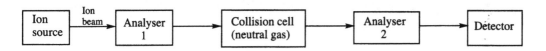

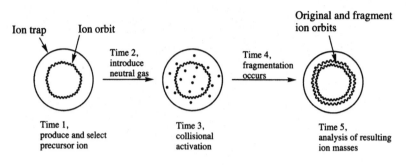

Figure 8.14. A schematic diagram showing how, within one zone of an ion trap, precursor ions can be selected (time 1), caused to collide with molecules of a neutral gas (times 2, 3) and the product ions from collision investigated later (times 4 and 5). The processes are separated *temporally*.

gained as the ions follow a linear path in the ion source but before they are injected into the analyser region (figure 8.15). In this type of system, soon after the sample solution to be examined has been nebulized into a fine spray, many ion/molecule collisions between species having little excess of energy tend to equalize any available excess (region 1, figure 8.15). As there are many more neutral particles than ionized ones, this redistribution of excess of internal energy leaves the ions with very little of the excess. Hence, these sorts of atmospheric pressure ionization sources-cum-inlets are characterized by the production of stable quasi-molecular ions, usually protonated molecules. However, before the stream of ions and gas is passed into the analyser of the mass spectrometer, it normally passes through at least one more region (region 2 in figure 8.15), where the pressure is very much lower and so multiple collisions are much reduced. The mean free path of the ions is increased and so, if they are now accelerated, when collision with a neutral gas molecule does occur the ion acquires extra internal energy, which may be enough for fragmentation to occur. Therefore, these sources can be operated in two modes, the first of which gives only stable quasi-molecular ions and the second of which gives these same ions plus fragment ions. The fragment ions are useful for structure elucidation.

The electrospray ionization/inlet system (section 3.9) is also characterized by the production of multiply charged ions. For small molecules these ions may have only one to four charges but, for large molecules like peptides and proteins, the number of charges on an ion may range up to, say, twenty or thirty. Because these ions are produced in some abundance under mild conditions and can be mass analysed by relatively simple spectrometers, the electrospray method has proved to be excellent for obtaining molecular mass information both for small molecules and for those with masses running into thousands. Again, a major problem has been to extract

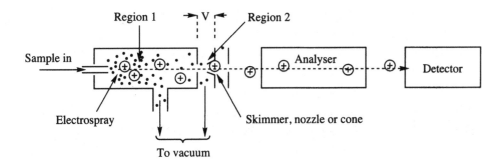

structural information at the same time as measuring relative molecular mass. This problem has been solved by collisionally activating the multiply charged ions as described in the above paragraph. There is a peculiar difficulty in this approach of using collisional activation with multiply charged ions. As was discussed in section 8.3.1 and shown in equation (8.10), fragmentation of multiply charged ions may give product ions with different numbers of charges. In the example given earlier (section 8.3.1) of the decomposition of one doubly charged, $[M - 2H]^{2+}$, ion in the mass spectrum of naphthalene (scheme (8.15)), *two* singly charged ions were formed.

Figure 8.15. A schematic diagram illustrating the operation of an electrospray ionization/inlet system in such a way that, after initial formation (in region 1) of multiply charged ions, these can be accelerated by an increase in the 'cone voltage' so as to collide energetically with neutral gas species present in region 2.

$$\text{Naphthalene (C}_{10}\text{H}_8) \xrightarrow{\text{EI}} \underset{\substack{\text{Doubly charged} \\ \text{fragment ion}}}{C_{10}H_6^{++}} \longrightarrow \underset{\substack{\text{Singly charged} \\ \text{fragment ions}}}{C_7H_3^+ + C_3H_3^+}, \qquad (8.15)$$

There is also evidence that Coulombic repulsion between similar charges enhances fragmentation into less highly charged fragments (section 8.3.1).

For multiply charged large molecules there are possibilities for many differently charged product ions following collisional activation. The point is illustrated simplistically in scheme (8.16*a*) in which the theoretical

$$N + M^{5+} \left\{ \begin{array}{l} f_1^{5+} + f_2^{0} \quad \text{and/or} \quad f_1^{0} + f_2^{5+} \\[4pt] f_1^{4+} + f_2^{1+} \quad \text{and/or} \quad f_1^{1+} + f_2^{4+} \\[4pt] f_1^{3+} + f_2^{2+} \quad \text{and/or} \quad f_1^{2+} + f_2^{3+} \end{array} \right\} \qquad (8.16a)$$

$$N + M^{5+} \longrightarrow N^+ + M^{4+} \left\{ \begin{array}{c} \\ \end{array} \right. \text{fragment ions} \qquad (8.16b)$$

fragmentation of a pentuply charged molecular ion, M^{5+}, is shown. At a minimum, just a single fragmentation mode of this ion could lead to ten differently charged fragments (f_1^+ to f_1^{5+} and f_2^+ to f_2^{5+}) and two neutral species (f_1^0 and f_2^0). Possibly all of these would be detected, making interpretation of the spectrum more difficult and less predictable, especially if it is borne in mind that this discussion has centred on only one bond cleavage. If *multiple* bond cleavages occur after collisional activation then the possible range of multiply charged fragment ions becomes very large, making the mass spectrum difficult to interpret. The possibility of charge exchange introduces yet another possible complication. The pentuply charged molecule may become quadruply charged by charge exchange with a neutral gas molecule (scheme 8.16(*b*)) and fragmentation can occur from this new quadruply charged species, which may exhibit different behaviour from that of any quadruply charged species originally present. Often, charge exchange can be eliminated entirely by choosing a collision gas with a high ionization energy (e.g., helium) but the spray issuing from the electrospray device contains solvent, usually having relatively low ionization energy in the region in which collisional activation occurs. Through use of a suitable drying ('curtain') gas after most of the solvent has been removed it should be possible to limit charge exchange (figure 8.15, region 2). The problem has been addressed partly through use of the high resolution and trapping capabilities of an FTICR instrument to select and examine the behaviour of single large, multicharged ions (Smith, Cheng, Bruce, Hofstadler and Anderson, 1994). The effect of collisional activation on multiply charged negative ions is described in a worked example of mass spectral interpretation later in this book (chapter 11, example F).

8.7.2.   *Surface-induced decomposition*

Collision-induced dissociation becomes increasingly less efficient as ion mass increases and therefore other methods for activating and fragmenting ions are needed. Activation of ions through collision with neutral gas molecules is thought to deposit somewhat less excess of internal energy in the ion than can be achieved by electron ionization at 70 eV. Especially for large molecules, which have many vibrators and rotators with which to spread out any excess of internal energy, these sorts of excess are not sufficient to induce significant decomposition. Accordingly, other means of activation of ions have been sought. Amongst these is surface-induced decomposition

(SID), in which an ion beam is directed at a metal surface (target) at a shallow, glancing angle and the scattered products of collision are examined. By arranging that the difference in potential between the ion source and the target is in the range of about 20–60 V these amounts of energy become available through the glancing collision of the ions with the surface of the target. Thus, MS/MS product ion studies can be carried out by colliding a beam of ions of known m/z value with the target and then scanning the resulting spectrum with a second sector or quadrupole. The total or original mass spectrum can be obtained by scanning the ion beam in the first mass selector and then collecting the separated ions in the second; for this operation, the potential on the target is raised so that the beam of ions is reflected without striking the target. It has been found that all ions in a mass spectrum do not behave similarly on impact; some ions appear to be particularly stable but others fragment more readily (Dekrey, Mabud, Cooks and Syka, 1985; Mabud, Dekrey and Cooks, 1985). For large kinetic energies of ions, energy transfers of up to 8–10 eV are possible and, unlike collisional activation with a gas or with electrons for which the energy distribution is rather wide, the distribution of excess of internal energy transferred on collision appears to be narrow (Dekrey, Mabud, Cooks and Syka, 1985). The sensitivity of MS/MS using this process is low compared with that of CID because of difficulties in collecting the scattered fragments. Collisions of ions having only a few tens of volts of kinetic energy are possible by slowing the ions having large kinetic energies and then deflecting them onto a surface. Both positive and negative ions can be activated but the post-collisional ion abundances are very dependent on the type of surface, any impurities on the surface, the collisional kinetic energy and the electron affinities of the ions involved. Most of the above experiments have been carried out with quadrupole or hybrid MS/MS instruments but similar studies have been effected with four-sector tandem mass spectrometers (Despeyroux, Wright, Jennings, Evans and Riddoch, 1992).

8.7.3. *Photon-induced dissociation*

Ions, like molecules, are capable of absorbing light energy. Just as molecules can be activated by electromagnetic radiation (e.g., light energy in photochemistry), ions can absorb photons to give excited ions having excesses of electronic, vibrational and rotational energies. If sufficient energy is absorbed, the irradiated ions will fragment. The process is photon-induced dissociation (PID) or simply photodissociation. The gas (number) density of the fast moving ions in the ion beam of a typical mass spectrometer is low and the cross-section for photon absorption is low. Therefore, it is necessary to use the intense light from laser beams to provide enough photons to activate a significant proportion of the ions. For absorption of photon energy, it is necessary that the ions should possess absorption bands at or near the wavelength of the laser beam. Because of this requirement, the sensitivities of different kinds of ions to PID vary depending on the nature of the ion. In fact, photon absorption by ions is so specific, compared with CID or SID, that PID can be used to investigate their 'absorption spectra' by examining the types and abundances of fragment ions found following irradiation of the ions over a range of UV/visible wavelengths. This sort of investigation has been carried out on magnetic/electric sector mass spectrometers for some time (Lubman, 1990) but only recently, with the more general availability of good array detectors (section 8.9.1), has the application of PID to analytical problems become possible.

With the low cross-sections of ions towards photon absorption compared with the high cross-sections for collisional processes, techniques that lengthen the time during which ions are exposed to a photon beam will improve their chances of absorbing photons and subsequently fragmenting. Therefore, unlike beam instruments (magnetic sectors or quadrupoles) in which the ion and photon paths coincide only once, ion trapping instruments can hold ions in orbits, which can be made to cross a photon beam continually, thereby increasing the chances of photon absorption. Studies with ion cyclotron resonance and ion trap instruments suggest that PID may be more effective than CID for masses greater than about 3000 daltons.

8.8.   MASS SPECTROMETRY/MASS SPECTROMETRY (TANDEM MS)

The introduction of ion activation processes, leading to induced fragmentation of normal ions, similar to the auto-decomposition of metastable ions, has led to the introduction and growth of the technique of mass spectrometry/mass spectrometry (MS/MS) or tandem MS. A simple layout for MS/MS is shown in figure 8.16, in which MS(1) and MS(2) can be simply

Figure 8.16. A diagrammatic outline of a system for MS/MS experiments. A collision cell is shown between two mass spectrometer sectors or instruments (MS 1 and MS 2).

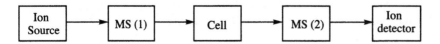

the electric and magnetic sectors of a conventional double-focussing mass spectrometer or they may be actually two connected sector mass spectrometers or they could be two quadrupole mass spectrometers or even a quadrupole and a sector mass spectrometer. Many variants are possible but only a few are in common use.

Consider an idealized mass spectrometric fragmentation of molecular ions ($M^{+\bullet}$) to give successive fragment ions (A^+, B^+, C^+) as shown in scheme (8.17); the ejected neutral particles are not shown. By the techniques

$$M^{+\bullet} \xrightarrow{\ a\ } A^+ \xrightarrow{\ b\ } B^+ \xrightarrow{\ c\ } C^+ \tag{8.17}$$

explained in section 8.1 and 8.2, metastable ions accompanying the processes (a)–(c) of fragmentation of normal ions do occur but the abundances of their ionic products of decomposition are rather low. By adding extra internal energy to normal ions in flight, as in MS/MS, the normal ions are induced to fragment as in the processes (a)–(c) and the same techniques that are used to examine metastable ions can be used to examine the induced decompositions. The major difference between simply examining metastable ions and MS/MS lies in the latter's activation of normal ions so that extra fragmentation is induced with a resultant increase in sensitivity for observation of composite mass spectrometric processes such as that shown in scheme (8.17). Therefore, the following discussion of MS/MS will be related to sections 8.1–8.6 on the examination of metastable ions. The various types of MS/MS instruments will be described as two sector, three sector, and so on. There seems nothing in the definition of the word 'tandem' that implies two or more sectors should be the same and, therefore, the terms MS/MS and tandem MS are used indiscriminately. With more sectors, more MS/MS variations become possible. Thus, two-sector instruments give MS/MS (or MS^2), three-sector ones can give both MS/MS (MS^2) and MS/MS/MS (MS^3) and so on. For FTICR and ion trap instruments, experiments with MS^n ($n = 1$–5) have been carried out in the one 'sector'.

As with metastable ions, in a fragmentation process such as that shown in equation (8.15), in which $M^+ \rightarrow A^+$, then M^+, A^+ may be referred to as parent and daughter ions respectively. Alternatively and more formally, M^+ and A^+ are *precursor* and *product* ions; this designation will be used here. A means of depicting MS/MS reactions has been developed and is illustrated in figure 8.17 for three common types of scan. In the *product scan* mode, a

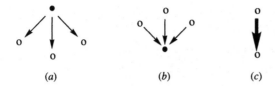

(a)   (b)   (c)

Figure 8.17. A method for depicting types of MS/MS scans. A filled circle (•) represents a fixed (set) mass, an open circle (o) a variable mass. The arrows represent transitions investigated. A heavy arrow is used to describe a neutral loss scan. Scheme (a) implies a product ion scan, (b) a precursor ion scan and (c) a fixed neutral loss scan.

precursor ion is selected, activated in a collision region and then the range of product ions is examined. In the *precursor ion scan*, the same sort of experiment is carried out but this time, a product ion is selected and all the precursor ions leading to it are detected. Precursor ion scanning on a four-sector instrument has been discussed in depth (Scrivens, Rollins, Jennings, Bordoli and Bateman, 1992). In a third popular mode, a *neutral loss scan* is carried out by which the precursor and product ions are separated in mass by a predetermined amount. For example, suppose that, in the reaction scheme of equation (8.17), the differences between ions M^+, A^+ and B^+, C^+ were 15 mass units each. By setting the two MS analysers to detect mass losses of 15, then M^+ and A^+ would be linked as would B^+ and C^+. Examples of these and other scans are given later. Discussions have appeared of ways for depicting the various scans, together with a survey of a range of some of the many possible types of scan (Louris, Wright, Cooks and Schoen, 1985; Creaser, 1988; Scrivens and Rollins, 1992). Examples of the various scans are given in the following sections and three common ones are depicted in figure 8.17.

8.8.1.   *Two-sector MS/MS*

Conventional magnetic/electric sector instruments. An arrangement of instrumentation for two-sector MS/MS has been outlined in figure 8.16. The layout, with gas cells, for a two-sector mass spectrometer is displayed in figure 8.18. It can be seen that collision cells (1–3) can be placed in the field-free regions (1–3) so that normal ions can be induced to fragment in the field-free regions where metastable ions fragment ordinarily. The sectors can be electric (E), magnetic (B) or quadrupolar (Q). In one of the simplest two-sector configurations (termed BE), there is a magnetic sector followed by an electric sector, with just the one collision cell (2) between them. The set-up is entirely comparable with that presented in figure 8.10 and MIKES analyses can be carried out. Alternatively, an EB arrangement is used. For example, a good method for examining the distribution of

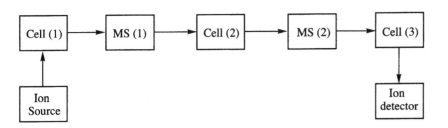

Figure 8.18. The schematic layout of sectors and collision cells for typical MS/MS experiments. Not all of the cells are necessarily used or even in place. Usually cell (2) is present.

oligomers in an organic polymer is by field desorption ionization, which normally reveals molecular ions but gives no details of structure because of the absence of fragmentation. In experiments with polystyrene, molecular ions produced by FD were collisionally activated at 10 keV in the first field-free region of an EB two-sector mass spectrometer, using helium as the target gas. By selecting one of the molecular ions as precursor and carrying out a B/E linked scan (section 8.3), a spectrum of product ions could be obtained (Danis, 1991). It was found that changes in the end-groups of the oligomers caused marked variations in the mass spectra of the product ions and enabled identification of the end-group. Figure 8.19 illustrates the effects for polystyrenes having *t*-butyl and vinyl end-groups. Similar results have been obtained with polyethylene glycols (Craig and Derrick, 1986; Lattimer, Muenster and Budzikiewicz, 1989).

Quadrupolar instruments. When quadrupole mass filters (Q) are used for mass selection and analysis, MS(1) and MS(2) are both of this type. A third quadrupole is used for the ion activation region (cell 2 of figure 8.20(*a*)) and, confusingly, this has been called a three-sector instrument. In fact, the first and third quadrupoles (Q1 and Q3 of figure 8.20(*a*)) are operated as normally for mass selection and analysis, having their usual RF and DC voltages applied. However, quadrupole (2) is used with only its RF voltage and is both a collision region and a means of reducing scatter in the ion beam after collision. Therefore, this 'three-sector' instrument is really a two-sector one from the viewpoint of MS/MS. A typical layout is shown in figure 8.20(*a*) which is entirely analogous to the triple-quadrupole instrument described in section 8.6 and is compared with a two-sector layout in figure 8.20(*b*).

Many MS/MS experiments are easier to carry out with a multiquadrupole instrument compared with magnetic/electric-sector spectrometers. Additional important influences on the promotion of quadrupoles for these

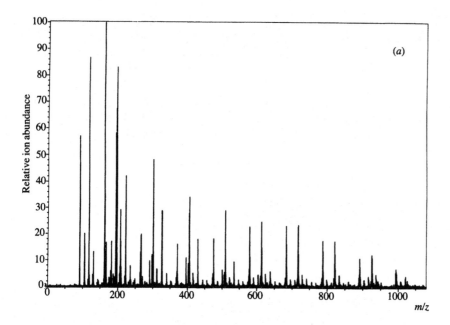

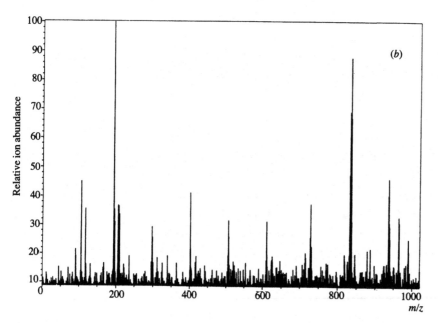

Figure 8.19. The product ion scans arising from polystyrenes having (*a*) *t*-butyl and (*b*) vinyl end groups. The major differences between the two spectra make end group identification easy. (Spectra are reproduced by kind permission of Dr. P.O. Danis, Rohm and Haas Company.)

uses have been improvements to their resolving power and their relatively lower cost. As discussed above, mass resolution in B/E or E/B scans can be quite low, sometimes falling to only hundreds or thousands, making exact identification of precursor or product ions more difficult as mass increases. Quadrupole operation offers the opportunity for at least unit mass separation for these ions.

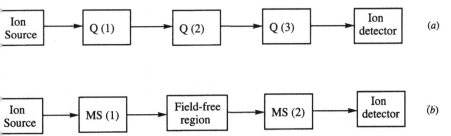

Figure 8.20. A simple comparison of layouts for (a) quadrupole and (b) sector instruments, showing their essential similarity.

The popular scan modes are very easy to carry out with the 'triple quad'. For a product ion scan, Q(1) is set to transmit one selected ion mass and the resulting single-mass ion beam is collisionally activated in Q(2); all the product ions arising in Q(2) are detected by scanning Q(3). There is no need for complicated B/E, B^2/E or $(B/E)(1 - E)^{1/2}$ linked scans. A precursor ion scan is just as easy. Quadrupole Q(3) is set to pass a selected product ion mass and Q(1) is scanned; when any precursor ion is activated by collision in Q(2), it fragments and the product ions are detected by Q(3). Finally, a constant neutral loss scan is carried out simply by off setting Q(1) and Q(3) by the number of mass units required and scanning each one.

Anthracyclines, of which doxorubicin (6) and epirubicin (7) compose a pair epimeric in the sugar residue, give relatively abundant $[M + H]^+$ ions by FAB; these fragment ions correspond to the aglycone and sugar portions and yield complete characterization of structure. However, distinguishing between epimers is not easy by simple FAB-MS but, by carrying out FAB-MS/MS in a triple quadrupole instrument at low collision energies, the two epimers could be distinguished easily. Protonated molecules for each of the epimers (6, 7) were mass-selected in Q(1), collisionally activated in Q(2) and the spread of fragment ions examined in Q(3). Comparison of the resulting

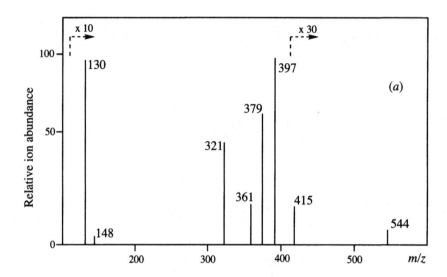

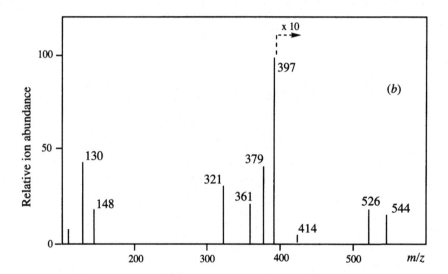

Figure 8.21. Product ion scans from $[M + H]^+$ ions of (*a*) doxorubicin and (*b*) epirubicin. The two epimers are clearly differentiated by relative abundances of various of the ions as, for example, at m/z 526 and 415. (Spectra are reproduced by kind permission of Dr. B. Gioia, Farmitalia Carlo Erba.)

mass spectra (shown in figure 8.21) obtained from Q(3) showed significant variations in both abundances and the presence of some m/z values for the individual epimers and these differences allowed the positive identification of the epimers (Gioia, Franzoi and Arlandini, 1991).

In a striking demonstration of the ability of MS/MS to deal directly with complex mixtures, without prior separation of components, a triple-quadrupole MS/MS instrument has been automated to examine phospho-lipids (**8**; figure 8.22) from a variety of biological sources. By carrying out

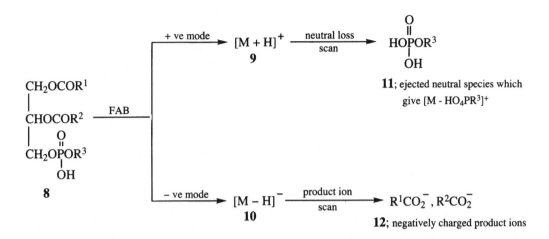

neutral loss scans with Q(1) and Q(3) on protonated molecules (9), the polar head group (11) could be identified for each component Product ion scans on negatively charged deprotonated molecules (10) characterized the fatty acyl components (12). The two lots of scans provided a complete phospholipid profile of the sample in a few minutes at picomole levels. Such profiles can be used for examining microbial contaminants of food and for identifying infections (Cole and Enke, 1991).

Ion kinetic energy in a quadrupole is much less than that of magnetic/electric-sector instruments, being of the order of a few to a few tens of volts compared with thousands of volts in the latter. Whilst this is instrumentally advantageous it does necessitate careful control of gas pressure in Q(2) for reproducible results. Ion transmission is not as high as for the E/B or B/E instruments, especially at high mass.

Figure 8.22. A schematic view of the identification of major components of a mixture of phospholipids (8). In (a), a neutral loss scan was used with (positive) protonated molecules (9) obtained by FAB in order to identify the range of phosphate end groups (11). In (b), negative ions (10), again from FAB, were used to determine the compositions of the fatty acid side chains (12).

8.8.2. *Three-sector instruments*

For reasons already given, the triple quadrupole (QQQ) has been described together with the two-sector instruments and will not be discussed in this section reserved for those instruments having a third sector to aid mass discrimination rather than ion activation.

Magnetic/electric sector instruments. As was discussed for metastable ions (section 8.3) and for two-sector MS/MS (section 8.8.1), the various linked scans (B/E, B^2/E, etc.) with EB or BE instruments give either poor product ion resolution or poor precursor ion resolution, making high-mass MS/MS or accurate mass measurement very difficult if not impossible. The situation

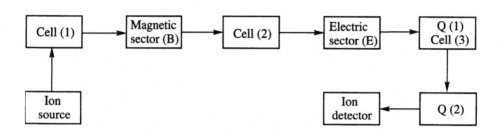

is made worse by scatter in the ion beam caused by collisional activation. To recover the high resolution inherent in BE or EB instruments, which is lost when operating in linked scan and MS/MS modes, an extra focussing sector can be added at relatively little extra cost, as described in section 8.6, to give an EBE instrument. The introduction of another field-free region accompanying the new sector introduces the possibility of even more varied MS/MS experiments. For example, it is possible to carry out a product ion scan on a mass-selected precursor ion and then carry out (sequential) product ion scans or neutral loss scans on the first product ions. These spectrometers provide a useful transition stage between the common two-sector and the much more expensive four-sector instruments. Three-sector instruments can operate as MS/MS (MS^2) combinations or as MS/MS/MS (MS^3).

Hybrid instruments. The advantages of B and E sectors can be combined with those of quadrupoles to give BEQ or EBQ three-sector combinations. As with the triple quadrupole which is really a two-sector MS/MS instrument with an intermediate quadrupole acting as a collision cell, the BEQ and EBQ arrangements usually contain a second quadrupole as a collision cell placed between the B and E sectors and final quadrupole mass analysers. These instruments are described as EBQQ or BEQQ but are, in fact, only three-sector from the viewpoint of mass discrimination and MS/MS studies. A typical layout of a BEQQ instrument is displayed in figure 8.23. As there are three mass or energy discrimination sectors, BEQQ hybrids can be used for either MS^2 or MS^3 experiments, using two of the collision cells. Generally, cells 2 and 3 are used as these afford good precursor and sequential product ion scans. There are two collision regimes. In cell 2, ions are moving with quantities of kinetic energy of a few kilo-electron volts whilst, for cell 3, they must be slowed to a few tens of volts of energy. An example of the excellent sensitivity and specificity that can be attained in trace analysis using MS/MS methods is afforded by the detection of the highly toxic and

Figure 8.23. One form of hybrid instrument in which a quadrupole (Q2) is coupled to a conventional magnetic/electric-sector instrument (MS(B) and MS(E)). Another quadrupole (Q1) is used as a collision cell.

$$ClCH_2OCH_2Cl^{+\bullet} \xrightarrow{-Cl^\bullet} ClCH_2OCH_2^+ \xrightarrow{-CH_2O} ClCH_2^+ \qquad (8.18)$$
$$\textbf{13; } m/z\ 114 \qquad\qquad m/z\ 79 \qquad\qquad m/z\ 49$$

carcinogenic compound, bis(chloromethyl)ether (BCME; **13**; scheme (8.18)). In its normal EI mass spectrum, BCME shows abundant fragment ions at m/z 49, arising by successive losses of Cl and CH_2O from the molecular ion. A safe working threshold for BCME in the atmosphere has been set at 1 ppb but, at this level, GC or single ion monitoring GC/MS have insufficient sensitivity and selectivity from other atmospheric impurities to be definitive. However, using a GC/BEQQ instrument with *single reaction monitoring* (SRM), BCME could be detected specifically even at 4 ppt. In this experiment, the effluent from GC was ionized but the first stages (BE) were set to transmit only ions at m/z 79. To differentiate these ions arising from BCME with any ions at m/z 79 from other atmospheric components, the ion beam was collisionally activated and the last quadrupole was set to accept only ions at m/z 49. Therefore, from all the ions formed from the GC effluent, only those taking part in the reaction, m/z 79 → m/z 49, were selected; all other fragmentation processes were rejected and hence the name, single *reaction* monitoring. As can be seen in figure 8.24, even at 4 ppt, the peak corresponding to the process, m/z 79 → m/z 49, stands out clearly from the background and provides the required degree of analytical specificity and sensitivity for detection of BCME (Blease, Scrivens and Morden, 1989). No peak corresponding to BCME could be observed by simple GC/MS because of the obscuring effect of larger, overlapping peaks arising from other pollutants.

Other types of hybrids have been described including quadrupole/magnetic/electric sectors (Glish, McLuckey, McBay and Bertram, 1986), quadrupole/time-of-flight (Glish and Goeringer, 1984; Glish, McLuckey and McKown, 1987) and quadrupole/Fourier-transform ion cyclotron resonance (McIver, Hunter and Bowers, 1985).

8.8.3. *Four-sector instruments*

By placing two magnetic/electric-sector spectrometers back-to-back, a four-sector tandem MS instrument can be assembled. Indeed, the two instruments are not infrequently arranged so that they can be used separately for MS or together for MS/MS studies. Generally EBEB or BEBE instruments have been used (figure 8.25) with five field-free regions so that

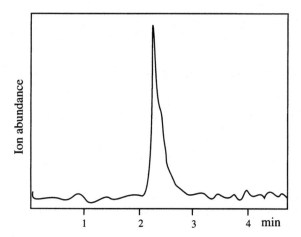

Ion abundance

1   2   3   4   min

Figure 8.24. The peak for *bis*(dichloromethyl)ether (13) emerging from a GC column at 2 min and subjected to single reaction monitoring, whereby the process m/z 79 → m/z 49 was monitored by MS/MS. This fragmentation is highly diagnostic for the ether and all the other components emerging from the GC column gave no response. In the ordinary gas chromatogram, the concentration of the ether was so low that it could not even be detected amongst ovelapping peaks arising from other substances eluting at or near the same time. (Adapted and produced in part by kind permission from Scrivens and Rollins (1992).)

experiments up to MS[5] type can be carried out. Because of dissipation of the ion beam through scattering in each collision region, the current sensitivity of MS[5] experiments is too low for anything but pure compounds (Tomer, Guenat and Deterding, 1988). When used for MS[2] experiments, then the precursor ion beam is usually selected by the first two sectors, with ion activation in collision cell 3; the final two sectors are used to record the product ion spectrum (figure 8.25).

An example of the fine tuning available on a multi-sector MS/MS instrument is provided by the collision-induced decomposition of renin substrate, a tetradecapeptide (**14**). The protonated molecule from the peptide was obtained by liquid phase secondary-ion mass spectrometry (LSIMS), using 25 keV Cs[+] ions, and was selected and focussed in the electric and magnetic sectors of the first mass spectrometer. After CID, the resulting spectrum was acquired on a second mass spectrometer with reversed geometry (EB). When helium was used as the collision gas (4.5 eV collision energy; laboratory frame), the resulting spectrum of the tetradecapeptide showed prominent ions corresponding to cleavage only along the peptide chain (figure 8.26(*a*)). However, with air as the collision gas (32 eV collision energy; laboratory frame), more extensive fragmentation involving the peptidic side chains was observed (figure 8.26(*b*)). The two sets of data could be used to provide amino acid sequence information and to identify types of amino acids present in the chain (Bordoli and Bateman, 1992).

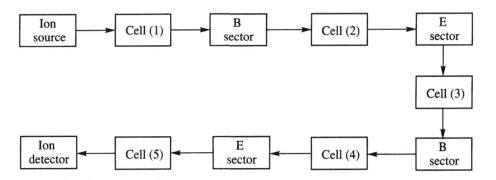

Figure 8.25. A set-up of a four-sector BEBE instrument showing how an MS[5] experiment could be carried out utilizing the five field-free regions represented by cells (1–5).

8.9. SUMMARY AND MISCELLANY OF MS/MS EXPERIMENTS

All of the methods used for examining metastable ions, including linked scanning, can be used to develop MS/MS experiments in which selected precursor ions are activated so that they fragment to give product ions. This approach constitutes simple MS/MS (or MS[2]). If more than two ion discrimination (mass or energy) sectors are used then multiple MS/MS type experiments can be carried out (MS[n]) and there may be big improvements in mass resolution in comparison with that for simpler two-sector instruments. The mass or energy discrimination sectors can be magnetic (B), electric (E) or quadrupole (Q) and many types of combination have been used or can be imagined, as with BE, EB, BEQ, QQQ, BEBE and so on. A mix of magnetic/electric sectors with quadrupoles is referred to as a hybrid (and see other types below). Mass resolution can be a problem with simpler MS/MS instruments and, in general, can be improved as the number of sectors increases.

Whilst MS/MS can be used on its own for examination of complex mixtures, it is frequently used along with GC or LC to improve specificity and sensitivity for identification of mixture components. Thus, GC/EI/MS/MS or LC/electrospray/MS/MS are commonly used to examine environmental problems, biochemical mixtures, drug contamination and so on. Another advantage of MS/MS lies in its use with mild ionization methods (CI, FI, FAB, electrospray, API, etc.), which were developed to allow ionization of thermally unstable molecules like peptides and proteins. Whilst successful for this purpose, the molecular or quasi-molecular ions produced also contain little or no excess of internal energy above their ground states and do not fragment. The appearance of abundant molecular ions procures excellent relative molecular mass data

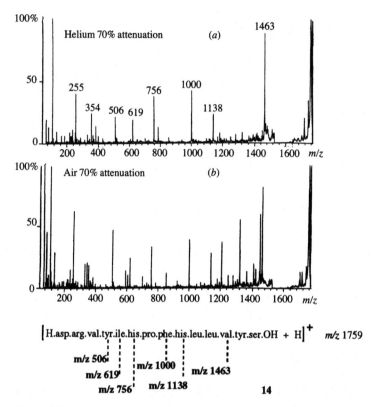

Figure 8.26. Product ion spectra from the protonated molecule of a tetradecapeptide (**14**) using (*a*) helium and (*b*) air as collision gas for activation. In (*a*), the fragments arise from the amide bonds between each amino acid and the next, and give the sequence of amino acids. In (*b*), the extra energy imparted by collision with air molecules causes further fragmentation of the amino acid side chains, thereby helping to identify them. (Spectra are reproduced by kind permission from Bordoli and Bateman (1992).)

but gives little or no information concerning molecular structure. By intentionally activating these molecular ions to cause them to fragment in MS/MS experiments, structural information is recovered. Furthermore, the ability to link precursor and product ions specifically means that the fragmentation paths for the molecular ions can be elucidated to give more specific structural information. Strategies for the examination of a wide range of biologically interesting substances by tandem MS/MS have been discussed in detail (Gross, 1994) and a brief overview of soft ionization and tandem MS methods for natural product research has appeared (Baldwin, 1995).

8.9.1. *Array detectors*

In mass spectrometry the main methods for ion detection are the Faraday cup (or cage), the electron multiplier, the photographic plate and the channeltron electron multiplier array. The first two techniques require the consecutive focussing of ions at a 'point' detector. Thus, whilst arrivals of ions at any one particular *m/z* value are being recorded, all the others that were generated in the ion source are being discharged and pumped away in some other region of the mass spectrometer. For any unit time, this waste means that the cup or multiplier does not provide the sensitivity of which the mass spectrometer could be capable. The ideal solution would be to collect all ions simultaneously at the detector, in which case, all ions produced in the source are, in theory, capable of arriving at the detector and the inherent sensitivity of the mass spectrometer is maintained. An early system for detection of ions used a photographic plate; all separated ions struck the plate simultaneously, producing line images from which mass measurement could be made (section 2.5). This method of detection is not popular because development and measurement of a photographic film means that results are not available 'instantaneously'. The loss of inherent sensitivity with point detectors has been accepted and partly counteracted by large gains in performance of commercially available electron multipliers. The advent of MS/MS investigations, particularly MS^3, MS^4 and MS^5 experiments in which many of the ions originally produced in the source are scattered and lost, has meant that further moves towards regaining the inherent spectrometer sensitivity have been sought. By arranging to place many small (8–25 μm diameter) electron multipliers (channeltrons) side-by-side, an array is formed, which is capable of simultaneous detection of ions as with the photo plate, together with having the instantaneous nature of the recording of results of a single electron multiplier. Discussion of the relative merits and uses of the various detectors has appeared (Boerboom, 1991) and a description of some of their uses in MS/MS has appeared (Boerboom, 1990; Scrivens and Rollins, 1992).

Array detectors give improvements in sensitivity of up to 100 times and improve the overall sensitivity of MS^n experiments. There is a limit set on mass resolution of about 500–5000, depending on the mass range being examined, due to the finite dimensions of each channeltron.

8.9.2. *Ion cyclotron resonance spectrometers and ion traps for MS/MS*

Most of this chapter has been concerned with MS/MS experiments carried out with 'conventional' magnetic or quadrupolar spectrometers, partly because this has been the main area of development and partly because the experiments are easier to describe and understand. Parts of this book are concerned with ion cyclotron resonance (ICR) and ion trap (IT) spectrometers and these require a somewhat different treatment of the *means* of acquiring MS^n data when compared with magnetic or quadrupolar spectrometers, but not of the interpretation of the spectra. Practical considerations with ICR and IT mass spectrometry mean that MS^n experiments tend to be somewhat time-consuming and limited to mass resolutions of less than 1000 but advances are to be expected (Johnson, Yost, Kelley and Bradford, 1990).

ICR instruments. Greater mass resolution in ICR cells can be attained with Fourier-transform (FT) techniques, giving rise to FTICR mass spectrometry (section 12.3). In ICR, ions are injected into a 'cell', where they move in 'circular' orbits under the influence of crossed electric and magnetic fields (see sections 2.4.5 and 12.3 for details). In the FTICR mode, a pulse of broad-band electromagnetic radiation (a range of RF frequencies) increases the velocities of all ions; measurement of the induced currents in the cell and transformation of these data from a time to a frequency domain provide m/z values. The FT mode gives much greater mass resolution than does the ordinary time domain ICR experiment. In an ICR cell, all or only selected ions can be ejected from the cell (lost to the walls) by supplying enough RF power of the right frequency. Unless ejected, the ions continue to circulate in the cell for milliseconds or longer ('trapped'), compared with the few microseconds flight-time in conventional beam mass spectrometers.

For a simple MS/MS precursor/product ion experiment, ions of a range of m/z values are injected into the ICR cell and then all ions other than those selected (e.g. molecular ions) are ejected from the cell by one pulse of RF irradiation. Target gas is allowed into the cell so that collisional activation and fragmentation occur and a final pulse of RF irradiation ejects all the product ions from the cell. Because ions are produced, collisionally activated and collected in the same space of an ion trap or FTICR cell the pressure in the cell is usually too high for good resolution of precursor and product ion masses. This problem has been circumvented in diffrent ways. In an unusual tandem combination of a quadrupole and an FTICR instrument, bromobenzene was introduced into the quadrupole and ionized to give bromobenzene molecular ions, which were injected into an FTICR cell held at low pressure. Collisional activation of the molecular ions gave m/z 78 product ions at a resolution of about 140 000, adequate for accurate mass determination of isobars (McIver, Hunter and Bowers, 1985; Wise, 1987). In a second method, two FTICR cells are used, separated from each other by a plate with a small hole in it and separately pumped. One cell is used for ionization and collisional activation of these precursor ions to give product ions. The pressure in this first cell may be relatively high, thereby affecting attainable resolution quite badly. However, if the product ions are passed through to the adjacent low-pressure cell, the m/z values for the product ions can be measured at high resolution, 200 000 at m/z 105 in this example (Wise, 1987). The measurements at low pressure and high resolution mean that MS^n ($n > 2$) experiments can be carried out and not just MS^2. The small size and design of an ICR cell make it particularly suitable for use with photon-induced dissociation in MS/MS studies, by which background gas pressure can be kept low because only a small hole large enough to admit a laser beam is needed in the side of the cell. In a typical MS/MS/MS experiment, $[M + H]^+$ ions from the tetrapeptide, gly.phe.ala.ala (OMe) were introduced into the analyser region of an FTICR cell; all other ions were ejected from the cell (scheme (8.19)). A first pulse of ultraviolet laser photon energy at 193 nm

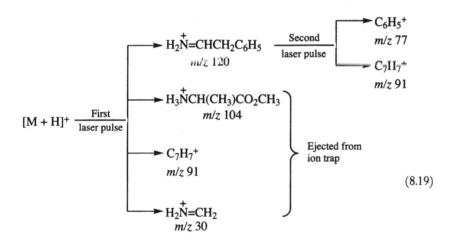

$$(8.19)$$

$$M = H_2NCH_2CONHCH(CH_2C_6H_5)CONHCH(CH_3)CONHCH(CH_3)CO_2CH_3$$

caused the $[M + H]^+$ ions to fragment to give ions at m/z 30, 91, 104 and 120. Again, all ions except m/z 120 were ejected from the analyser and then a second pulse of laser photon energy was applied. This second pulse caused the ions at m/z 120 to fragment to give ions at m/z 77 and 91. The added photon energy (6.42 eV) was quite large compared with collision-induced dissociation energies and the MS/MS experiment was highly efficient since the ions under investigation were trapped in the analyser until required at each stage (Bowers, Delbert and McIver, 1986). The high efficiency for

MS/MS of ion trap mass spectrometers (tandem in time) compared with a triple quadrupole (tandem in space) has been demonstrated for phosphonate esters (Johnson, Yost, Kelley and Bradford, 1990).

Ion traps. Although ICR instruments trap ions for extended periods compared with conventional sector spectrometers, the term ion trap has come to be used almost entirely for types of spectrometers that can be thought of as 'three-dimensional' quadrupoles. The electric fields in a standard quadrupole assembly constrain ions into stable two-dimensional motion between the poles (rods) if the required voltages and frequencies are correct for any particular m/z value (section 2.4.2); an ion travels in the third dimension along the central axis of the rods only by virtue of the small amount of kinetic energy with which it is injected from the ion source into the pole assembly. By having an arrangement of two end-cap electrodes, one on either side of a third annular electrode and all of hyperbolic cross-section, ions injected into the space between the three electrodes can be constrained (trapped) into complex circular motion by application of suitable DC and RF electric fields. To obtain a mass spectrum from the trapped ions, the amplitude of the RF field must be increased so that ions are ejected through the end-caps in proportion to their m/z value and the increase in amplitude of motion. Initial ionization of substances can be by any of the usual means. In an MS/MS experiment, the sample is ionized, all of the ions are trapped and then all of the ions except a selected precursor are ejected. A 'bath' gas, usually helium, is used to damp the motion of ions in the trap and, after a short time, depending on the pressure in the trap, the selected precursor ions will have collided with gas molecules and will have been activated enough to fragment. Usually, the selected precursor ions are given more kinetic energy ('tickled') through use of an extra RF voltage applied to the end-caps so that total collision energy can be increased and a spectrum of product ions can be obtained. This tickle voltage needs to be carefully controlled to match the secular motion of the ions. This fine tuning can be avoided by adjusting RF and DC voltages in the trap so that any trapped ions are moved to the boundary of the stability diagram (section 2.4), *viz.,* the ions are on the point of being ejected from the trap and are moving with much more kinetic energy. Collisions with neutral gas molecules then give fragmentation. The resulting 'boundary activated dissociation' (BAD) spectra are like collisionally activated dissociation spectra but are easier to achieve (Paradisi, Todd,

Traldi and Vettori, 1992). As with FTICR intruments, ions held in an ion trap may be induced to dissociate through activation by absorption of laser light energy. The efficiency of this process varies with the frequency of the irradiating laser light and with the light absorption characteristics of the ions (section 9.6.2). Generally, photodissociation does not appear to impart as much excess of internal energy as does collisional activation but there are exceptional cases (Creaser, McCoustra and O'Neill, 1991).

As an example of ITMS, figure 8.27 shows mass spectra of vanadyl etioporphyrin III (**15**). At small amplitudes of the RF voltage, only protonated molecular ions are formed after injection of a solution of the porphyrin (**15**) through an electrospray inlet (figure 8.27(*a*)). By increasing the RF

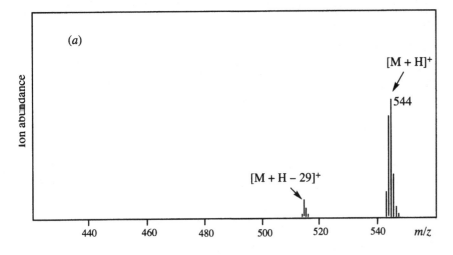

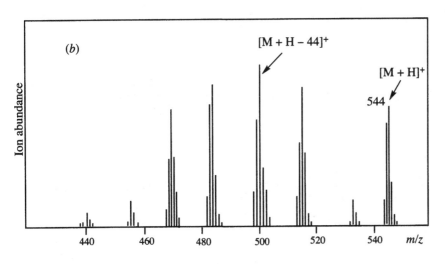

Figure 8.27 The protonated molecule of a porphyrin (**15**) exhibits very little fragmentation (mainly the loss of 29 mass units; ethyl radical) in spectrum (*a*). After collisional activation, some of the protonated molecules fragment to give much more structural information through formation of fragment ions by successive ejections of ethyl and methyl radicals, as with the loss of 44 mass units from the protonated molecule (*b*). (Spectra are adapted and produced in part with kind permission from Van Berkel, McLuckey and Glish (1991). Copyright (1991), American Chemical Society.)

15

amplitude, these protonated molecules are given extra kinetic energy such that, when they collide with molecules of the helium bath gas, induced dissociation occurs to give fragment ions (figure 8.27(*b*)). These MS/MS experiments were used to correlate the mass spectral fragmentation of a range of unmetallated and metallated porphyrins (Van Berkel, McLuckey and Glish, 1991).

Ion traps are amenable to MS^n experiments because collection efficiencies for multiple collisions are high so that initial ion abundances are not drastically attenuated at each MS stage. For example, 90 per cent fragmentation efficiency at each stage of an MS^4 experiment means that about 73 per cent of initial ions are finally detected; this may be compared with a 50 per cent fragmentation efficiency at each step in a typical sector experiment, leading to a final 13 per cent detection rate for equivalent experiments.

9 Theory of mass spectrometry

9.1. INTRODUCTION

A thorough understanding of mass spectrometric processes is severely hampered by poor knowledge of the structures and electronic states of ions. The fragmentation of ions depends on their structures and the excess of internal energy that they contain and this last is governed mostly by three factors: the thermal energy in the molecule immediately prior to ionization, the energy gained during ionization and the subsequent environment of the ion, which determines whether or not it collides with molecules. These energies are depicted schematically in figure 9.1, from which it can be seen that ions formed by removal of an electron ($M^{-\bullet}$ in figure 9.1(a)), addition of a proton ($[M + H]^+$ in figure 9.1(b)) or addition of an electron ($M^{-\bullet}$ in figure 9.1(c)) require different energy inputs to achieve minimum energy levels of ionization. Such ions have no excess of internal energy and will be stable towards fragmentation. However, ions can gain extra internal energy in several ways, *some* of which are shown in figure 9.1(d), and this extra energy (excess of internal energy) may well be sufficient to induce decomposition not only in the first-formed ions (primary fragmentation) but also in those arising from this initial fragmentation (secondary fragmentation). The very act of heating a sample to vaporize it into the ion source adds extra thermal energy (E_v) to the molecules even before the ionization step; additionally, some sources such as those for EI are usually hot, so that collision of the sample molecules with the walls of the source before ionization also adds further excess of thermal energy. For ionization methods that require initial vaporization of the sample, then, even before the ionization step, the molecules are thermally excited; ionization quires a further amount of energy (E_i) and produces an excited ion, $\overset{*}{M}{}^{+\bullet}$, having an excess of internal energy. The excess may be increased even further during the ionization step itself, depositing a further E_e units of excess of internal energy. Thus, compared

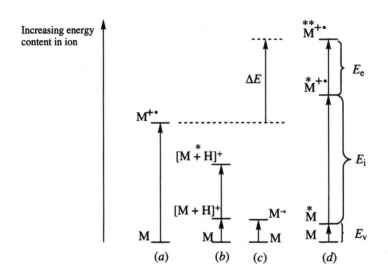

Figure 9.1. A diagrammatic representation of how excess of internal energy can arise in ions. In (a), molecules (M) are ionized by electrons with the minimum amount of energy need to remove an electron and give the ground-state cation-radical ($M^{+\bullet}$). Similarly, addition of a proton (b) or an electron (c) yields ions $[M + H]^+$ and $M^{-\bullet}$ respectively without excess of internal energy. Collisional or other activation is necessary to give, for example, the activated ion, $[M \overset{*}{+} H]^+$. In (d), the molecule (M) must first be heated to vaporize it (energy E_v), giving a vibrationally excited molecule, $\overset{\bullet}{M}$, and then it is ionized to a vibrationally excited state, $\overset{*}{M}{}^{+\bullet}$, with ionization energy, E_i, or if more energy, E_e, is deposited, to a much more excited state, $\overset{**}{M}{}^{+\bullet}$. Compared with ($a$), the ions produced in (d) have an excess of internal energy, ΔE.

with the simple ground-state process, $M \rightarrow M^{+\bullet}$, after ionization an excited-state ion can result, which has a considerable excess of energy (ΔE, figure 9.1). For the protonated molecule ($[M + H]^+$ in figure 9.1), collisional or other activation after its initial formation may produce an excess of internal energy in it. The key to producing either a 'stable' ion species or one that fragments depends on controlling the excess of internal energy. This is done in several ways, which are discussed below and in other parts of this book. For example, stable protonated molecules, which are excellent for giving relative molecular mass information, but give little structural information, can be induced to fragment through energetic collision with other molecules or through absorption of photons; the induced decomposition gives the required structural data. As discussed in chapters 1 and 8, some ions have intermediate excesses of internal energy such that decomposition does take place on the time scale of the mass spectrometric analysis but in regions lying between the ion source and the detector; these are the metastable ions.

Despite inadequate knowledge of ion structures, many generalizations may be made, which are useful for understanding mass spectrometric fragmentations of compounds, particularly those resulting from electron ionization for which most fundamental research has been done.

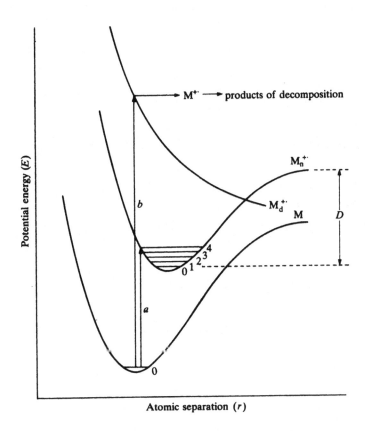

Figure 9.2. Potential energy (Morse) curves for a diatomic molecule (M) ionized to either a non-dissociative ($M_n^{+\cdot}$) or dissociative state ($M_d^{+\cdot}$) of the ground-state ion by vertical transitions a and b. D is the dissociation energy for ions in the non-dissociative state.

9.2. ENERGY STATES RESULTING FROM ELECTRON IONIZATION AT LOW GAS PRESSURES

Consider a diatomic molecule (M) ionized by an electron to yield a molecular ion $M^{+\cdot}$. The vibrational energy of both the molecule and the molecular ion may be represented in the usual way by potential energy (Morse) curves as shown in figure 9.2. For the moment, it is only necessary to consider the zeroth vibrational energy level for M because, even at 500 K, the higher vibrational levels are sparsely populated. However, it is necessary to consider the vibrational energy levels, $v = 0,1,2,...$ in the ion $M^{+\cdot}$. Since ionization is much faster than bond vibration (10^{-15} as against 10^{-12} s) then, by the Franck–Condon principle, there will be a vertical transition between states of M and $M^{+\cdot}$. The minima of the two potential curves for M and $M^{+\cdot}$ are unlikely to coincide so that a vertical transition during ionization will leave $M^{+\cdot}$ in a vibrational state above its zeroth (vibrational ground state) level. For example, transition a in figure 9.2 produces the ion in the vibrational state $\nu = 4$ and therefore the ion has an excess of internal rovibrational energy simply because of the change of state. A vertical transition may occur to a vibrational level in $M^{+\bullet}$ above its dissociation limit and, in this case, the ion would contain energy in excess of the dissociation energy (D). In this case, $M^{+\bullet}$ would fragment as soon as it began to vibrate. It is also possible for the ground state of the ion $M^{+\bullet}$ to be completely dissociative (transition b, figure 9.2) so that decomposition of the ion follows immediately after its formation. This is the reason why molecular ions, $M^{+\bullet}$, of molecules such as trichloro- and tetrachloromethane are never observed; immediately after formation, the molecular radical-cation begins to vibrate from a completely dissociative state and simply falls apart to give a halogen atom and a carbocation:

$$CCl_4 + e^- \xrightarrow{\text{Ionization}} CCl_4^{+\bullet} + 2e^- \xrightarrow[\text{fragmentation}]{\text{Immediate}} CCl_3^+ + Cl^\bullet + 2e^-$$

Following electron ionization then, from a number of molecules M (a sample present in the ion source), ions $M^{+\bullet}$ are produced having different amounts of internal energy proportional to the transition probabilities between M and $M^{+\bullet}$, the energy of the ionizing electron and the closeness of approach of the electron to a molecule. Also, some 'hot' molecules (M) are ionized from vibrational levels above their zeroth levels and the ions formed will again contain different amounts of excess of internal energy (figure 9.1). It is important to remember that this distribution of energies is not a 'Maxwell–Boltzmann' distribution of thermal energies because the latter applies to collisionally equilibrated species and, in electron ionization carried out under high-vacuum conditions, the ions are not so equilibrated because ion/ion and ion/molecule collisions are rare. Once formed, each ion behaves as an isolated entity containing its own specific amount of excess of energy, which, if in excess of the ground state energy, may be sufficient to cause fragmentation. *It is the possibility of forming more stable species by using up the excess of internal energy that leads to decomposition by bond dissociation and/or to reorganization (rearrangemnt) followed by bond cleavage.* This explains why commonly observed fragmentations of ions lead to ejection of such neutral species as CO_2 and H_2O having large negative heats of formation.

In complex molecules, the simple two-dimensional potential energy curves discussed above must be replaced by multidimensional potential energy *surfaces*, but similar principles to those used for the discussion of

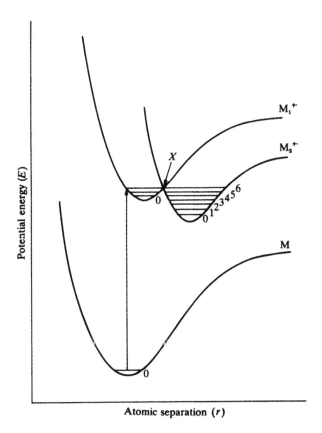

Figure 9.3. An ionizing transition from molecule, M, to ion state, $M_1^{+\cdot}$, followed by radiationless transition at X to ion state, $M_2^{+\cdot}$.

diatomic molecules may be applied. For a species that has various vibrational modes, each mode will be represented by a potential energy surface. If the surfaces cross, similar to the crossing of two potential energy curves shown in figure 9.3, a radiationless transfer of energy to a lower energy state can occur. Consider two vibrational states, $M_1^{+\cdot}$ and $M_2^{+\cdot}$, of the ion $M^{+\cdot}$ in figure 9.3, which cross at some point X. Ionization by vertical transition from the $v = 0$ level of the molecule to the $v = 2$ level of the ion state $M_1^{+\cdot}$ can occur, but the two states, $M_1^{+\cdot}$ and $M_2^{+\cdot}$, coincide at X and crossing from state $M_1^{+\cdot}$ to $M_2^{+\cdot}$ may take place. In state $M_2^{+\cdot}$, the internal energy transferred from $M_1^{+\cdot}$ may be above the dissociation energy so that the ion fragments. In complex molecules there are many such energy surfaces by which internal transfer of energy may occur and so, after ionization, equilibration of any excess of internal energy in the ion can take place rapidly by these surface-crossing processes. The presence of closely spaced rotational energy levels associated with the vibrational levels means that, in polyatomic molecules, surface crossing becomes easy and some vibrational

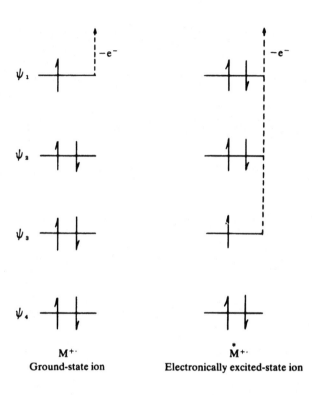

Figure 9.4. Orbital energy levels showing ionization to the ground-state molecular ion, $M^{+\cdot}$, and an electronically excited-state ion, $\overset{\cdot}{M}{}^{+\cdot}$.

M$^{+\cdot}$
Ground-state ion

$\overset{\cdot}{M}{}^{+\cdot}$
Electronically excited-state ion

energy is converted into rotational energy and vice versa. However, despite such energy transfer and equilibration into all the oscillators in the ion, there may be insufficient excess of energy to cause fragmentation and this results in a stable ion. If a relatively small amount of excess of internal energy is equilibrated throughout a large molecular ion by internal state crossing then, on a statistical basis, each of the oscillators in the ion may well have little or no excess of vibrational energy and bond breaking will not take place or will be very slow, awaiting statistical accumulation of sufficient energy in one bond so as to cause it to break. It is worth stressing here again that this discussion is concerned entirely with high-vacuum conditions such that ions, once formed, continue to exist and behave as *isolated* species, *viz.*, they undergo no collisions with other species and they are not subjected to other methods of adding or removing any excess of internal energy. Ions that do collide with other species after formation, as in chemical ionization, are discussed later. Various complications to this basic approach to ionization and fragmentation of isolated species must be considered.

The only ionization discussed so far corresponds to removal of the most loosely bound electron to give a vibrationally excited ion in its

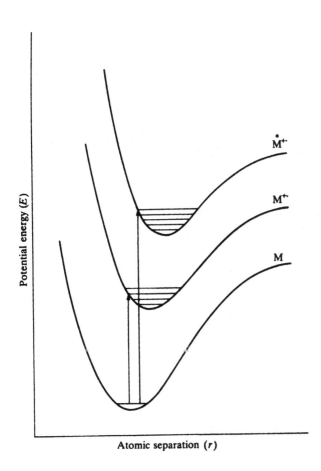

Figure 9.5. Ionizing transitions from the molecule M to the ground-state ion $M^{+\bullet}$ and to an excited-state ion, $\dot{M}^{+\bullet}$, corresponding to the electron configurations shown in figure 9.4.

electronic ground state, corresponding to the first ionization energy. At ionizing energies greater than this minimum, it is possible to remove electrons from deeper orbitals to give *electronically* excited ions. Figure 9.4 shows the electron configuration of a ground-state ion (removal of an electron from orbital ψ_1) and an excited-state ion (removal of an electron from ψ_3). Figure 9.5 represents the formation of these ions from the molecule by vertical ionizing transitions. In figure 9.5, the curve for $M^{+\bullet}$ represents the ground state ion and $\overset{*}{M}^{+\bullet}$ the electronically excited ion. The energetically excited ion may lose or redistribute its energy by two main processes: (i) internal electron redistribution with concomitant *emission* of radiation to give a ground-state ion and (ii) radiationless potential surface crossings, as in figure 9.3, whereby the electronic energy in the excited state ion can be *transformed* into excess of vibrational energy in the ground-state ion. Thus, either by radiation of energy or by radiationless transfer of energy, the excited-state ion can revert eventually to an electronic ground-state ion having an excess of vibrational and rotational energy. There are many possible excited states of ions that can be reached by electron ionization. Energy conversion by surface crossing does not necessarily lead simply to a ground-state ion since, if there is sufficient energy to cause decomposition from any excited first-formed or intermediate state, then decomposition will occur from that state. Because the time for a bond vibration (about 10^{-13} s) is short compared with the maximum lifetime of an excited ion (about 10^{-8} s), many bond vibrations occur with the possibility of radiationless state crossing within this lifetime. However, the lifetime of an excited state ion (about 10^{-8} s) is considerably shorter than the time (about 10^{-5}–10^{-6} s) that an ion spends in the ion source and in flight to the detector. Therefore, in conventional electron ionization mass spectrometry, there is ample time for the excess of electronic energy in an ion to be converted into an excess of vibrational energy in a lower electronic state. As a general working hypothesis in mass spectrometry, it is usually supposed that this less electronically excited state from which most reactions (fragmentations) occur is the electronic ground state of an ion that is vibrationally and rotationally excited. The process may be represented schematically as follows for isolated molecules M interacting with ionizing radiation and fragmenting to yield ions found in the mass spectrum:

$$n.\text{M} + \text{energy} \xrightarrow{\;10^{-15}\text{–}10^{-16}\text{ s}\;} a.\text{M}^{+\bullet} + b.\overset{*}{\text{M}}_1^{+\bullet} + c.\overset{*}{\text{M}}_2^{+\bullet} + \ldots\ldots + ne^-$$

$\downarrow$ $< 10^{-7}$ s (internal energy conversation)

$$a.\text{M}^{+\bullet} + b.\text{M}_1^{+\bullet} + c.\text{M}_2^{+\bullet} + \ldots\ldots$$

$\downarrow$ 10^{-10}–10^{-5} s (fragmentation)

$$p.\text{M}^{+\bullet} + q.\text{A}^+ + r.\text{B}^+ + \ldots\ldots$$

$\downarrow$ (ions accelerated from ion source)

mass spectrum

$(\overset{*}{M}_1{}^{+\cdot})$, $(\overset{*}{M}_2{}^{+\cdot})$ and so on represent electronically and rovibrationally excited molecular ions; $M^{+\cdot}$, $M_1{}^{+\cdot}$ $M_2{}^{+\cdot}$ etc. are rovibrationally excited molecular ions; A^+, B^+, etc., are fragment ions. This scheme illustrates how, by collision with electrons in an evacuated ion source, an assembly of n molecules first forms a distribution (a, b, c ...) of ground and excited state molecular ions. Some short time later, *and mostly within the ion source*, this distribution is transformed into a new distribution (p, q, r ...) of molecular and fragment ions. These changes give the final m/z distribution of ions and their abundances, which are analysed to give a mass spectrum. The above scheme should be modified slightly to accommodate field ionization, in which the ions spend a very much shorter time (10^{-11}–10^{-9} s) in the ion source.

Removal of one electron from a molecule leaves an ion in the doublet state (one unpaired spin, so that $2S + 1 = 2$, where S is the total spin). During ionization by electrons only the total spin of the system (electron plus molecule) needs to be conserved so that an ion may also be formed even in its quadruplet state (three unpaired spins). Just as triplet states have longer lifetimes than singlet states, so it is likely that quadruplet states have longer lifetimes than doublet states. Hence, fragmentation may take place from relatively long-lived excited-state ions.

Broadly, a picture emerges of electron ionization producing molecular ions in different energy states, rapid internal energy conversion to produce ions with individual amounts of excess of energy, and fragmentation taking place at different rates depending on the electronic state of and excess of vibrational energy in each ion. Within each ion, the excess of internal energy is considered to be distributed statistically amongst all the vibrational and rotational modes. The distribution of internal energies amongst the ions may be represented as in figure 9.6.

Some ions have insufficient energy to fragment (figure 9.6) and the proportion of them increases as the energy of the ionizing radiation decreases to the minimum ionization energy. For this reason, reducing the electron beam energy in an electron ionization source reduces the degree of fragmentation observed in the mass spectrum. It is possible to consider a mean or average effect when discussing the total behaviour of many such ions, but this situation of an average over many isolated species reacting independently of one another is quite unlike the mean (equilibrated) effect resulting from collisional processes in solution or normal gas-phase chemistry. In solution chemistry, collision processes leave

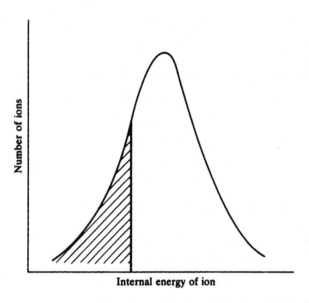

Number of ions

Internal energy of ion

Figure 9.6. The distribution of internal energies in ions at any one m/z value. The true shape of the curve would be governed by the cross-sections for all the various ionizing transitions. The hatched portion indicates those ions with insufficient energy to fragment. Other ions (the non-hatched portion under the curve) fragment at rates partly governed by their internal energies, the rates increasing as the excess of internal energy increases.

molecules with sufficient energy to react or decompose. In electron ionization mass spectrometry, the ion initially produced does not normally collide with other species and usually decomposes endothermally if sufficient energy has been transferred during ionization. As remarked earlier, these considerations need to be modified for other modes of ionization.

9.3. ENERGY STATES RESULTING FROM IONIZATION AT HIGHER GAS PRESSURES

The previous section has dealt with the energy states resulting from ionization under low-pressure conditions, typically 1.3–0.13 mN m^{-2} (10^{-5}–10^{-6} Torr), when the initially formed ions do not collide with other molecules before their extraction from the ion source. Under these circumstances, any excess of energy provided during ionization can lead to fragmentation. For other modes of ionization, gas pressure conditions in the ion source are quite high, ranging from about 0.2 N m^{-2} for chemical ionization (CI) to 133 N m^{-2} for atmospheric pressure ionization (API). The collisional processes leave newly formed ions with at least some excess of vibrational energy, although this is not considered to be very large. At these higher source pressures, the mean free path of any newly formed ion is very short so that the ion suffers multiple collisions with other (neutral) sample or reactant (CI) gas molecules before it enters the low-pressure regions of the analyser section of the mass spectrometer. Because the kinetic energies of the newly formed ion and any surrounding neutral gas molecules will be similar, collision leads to the species with an excess of internal energy passing some of this onto the other. Since there are many more neutral gas molecules without an excess of internal energy compared with relatively few ions with an excess, the result of multiple collisions of the first-formed ions with neutral molecules means that the ions rapidly lose much of their excess of internal energy to molecules of the surrounding gas. This excess is eventually dissipated as heat through collision of neutral molecules with the walls of the mass spectrometer. Thus, at higher gas pressures, a short time after formation, a newly formed ion will have lost most of any excess of energy that it gained during ionization and there is no or very little excess available for endothermic fragmentation. The ion may be said to be collisionally equilibrated. Equation (8.13) shows that, if an ion suffers little acceleration before it collides with a neutral gas molecule so that E_{LAB} is very small and, if the mass of the ion equals that of the neutral species ($m_g = m_i$) then the resulting collision energy, E_{CM} ($= E_{LAB}/2$), will be small and little translational energy is transferred from one species to the other due to the energy of collision alone. Therefore, the collisional process can be considered to be one in which excess of rovibrational energy is transferred from a relatively few ions to many surrounding neutral molecules.

Ions formed under high-pressure conditions are notable for their lack of fragmentation, unlike ions formed under low-pressure conditions, which can fragment readily. Molecular or quasi-molecular ions formed in this way are stable enough to arrive at the ion detector without any fragmentation having occurred. This lack of much or any excess of internal energy and the subsequent lack of fragmentation in molecular ions produced under such conditions has led to these methods of ionization being referred to as 'soft' or 'mild'. It might be noted here that *monatomic* neutral gas molecules have no rotational or vibrational modes and any energy transferred to them must appear as a change in their electronic or kinetic energies. Mass spectra of substances ionized at low pressure (EI, FI) are characterized by an abundance of fragment ions; sometimes, *molecular* ions formed by EI decompose so easily and rapidly that none reaches the detector of the spectrometer and none is observed in the spectrum so that no molecular mass information is available. On the other hand, mass spectra of substances formed under high-pressure conditions (CI, API) are dominated by molecular or quasi-molecular ions and few fragment ions are observed. For mass spectrometry, these opposing characteristics are not necessarily beneficial. With excessive fragmentation, absence of or uncertainty about molecular ions prevents or hinders measurement of molecular mass, one of the great virtues of mass spectrometry. On the other hand, an absence of fragmentation removes structural information, another of its virtues. This problem is neatly solved by using combinations of methods. Thus, alternate EI (much fragmentation) and CI (little fragmentation) can be carried out (ACE) by successively reducing and increasing gas pressure in the ion source, the CI mode being used to confirm molecular mass. Molecular ions formed under API conditions, which would not normally fragment, can have their excess of internal energy increased by accelerating them through a cell containing a gas at relatively high-pressure, as in tandem MS in which highly energetic (E_{CM} large) ion/molecule collisions cause fragmentation of the ions so as to yield structural information (see section 8.7.1).

9.4. ENERGY STATES RESULTING FROM IONIZATION IN CONDENSED (LIQUID) MEDIA

In the previous two sections, ionization has been discussed as taking place after vaporization of a molecule into a low- or high-pressure gas phase (EI, CI, FI). However, other ionization techniques utilize the condensed liquid phase, as with API (thermospray, plasmaspray, electrospray), SIMS, FAB and FD. Sample molecules in such liquid phases at room temperature have little or no excess of internal energy and the different ionization processes also tend to impart little extra so that fragmentation of molecular or quasi-molecular ions is greatly reduced. In API, ions are formed by a mixture of solvent evaporation and low-energy collision processes, both of which afford ions having little excess of energy. In FAB and SIMS, ionized species are vaporized from a liquid solution (matrix) mostly by momentum transfer from a primary atom or ion beam; ions formed at or near the surface of the liquid are rapidly thermally equilibrated and form few fragment ions after extraction from the ion source. In field desorption, ions formed in the condensed phase at the surface of the emitter by electron tunnelling leave with little excess of vibrational energy. In all of these ionization modes, ions are formed with little or almost no excess of rotational and vibrational energy let alone excess of electronic energy. Indeed, cluster ions are often observed in which one proton may hold together two, three, four or more molecules, $[M_n + H]^+$ (n is an integer). Further discussion of ion formation by 'mild' methods can be found particularly in chapter 3. In such highly condensed phases, immediately after ion formation, multiple low-energy collisions lead to distribution to surrounding neutral species of any excess of internal energy gained through ionization. Thus, in these mild methods of ionization, sample molecules do not have to be heated to vaporize them, the newly formed ions have little excess of internal energy from the ionization step and what excess there is is soon dissipated through low energy collisions with surrounding molecules. Ions produced by these methods show little or no tendency to fragment.

9.5. ENERGY STATES RESULTING FROM COLLISIONAL AND
OTHER ACTIVATION

When two molecular species collide, vibrational and rotational modes in each are distorted. At low collisional velocities (E_{CM} small, equation (8.13)), these distortions are relatively minor and the two species 'bounce' away from each other; the total momentum of the two is the same before and after the collision, which is then said to be *elastic*. Whilst collisions between atoms, which have no vibrational or rotational modes, is necessarily elastic, collisions between atoms and molecules or ions and molecules or molecules and molecules is unlikely to be truly elastic and some of their total kinetic energy is converted into extra vibrational or rotational energy in one or both colliding entities. Viewed very simply, the arrival of a colliding molecule end-on to a vibrational mode in a bond of a second species will result in some of the total kinetic energy of the two colliding species being changed into vibrational and rotational energy; such a collision is *inelastic*. A collision at an angle to a bond will impart also some rotational energy. As the collision energy increases, more and more kinetic energy is transformed into excess of vibrational and rotational energy in the colliding species and, at high translational energies, even electronic excitation occurs. These inelastic collisions become more likely as collisional speeds increase (up to a limit) such that the actual collision itself leads to severe distortion of vibrational, rotational and electronic modes. In simplistic terms, the process can be likened to dropping a glass globe onto a hard floor; at low closing speeds (small drop height), the globe hits the floor, distorts, regains its shape and bounces off (total momentum preserved). At high closing speeds (big drop height), the globe hits the floor, distorts severely, the distortion leads to cracking and some of the kinetic energy is converted into thermal energy; the original momentum is shared amongst the resulting fragments. In ion sources for chemical ionization or atmospheric pressure ionization, collision speeds are relatively low and much of the total kinetic energy is preserved but there is some change in vibrational energy in that vibrationally excited ions collide with ground-state neutral entities and transfer (off-load) some of their excess at each encounter. Even more simply, a proton may be passed from an excited state ion onto a ground-state neutral species, thereby forming a ground-state protonated molecule and leaving a vibrationally excited neutral entity (reaction (9.1)):

$$[M + H]^+ \quad + \quad M \quad \longrightarrow \quad M \quad + \quad [M + H]^+$$

| energetically excited protonated molecule | ground-state molecule | energetically excited neutral molecule | ground-state protonated molecule | (9.1) |

If collisional speeds are increased by, for example, accelerating an ion through potentials of tens or thousands of volts before it reaches a neutral collision gas, then part of the total kinetic energy so gained will be converted into vibrational, rotational and even electronic energy in the ion; this extra energy may be sufficient to cause the impacting ion to fragment, much as it would do after electron ionization. The process is known as *collisional activation* or *collisionally induced decomposition* (section 8.7.1). For some molecules or atoms, collision may even lead to charge-exchange processes in which electrons are passed from one species to another (see below).

Although collisional effects are an intrinsic part of chemical ionization and other mild ('soft') ionization techniques, there being relatively few ions present in a bath of neutral molecules means that energy transfer is almost one way, *viz.*, from the ions to the neutral species so that the ions end up with little excess of energy and do not fragment. Because this lack of fragmentation gives no structural information, it is now common to add more energy to such stable ions to cause them to fragment, as in MS/MS (chapter 8). Ions can be accelerated to collide with a collision gas, as described above, but it is possible to add the required extra energy in other ways. For example, an ion can be irradiated with photons from laser sources or collided with a metal plate or even passed through an electron beam. Further details on these variations appear in chapter 8, under MS/MS techniques.

9.6. FORMATION OF IONS

9.6.1. *Electron ionization*

Electrons accelerated through a potential of V volts have a 'de Broglie' wavelength of $(1.5/V)^{1/2}$ nm, e.g. 75 V electrons have a wavelength of 0.14 nm corresponding to short-wavelength radiation. During the approach of an electron to a molecule, the impacting electron waves and the electric field of the molecule mutually distort one another. The distorted electron wave can be considered to be composed of many different sine waves and some of these component waves will be of the correct frequency (energy) to interact with molecular electrons. Thus, the passage of an electron through or close approach of an electron to a molecule may lead to electronic excitation in the molecule by promoting an electron from a lower to a higher orbital (this is the same effect as in ultraviolet spectroscopy). Similarly, a molecular electron may be promoted to an outer orbital (compare this with vacuum ultraviolet Rydberg bands) or an electron may be ejected from the molecule altogether to leave a positive ion. Direct attachment of an electron to a molecule to give a stable radical-anion is of low probability because the translational energy of an electron attaching itself to a molecule must be taken up as excess of internal energy in the new radical-anion. Usually, this excess of energy leads either to the electron being 'shaken' off again or to fragmentation of the radical-anion. To form stable radical-anions by direct electron attachment, it is necessary to have sufficient gas pressure for 'third-body' collisions to occur so as to remove sufficient excess of energy that the radical-anion does not fragment. Formation of negative ions directly by electron ionization normally requires higher gas pressures so as to improve the chances of three-body collisions. Production of negative ions is also improved if the sample molecules are highly electronegative (molecules with high electron affinities), just as formation of positive ions is eased by electropositive materials (ones with low ionization energies). Thus, fluorinated, chlorinated and nitrated compounds are particularly effective in forming negative ions and, for these types of compound, the efficiency of ionization (and therefore limit of detection) in the negative mode is better than that in the positive mode. The improvement in ionization efficiency gained at higher gas pressures is discussed more generally in chapter 3. Other electronic effects in the molecule due to reaction with an electron are also observable but have little direct use in analytical mass spectrometry and so there is insufficient space to deal with them here.

The commonly used term 'electron impact' is misleading because an electron is so small in molecular terms and there is comparatively so much empty space in a molecule that it would have difficulty 'hitting' any part of a molecule that it met. It is better to think of the electron as passing close to or even through a molecule rather than of any 'impact' taking place. It is recommended that the term electron impact ionization be discontinued in favour of the more apposite one, electron ionization.

The effects of the electron on the molecule are experienced as it approaches and there is some distortion of the molecule during the transition to an ion, i.e., the ionizing transitions are not strictly vertical and allow more states to be reached than by photo-ionization. Also, relative to photo-ionization, more spin states of the ion are accessible by electron ionization. Therefore, in comparison with both field ionization and photo-ionization, ionization by electrons at the conventionally used energy of 70 eV results in the deposition of an excess of energy of about 3–5 eV into the initially formed molecular ions, which is enough to cause bond breaking (fragmentation). With this excess of energy, fragmentation is fast and most ions that are going to decompose will have done so within a microsecond, which is the usual approximate residence time of an ion in an electron ionization source. By reducing the electron energy well below the conventional 70 eV to, say 15 eV, much less excess of internal energy or almost none is deposited in the ion following ionization so that fragmentation is greatly reduced and electron ionization spectra become more comparable with photo-ionization and field ionization spectra. The penalty that has to be paid for this reduction in fragmentation is a greatly reduced efficiency in actual ionization so that the numbers of molecules that are ionized decreases and, therefore, so does the sensitivity of the method.

9.6.2. *Photon ionization*

The wavelength of radiation required for photon ionization (photo-ionization) is about 100 nm, which is much greater than normal molecular dimensions and means that a molecule is subjected effectively to a uniform electromagnetic field. In quantum mechanical terms, only the effect of the radiation on the molecule need be considered, unlike the case of electron ionization, in which the wavelength of the radiation and the molecular dimensions are similar and mutual quantum effects must be considered. Because the electric field of the photons is essentially uniform over the whole molecule, almost no molecular distortion occurs during ionization. The ionizing transitions are very fast (10^{-16} s) and are vertical in accord with the Franck–Condon principle. Also, because of spin conservation from molecule to ion, fewer states are accessible than by electron ionization. In comparison with the latter, somewhat less vibrational energy is deposited in a molecular ion after absorption of a photon and, like field ionization spectra, photo-ionization spectra yield abundant molecular ions and relatively few fragment ions.

Provided that the energy ($h\nu$) of the incident photons is known, the excess of energy (E^*) deposited in the ion during ionization is given exactly by the expression $E^* = h\nu - I$, where I is the ionization energy. Thus, it is possible to vary the excess of energy in the molecular ion in a controlled way by closely controlling the energy (wavelength) of the incident photons.

The lifetimes of photo-excited states are normally quite short and any ion that has absorbed a photon will have dissipated the acquired energy within a few nanoseconds or less. If lasers are used to photo-excite ions then more than one photon may be absorbed within a very short space of time because of the intensity of light in a laser beam. Thus, multiphoton absorption can be observed and this leads to decomposition of ions from energy states that would be inaccessible were only one photon absorbed at the same wavelength. Just as neutral sample molecules in an ultraviolet spectrometer yield a spectrum when light of different wavelengths is absorbed (figure 9.7(a)), ions when irradiated with ultraviolet light of a range of wavelengths can be shown to absorb the light preferentially. In this case, the absorption cannot be observed directly because the density of ions in an ion beam is too low for direct measurement of the amount of light absorbed but, since the ion reacts to absorption of quanta of light energy by fragmenting, the range of fragment ions produced as the wavelength of light is changed gives an 'action spectrum'. This is derived by plotting the degree of fragmentation against the wavelengths of light used to irradiate the ions. A simple characterization of an action spectrum is shown in figure 9.7(b). This principle is used to activate 'stable' ions to cause them to fragment as an alternative to collisional activation. For example, protonated molecules can be formed in one mass spectrometer or sector and then irradiated with light to cause them to

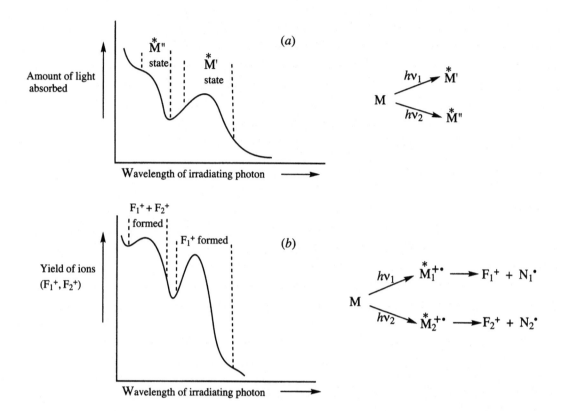

fragment; the range of fragment ions is examined in a second mass spectrometer or sector (MS/MS – see chapter 8). Instead of sector-based MS/MS, ion traps and ion cyclotron resonance may be used in a similar way, *viz.*, ions are first formed in, say, a trap and are then irradiated with light after which the range of fragments in the trap is examined.

9.6.3. *Field ionization*

If a sharp edge or fine wire (called an emitter) is maintained at a high electrical potential, the small radius of curvature at the edge or along the wire produces an intense electrical field. In such a field, the molecular orbitals of any molecule vaporized onto the emitter are distorted and the potential energy barrier to transfer of an electron from the molecule to the emitter is reduced considerably. Under these conditions, quantum tunnelling of an electron occurs, converting the molecule into a radical-cation, which is promptly repelled by the high positive electric potential. By use of a negative potential, negative ions can be formed similarly. Field ionization spectra,

Figure 9.7. A simplified comparison of an ultraviolet absorption spectrum (*a*) and an ultraviolet 'action spectrum' (*b*). In (*a*), light of specific wavelengths (hv_1, hv_2) is absorbed by molecules (M) to give photo-excited molecular states, $\overset{*}{M}'$ and $\overset{*}{M}''$ the spectrum is obtained by measuring the amount of light absorbed at each wavelength. In (*b*), light of specific wavelengths (hv_1, hv_2) is absorbed by ions ($M^{+\cdot}$) to give photo-excited states $\overset{*}{M}_1{}^{+\cdot}$ and $\overset{*}{M}_2{}^{+\cdot}$; the 'spectrum' is obtained by measuring the fragment ion (F_1^+ and F_2^+) yields at each wavelength.

discussed more fully in section 3.7, are characterized by the presence of abundant molecular ions and few fragment ions.

The emitter electrode is normally held at a high electrical potential, positive if positive ions are to be investigated and *vice versa* for negative ions. Since the repulsion between two like-charged species in an intense electric field is large and the emitter has large mass compared with an ion then the latter, once formed, is propelled very rapidly away from the emitter. The promptness of this process and the high velocities attained by the ions mean that they arrive at the ion detector in much shorter times than those of ions in more conventional instruments. Typically, the time between formation of the ion and its detection is of the order of 10^{-13}–10^{-11} s. This is very little time for much fragmentation to occur, especially considering the small amounts of excess of internal energy carried by these ions. Thus, although such ions are formed by addition or subtraction of an electron as in electron ionization, the energy regimes and time scales in FI are such that little or even no fragmentation is observed and the technique is good for obtaining molecular mass information. It has also been used extensively to investigate early stages of fragmentation, facilitated by the short time scales (see chapter 3).

9.6.4. *Chemical ionization*

This involves the collision of two bodies, an ion and a molecule, which are massive compared with an electron. The collisional interaction is slower, allowing equilibration of energy between the two colliding species. The necessary ions are formed initially by electron ionization of a *reactant (reagent) gas* (A) at relatively high pressures in the ion source. The resulting reactant gas ions, A^+, collide with neutral molecules (A) and lose much of their excess of energy in the process. These thermally equilibrated ions (A^+) collide with molecules, B, of any sample under investigation when a chemical reaction occurs in which B is ionized – hence the term 'chemical ionization'. Usually, this reaction entails addition or subtraction of charged entities to or from molecules B to give 'quasi-molecular ions'. For example, if the reactant gas contains H_3^+ ions, formed from H_2 by electron ionization at moderate gas pressures (equation (9.2a)), a chemical gas phase reaction can occur with sample molecules B as shown in equation (9.2b), in which proton transfer affords the protonated molecule, $[B + H]^+$. During this transfer of a proton, one bond to hydrogen in the reactant gas ion is broken as one bond forms to the same hydrogen from the sample molecule. The net energy

$$\text{H}_2 + \text{e}^- \rightarrow \text{H}_2^{+\bullet} + 2\text{e}^- \text{ (electron ionization)}$$
$$\text{H}_2 + \text{H}_2^{+\bullet} \rightarrow \text{H}_3^+ + \text{H}^\bullet \text{ (hydrogen abstraction)}$$

$$(9.2\ a)$$

$$\text{H}_3^+ + \text{B} \rightarrow \text{H}_2 + [\text{B} + \text{H}]^+ \text{ (proton transfer)} \qquad (9.2\ b)$$

transfer is small because there is usually little difference in energy between the bond being broken and that being formed. The sample molecule receiving a proton then has little extra energy and that which there is may be reduced further by collision of the newly produced protonated molecule, $[\text{B} + \text{H}]^+$, with other neutral reactant gas molecules (A). Chemical ionization spectra are characterized by stable molecular or quasi-molecular ions and very little fragmentation. Because there are many fewer sample molecules compared with reactant gas molecules, the chances of direct electron ionization of the sample are low but the chances of proton transfer are high. By suitable choice of reactant gas, negative ions can be produced efficiently (section 3.4). Chemical ionization is described in greater detail in section 3.3.

9.6.5. *Field desorption ionization*

This technique is identical to that of field ionization except that the substance under investigation is precoated as a thin film onto the emitter instead of being vaporized onto it; the emitter may itself be heated by passing an electric current through it to help desorb the ions. Whereas initial ionization again occurs via electron tunnelling to give ions ($\text{M}^{+\bullet}$) having little excess of energy, as with field ionization, repulsion of these ions from the emitter on their way to the mass analyser causes them to collide with co-adsorbed neutral species (M) so that chemical ionization-like processes occur resulting in the formation of quasi-molecular ions, frequently $[\text{M} + \text{H}]^+$ and sometimes clusters $[\text{M}_n + \text{H}]^+$. Like simple field ionization spectra, field desorption spectra have few fragment ion peaks. Through use of a negative potential on the emitter, negative ions can be produced.

9.6.6. *Charge exchange*

During collision between two molecules or an ion and a molecule, charge exchange (charge transfer) may occur as illustrated by equation (9.3), in which neutral caesium atoms and phosphorus trichloride molecules afford positive and negative ions, respectively:

$$Cs + PCl_3 \rightarrow Cs^+ + PCl_3^- \qquad (9.3)$$

Charge exchange can be effected by forming ions in the ion source of a mass spectrometer and directing them through a region containing a neutral gas. Ion/molecule collisions occur and, in suitable cases, charge transfer produces new ions, as exemplified in equation (9.4), in which initially formed I^- ions yield $BrCl^-$ ions after collision with neutral BrCl. Ions formed by these methods are frequently difficult to produce by direct ionization. Charge exchange can be observed in tandem MS when collisional activation is used (section 8.7.1).

$$I^- + BrCl \rightarrow I^\cdot + BrCl^{-\cdot} \qquad (9.4)$$

9.6.7. *Ion structures*

It is essential to remember that, although the structure of a molecule put into a mass spectrometer may, at least in principle, be determined by other physical methods (e.g. X-ray analysis), after ionization it is not correct simply to assume that the ion has the same structure as the original molecule. Immediately after the ionization step, the original molecule continues to vibrate as an ion and this vibration, together with the different content and distribution of electrons in the ion compared with the original molecule, may lead rapidly to a structure quite different from that of the original molecule. There is ample experimental evidence to suggest that, frequently, rearrangement does occur rapidly after ionization, hydrogen rearrangement being particularly easy. For example, the EI mass spectra of simple isomeric alkenes are very similar because extensive rearrangement leads to common ion structures before fragmentation can occur. The importance of excess of internal energy in leading to rearrangement of an ion before its fragmentation is discussed in more detail in sections 9.7 and 10.9.

In the cases of ions being formed by mild ionization techniques, it is more probable that the structure of an ion is similar to that of the original molecule before ionization. For example, addition of a proton to a molecule that has not been heated above or much above ambient temperatures to give a protonated molecule, $[M + H]^+$, as in API, is likely to proceed without significant change of structure. However, it should be remembered that this is conjecture and, in the absence of the solvent that is present in normal solution phase chemistry, such gas phase protonated molecules may rearrange.

9.7.   THEORIES OF FRAGMENTATION RATES

A quantitative treatment of mass spectrometric fragmentation has been
attempted using mathematical expressions, which statistically distribute the
available excess of internal energy in a newly formed ion amongst all its
oscillators and rotators. Use of this quasi-equilibrium theory (QET) allows
the rate of decomposition of an ion through cleavage of any one particular
bond to be calculated for any given excess of internal energy (Rosenstock,
Wallenstein, Wahrhaftig and Eyring, 1952; Vestal, Wahrhaftig and Johnston,
1962). An almost identical approach is used in RRKM (Ramsberger, Rice,
Kassel and Marcus) theory (Marcus, 1952), which was first applied to the
calculation of chemical reaction rates from the distribution of energy
amongst the oscillators and rotators in the transition state. Such calculations
are particularly applicable to ion/molecule collisions in which the collision
complex can be approximated to a transition state. As a further elaboration,
consideration of phase space, the six-dimensional space necessary to
describe the positions (x, y, z) and momenta (mv_x, mv_y, mv_z) of the colliding
species and/or fragments from decomposition gives a total description of
the distribution of energy into various reaction (decomposition) channels
(Light, 1967). For all but very simple molecules, these mathematical
approaches are too time-consuming to be evaluated without the aid of a
computer and, even with the aid of a powerful computer, there is still the
problem of the assumptions made concerning likely structures for transition
states and the products of fragmentation, let alone the assumptions made
about the structure of the initially formed ion. The actual distribution of
energies (energy deposition function) in the initially formed ion is also a
problem since these are not usually known and are frequently represented
by some simple square wave function. Despite such difficulties, the quasi-
equilibrium and RRKM theories have had partial success in predicting fragmentation patterns for some relatively simple molecules. Additionally,
simplified versions of these theories have provided rational explanations for
commonly observed features of fragmentation (see later).

Rates of fragmentation calculated from RRKM theory with allowance for
phase space have been used to compare predicted and observed reaction
rates in ion/molecule reactions (Safron, Weinstein, Herschblach and Tully,
1972; Marcus, 1975; Klots, 1976; Chesnavich and Bowers, 1978). Some of the
success achieved may only be apparent because, in applying the theories, for

the reasons cited above, where there is disagreement between theoretical and observed behaviour, it is uncertain whether this is due to unwarranted assumptions about the application of the theory, about the ion structures or about the energy deposition functions. Nevertheless, as knowledge of ion structures improves, comparison of predicted and observed rates of fragmentation is leading to a better understanding of the finer details of fragmentation processes and these refined treatments become more important. The advent of larger, faster computers means that it is feasible to examine a range of assumptions to determine which give results more in keeping with actuality.

There is a simplified version of the QET equations, which is useful for discussing mass spectra in a semi-quantitative fashion and can be used with complex molecules without the need of a computer. This expression (9.5) is not applicable at very low and very high excesses of internal energy in a reacting species. In this simplified expression, k is the rate constant for decomposition of an ion, v is a frequency factor, E the excess of internal energy, E_0 the energy of activation for the decomposition and N the number of oscillators in the ion:

$$k = v[(E - E_0)/E]^{N-1} \qquad (9.5)$$

Effectively, the frequency factor is a measure of the entropy factor in the decomposition. For example, for a simple bond cleavage it is put equal to the bond-vibration frequency, which may be obtained approximately from a bond stretching frequency in an infrared spectrum. For a rearrangement reaction, a particular spatial disposition of atoms is probably required and it may be expected that v would be at a lower frequency than for simple bond cleavage (Field and Franklin, 1957). The number of oscillators (N) in a molecule containing n atoms is given by the expression ($3n - 6$); often it is found that not all the oscillators appear to be equally effective, in which case a calculated value for k is arrived at which is too large. For this reason, the number of oscillators is usually divided empirically by three for better results. The part of the expression (9.5) containing the energy terms, ($E - E_0)/E$, is a dimensionless quantity and energy can be considered in any convenient units, e.g., electron-volts.

The simplified rate equation (9.5) has been used to discuss the fragmentation of molecules in a general sense. As examples of its use, the following may be mentioned: the greater the excess of energy deposited in an ion during the ionization step, i.e., $(E - E_0)/E$ increases, the faster an ion decomposes, eventually approaching a rate equal to that of a bond-stretching vibration. For two alternative fragmentations of an ion that have similar activation energies, the rates are proportional to the frequency factors, i.e., $k_1/k_2 = v_1/v_2$; if the frequency factors are similar, the rates of fragmentation are approximately proportional to the activation energies; comparison of the rate for a simple bond-fission reaction with the rate for fission accompanying rearrangement suggests that the rearrangement process may become more prominent as the internal energy of the ion decreases (section 10.9). It should be noted that these rate constants (k) refer to individual ions with internal energy (E) and that they are not equivalent to the rate constants found in solution chemistry, which are the result of an equilibration process amongst many molecules. However, it is possible to consider in a qualitative sense the behaviour of an ion having an 'average' of the excesses of energy of many similar ions and then to extrapolate this mean behaviour to all the other ions of the same structure but slightly different internal energies.

When the *simplified* QET equation is used for a process such as $A^+ \rightarrow B^+$, subsequent fragmentation of B^+ is sometimes ignored. Particularly at greater excesses of internal energies, these further fragmentations are important and the abundance of ion B^+ (or A^+) cannot be related only to its rate of formation (or decomposition). As stressed in section 1.2, the abundance of an ion in a mass spectrum depends upon *both* its rate of formation and its rate of decomposition, and care should be exercised in the use of the QET equation if observed ion abundances are to be related to calculated rates of formation or decomposition. Abundances of ions considered theoretically in this way are unlikely to reflect very closely the observed abundances unless some account is taken also of the distribution of excess of energy amongst the ions. The actual distribution of excess of energy (the energy deposition function) may be such that the use of an 'average' excess, as suggested above, would lead to gross differences between theory and practice. The actual distribution is difficult to determine experimentally and often a very simple function is chosen for calculations using the QET or RRKM equations. When energy deposition functions approximate to the true distribution obtained experimentally (such as from photoelectron or photo-ion spectra), agreement or contrast between predicted and observed rates of fragmentation is more significant.

9.8.   THERMOCHEMICAL ARGUMENTS

Measurement of the minimum energy required to ionize a molecule (the *ionization energy, I*) or to cause the appearance of fragment ions (the *appearance energy, A*) can provide valuable thermochemical data. For a molecule, M, giving first a molecular ion, $M^{+\cdot}$, and then a fragment ion, F^+, with ejection of a neutral species, $N^\cdot$, the ionization and appearance energy are related to heats of formation (ΔH_f) as shown in equation (9.6). Thus, measurement of I or A can yield values for heats of formation of ions or neutral species, provided that the remaining required data are available. There are uncertainties about some of the values obtained. These uncertainties stem both from difficulties in the measurements themselves and from relatively poor knowledge of ion structures (Howe, 1973). However, the same considerations can be used differently so that knowledge of ion structures is improved by using thermochemical data. In the example shown in equation (9.7), two different molecules, M_1 and M_2, fragment to produce ions, F^+, of the same *composition* but not necessarily the same *structure*. By measurement of appearance energies, A_1 and A_2, and insertion of heats of formation for M_1, M_2, N_1 and N_2, the heat of formation of F^+ can be compared for the two reactions. If the value found for $\Delta H_f(F^+)$ is the same for the two fragmentations, it provides some positive evidence that the structure of ion F^+ is the same in each case. Similarly, a gross difference in the values found for $\Delta H_f(F^+)$ would strongly suggest that the ions, F^+, from each fragmentation have different structures, even though they had the same elemental composition. Tables are available listing the known heats of formation of ions (Rosenstock, Draxl, Steiner and

$$M \rightarrow M^{+\cdot} + e; \qquad\qquad I = \Delta H_f(M^{+\cdot}) - \Delta H_f(M)$$

$$M \rightarrow M^{+\cdot} + e \rightarrow F^+ + N + e; \qquad A = \Delta H_f(F^+) + \Delta H_f(N) - \Delta H_f(M)$$

$$\therefore A = I + \Delta H_f(F^+) + \Delta H_f(N) - \Delta H_f(M^{+\cdot})$$

$$(9.6)$$

$$M_1 \rightarrow F^+ + N_1 + e^-; \qquad \text{appearance energy}, A_1$$

$$M_2 \rightarrow F^+ + N_2 + e^-; \qquad \text{appearance energy}, A_2$$

$$\Delta H_f(F^+) = A_1 + \Delta H_f(M_1) - \Delta H_f(N_1)$$

$$\Delta H_f(F^+) = A_2 + \Delta H_f(M_2) - \Delta H_f(N_2)$$

$$(9.7)$$

Herron, 1977 and see references below), but these sorts of thermochemical arguments have been enormously improved by advances in the calculation of heats of formation both of ions and of molecules, either by empirical or by *ab initio* methods (see below). As an example of this sort of approach, for the reactions shown in equations (9.7), it is now possible to calculate the expected heat of formation of an ion after assuming a structure. Thus, the experimental result(s) for the heat of formation of an ion, F^+, can be compared with predicted (calculated) heats of formation. Close agreement between experimental and calculated values would suggest a possible structure for the ion, F^+.

These types of thermochemical argument can be used to predict the likely course of fragmentation. For two possible competing decompositions of a molecular ion ($M^{+\cdot}$), two fragment ions (F_1^+ and F_2^+) with two corresponding neutral species (N_1^0 and N_2^0) may be envisaged as in equation (9.8). The heats of reaction ($\Delta H_R 1$ and $\Delta H_R 2$) are given by the differences in the summed heats of formation. Now, the various heats of formation can all be calculated by empirical or *ab initio* methods after assuming structures for the various ions and neutral species. If entropic factors are not significantly different in the two possible reactions then the free energies of reaction may be equated to the enthalpy changes, $\Delta H_R 1$ and $\Delta H_R 2$. If the two heats of reaction are significantly different, the preferred mode of fragmentation can be predicted to proceed along the most endothermic pathway (the difference, $|\Delta H_R 1 - \Delta H_R 2|$, should be the most positive). By comparing whether or not this prediction closely fits with experimental observation, the assumed structures for the various species may be supported or rejected.

$$M^{+\cdot} \quad \begin{array}{l} \nearrow \ F_1^+ \ + \ N_1^0 \\ \\ \searrow \ F_2^+ \ + \ N_2^0 \end{array}$$

$$(9.8)$$

$$\Delta H_R 1 \ = \quad \Delta H_f(F_1^+) + \Delta H_f(N_1^0) \ - \ \Delta H_f(M^+)$$

$$\Delta H_R 2 \ = \quad \Delta H_f(F_2^+) + \Delta H_f(N_2^0) \ - \ \Delta H_f(M^+)$$

$$\Delta H_R 1 - \Delta H_R 2 \ = \ \{\Delta H_f(F_1^+) - \Delta H_f(F_2^+)\} \ + \ \{\Delta H_f(N_1^0) - \Delta H_f(N_2^0)\}$$

As just discussed, there is a need to know the heats of formation of more ions and molecules than have been measured and many schemes have been devised for predicting them. With molecules for which structures are generally known or can be elucidated, these calculated values are now mostly as accurate as measured values and can be used with considerable confidence. For ions, the situation is not quite so straightforward, partly because of the assumptions that have to be made for all but the simplest of structures. However, good schemes have been formulated for estimating both neutral and ionic heats of formation (Jolly and Gin, 1977; McKelvey, Alexandratos, Streitwieser, Abboud and Hehre, 1976; Sanderson, 1982; Pihlaja, Rossi and Vainiotalo, 1985) and *ab initio* molecular orbital calculations can provide surprisingly detailed information (Bentley, 1979; Radom, 1992; Smith and Radom, 1993). The use of these calculated heats of formation greatly extends the degree of speculation that can be applied to mechanisms of ion fragmentation. The use of thermochemical arguments is further exemplified in section 9.10.

9.9. ION LIFETIMES

Discussion of theoretical aspects of mass spectrometry has so far been concerned with the rates at which ions fragment in relation to the excesses of internal energy that they contain and their structures, and with predictions on the direction that fragmentation might take based on rate equations and thermochemical arguments. It should be remembered that a mass spectrometer samples a time 'window', which is determined by the difference between the times of formation and detection of an ion in the instrument. In turn, these times depend upon the construction and operation of the instrument. In sector instruments, a typical time window might be a few microseconds but, in an ion trap or an ion cyclotron resonance spectrometer, the time window may be several or tens of milliseconds and in the case of field ionization only a few picoseconds. Thus, the appearance of a mass spectrum depends not only on the behaviour of ions after their formation with various amounts of excess of internal energy but also on the time window sampled. For example, in field ionization, the time spent by an ion in the ion source after formation is very short (10^{-12} s) and, because of its rapid acceleration from the source, its flight time to the detector is also short. Therefore, very fast decompositions can be examined (the limiting rate of any decomposition is equal to a bond vibration frequency and is usually of the order of 10^{-13}–10^{-10} s^{-1}). This time window has led to the development of *field ion kinetics*, in which the fragmentation behaviour of ions can be examined soon after formation. For greater detail on this field of research, the reader is referred to excellent reviews (Beckey, 1961; Derrick, 1977).

On the other hand, ions that contain very small amounts of excess of energy or that undergo extensive rearrangement before fragmentation will decompose much more slowly, in say 10^{-7}–10^{-6} s, a time window more suited to sector instruments or ion traps. These long-lived (meta-stable) ions provide structural information on ions with small, but ill-defined, excess of internal energy. The uses and investigation of metastable ions (Cooks, Beynon, Caprioli and Lester, 1973) are discussed more fully in chapter 8. By 'tuning' a mass spectrometer to these different time windows, valuable information on fragmentation or structure of ions can be obtained. From a practical, analytical viewpoint, observed mass spectra may differ in appearance, depending on the excess of internal energy in the newly formed ion and on the time window needed for examining the spectrum of m/z values. For this reason it is essential to state clearly in any publication, which instrument has been used and under what conditions it was operated. The success of library matching programmes is critically dependent on being able to compare spectra in a library with spectra for unknowns obtained under similar conditions. This is one of the major reasons why perfection in 'fit' values is rare and why it is acceptable to deduce that a less than perfect fit with a library spectrum might still correctly identify an unknown substance.

9.10. QUALITATIVE THEORIES

Discussion of mass spectrometric fragmentation for structural analysis must be carried out in terms of the structure of the intact molecule before ionization because the structures of ions are generally unknown. Fragmentation of a molecular ion is described with minimum structural change at each step. For example, the loss of $CH_3O^{\bullet}$ from methyl benzoate can be described as a simple bond cleavage, without knowledge of the actual structures of the ions involved, simply by considering the minimum of structural change during ionization and decomposition that is necessary to remove a fragment of the composition CH_3O:

$$[C_6H_5CO\text{-}OCH_3]^{+\bullet} \longrightarrow C_6H_5CO^+ + CH_3O^{\bullet}$$

In the absence of knowledge of ion structures, this empirical approach is valid and provides a convenient way of trying to bring together into a reasonably simple scheme the diverse behaviour of complex molecules after ionization. Conversely, loss of a unit, having an *elemental* composition, CH_3O, from a substance of unknown structure may be taken to imply that a *structural* unit, CH_3O, must be present, without any knowledge of ion structures other than their elemental compositions. Of course, the elemental unit, CH_3O, may actually be derived from the structural unit, CH_2OH, or from other structural units capable of eliminating these elements as one entity. Thus, for structural analysis of unknown compounds by mass spectrometry, it is necessary to investigate and collate the mass spectra of many compounds of known structure and to apply this knowledge empirically to the analysis of an unknown substance. This empirical approach is common in organic chemistry and has proved extremely useful in ultraviolet, infrared and nuclear magnetic resonance spectroscopy. Because the actual structures of ions are unknown, it is not possible to discuss adequately the *mechanisms* of mass spectrometric fragmentations, and still less the intervention of intermediates and transition states. Nevertheless, 'mechanisms' for mass spectrometric fragmentation are frequently published and sometimes supported by evidence from isotopic labelling experiments or other approaches. Whether or not such mechanisms are ever proved correct or incorrect has no effect on the value of an empirical approach to mass spectrometry for structural analysis. Two qualitative theories of mass spectrometry that have proved of considerable value as aids to structure determination are described. The main attribute of these theories is that they classify the many types of mass spectrometric fragmentation into a relatively few categories of behaviour. Knowing these categories, it is easier to discuss or interpret mass spectrometric fragmentation of new compounds of known or unknown structure.

9.10.1. *Stabilities of fragmentation products*

For the fragmentation of an ion to take place it must possess an excess of internal energy sufficient to exceed the activation energy for the reaction considered (E_f in figure 9.8). It has been demonstrated for many fragmentation

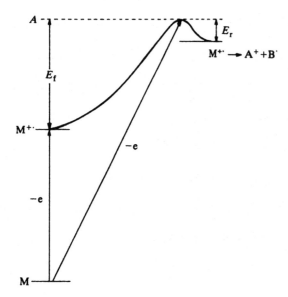

Figure 9.8. An energy diagram showing the relationship between the activation energies (E_f and E_r) for the forward and backward reactions in the decomposition of the ion ($M^{+\cdot}$) to fragments (A^+ and $B^{\cdot}$). The appearance energy for the process is marked (A).

reactions of ions, and is believed to be generally approximately correct, that the activation energy for the reverse reaction (E_r) is almost zero or at least very small. By Hammond's postulate (Hammond, 1955) for reaction kinetics, it may be supposed, for such endothermic fragmentations, that the transition state will resemble the products of reaction and, therefore, that the stabilities of the products will, to an approximation, determine the course of the frag-

$$[C_6H_5COOCH_3]^{+\cdot} \Bigg\langle \begin{array}{l} {}_{\nearrow}C_6H_5CO^+ + CH_3O^{\cdot} \qquad\qquad (9.9a) \\ {}_{\searrow}C_6H_5{}^+ + CH_3OCO^{\cdot} \qquad\qquad (9.9b) \end{array}$$

mentation. Consider two alternative decompositions (9.9a and b) of the molecular ion of methyl benzoate:

Figure 9.9 shows the relative stabilities of various species with respect to the molecular ion. If the transition state does resemble the products ($E_r \approx 0$) and if the frequency factors (equation (9.5)) are similar, then reaction (9.9a) will be favoured over (9.9b) as is found experimentally. For the proper use of this method, it is necessary to take account of the heats of formation of all the products of fragmentation. When the ejected neutral particle is a stable molecular species like H_2O or CO_2, its large negative heat of formation should not be ignored, since it will lead to enhanced fragmentation along the pathway to H_2O or CO_2 in preference to others that might be possible. By such reduction in the energy required for the process compared with similar

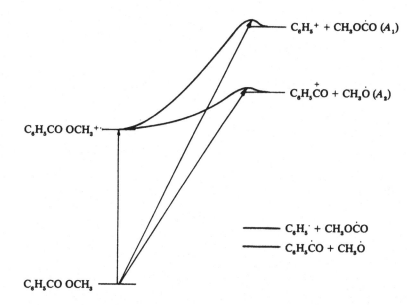

Figure 9.9. An energy diagram showing the relationship between competing fragmentation processes from one state of the molecular ion of methyl benzoate. The appearance energies (A_1 and A_2) relate to the two reactions in scheme (9.9).

processes in which the ejected neutral fragment does not have a large negative heat of formation, one decomposition mode becomes predominant and may mask the presence of other functional groups. Quite often only the stabilities of the ionic products of fragmentation are considered and the heats of formation of any ejected neutral species are often ignored. For example, the fragmentation of an amine, $RCH_2NR'_2$, *beta* to nitrogen is ascribed to 'resonance stabilization' of the positive charge in the fragment ion, $[CH_2NR'_2]^+$; the 'stability' of the neutral reaction product $R^\bullet$ is (incorrectly) not taken into account (equation (9.10)).

$$[R \overset{\backslash}{\diagup} CH_2NR'_2]^{+\bullet} \rightarrow R^\bullet + \overset{+}{C}H_2NR'_2 \leftrightarrow CH_2 = \overset{+}{N}R'_2 \qquad (9.10)$$

[For $E_r = 0$, then $E_f = \Delta H_f(R^\bullet) + \Delta H_f(\overset{+}{C}H_2NR'_2) - \Delta H_f(RCH_2NR'_2)$]

The 'wiggly' line in equation (9.10) is drawn to indicate which bond is considered to break in the process. A charged species is drawn as a structure inside square brackets. An even-electron ion is marked by a ($+$) sign outside the brackets and an odd-electron species, a radical-cation, by a ($+\bullet$) sign; negative species may be radical-anions ($-\bullet$) or anions ($-$). Many types of mass spectrometric fragmentation have been classified by Biemann (1962) into a few general schemes, some of which are outlined in table 9.1.

These classifications are empirical, imply actual knowledge of ion structures, and consider fragmentation in terms of the structure of the original molecule. Two examples will suffice to illustrate the argument.

Table 9.1 *Examples of empirically classified fragmentation modes (Biemann, 1962).*[a]

Type A$_3$:

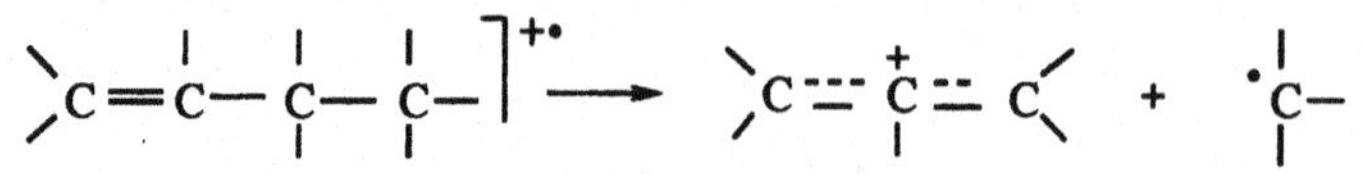

Type A$_5$:

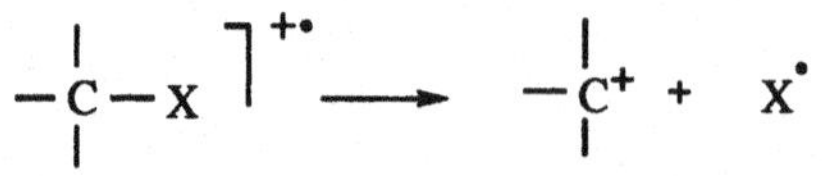

Type B:

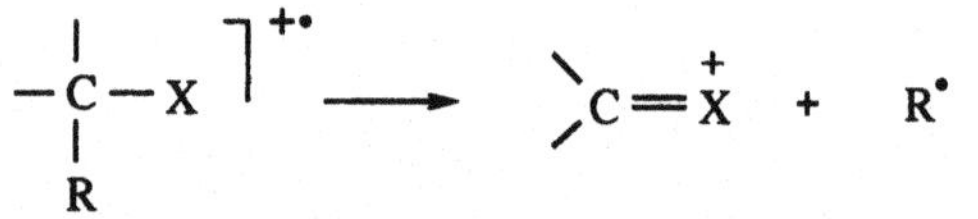

Notes:

[a] These and other reaction types are classified slightly differently in this book; the Biemann classification was the first, and still valid comprehensive effort in this way of approach to reaction paths in mass spectrometry. Type A$_3$ represents 'allylic' cleavage; Type A$_5$ and B represent cleavage adjacent to the α-carbon atom.

(i) n-Alkane molecular ions decompose extensively to give a series of fragment ions $[C_nH_{2n+1}]^+$ separated by 14 mass units:

$$[C_nH_{2n+1}]^{+\bullet} \longrightarrow CH_3^+, C_2H_5^+, C_3H_7^+, \ldots$$

However, a branched alkane shows enhanced or preferred bond fission adjacent to the branching point and this has been ascribed to the increased stability of secondary or tertiary cations compared with the primary ones generated from straight-chain alkanes:

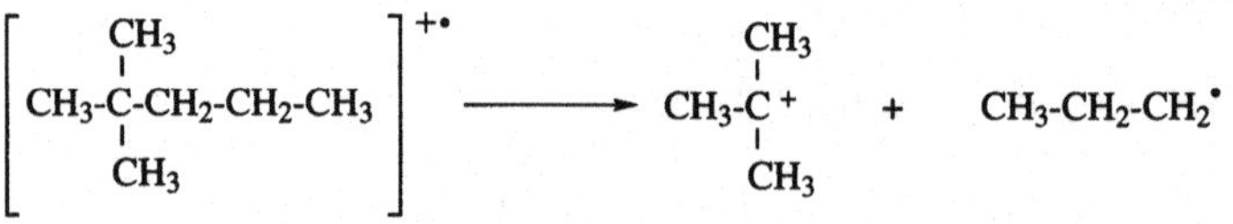

(ii) Bond cleavage of the cinnamyl compound (1) takes place pre-dominantly at the position shown and this has been thought to be due to the stability of the resulting cinnamyl cation (2). The subsequent loss of H_2 from this cation may be ascribed to cyclization with formation of a stabilizing indenyl cation (3):

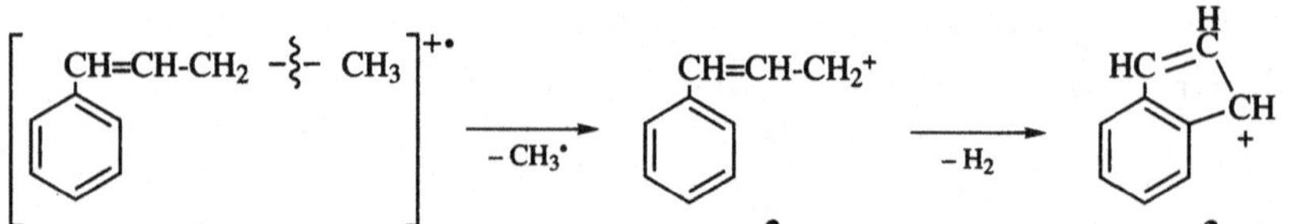

Whether or not the ion structures are correct makes no difference to this empirical approach based on the mass spectrometric behaviour of many known series of compounds. It is a useful aid for summarizing various types of mass spectral fragmentation so that this knowledge can be applied more readily to the interpretation of mass spectra of unknown substances.

9.10.2. *Charge localization*

An alternative theory of mass spectrometric fragmentation applied extensively by Djerassi (Budzikiewicz, Djerassi and Williams, 1967) supposes that, after ionization, the charge on the molecular ion may be considered to be localized at some particular place in the ion. For an aliphatic amine, $RCH_2NR'_2$, this site of charge localization is the nitrogen atom because one of the 'lone-pair' electrons on nitrogen would be easiest to remove. The charge site is then considered to 'trigger' fragmentation by one- or two-electron shifts:

$$R - CH_2 - \overset{\cdot\cdot}{N}R'_2 \xrightarrow{-e^-} R - CH_2 - \overset{+\cdot}{N}R'_2 \longrightarrow R^\cdot + CH_2 = \overset{+}{N}R'_2$$

A double-headed arrow indicates a two-electron shift and a single-headed arrow, or 'fish-hook', a one-electron shift. Charge stabilization by the product ion is therefore a 'bonus' making the reaction more favourable. It has also been argued that it is specifically the presence of the radical site that is the driving force for the reaction (McLafferty and Turecek, 1993). To apply the charge localization treatment it is only necessary to decide where the charge is localized (in fact, this may not be at all obvious in a complex molecule). It is unnecessary to classify fragmentations by type, although this is often done, and therefore the method provides both an *aide-memoire* for collecting large amounts of information on mass spectrometric fragmentation behaviour and a means of assessing the possible direction of this fragmentation in compounds for which mass spectra have not been determined. Similarly, using these same principles for the interpretation of the mass spectra of compounds of unknown structure affords a convenient guide when deciding on alternative possible structural features. The losses of an ethyl radical from the epoxide (**4**) and of C_2H_2O from acetanilide (**5**) are used to illustrate the method:

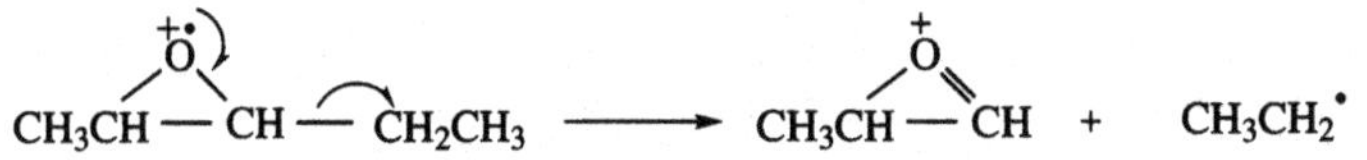

In these cases, the charge-site is the atom or group of lowest ionization energy and bond cleavage is shown as occurring adjacent to these sites so as to yield 'stable' cations.

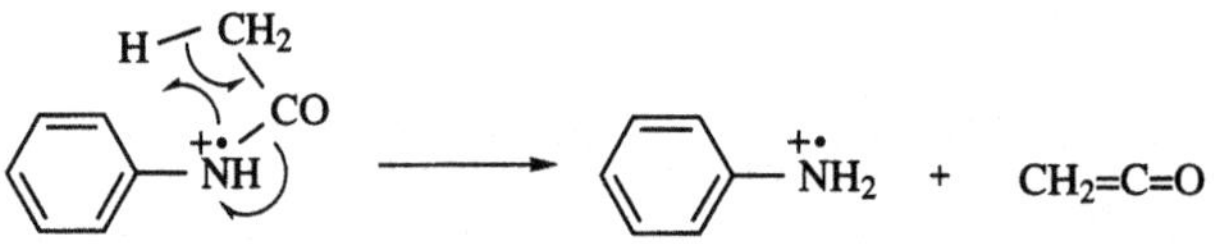

For relatively stable entities such as protonated molecules, the site of protonation has been examined in various ways. It appears that the protons mostly populate those sites of greatest basicity, which might have been expected. What is unexpected are the changes in basicity observed for molecules in the gas phase compared with the same molecules in solution. Thus, in solution, a benzenoid ring would be considered much less basic than an amino group so that protonation is entirely on the latter species. In the gas phase, this is not necessarily true. For example, in aniline, the site of protonation is thought to be mainly in the aromatic ring and not on the nitrogen.

9.10.3. *Comments on the two qualitative theories*

For a theory to be worthy of the name, it should have a truly predictive capability. This criterion does not apply to either of the qualitative theories just discussed. Neither is truly predictive because both rest heavily on argument by analogy or comparison with previous experience. From a theoretical point of view, the first hypothesis, based on Hammond's postulate of the transition state, is possibly more satisfying, albeit at one time less fashionable. The second theory, which requires a charge-localized site to trigger fragmentation, would appear to be opposite to Hammond's postulate. There is no reason to suppose that triggering is necessary for fragmentation along any particular pathway. Electrons in a molecule are held in molecular orbitals and removal of one electron affects the whole structure to various degrees. Even 'lone-pair' electrons are not strictly localized on atoms in ions, although this is a good approximation for ground-state properties of organic molecules (Dewar and Worley, 1969). For example, the 'lone-pair' electrons on the oxygen atom in methanol have been estimated to be about 25 per cent delocalized over the molecule. Perhaps these considerations can lead to a fusion of the two qualitative theories of mass spectrometric fragmentation. Removal of an electron from a molecular orbital will change the various bond vibration frequencies and strengths to different extents in different parts of the ion. It is only necessary for sufficient vibrational energy to be concentrated along the reaction coordinate for fragmentation to occur. If the electron deficiency in the ion is spread out over a 'molecular' orbital it may nevertheless be concentrated at certain parts in the ion (charge localized). Many interpretations of charge localization really use product stability arguments to decide the direction that fragmentation will take and, from this, it has appeared that charge localization leads to 'triggering' of fragmentation. Therefore, triggering often appears as a deduction from charge localization theory when, in fact, product stability has already been invoked to explain preferred modes of cleavage. By way of illustrating this point, two competing fragmentations of compounds of the type C_2H_5X may be considered, as shown in scheme (9.11):

$$CH_3 \text{—} CH_2 \text{—} \overset{\cdot+}{X} \longrightarrow CH_3^{\cdot} + CH_2 = \overset{+}{X}$$

or

$$CH_3 \text{—} CH_2 \text{—} \overset{\cdot+}{X} \longrightarrow C_2H_5^{+} + X^{\cdot}$$

charge localization
description of processes

actual fragmentation
path determined   (9.11)
by product stabilities

Charge localization alone cannot determine, which cleavage in the very simple C_2H_5X compounds (scheme (9.11)) is more favourable, but product stability arguments can. In fact, although bond strengths have been affected in different ways on ionization, Hammond's postulate may still be applied and the stabilities of products will largely determine the direction of fragmentation in the absence of a large activation energy for the reverse reaction and not forgetting the effects of frequency factors. Therefore, it may be suggested that charge localization can indicate which bonds in a molecule are most likely to be affected on ionization and that product stability arguments will then mostly determine the actual direction of fragmentation. From this point of view, the two theories are complementary and their conventions of the 'wiggly line' and 'fish-hooks' or 'arrows' are used somewhat indiscriminately throughout this book as a means of 'electronic book-keeping'.

10 Structure elucidation

10.1. CLASSIFICATIONS OF MASS SPECTRA

Electron ionization mass spectra of many thousands of compounds of known structure have now been determined and collated and are available as 'libraries' of mass spectra. Published volumes of mass spectra, the contents of which are usually in the form of normalized tables of mass spectra, are available from various sources (section 12.5). If the mass spectrum of a substance of unknown structure has been obtained, it is useful to be able to sort through compilations of published spectra in order to compare them with it. In this comparative way, it may be possible to identify the unknown compound without needing to understand the mass spectrum at all. Even if the structure of an unknown substance can be deduced from its mass spectrum, the compilations are useful for confirming that structure. The task of searching manually through large compilations is tedious but preferable to looking through scattered research literature. Fortunately, searching through large compilations of mass spectra is a task ideally suited to a computer and automatic library searching has already been discussed (section 4.4.3). Of course, the classification of mass spectra in this way is of limited use if the mass spectrum of the substance to be identified has not been determined previously. In such circumstances, the mass spectrum must be interpreted by rationalizing the fragmentation pattern or subjected to computerized spectral interpretation (section 4.4.3). The remainder of this chapter is concerned largely with interpretation through rationalization of mass spectra.

10.1.1. *The functional group approach*

Empirical classifications of data based on the functional group approach, which is so successful in organic chemistry, have been equally successful in organizing the modes of mass spectrometric fragmentation. For example, the fragmentation reactions of aliphatic ketones may be classified together and any fragmentations that appear to be common features of the presence of a ketone group are noted particularly. If an unknown substance is suspected of being a ketone, then its mass spectrum can be examined for those features of fragmentation pathways that are characteristic of other ketones. Classification by functional group makes examinations like this much easier. Frequently, the effects of changes in molecular structure on the fragmentation behaviour of the functional group are included in the classifications and provide very useful extra information. For example, a comparison of the effects on fragmentation of a ketone group in ring and straight-chain compounds can help to decide the environment of a ketone group in an unknown substance.

The functional group approach has its limitations. Unlike many other methods, such as infrared spectroscopy, in which the effect of the functional group can be essentially isolated from the remainder of the molecule outside its immediate environment, mass spectrometric fragmentations are really a set of chemical reactions affecting the whole ion. Therefore, there may not be sufficient difference between the effects of functional groups to allow a positive identification to be made. In particular, changes in the rest of the structure can produce large changes in the 'standard' effects of functional groups. The molecule may contain more than one functional group and, if these interact in the fragmentation process, then the effect could be to mask the 'standard' effects of each of the groups. These problems are not exclusive to mass spectrometry but occur frequently in chemistry and illustrate the inherent dangers of relying exclusively on one particular technique for providing an answer to a problem.

There is neither the scope in, nor is it the intention of, this book to provide more than this summary of the functional group approach to the interpretation of mass spectra. In general, the scheme is very close to similar ones long familiar to chemists in infrared, ultraviolet and nuclear magnetic resonance spectroscopy, and in the classification of reactions of compounds. Accurate mass measurement of the molecular ion in a mass spectrum will give the molecular formula of the compound and this in itself often affords a good indication of the type of functional group present. Several excellent books have been published, which classify and discuss mass spectra in detail on the basis of functional groups and the reader is recommended to read them for further information (Budzikiewicz, Djerassi and Williams, 1964a, 1964b, 1967; Reed, 1966; Beynon, Saunders and Williams, 1968; Hill, 1972; McLafferty and Turecek, 1993). The behaviour of organic compounds during mass spectrometry, classified according to functional groups, has been well reviewed (Bowie, 1971, 1973, 1975, 1977, 1979, 1984, 1985, 1987; Bowie, Trenerry and Klass, 1981; O'Hair and Bowie, 1989). The behaviour of inorganic compounds has also been recorded (Spalding, 1979). These reviews cover the reactions of both positive and negative ions.

10.1.2. *Observation of characteristic m/z values*

It is often observed that a certain feature of molecular structure will give rise to a characteristic peak in the mass spectrum such that the presence of an ion at that m/z value may be taken as good evidence for the feature. This approach is not the same as that based on functional groups since the latter concentrates on fragmentation pathways whereas one based on characteristic m/z values relies on the identification of one or two particular peaks. Thus, an ion at m/z 43 is often evidence for the presence of the grouping CH_3CO in a molecule; it is of course better to determine the elemental composition of the ion at m/z 43 by accurate mass measurement to support the diagnosis. The presence of an ion at m/z 91 ($C_7H_7^+$) is almost classical in mass spectrometry and usually indicates the presence of either a benzylic residue in the molecule or a structure that readily rearranges to give the $C_7H_7^+$ ion. Similarly, an ion at m/z 30 (CH_4N^+) strongly indicates an amine, particularly a primary one.

Lists of such characteristic peaks have been published and they may be used in much the same way as, for example, nuclear magnetic resonance correlation charts (McLafferty, 1963; Beynon, Saunders and Williams, 1968; McLafferty and Turecek, 1993). Just as peaks near $\delta = 7$–8 ppm in a proton magnetic resonance spectrum would suggest the presence of an aromatic structure, so too would the presence of an ion at m/z 91 in the mass spectrum. However, in general, these compilations of characteristic peaks do not have the same usefulness as the correlation charts used in ultraviolet, infrared and nuclear magnetic resonance spectroscopy because, in mass spectrometry, rearrangement reactions accompanying fragmentation are common, i.e. the original molecule may be considerably disturbed on ionization. For example, although an ion of composition C_2H_3O at m/z 43 suggests the CH_3CO grouping in a molecule, it is found that many oxygen-containing compounds without this grouping also give abundant ions at this m/z value. The straight-chain ketone $C_2H_5COC_5H_{11}$ has its base peak due to $C_2H_3O^+$ $(m/z\ 43)$. In this case, the use of tables of characteristic peaks would lead to an inference of a methyl ketone, whereas in fact the ion is a product of a complex rearrangement. As a mass spectrometrist gains experience from interpreting mass spectra, so the usefulness of characteristic peaks increases.

This method of searching for characteristic ions is particularly valuable when closely similar sets of compounds are examined. For example, an ion at $m/z\ 105$ $(C_7H_5O^+)$ is prominent in mass spectra of molecules containing the grouping C_6H_5CO and, on substitution into the benzene ring, the corresponding ion appears at a different m/z value. Thus, benzophenone yields abundant ions at $m/z\ 105$, but 3-chloro-3'-methylbenzophenone gives analogous ions at $m/z\ 119$ $(C_8H_7O^+$ or CH_3—$C_6H_4CO^+)$ and $m/z\ 139,\ 141$ $(C_7H_4OCl^+$ or Cl—$C_6H_4CO^+)$:

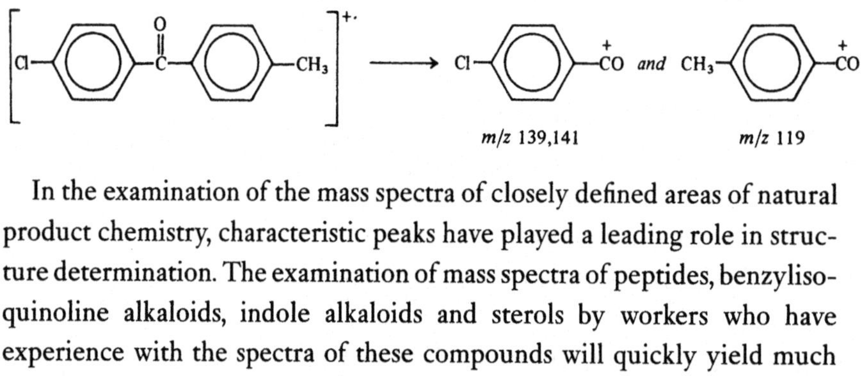

In the examination of the mass spectra of closely defined areas of natural product chemistry, characteristic peaks have played a leading role in structure determination. The examination of mass spectra of peptides, benzylisoquinoline alkaloids, indole alkaloids and sterols by workers who have experience with the spectra of these compounds will quickly yield much valuable information. Therefore, when closely related compounds are being examined it is worthwhile searching their mass spectra for peaks that are characteristic of particular common groupings, so that these groupings may be recognized in the mass spectra of similar compounds of unknown structure. For example, benzylisoquinoline alkaloids (1) give a prominent, characteristic ion (2) in their mass spectra, which is due to ejection of the benzylic residue; trimethylcoclaurine $(R_1 = R_2 = R_3 = CH_3)$ yields the ion (2) at $m/z\ 206$.

Observation of a characteristic m/z value in the mass spectrum of a substance of unknown origin should be treated initially with caution. The cautious interpreter, knowing that all aeroplanes have wings, would not deduce that a bird is an aeroplane!

10.2. EXAMINATION OF THE MASS SPECTRUM

The first step in examining an EI mass spectrum is usually to decide which peak, if any, represents the molecular ion. The spectrum is examined for significant peaks at highest m/z values; the point at which these peaks occur is the potential molecular ion region. In assigning the molecular ion peak it is worth recalling that it is accompanied by peaks representing isotopes, one or more mass units greater ($[M + 1]^{+\cdot}$, $[M + 2]^{+\cdot}$, etc.). For chlorine, bromine and sulphur the $[M + 2]^{+\cdot}$ ions are abundant compared with the $[M + 2]^{+\cdot}$ ions of similar compounds without these heteroatoms. Figures 10.1(a) and (b) show, respectively, examples of the molecular ion regions for a typical hydrocarbon, with significant peaks up to $[M + 2]^{+\cdot}$, and for a dihalo-compound with peaks up to $[M + 5]^{+\cdot}$ ions. Often ions $[M - 1]^{+}$ and sometimes $[M - 2]^{+\cdot}$, $[M - 3]^{+}$, etc. are observed owing to losses of hydrogen from molecular ions. Figure 10.1(c) illustrates the molecular ion region of 2-styrylpyridine, which looks very similar to that of the hydrocarbon in figure 10.1(a.). However, the most abundant ion in the nitrogen compound is the $[M - 1]^{+}$ peak and not the molecular ion. Sometimes these losses of hydrogen are so marked as to lead to the wrong assignment of the molecular ion because the $[M - 1]^{+}$ ion is so much more abundant than the molecular ion. If such losses of hydrogen are suspected it is worthwhile examining the spectrum at lower ionizing energies or recording the mass spectrum using a mild method of ionization, such as chemical ionization or fast atom bombardment, or observing any metastable ions that fragment by losses of hydrogen. Figure 10.1(d) shows the molecular ion region in biphenyl with the marked loss of up to four hydrogen atoms from the molecular ion.

Difficulties are also encountered when molecular ions are of low abundance or even nonexistent so that it becomes impossible to determine directly the relative molecular mass of the compound. Sometimes the

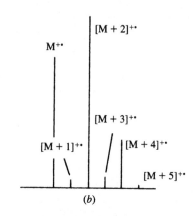

(a)

(b)

Figure 10.1. A comparison of the molecular ion regions of (*a*) phenylethyne, (*b*) 1-bromo-2-chlorobenzene, (*c*) 2-styrylpyridine and (*d*) biphenyl.

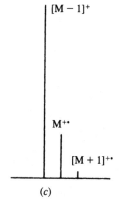

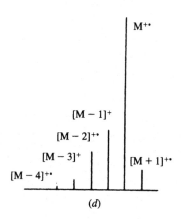

(c)

(d)

difficulty can be resolved by examining the fragment ions and deducing the molecular mass. It is more satisfying to re-examine the mass spectrum after ionization with low electron beam energies, or after chemical ionization, or after ionizing the substance with one of the very mild techniques like electrospray, since these methods often afford abundant molecular or quasi-molecular ions when, under normal electron-ionization conditions, only uncertain ones are found, or none at all.

To be due to molecular ions, a candidate peak must pass several criteria. Firstly, the molecular ion peak must be that of largest mass in the spectrum, apart from those of natural isotopes. However, it is reasonable to expect some 'background' ions of relatively low abundance, for example, ions derived from vacuum pump oil, from impurities such as plasticizers in samples, from the liquid matrix in techniques like fast atom bombardment and electrospray, and from column bleed in combined gas chromatography/mass spectrometry. Impurity ions may extend to high mass and complicate the assignment of the molecular ion peak but, with experience, a mass spectrometrist can usually distinguish such signals without too much trouble. Secondly, in the analysis of organic compounds, the elements most commonly met (C, H, N, O, S, Cl, Br, etc.) very rarely give rise to losses of 5–14 and 21–25 mass units, inclusively. If a candidate peak for the molecular ion shows such losses, it is most unlikely to be correctly assigned. On the other hand, losses of H$^{\bullet}$ and CH$_3$$^{\bullet}$ (1 and 15 mass units) are commonly observed for molecular ions of organic compounds. Thirdly, molecular ions must contain all of the elements that are in evidence elsewhere in the spectrum. For instance, if the characteristic isotope ratio for chlorine is seen in some lower mass ions, the molecular ion region must show it too. In high-resolution mass spectra that provide elemental compositions of ions, fragment ions cannot contain greater numbers of any element than does the true molecular ion. Lastly, if data from metastable ions implicate the candidate peak as being due to product ions of some higher mass species, then it cannot be the true molecular ion peak.

If the relative molecular mass or molecular formula of the compound is the only requirement, then deciding on the molecular ion will be the sole purpose of examining the spectrum. Even so, this can still lead to much valuable information. As has been shown above, the very appearance of the molecular ion region can furnish information. It is also useful to remember that an *odd* number for the relative molecular mass indicates an *odd* number of nitrogen atoms in the compound. Abundant [M + 2]$^{+\bullet}$ ions would suggest the presence of chlorine or bromine, and less abundant ones the presence of sulphur. An approximate idea of the number of carbon atoms in the molecule can be gained from the relative heights, h and h', of the M$^{+\bullet}$ and [M + 1]$^{+\bullet}$ peaks, respectively. Since ^{13}C is present in about 1.1 per cent natural abundance in organic compounds, the approximate number of carbon atoms is given by $100h'/1.1h$. For example, in the mass spectrum of naphthalene, which has ten carbon atoms, $h'/h = 10.9/100$ and $(100 \times 10.9)/(1.1 \times 100) = 10$ to the nearest integer. The method is not very accurate for more than ten or twelve carbon atoms and may yield highly erroneous results if the compound contains nitrogen (because of the contribution of ^{15}N to the [M + 1]$^{+\bullet}$ peak), if the compound has a strongly basic group that is prone to protonation, giving [M + H]$^+$ ions, or if there are overlapping peaks due to impurities. For large numbers of carbon atoms in a molecule,

Table 10.1. *Some elemental compositions at nominal mass 100.*

Composition	Accurate mass
$C_3H_6N_3O$	100.0511
$C_4H_8N_2O$	100.0637
$C_5H_{10}NO$	100.0762
$C_6H_{14}N$	100.1126
C_7H_{16}	100.1251
C_2F_4	99.9936
$C_2H_4OSi_2$	99.9700

the $[M + 1]^{+\cdot}$ ions assume greater prominence. For more than 90 carbon atoms, the abundance of $[M + 1]^{+\cdot}$ ions and therefore its ion peak exceeds that of $M^{+\cdot}$ ions (section 12.2.3).

To obtain the molecular formula of a compound, it is necessary to measure the accurate mass of the molecular ion. Knowing the accurate mass, the molecular formula can be obtained from tables or by calculation (Beynon and Williams, 1963; Lederberg, 1964). Most computer workstations designed for use with a high-resolution mass spectrometer have a pro-gramme for determining elemental compositions from measured values of accurate masses. Table 10.1 shows some of the elemental compositions pos-sible at m/z 100. If the measured accurate mass of the molecular ion of an unknown compound was 100.0635, it can be seen from table 10.1 that this mass corresponds very closely to that calculated for $C_4H_8N_2O$. The error between calculated and observed masses is often reported in parts per million (ppm). In the example cited, the difference between calculated and observed values is 0.0002 mass units, i.e. the error is two parts in 1 000 637, which is about 2 ppm. Errors of up to ± 10 ppm are usually acceptable but the smaller the error, the more certain the molecular formula. It may be mentioned here that accurate mass measurement is not a substitute for ele-mental analysis, although the two complement each other very well. The mass spectrometer yields the exact molecular formula but may give little indication of the purity of a compound. The elemental analysis provides a cross-check on the purity of a compound and on its molecular formula.

Having obtained a molecular formula, it may then be used to calculate the number of double-bond equivalents in the compound. For the commoner elements met in organic chemistry, this may be done very simply by substituting CH_3 for each halogen, CH_2 for each oxygen or sulphur atom, and CH for each nitrogen. For example, carrying out these substitutions on the molecular formula C_6H_6ONCl gives C_9H_{12}; an alkane with nine carbon atoms would have a molecular formula C_9H_{20}. The difference between these formulae, $C_9H_{20}-C_9H_{12}$, is eight hydrogen atoms and therefore there are four double-bond equivalents in C_6H_6ONCl.

During the initial examination of a mass spectrum, its general appearance is frequently a guide to the nature of the compound under investigation. A spectrum exhibiting many fragment ions increasing in abundance towards low m/z values will suggest a predominantly aliphatic structure whereas a spectrum with few fragment ions, abundant molecular ions and doubly charged ions will suggest an aromatic structure. Figure 10.2 illustrates this point by comparing the EI mass spectra of naphthalene and pentanal.

The next step in examining the mass spectrum is to note the major

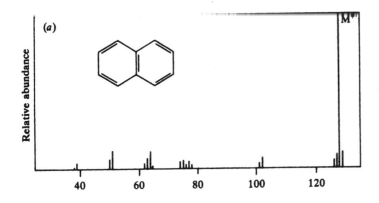

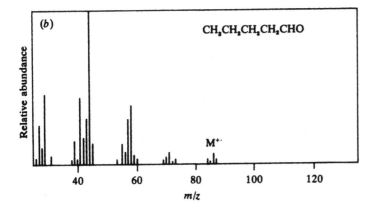

Figure 10.2. A comparison of mass spectra of (a) an aromatic (naphthalene) and (b) an aliphatic compound (pentanal).

fragment ions and attempt to elucidate the main fragmentation pathways. The major fragment ions may occur at 'characteristic' m/z values (see above) and give immediate information. In the absence of nitrogen atoms, fragment ions occurring at odd-numbered m/z values are even-electron species resulting mainly from simple bond fission, whereas fragment ions at even-numbered m/z values are odd-electron species produced by multiple bond cleavage, suggesting that rearrangement may have occurred. In this latter case, extra care must be taken in interpreting the spectrum because some of the fragment ions may well arise from rearranged species. Nonetheless, some rearrangements accompanying fragmentation are highly characteristic of certain structural features. Hence, they are particularly worthy of attention during interpretation.

It is best to check supposed fragmentation pathways by looking for the confirmatory fragmentation of metastable ions in the routine spectrum. For a mass spectrum recorded on photographic paper, recall that the position of the apparent mass of the product ion peak (m^*) for metastable ion decomposition is given by the expression $(m_2)^2/m_1$, where m_1 is the mass of the normal precursor ion and m_2 the mass of the normal product ion. The mass spectrum of ethyl benzene ($M^{+\cdot}$ at $m/z = 106$) shows large peaks for $[M - 15]^+$ at $m/z\,91$ and $[M - 41]^+$ at $m/z\,65$. The presence of peaks at m/z 78.1 and m/z 46.4 arising from metastable ions proves that one fragmentation path at least is $M^{+\cdot} \rightarrow [M - 15]^+ \rightarrow [M - 41]^+$ because the calculated values for the masses of the precursor and product ions, m_1 and m_2, for the metastable ion data are m/z 106 and 91 ($91^2/106 = 78.1$) and m/z 91 and 65 ($65^2/91 = 46.4$). Even with photographic recording the evidence for metastable ions for suspected fragmentation reactions may not be visible in the routine spectrum and certainly will not be when the spectrum has been acquired by computer. It is then necessary to look for metastable ion decomposition by the specialized techniques described fully in chapter 8.

Having determined the relative molecular mass and the main features of the fragmentation of the compound under investigation, it may be possible to make tentative or even definite structural assignments. Comparison of the mass spectrum with those in a library may allow immediate identification of the compound (section 10.1). At this stage, it is advisable to assemble any other available physical data on the compound. For example, interpretation of a proton magnetic resonance spectrum will benefit from a knowledge of the molecular formula, which gives the exact number of protons present and also indicates the presence of particular heteroatoms. The ultraviolet spectrum will yield information on the likely degree of unsaturation so that the number of double-bond equivalents obtained from the molecular formula may indicate whether there are any rings, and if so how many. The infrared spectrum will also often give information on the state of saturation of the molecule and may give very valuable clues concerning the presence of carbonyl, hydroxyl and other groups. If it is known that such groups are present, it is possible to examine the mass spectrum in conjunction with the known fragmentation behaviour of these functional groups (section 10.1). If the infrared spectrum indicates the presence of a hydroxyl group, then reference to the fragmentation reactions of alcohols in general may reveal similar fragmentations in the spectrum of the unknown substance. It is very much a case of pooling all available information, using it logically and calling on experience.

It may not be possible to proceed any further with the interpretation without knowledge of the elemental compositions of the fragment ions. Loss of 27 mass units from the molecular ion of C_6H_6ONCl may be ejection of HCN or $C_2H_3 \cdot$ and accurate mass measurement on the fragment ion will determine which it is. A simple mass spectrum containing few fragment ions requires only a little extra work to obtain the accurate masses of these ions and hence their elemental compositions. On the other hand, a complex mass spectrum requires many accurate mass measurements and it should be remembered that any peak at an integer m/z value may represent ions of different compositions. Therefore, at high resolution, some peaks will appear as doublets or triplets. A long list of elemental compositions corresponding to the most abundant ions in a mass spectrum is difficult to sift for information that would be of help in structural work. The problem has been recognized especially following the application of computers to accurate mass measurement (section 4.4.1) since this technique generates a great deal of information, which must be examined in detail. A method of assembling all of the data into some order is through the use of an *element map*. An example of one common type of these maps is given in figure 10.3 and relates to the mass spectrum of a natural product, genipin. The ions are arranged in an array of increasing complexity of elemental compositions from left to right, and with increasing m/z values from top to bottom. Thus, the molecular ion is found on the right of the map; doublet peaks are marked (d). The sizes of the peaks are not given in figure 10.3, but this information is

m/z	C_xH_y	m/z	C_xH_yO	m/z	$C_xH_yO_2$	m/z	$C_xH_yO_3$	m/z	$C_xH_yO_4$	m/z	$C_xH_yO_5$
				60	$C_2H_4O_2$						
				61	$C_2H_5O_2$						
		59	C_3H_7O	71	$C_3H_3O_2$(d)	103	$C_4H_7O_3$(d)				
		70	C_4H_6O	83	$C_4H_3O_2$(d)						
69	C_5H_9	81	C_5H_5O	100	$C_5H_8O_2$						
71	C_5H_{11}(d)	82	C_5H_6O								
		83	C_5H_7O(d)								
77	C_6H_5	94	C_6H_6O	109	$C_6H_5O_2$(d)	128	$C_6H_8O_3$				
78	C_6H_6	95	C_6H_7O			129	$C_6H_9O_3$				
79	C_6H_7	96	C_6H_8O								
80	C_6H_8										
90	C_7H_6	105	C_7H_5O	124	$C_7H_8O_2$	139	$C_7H_7O_3$				
91	C_7H_7	106	C_7H_6O								
92	C_7H_8	107	C_7H_7O								
93	C_7H_9	108	C_7H_8O								
		109	C_7H_9O(d)								
102	C_8H_6	119	C_8H_7O	135	$C_8H_7O_2$						
103	C_8H_7(d)	120	C_8H_8O	137	$C_8H_9O_2$						
		121	C_8H_9O								
		130	C_9H_6O	147	$C_9H_7O_2$	165	$C_9H_9O_3$				
		131	C_9H_7O	148	$C_9H_8O_2$	166	$C_9H_{10}O_3$				
				149	$C_9H_9O_2$	167	$C_9H_{11}O_3$				
				151	$C_9H_{11}O_2$						
				152	$C_9H_{12}O_2$						
				158	$C_{10}H_6O_2$	176	$C_{10}H_8O_3$	194	$C_{10}H_{10}O_4$		
				159	$C_{10}H_7O_2$			195	$C_{10}H_{11}O_4$		
				161	$C_{10}H_9O_2$	179	$C_{10}H_{11}O_3$				
				162	$C_{10}H_{10}O_2$	180	$C_{10}H_{12}O_3$				
				163	$C_{10}H_{11}O_2$						
						190	$C_{11}H_{10}O_3$	208	$C_{11}H_{12}O_4$	226	$C_{11}H_{14}O_5$

Figure 10.3. An element map of genipin. The ions are arranged in order of increasing complexity from left to right and top to bottom. Doublet peaks are marked (d).

frequently included also on element maps. Other methods of element mapping have been proposed.

Inspection of the map in figure 10.3 shows, for example, that the ion at m/z 129 cannot arise from m/z 147 (difference of 18 mass units, which could have been due to H_2O) because the former has one more oxygen atom than the latter. The minimum number of carbon atoms associated with all five oxygen atoms is eleven; with three oxygen atoms it is four (m/z 103). The column headed C_xH_yO in the element map shows that nine carbon atoms

3

are associated with one oxygen (m/z 131). The formula for genipin (**3**) shows which atoms these must be. Thus, if the structure of genipin had been unknown, significant structural features could have been gleaned from the map.

These maps provide a convenient means of presenting all of the information relating to the elemental compositions of the ions in a spectrum but usually do not display data for metastable ions. With the advent of automatic recording of metastable ions in the first and second field-free regions (chapter 8), the use of *metastable-ion maps* in conjunction with element maps is likely to increase.

10.3. MODIFICATION OF MASS SPECTRA THROUGH INSTRUMENTAL PARAMETERS

After initial examination of an electron ionization mass spectrum, it may be desirable to modify it by changing the ionizing energy, the temperature of the ion source, or indeed the method of ionization itself. A complex EI spectrum can be simplified by using electrons of lower energy for ionization. A change in the appearance of a spectrum with time or the presence of unexplained peaks may be due to thermolysis of the sample, or to the sample being a mixture. Changing the temperature or trying a different (milder) ionization technique will shed some light on these possibilities. Modification of mass spectra by the use of chemical derivatives is discussed separately (see chapter 6).

10.3.1. *Ionizing energy*

Field ionization and photo-ionization sources usually operate at a single ionizing energy, for example the 21.21 eV helium line for the latter, but electron ionization sources have continuously variable electron-beam energies.

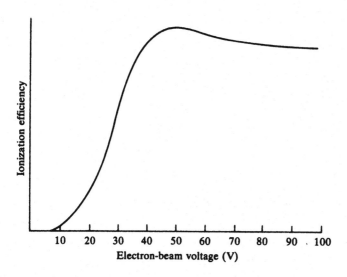

Figure 10.4. A typical curve illustrating the variation in the yield of ions from a compound with increasing electron-beam voltage.

Electrons from a hot filament are accelerated through a potential (V), which ranges from about 5 to 100 V. By convention, *standard* mass spectra are obtained with 70 V electrons because, at this energy, the yield of ions is near maximum and almost constant over a small voltage range. A typical graph showing ion yield as a function of electron energy is given in figure 10.4. A maximum yield of ions is often obtained at electron energies of 50 rather than 70 V but the former energy is not used for standard work because, away from the rather flat maximum near 70 volts, small variations in voltage can lead to pronounced changes in the appearance of the mass spectrum. There is little significance attached to the conventional figure of 70 V for the energy of the ionizing electrons and a mass spectrum can be obtained at any energy above the ionization energy of the compound examined. However, because ion *yield* increases with increasing electron energy up to about 70 V then, for greatest sensitivity, it is better to measure spectra at 50–70 V.

The ionization energies of most compounds lie between 7 and 11 eV and, from the ionization energy up to an electron-beam energy of about 20 V, not only does the yield of ions increase but also the *appearance* of the mass spectrum changes appreciably. At, and just above, the ionization energy, only molecular ions are formed, provided that the molecular ion is not in a dissociative state. As the electron-beam energy is increased more rotational and vibrational energy is deposited in the molecular ion by internal energy conversion processes outlined earlier (section 9.2). With a sufficient excess of internal energy, the molecular ion fragments and this fragmentation

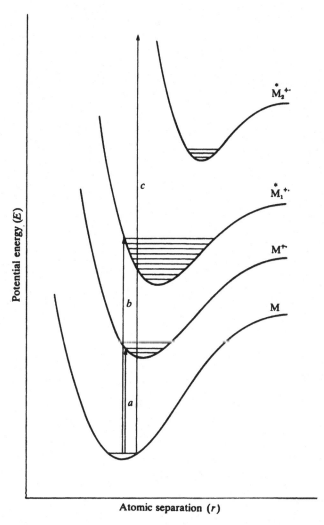

Figure 10.5. Morse curves showing ionizing transitions to excited states $[M_1^{+\bullet}]^*$ and $[M_2^{+\bullet}]^*$ for a molecule M.

becomes increasingly complex as the initial excess increases. Increasing ionizing electron energy means that higher and higher excited states of an ion (for example, $[M_1^{+\bullet}]^*$ in figure 10.5) can be reached by vertical transitions from the molecule M. At some point, vertical transitions to highly excited states of the ion (for example, $[M_2^{+\bullet}]^*$) are not possible and no more energy can be transferred from the ionizing electron. This point is not reached abruptly and maximum transfer of energy from the ionizing electron is reached relatively slowly. Thus, as the energy of the ionizing electron is increased, the energy that is deposited in a molecule on ionization increases rapidly at first and then much more slowly after about 20 eV. At the same time, the *probability* of ionization continues to increase so that the actual *yield*

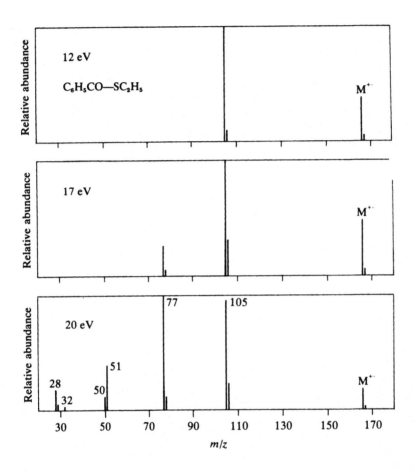

Figure 10.6. Changes in the mass spectrum of ethyl thiolbenzoate with increasing electron-beam energy.

of ions continues to increase from about 20 to about 50 eV although the maximum transfer of energy occurs at around 20 eV. Therefore, although more and more ions are formed up to 50–70 eV, little extra fragmentation is observed after 20 eV. In fact, experimental evidence suggests that only about 5–6 eV of energy above the ionization energy of a compound are transferred by 70 V electrons. The effect on the appearance of the mass spectrum is as follows. At the ionization energy, molecular ions are formed; up to about 20 eV the mass spectrum becomes increasingly complex as more and more energy is transferred during ionization, leading to increased fragmentation; after 20 eV and up to 50–70 eV the *appearance* of the mass spectrum changes little but the total *yield* of ions increases, leading to greater sensitivity in the mass spectrometer. Figure 10.6 illustrates changes in the mass spectrum of ethyl thiolbenzoate with increasing electron beam energy.

Conversely, a standard 70 V EI mass spectrum can be simplified

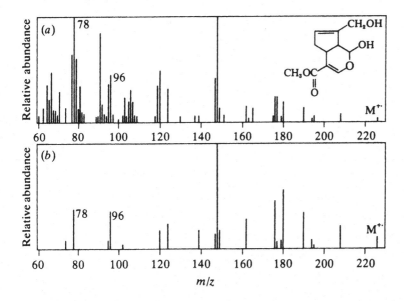

Figure 10.7. Mass spectra of genipin at (*a*) 70 and (*b*) 10 eV. The lower energy spectrum is much simpler and peaks at m/z 78 and 96 remain prominent.

considerably by measuring it at electron-beam energies below 20 V. Simplification of a complex mass spectrum in this way may lead to a better understanding of the fragmentation paths. Figure 10.7 shows the effect on the mass spectrum of genipin of reducing the ionizing electron-beam energy from 70 to about 10 V. The spectrum is greatly simplified and two ions at m/z 78 and 96 of particular importance for its interpretation 'emerge' from the clustered fragment ions at low m/z values (typical of aliphatic compounds) in the higher energy spectrum.

The simplifying effect of reducing the electron-beam energy may be used in conjunction with reduction of the temperature of the ion source (see below) to afford even simpler electron ionization mass spectra.

10.3.2. *Temperature*
The thermal energy imparted to a sample by high temperatures in the inlet system or ion source may have profound effects on the mass spectrum. The effects observed may result from actual changes in molecular structure or may be caused by the excess of vibrational and rotational energy gained by the molecule before it is ionized (section 9.1).

It is often necessary to heat a sample to achieve a great enough vapour pressure in the ion source for mass spectrometry. On heating, a compound may react uni- or bimolecularly to give new compounds. The mass spectrum of the sulphonylhydrazide (**4**) was complicated by the presence of the

$$(10.1)$$

azine (5) formed from it thermally (reaction (10.1)). Effects of this kind are more often found with heated inlet systems operating at relatively high pressures (10^{-2} Torr), particularly if they contain metallic parts that can catalyse reactions. Normally, these effects of temperature on molecular structure are far less severe in direct inlet systems, which operate at a lower pressure (10^{-6} Torr) and therefore require less heat for vaporization. A direct insertion probe that may be heated or cooled independently is a good device for getting a sample into the ion chamber in a controlled manner. Although bimolecular reaction is much less likely at the low pressures employed, unimolecular reaction can still take place inside the ion chamber.

The temperature of the ion source can affect a mass spectrum markedly other than by actual chemical reaction taking place. On entering the ion source, molecules are not necessarily ionized immediately but, because of the low vapour pressure (long mean free path), they can collide with the walls of the source many times before ionization. Even if there are no chemical effects at each wall collision, equilibration of thermal energy occurs. The ion sources of electron ionization sources typically operate at temperatures between 150 and 250°C so that a molecule can gain a lot of thermal energy before crossing the electron beam and being ionized; this thermal energy must be added to that imparted during ionization and the molecular ion contains excesses of energy from the hot source and from electron ionization. In practice, it is found that mass spectra of aromatic compounds that have fairly rigid molecular structures show little evidence of this thermal effect but aliphatic structures frequently show greatly enhanced fragmentation because of it. Figure 10.8 compares the partial mass spectrum of a peptide at two temperatures of the ion source and it can be seen that the size of the molecular ion peak is greatly reduced at the higher temperature. Ion

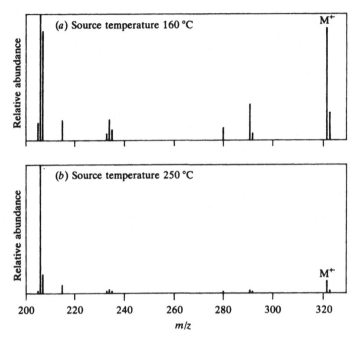

Figure 10.8. A partial mass spectrum of the peptide $C_6H_5CH_2OCO$-Val-Gly-OMe, at (*a*) 160°C and (*b*) 250°C. Both spectra are normalized with respect to *m*/*z* 206.

sources are available that can be cooled externally, thereby giving some control over the temperature to which a sample is subjected.

The simple thermochemical arguments given above are generally applicable to compounds that are not particularly susceptible to thermal decomposition. For thermally labile compounds (often large natural products), the activation energy for decomposition is lower than that for vaporization and hence at lower temperatures only decomposition occurs. At higher temperatures, volatilization and decomposition are competing processes. Once volatilized, a molecule escapes the region around the direct insertion probe, which is dense with sample molecules and within which bimolecular interactions can occur. The molecule then enters the low-pressure region of the ion source, where the chances of decomposition are lower because collisions are so rare. Therefore, if the rate of heating of the sample is very rapid (between ten and several thousand degrees per second), non-equilibrium conditions obtain and the sample molecules spend less time at a temperature at which decomposition is dominant and collisions are common. In effect, the sample is volatilized into the low-pressure region before it has time to decompose in the higher pressure region. Reactions such as those in scheme (10.1) may also be suppressed under such conditions. Rapid heating normally requires a modified direct insertion probe and is successful in obtaining electron ionization mass spectra of thermally labile compounds of low volatility such as large peptides, carbohydrates and even some salts (see section 3.6). Alternatively, an inlet can be used that does not require the sample to be vaporized thermally, as with FAB or electrospray mass spectrometry (section 10.3.3).

10.3.3. *The method of ionization*

The appearance of a mass spectrum depends markedly on the method of ionization (see figures 3.1, 3.2 and 3.6), so the use of another method of ionization provides collaborative or complementary information. If the EI mass spectrum indicates that an unknown sample is a polyhalogenated compound, then negative-ion chemical ionization is likely to yield useful information. For a natural product that is too labile thermally to give a useful electron ionization, chemical ionization or the field ionization mass spectrum, the various mild techniques that do not require heating, such as fast atom bombardment, laser desorption, or electrospray mass spectrometry, will provide better analytical information. Such methods of ionization and their applications are discussed in chapter 3.

10.4. POSTULATION OF ION STRUCTURES

In interpreting a mass spectrum it is convenient to postulate ion structures to explain the observed fragmentation pathways. Generally, the actual structures of ions in a mass spectrometer are not known, although the elemental compositions may be, and hence there is a need to postulate structures. If the compound investigated has a known structure, the mass spectrum is often interpreted by assuming a minimum of structural change at each fragmentation step. To illustrate this point, part of the fragmentation of acetanilide may be considered in three alternative ways (scheme (10.2*a*, *b*, *c*)). The first way (*a*) shows ionization and fragmentation in terms of m/z values and gives

$$(10.2)$$

little information, but it is verifiable. The second (*b*) gives the same pathway in terms of elemental compositions of the ions; this affords much more information since the compositions of the ions and of the unit lost at the fragmentation step are known. Assuming that accurate mass measurements have been completed, this depiction represents experimental fact. The third way (*c*) adds postulate to experimental fact by assuming or postulating the structures of the molecular and fragment ions as well as the neutral species to be as shown, i.e. with minimum change from the original molecule. It is not a straightforward task to confirm these ionic structures. They may or may not be correct, but this is quite immaterial to the convenience of describing fragmentation pathways in this way.

An added advantage of postulation of ion structures appears when the mass spectra of groups of compounds are compared for similar fragmentation behaviour since this becomes much more obvious from structural relationships than from elemental composition relationships. It is easier to see the similarity in behaviour between diethylmethylamine (**6**) and 1-pentylamine (**7**) from the postulated ion structures (scheme (10.3*a*)) than from the corresponding elemental compositions (scheme (10.3*b*)).

The postulation of ion structures does cause difficulty in discussing thermochemical aspects of mass spectrometry. If it is necessary to postulate or assume an ion structure, then that postulate or assumption is built into any thermochemical result achieved and therefore introduces uncertainty. Discussion of fragmentation mechanism is also clouded by those assumptions. It is one thing to postulate ion structures as an aid to describing fragmentation but it is another thing altogether to discuss detailed aspects of fragmentation mechanisms in terms of assumed structures!

As an aid to the rationalization of ionic behaviour, fragmentation is often shown mechanistically using curly arrows and fish-hook arrows. Indeed,

$$
\begin{aligned}
&\left. CH_3CH_2N(CH_3)CH_2CH_3 \right]^{+\cdot} \longrightarrow CH_3^{\cdot} + CH_2{=}\overset{+}{N}(CH_3)CH_2CH_3 \\
&\hspace{2.5cm} \textbf{6} \\[4pt]
&\left. CH_3CH_2CH_2CH_2CH_2NH_2 \right]^{+\cdot} \longrightarrow CH_3CH_2CH_2CH_2^{\cdot} + CH_2{=}\overset{+}{N}H_2 \\
&\hspace{2.5cm} \textbf{7}
\end{aligned}
\right\}(a)
$$

$$(10.3)$$

$$
\begin{aligned}
&C_5H_{13}N^{+\cdot} \longrightarrow CH_3^{\cdot} + C_4H_{10}N^{+} \\[8pt]
&C_5H_{13}N^{+\cdot} \longrightarrow C_4H_9^{\cdot} + CH_4N^{+}
\end{aligned}
\right\}(b)
$$

Table 10.2. *Bond strengths in ethane and the ethane radical-cation.*

	D_{C-C}	D_{C-H}
C_2H_6	83 (347)	96 (402)
$C_2H_6^{+\bullet}$	44 (184)	26 (109)

Bond strength (D) is given in kcal mol^{-1} and, in parentheses, in kJ mol^{-1}.

such formulations were used in the previous chapter. Frequently, there is experimental evidence for the reaction mechanisms depicted but, given the uncertainty surrounding many ion structures, many of these mechanisms must also remain tentative. However, even if some of the mechanistic details are shown eventually to be wrong, it does not diminish their value for rationalizing fragmentation. It is reasonable to speculate on and draw fragmentation mechanisms just so long as (i) they are shown to be consistent with known mass spectrometric behaviour and hence beneficial to interpretation, and (ii) the tentative nature of any mechanisms is borne in mind.

The usefulness of postulating ion structures and mechanisms to describe fragmentation pathways also applies in a reverse sense for structural work. Knowing a fragmentation pathway, one can begin to postulate structures that might behave in this way and hence ideas can be formed regarding the structure of an unknown compound. For example, knowing that a molecular ion ejects $C_2H_3O_2$ (59 mass units), one can postulate a $COOCH_3$ (methyl ester) grouping in the original molecule without knowing the actual structures of the ions and the mechanism of the fragmentation.

10.5.   FRAGMENTATION OF HYDROCARBONS

The carbon–carbon bonds in alkanes are weaker than the carbon–hydrogen bonds. In the molecular ion of an alkane, assuming it to have a similar structure to that of the molecule before ionization, the strengths of C–C and C–H bonds are greatly reduced. Table 10.2 compares the C–C and C–H bond strengths in ethane and the ethane ion.

If all of the C–C bonds in the molecular ion of an alkane have similar strengths it might be expected that alkane ions would be cleaved at them almost indiscriminately. Figure 10.9 shows the mass spectrum of decane, in

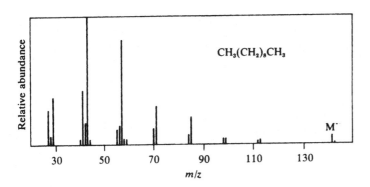

Figure 10.9. A typical electron ionization mass spectrum of an alkane, decane.

which there can be seen bunches of ions approximately 14 mass units apart extending down from the molecular ion. The differences of 14 mass units are not due to elimination of units of CH_2 (an energetically unfavourable process) but rather to fragmentation at different places in the molecular and fragment ions (scheme (10.4)). For an alkane, fragment ions appear at m/z values corresponding to the compositions C_nH_{2n+1}. This simple view of the

$$[CH_3 \!-\! CH_2 \!-\! CH_2 \!-\! CH_2 \!-\! - - - - - - -]^{+\cdot} \longrightarrow [C_nH_{2n+1}]^+$$

$$\tag{10.4}$$

−15 −29 −43 −57

14   14   14

mass spectra of alkanes is complicated by rearrangement processes and excess of internal energy in the ions. The excess of energy causes rapid fission of the C–C bonds in the ions resulting in a considerable abundance of ions of small mass near m/z 29, 43 and 57 and a rapid falling off in the sizes of the peaks at higher m/z values up to the molecular ion, which is generally of very low abundance. The excess of energy imparted to the molecular ion may be reduced by measuring the mass spectrum at low electron-beam energies and by using a cooled ion source. Under these conditions the mass spectrum is much more uniform with all fragment ions of similar abundance and without the bunching at low m/z values. Alternatively, chemical ionization could be employed because, in the chemical ionization mass spectra of alkanes, the bunching at low m/z values is not so pronounced and quasi-molecular ions, $[M - H]^+$, usually account for

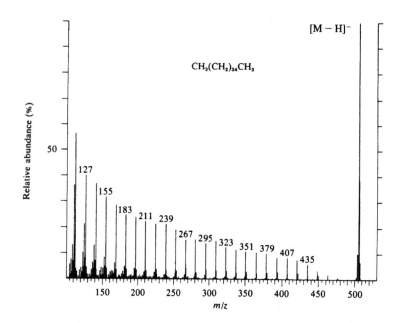

Figure 10.10. A typical chemical-ionization mass spectrum of an alkane, hexatriacontane. Isobutane was used as the reactant gas.

the base peak of the spectrum. The chemical ionization spectrum of hexatriacontane, $C_{36}H_{74}$, obtained by using isobutane as a reactant gas illustrates these features (figure 10.10). The fragment ions, $C_nH_{2n+1}^+$, are due to losses of C_2H_4, C_3H_6, C_4H_8, etc., from quasi-molecular and fragment ions.

Groups of peaks rather than single peaks, 14 mass units apart, are found in the mass spectra of alkanes because elimination of H_2 from various fragment ions can take place. Thus, the ions at m/z values corresponding to $C_nH_{2n+1}^+$ are accompanied by less abundant ions of $C_nH_{2n-1}^+$. Similar losses of hydrocarbon entities yield $C_nH_{2n}^{+\cdot}$ ions.

The supposition that product stability governs the mode of fragmentation (section 9.10) receives some support from the mass spectra of branched-chain alkanes, which show enhanced bond breaking at the branching points. The increased fragmentation at these points has been ascribed to the increasing stability of primary, secondary and tertiary carbocations. Figure 10.11 shows the mass spectrum of 3,3-dimethylhexane compared with that of its straight-chain isomer, octane. There are increased abundances of ions at m/z 71, 85 and 99 corresponding to bond scission adjacent to the tertiary carbon centre in the branched chain isomer (scheme (10.5)). The phenomenon is useful for determining branching points in alkanes.

Cycloalkanes give spectra very similar to those of linear alkanes except that the molecular ions are more abundant. Once a cycloalkane ring has

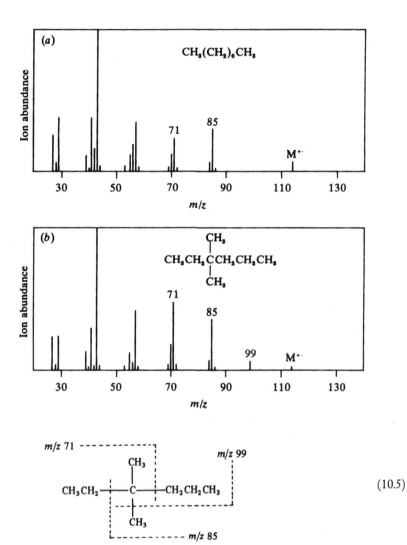

Figure 10.11. A comparison of mass spectra of (*a*) a straight-chain hydrocarbon, octane, and (*b*) its branched-chain isomer, 3,3-dimethylhexane. Note the enhanced abundance of ions at m/z 71, 85 and 99 in spectrum (*b*).

broken, the residual ion behaves like the ions of linear alkanes, again giving bunches of ions 14 mass units apart; for cycloalkanes, principal fragment ions occur at m/z values corresponding to $C_nH_{(2n + 1 - 2r)}^+$, where r is the number of rings.

The introduction of a double bond into an alkane lowers the ionization energy but does not alter the appearance of the mass spectrum greatly apart from an apparently preferred cleavage at allylic positions as shown in scheme (10.6). This latter cleavage has been ascribed to resonance stabilization of the resulting ion.

A cycloalkene with the double bond in a six-membered ring exhibits the

$$\left[-\overset{|}{\underset{|}{C}}-\overset{|}{\underset{|}{C}}-\overset{|}{C}=\overset{|}{C}- \right]^{+\cdot} \longrightarrow -\overset{|}{\underset{|}{C}}\cdot \; + \; \Big\rangle \overset{+}{C}\cdots\overset{|}{\underset{|}{C}}\cdots \overset{}{C}\Big\langle \qquad (10.6)$$

mass spectrometric analogue of the retro-Diels–Alder reaction. The molecular ion splits into two in accord with scheme (10.7), the charge appearing on either fragment:

$$ (10.7) $$

Thermochemical argument suggests that the charge will appear on the fragment with the lower ionization energy. If the ionization energies of the fragments are widely different then one of them will take the charge almost exclusively, but if the ionization energies are similar then either fragment may take the charge and appear as an ion in the mass spectrum. Figure 10.12 illustrates the retro-Diels–Alder process in the mass spectrum of cyclohexene and the larger molecular ion peak found with cyclic compounds.

There is evidence that, after electron ionization, migration of the double bond occurs in alkenes. The net practical result is that alkene isomers frequently show very similar if not identical mass spectra. Figure 10.13 shows that the isomeric butenes yield very similar mass spectra. Since it can be difficult to reproduce spectra exactly, especially between different laboratories, an identification of any one of the butenes from its mass spectrum alone would be questionable. It is generally not easy to locate a double bond by mass spectrometry of the alkene itself but it may be located by 'fixing' the bond in a derivative (section 6.3.1, scheme (6.7)). Alternatively, location of double bonds in alkenes may be brought about by chemical ionization with methyl vinyl ether as reactant gas (section 3.3).

With increasing numbers of double bonds, alkenes give larger molecular ion peaks and show less of the 'random' bond cleavage that is characteristic of alkanes and simple alkenes. Terpenoid molecules frequently fragment by loss of C_5 units or multiples of these. The long-chain terpenoid alcohol, solanesol, breaks down in this way in the mass spectrometer, the favoured

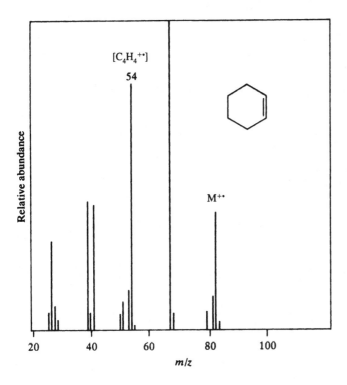

Figure 10.12. The mass spectrum of cyclohexene showing large peaks for the molecular ion and the ion at m/z 54 due to a retro-Diels–Alder reaction.

positions of bond breaking being allylic (scheme (10.8)). The elimination of 43 mass units ($C_3H_7^{\bullet}$) is commonly observed in terpenoid compounds containing isopropyl, isopropenyl and isopropylidene groups; in the latter two cases, double-bond migration must occur.

$$\left[\mathord{\text{⋀⋀⋀}}\left(\mathord{\text{⋀⋀}}\right)_7 \mathord{\text{⋀⋀⋀}}_{OH}\right]^{+\bullet} \longrightarrow \left[\mathord{\text{⋀⋀⋀}}\left(\mathord{\text{⋀⋀}}\right)_n \mathord{\text{⋀⋀}}CH_2\right]^{+} \quad (10.8)$$

$$n = 0 - 7$$

It was mentioned earlier that mass spectra of aromatic molecules are characterized by few fragment ions and abundant molecular ions. Benzenoid hydrocarbons often afford ions at m/z 91 ($C_7H_7^{+}$), which can be very abundant, particularly when benzylic groups are present (reaction (10.9)). The $C_7H_7^{+}$ ion may retain the benzylic structure or, in some cases, it is known to

$$\left[\mathord{\text{⬡}}\text{—}CH_2\text{—}CH_2CH_3\right]^{+\bullet} \longrightarrow C_2H_5^{\bullet} + C_7H_7^{+} \left[\mathord{\text{⬭}}^{+}\right] \quad (10.9)$$

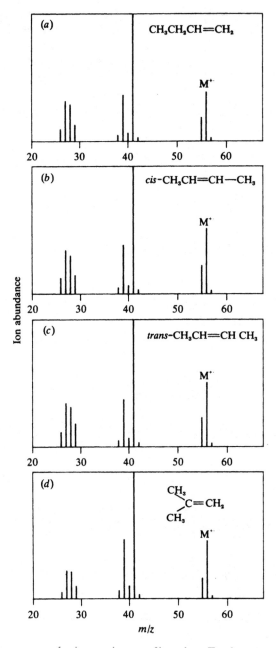

Figure 10.13. Mass spectra of the alkene isomers (*a*) but-1-ene, (*b*) *cis*-but-2-ene, (*c*) *trans*-but-2-ene and (*d*) 2-methylprop-1-ene.

rearrange to the isomeric tropylium ion. For interpretational purposes, it is important to recognize the stability of the $C_7H_7^+$ ion; it is less important to distinguish the main two structural possibilities. It is prudent to examine the mass spectrum of a suspected aromatic compound for an ion at $m/z\ 91$ but it should not be assumed that its presence necessarily implies a $C_6H_5CH_2$ unit in the original molecule. Ions at m/z 91 are frequently observed in the mass spectra of compounds that do not possess a benzyl group; *tert*-butylbenzene, 1,4-diphenylbutadiene, and camphene (structures **(8)**, **(9)** and **(10)**) each

give a large peak at m/z 91, which must result from rearrangement. However, an ion at m/z 91 of composition C_7H_7 is usually a good guide to the aromatic or benzenoid nature of a molecule. Rearrangement processes in the mass spectra of aromatic compounds are often more *obvious* and widespread than with aliphatic compounds and this is probably due to the fact that, with increased stability in the molecular ion, there is time for the excess of energy that it contains to rearrange the structure before cleavage occurs. The presence of heteroatoms frequently makes these reactions more apparent than in hydrocarbons. The consecutive losses of two CO entities from anthraquinone is an example of one such rearrangement (scheme (10.10); see section 10.9 for other examples).

$$\xrightarrow{-\text{CO}} C_{13}H_8O^{+\cdot} \xrightarrow{-\text{CO}} C_{12}H_8^{+\cdot} \quad (10.10)$$

10.6. PRIMARY FRAGMENTATIONS OF ALIPHATIC HETEROATOMIC COMPOUNDS

The effects of heteroatoms on the mass spectra of organic molecules may be loosely grouped by regarding the heteroatom either as a simple 'saturated' substituent in an alkane or as an 'unsaturated' substituent. Saturated substituents include the halogens, hydroxyl, amino and sulphydryl groups and are characterized by having lone-pair (non-bonded) electrons. Unsaturated substituents include carbonyl and nitrile groups and have both π-electrons

Table 10.3. *Common 'saturated' constituents (X) in R–X compounds (R = alkyl).*

X	X
F	SH
Cl	SR
Br	NH_2
I	NHR
OH	NR_2
OR	$MetR_n$ (Met = metal atom)

and lone-pair electrons. Double bonds and aromatic rings may also be included in this category.

Compounds with saturated heteroatomic substituents may be designated R—X, the group X being any one of the common ones listed in table 10.3. Some of the more important features of the mass spectrometric fragmentation of these compounds can be grouped under a few headings.

(i) Simple bond cleavage adjacent to the heteroatom. This is observed with the charge remaining on the hydrocarbon fragment:

$$R \overset{}{\underset{}{\Big\rceil}} X \Big]^{+\cdot} \longrightarrow R^+ + X^\cdot$$

The mass spectra of 1-chloro- and 1-bromopropane (figure 10.14) show prominent losses of halogen. Note the change in the isotope pattern between the molecular ion and the first fragment ion containing no halogen. The loss of halogen is so obvious from the isotope patterns and the number of mass units lost (35 and 37 for Cl; 79 and 81 for Br) that accurate mass measurement is unnecessary.

The carbon–halogen bond strengths in the molecules CH_3–X (X = halogen; table 10.4) decrease from the very strong C–F bond to the relatively weak C–I bond. However, in the ions CH_3–$X^{+\cdot}$ the bond strengths are greatly reduced for X = F, Cl and Br, but for the iodo compound the bond strength in the ion is greater than that in the molecule. The C–Cl and C–Br bonds in the ions are relatively weak and loss of Cl and Br from the

Table 10.4. *Relative bond strengths in some* CH_3X
molecules and molecular ions for the processes $CH_3X \rightarrow$
$CH_3^{\cdot} + X^{\cdot}$ *and* $CH_3X^{+\cdot} \rightarrow CH_3^{+} + X^{\cdot}$.

X	D_{CH_3-X}	$D_{CH_3-X^{+\cdot}}$
F	118 (494)	48 (201)
Cl	81 (339)	49 (205)
Br	67 (280)	51 (213)
I	53 (222)	61 (255)
OH	90 (377)	67 (280)
OCH₃	77 (322)	74 (310)
SH	67 (280)	77 (322)
SCH₃	75 (314)	102 (427)
NH₂	80 (335)	91 (381)

Notes:

Bond strength (D) is given in kcal mol⁻¹ and, in
parentheses, in kJ mol⁻¹.

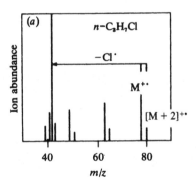 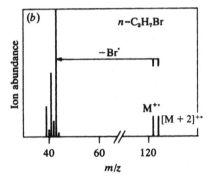

Figure 10.14. Mass spectra of (*a*) 1-chloropropane and (*b*) 1-bromopropane. In each spectrum, elimination of the halogen changes the isotope distribution considerably. Peaks above m/z 45 in the spectrum of 1-chloropropane have been scaled up five times.

molecular ions of organic chlorides and bromides is generally observed. Although the C–F bond strength is similar to that of C–Cl and C–Br in the ion, simple loss of fluorine from organic fluorides is not commonly observed because of the ease of competing reactions (see below and table 10.5).

Table 10.4 gives the bond strengths for several CH_3–X compounds. The order of bond strengths in the molecule is F > OH > Cl > NH₂ > SMe > SH, Br > I but in the molecular ion this order changes to SMe > NH₂ > SH > OMe > OH > I > Br > Cl > F (Reed, 1962). Note particularly how the C–I, C–S and C–N bond strengths increase in going from the molecule to the molecular ion, possibly due to the fact that iodine, sulphur and nitrogen can stabilize a positive charge much better than can fluorine,

Table 10.5. *Bond dissociation energies for ions,* $C_2H_5X^{+\cdot}$.

Ion from	D_{C-C}	D_{C-X}	D_{C-H}
C_2H_6	44 (184)	26 (109)	26 (109)
C_2H_5F	17 (71)	29 (121)	11 (46)
C_2H_5Br	39 (163)	30 (125)	45 (188)
C_2H_5OH	24 (100)	49 (205)	24 (100)
C_2H_5SH	52 (217)	56 (234)	60 (251)
$C_2H_5NH_2$	31 (130)	66 (276)	29 (121)
C_2H_5CN	48 (201)	36 (150)	56 (234)
$C_2H_5OCH_3$	30 (125)	53 (222)	35 (146)
$C_2H_5SCH_3$	59 (247)	74 (309)	68 (284)

Notes:

Bond dissociation energy (D) is given in kcal mol^{-1} and, in parentheses, in kJ mol^{-1}.

chlorine, bromine and oxygen. Unlike the other halogens, in the positive-ion spectrum iodo-compounds frequently fragment with the charge retained predominantly on iodine to give an ion at m/z 127. In negative-ion spectra, fluorine and chlorine compounds produce F^- and Cl^- ions.

Elimination of OH·, SH· and NH_2· from alcohols, thiols and amines is observed to a slight extent but, particularly with amines, other processes are more important. Cleavage of metal–carbon bonds in simple organometallic compounds is frequently so easy as to lead to low abundances of the molecular ions. One of the most abundant ions in the mass spectrum of tetra-ethylgermanium is produced by elimination of an ethyl radical from the molecular ion (figure 10.15):

$$[(C_2H_5)_4Ge]^{+\cdot} \xrightarrow{-C_2H_5} [(C_2H_5)_3Ge]^+ \xrightarrow{-C_2H_4} [(C_2H_5)_2GeH]^+ \xrightarrow{-C_2H_4} [C_2H_5GeH_2]^+$$

$$m/z\ 190 \qquad\qquad m/z\ 161 \qquad\qquad m/z\ 133 \qquad\qquad m/z\ 105$$

(ii) Simple fission adjacent to the heteroatom with simultaneous hydrogen transfer. This is commonly observed for R–X compounds. Simple cleavage of a molecular ion gives an even-electron species, but cleavage with hydrogen transfer involves more than one bond breaking and gives odd-electron ions:

$$H\!-\!\!-\!\!-A\!\!\stackrel{}{\diagdown}\!\!X\Big]^{+\cdot} \longrightarrow X^{\cdot} + H\!-\!\!-\!\!-A^+ \text{ (Even-electron species)}$$

$$H\!\!\stackrel{}{\diagup}\!\!A\!\!\stackrel{}{\diagdown}\!\!X\Big]^{+\cdot} \longrightarrow HX + A^{+\cdot} \quad \text{(Odd-electron species)}$$

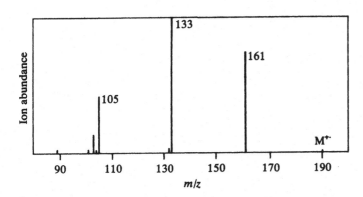

Figure 10.15. A partial mass spectrum of tetraethylgermanium showing large abundance of $[M - C_2H_5{}^{\bullet}]^+$ ions at m/z 161.

Whereas loss of $F^{\bullet}$ from fluoro-compounds is not generally observed, the loss of HF is more common, perhaps because of its particularly favourable negative heat of formation. The elimination of HCl from chloroalkanes is more favourable than the elimination of HBr from bromoalkanes.

Although the ejection of $OH^{\bullet}$ from alcohols can be observed in their mass spectra, the elimination of H_2O is usually more common, especially when a suitable hydrogen donor site is available. Deuterium labelling has shown that loss of H_2O from alkyl alcohols predominantly involves a hydrogen on a γ- or δ-carbon atom (scheme (10.11)) unless steric constraints dictate other-

$$
\begin{array}{c}
\underset{HO\cdots\cdots\cdots H}{C-C-C-C}^{+\bullet} \longrightarrow [M-H_2O]^{+\bullet} \qquad (10.11)
\end{array}
$$

wise. A similar process is observed in thiols, which eject H_2S, but amines do not readily eliminate NH_3, possibly due to the stability of the ammonium ion making an alternative fragmentation route more favourable (scheme (10.12); and see category (iii) below).

$$
\begin{array}{c}
\underset{H_2N\cdots\cdots\cdots H}{C-C-C-C}^{+\bullet} \longrightarrow \underset{H_3N^+}{C-C-C-C^{\bullet}} \not\longrightarrow [M-NH_3]^{+\bullet} \qquad (10.12)
\end{array}
$$

Because of the ease with which H_2O can be ejected from alcohols after ionization, the molecular ions of alcohols are usually of low abundance. In such cases, field ionization and chemical ionization methods are advantageous since these techniques normally yield abundant molecular or quasi-molecular ions. Thus, by comparing electron ionization spectra with those obtained by either field or chemical ionization, details of the molecular ion region become easier to interpret. Alternatively, alcohols may be derivatized to compounds likely to give molecular ions by electron ionization; alkylation, trimethylsilylation or oxidation may serve this purpose. The transfer of hydrogen to hydroxyl during elimination of H_2O from alcohols seems to involve mainly a six-centre transition state. Deuterium labelling of halogen compounds suggests that a five-membered transition state is most likely during the elimination of hydrohalide. The elimination of ROH from ethers, i.e. cleavage with hydrogen transfer (scheme (10.13)), is not abundant, perhaps because of the greater ease of alternative reactions, as indicated below. The corresponding reaction in organometallic compounds affords [metal + H]$^{+\cdot}$ ions and these are generally found in their mass spectra.

$$\left[\begin{array}{c}C-C-C-C \\ | \qquad\qquad | \\ RO\cdots\cdots\cdots H\end{array}\right]^{+\cdot} \longrightarrow [M-ROH]^{+\cdot} \text{ (low abundance)} \qquad (10.13)$$

(iii) Simple cleavage of groups, other than X, from the α-position (α-cleavage). This is a very common reaction of saturated substituent compounds (scheme (10.14)). The ejected radical species may be hydrogen or an alkyl radical. It is

$$\left[R-\overset{|}{\underset{|}{C}}\overset{\alpha}{-}X\right]^{+\cdot} \longrightarrow R^{\cdot} + {>}\overset{+}{C}-X \longleftrightarrow {>}C{=}\overset{+}{X} \qquad (10.14)$$

$$\left[H-CH_2-OH\right]^{+\cdot} \longrightarrow H^{\cdot} + \overset{+}{C}H_2-OH \longleftrightarrow CH_2{=}\overset{+}{O}H \qquad (10.15)$$

thought that stabilization of the positive charge by the group X makes the process favourable. This type of fragmentation is more prominent in those compounds having a substituent (X) that can more readily stabilize the charge, as with the $NR'R''$ group. Likewise, it has been shown by deuterium labelling that elimination of hydrogen from methanol takes place from the carbon atom and not from the oxygen (scheme (10.15)).

The same process can be observed in ethers, although elimination of an alkyl radical is normally more prominent. The radical may be ejected from either side of the oxygen atom in ethers (scheme (10.16)); the larger group (R or R') appears to be ejected more favourably. For example, the loss of a propyl radical from ethyl 1-butyl ether is more in evidence than the corresponding loss of a methyl radical (figure 10.16). However, examination of the

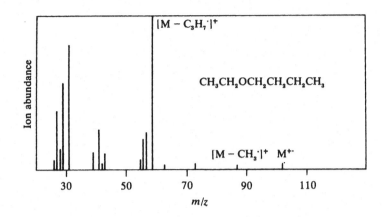

Figure 10.16. A mass spectrum of ethyl 1-butyl ether obtained at an electron beam energy of 70 eV. The peak at m/z 59, corresponding to loss of propyl radical from the molecular ion, is much larger than that at m/z 87, corresponding to loss of methyl radical.

$$R\text{—}CH_2\text{—}O\text{—}CH_2\text{—}R'\rceil^{+\cdot} \quad\begin{array}{c} \nearrow R^\cdot + \overset{+}{CH_2}\text{—}O\text{—}CH_2R' \longleftrightarrow CH_2\overset{+}{=}O\text{—}CH_2R' \\ \\ \searrow R''^\cdot + RCH_2\text{—}O\text{—}\overset{+}{CH_2} \longleftrightarrow RCH_2\text{—}\overset{+}{O}=CH_2 \end{array} \qquad (10.16)$$

mass spectra of ethers at reduced electron ionization energies suggests that the favourable loss of the larger alkyl fragment may be more apparent than real. At low electron energies, subsequent decomposition of the initially produced fragment ions is reduced and the results then suggest that the smaller alkyl fragment is lost more readily. It is a timely reminder of the importance of the initial internal energy in the molecular ion in determining subsequent fragmentation rates (sections 1.1 and 9.7) and that the abundance of any one ion is a balance between its rates of formation and decomposition.

Similar α-cleavage reactions are found in thioethers and are particularly marked in amines, in which the nitrogen atom can stabilize the positive charge as an immonium species (scheme (10.17)):

$$[CH_3CH_2N(CH_3)_2]^{+\cdot} \rightarrow CH_3^\cdot + \overset{+}{CH_2}\text{—}N(CH_3)_2 \longleftrightarrow CH_2=\overset{+}{N}(CH_3)_2 \qquad (10.17)$$

In an amine $R_SCH_2N(R)CH_2R_L$ with small and large alkyl groups R_S and R_L, respectively, the relative importance of ejection of $R_S^\cdot$ compared with $R_L^\cdot$ increases with decreasing energy of the ionizing electrons, as with ethers.

The two types of simple bond cleavage discussed in (i) and (iii) both involve fission adjacent to the α-carbon atom and are competitive processes. In table 10.5 are listed the dissociation energies for α-cleavage of C–C, C–H and C–X bonds in ions of the type $C_2H_5X^{+•}$ (X = H, F, Br, OH, NH_2, SH, OCH_3, SCH_3 and CN). For two competing simple bond fissions in the ground-state ion, the simplified quasi-equilibrium theory suggests that their relative rates of fission (k_1 and k_2) for any internal energy (E) are given by

$$k_1/k_2 = (v_1/v_2)\{[1 - (E_1/E)]/[1 - (E_2/E)]\}^{N-1}$$

where v_1 and v_2 are frequency factors, N the number of oscillators and E_1 and E_2 the energies of activation (section 9.7). Approximating this expression through the binomial expansion and putting $v_1 \approx v_2$ (for simple fission processes, the frequency factors may be equated with the bond vibration frequencies, which, for the reactions considered here, are not likely to be sufficiently different to upset the main argument developed by assuming their equality) gives

$$\ln (k_1/k_2) \approx [(N - 1)/E](E_2 - E_1)$$

That is, for two simple cleavage processes from a molecular ion, the relative reaction rate constants leading to initial fragmentations are proportional to the difference between the energies of activation for the two processes. In the absence of significant activation energies for the reverse reactions, the energies E_1 and E_2 may be equated (section 9.8) to the bond dissociation energies in the ion, as listed in table 10.5. Hence, inspection of table 10.5 should yield at least a qualitative guide to the expected mass spectrometric fragmentation of compounds C_2H_5X, i.e., the bond strength difference when equated with $E_2 - E_1$ should give a semi-quantitative measure of ion abundances in their mass spectra. For comparison, the mass spectra of these compounds are recorded in figure 10.17. Striking relationships may be observed immediately and some of these are annotated here.

(a) From table 10.5 it can be seen that all of the bond dissociation energies are low for $C_2H_5F^{+•}$ and that C–H bond cleavage is particularly easy. Hence, it might be expected that, with all bond dissociation

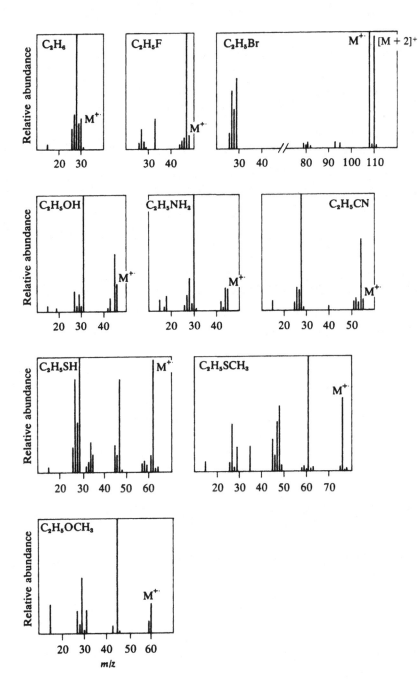

Figure 10.17. Mass spectra of compounds C_2H_5X (X = H, F, Br, OH, NH$_2$, CN, SH, SCH$_3$, OCH$_3$).

energies being low, fluoroethane would not show abundant molecular ions and that the $[M - 1]^+$ peak would be most prominent. The mass spectrum shows that this is the case and confirms that loss of F· is an unimportant process, in keeping with the C–F bond being the strongest of a set of weak ones.

(b) By way of contrast to fluoroethane, in bromoethane all the bonds are moderately strong with the C–X (X = Br) being the weakest, the C–C bond stronger and the C–H bond the strongest. Therefore, it might be expected that abundant molecular ions would be observed and that the most prominent process would be the loss of Br⋅. The mass spectrum is in keeping with this deduction.

(c) The mass spectrum of ethane thiol, C_2H_5SH, shows a large molecular ion peak with C–C and C–SH bond cleavage of similar importance. Table 10.5 shows that the bonds in the $C_2H_5SH^{+⋅}$ ion are all rather strong and that C–C and C–SH cleavage reactions are almost equally easy.

(d) In ethanol and ethylamine, the bond dissociation energies for C–C and C–H fissions are very much less than for C–X fission. Therefore, these substances should give moderately abundant molecular ions with marked losses of hydrogen and methyl radicals. Examination of their mass spectra bears out this argument.

Table 10.5 may be used to 'predict' the mass spectra of the simple compounds C_2H_5X. Because the compounds are so simple, further decomposition of the initial fragment ions is greatly reduced. The abundance of an ion in a mass spectrum depends upon its rate of formation and upon its rate of decomposition (section 1.1) but, in the case of C_2H_5X compounds, further decomposition of the initial fragment ions is negligible and their rates of formation determine their abundances. The general principles apparent from an examination of table 10.5 carry through to a large extent in more complex molecules and serve to provide a thermochemical understanding of many fragmentation processes. In the remainder of this section on fragmentation reactions, reference to the bond dissociation energies given in table 10.5 will prove of value.

It is interesting to note that the C–C bond dissociation energy in ethane is comparatively large and greater than for C–H bond fission; these relative strengths are reflected in the appearance of the mass spectrum, which exhibits a facile loss of hydrogen from the molecular ion. However, as the alkyl chain length increases in alkanes, C–C bond dissociation energy decreases more rapidly than C–H bond dissociation energy. Table 10.6 illustrates this effect for loss of a methyl radical from straight-chain alkane molecular ions; for butane the relative strengths of C–H and C–C bonds are very

Table 10.6. *C–C bond dissociation energies in n-alkane radical-cations*[a].

Ion from	D_{C-C}
C_2H_6	44 (184)
C_3H_8	30 (126)
C_4H_{10}	21 (88)
C_5H_{12}	15 (63)

Notes:

[a] Calculated for reactions, $C_nH_{2n+1}CH_3^{+\bullet} \rightarrow C_nH_{2n+1}^{+}$ + $CH_3^{\bullet}$, in kcal mol^{-1} and, in parentheses, in kJ mol^{-1}.

similar and much less than the values in ethane and most of the other compounds in table 10.5. In long-chain alkyl compounds, $C_nH_{2n+1}X$, this reduction of the C–C bond strength, coupled with the greater statistical probability of the excess of energy in the ion appearing in a C–C bond, may be expected to lead to an increasing importance of general fragmentation of the alkyl chain; such an effect is observed in the mass spectra of these compounds. As the chain length increases, the specific fragmentations associated with the heteroatom, X, decrease in importance and extensive fission of the alkyl chain is observed.

(iv) Bond breaking at positions other than that adjacent to the α-carbon atom. This is observed in the mass spectra of aliphatic heteroatom compounds. As the proportion of hydrocarbon in aliphatic heteroatom compounds increases, the influence of the heteroatom on the appearance of the mass spectrum decreases; other bond-breaking processes become more prominent. Thus, in the long-chain compound shown in scheme (10.18), cleavages such as those at the positions γ, δ... increase in importance as R increases in length:

$$\left[R\text{---}CH_2\text{---}CH_2\text{---}CH_2\text{---}CH_2\text{---}X \right]^{+\bullet}$$

$$\nearrow RCH_2CH_2CH_2CH_2^{+} + X^{\bullet}$$

$$\rightarrow RCH_2CH_2CH_2^{\bullet} + CH_2{=}\overset{+}{X}$$

$$\searrow RCH_2CH_2^{+} + \dot{C}H_2CH_2X$$

$$\searrow RCH_2^{+} + \dot{C}H_2CH_2CH_2X$$

(with positions labelled δ, γ, β, a)

(10.18)

In chloroalkanes and bromoalkanes, $C_nH_{2n+1}X$, this simple bond breaking may also lead to particularly abundant ions, $C_nH_{2n}X^+$, for $n = 3, 4, 5$. The propensity for this reaction shown by these compounds has been attributed to the formation of cyclic structures (reaction (10.19)) similar to those found

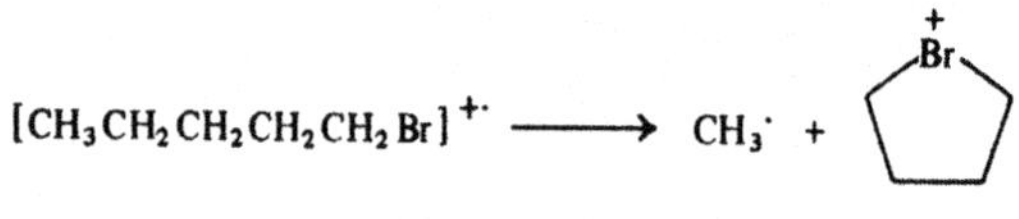

$$[CH_3CH_2CH_2CH_2CH_2Br]^{+\cdot} \longrightarrow CH_3^{\cdot} + \overset{\overset{+}{Br}}{\bigcirc} \qquad (10.19)$$

in solution chemistry. The abundance of ions formed by this process decreases with chain branching in the alkyl moiety and the reaction is not prominent in fluorine or iodine compounds.

A similar increased abundance of $C_nH_{2n}X^+$ ions at $n = 3, 4, 5$ is found in amines, but not alcohols or thiols. Unlike the halogen compounds, amines give more abundant ions at $n = 5$ (reaction 10.20), which has been ascribed

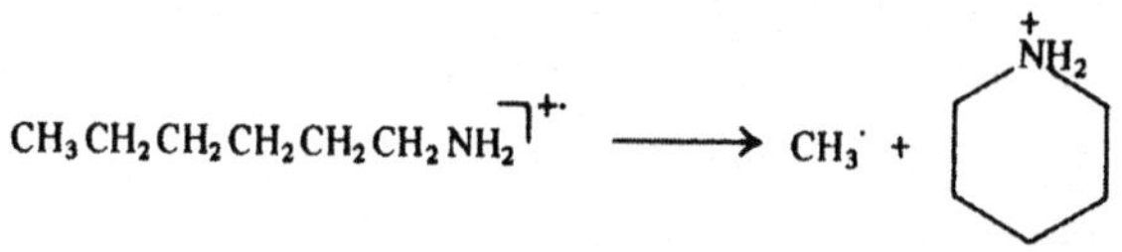

$$CH_3CH_2CH_2CH_2CH_2CH_2NH_2{}^{\rceil +\cdot} \longrightarrow CH_3^{\cdot} + \overset{\overset{+}{NH_2}}{\bigcirc} \qquad (10.20)$$

to a preference for six-membered ring formation (compare with losses of HX from R–X compounds in which five- and six-membered transition states appear to be involved). The commonest 'unsaturated' substituents of alkanes are the carbonyl functional groups –CO–, –CHO, –COOH, –COOR and –CONR$_2$, together with the sulphoxide, sulphone, phosphate and nitrile groupings.

The mass spectra of ketones are dominated by bond breaking adjacent to the α-carbon atom (α-cleavage) similar to reaction scheme (10.14) shown above for saturated substituents. The comparable breaking of the carbon–heteroatom bond, as in (i) above, is not observed presumably because of the increased strength of C=O bonds as against C–OH:

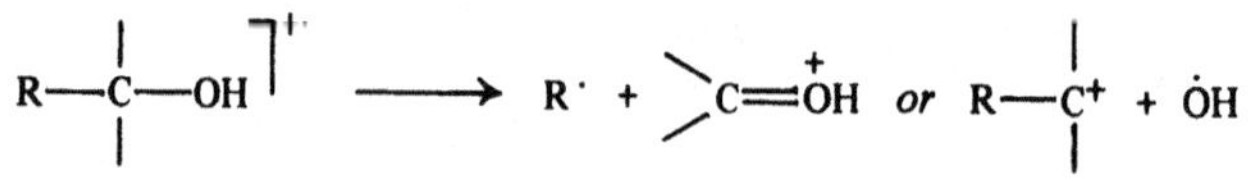

$$R{-}\overset{|}{\underset{|}{C}}{-}OH {}^{\rceil +\cdot} \longrightarrow R^{\cdot} + {\Large>}C{=}\overset{+}{O}H \ \ or \ \ R{-}\overset{|}{\underset{|}{C}}{}^{+} + \overset{\cdot}{O}H$$

but

$$R{-}\overset{|}{C}{=}O {}^{\rceil +\cdot} \longrightarrow R^{\cdot} + {-}C{\equiv}\overset{+}{O} \ \ and \ not \ R{-}\overset{|}{C}{}^{+\cdot} + O$$

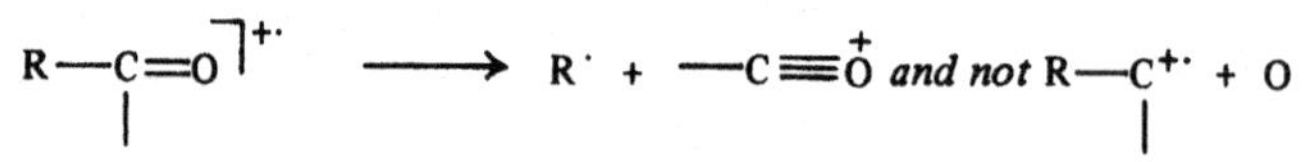

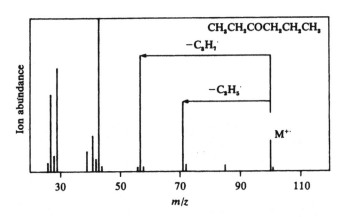

Figure 10.18. Mass spectra of ethyl 1-propyl ketone showing that the ion abundance at m/z 57 corresponding to loss of $C_3H_7^{\cdot}$ is greater than that at m/z 71 corresponding to loss of $C_2H_5^{\cdot}$ at 70 eV.

The cleavage adjacent to carbonyl is reminiscent of the photolytic decomposition of ketones and aldehydes. The large degree of resonance stabilization of the positive charge in the fragment ion is said to be responsible for the ease of fission at the α-position. As with α-cleavage of saturated substituent groups, cleavage adjacent to the carbonyl group appears to lead preferentially to the ejection of the largest radical from the molecular ion in 70 eV spectra. The α-cleavage process is frequently followed by elimination of CO from the fragment ion. The mass spectrum of ethyl 1-propyl ketone (figure 10.18) shows the elimination of both ethyl and propyl groups followed by ejection of CO in each case (scheme (10.21)). The pathways illustrated in scheme (10.21) may also be rationalized in charge localization terms with curly arrows (scheme (10.22)):

$$
\left. C_2H_5 - \underset{\underset{O}{\|}}{C} - C_3H_7 \right]^{+\cdot}
\begin{array}{c}
\xrightarrow{-C_3H_7^{\cdot}} C_2H_5CO^+ \xrightarrow{-CO} C_2H_5^+ \\[2em]
\xrightarrow{-C_2H_5^{\cdot}} C_3H_7CO^+ \xrightarrow{-CO} C_3H_7^+
\end{array}
\right\} \quad (10.21)
$$

$$
\left.
\begin{array}{l}
C_2H_5 - \underset{C_3H_7}{\overset{\cdot+}{C}} \!\! = \!\! O \xrightarrow{-C_3H_7^{\cdot}} C_2H_5 - C \!\! \equiv \!\! \overset{+}{O} \xrightarrow{-CO} C_2H_5^+ \\[2em]
C_2H_5 - \underset{C_3H_7}{\overset{+}{C}} \!\! = \!\! O \xrightarrow{-C_2H_5^{\cdot}} C_3H_7 - C \!\! \equiv \!\! \overset{+}{O} \xrightarrow{-CO} C_3H_7^+
\end{array}
\right\} \quad (10.22)
$$

In aldehydes, one of the groups adjacent to the carbonyl is simply hydrogen, and elimination of atomic hydrogen from these compounds is observed:

$$CH_3CH_2C\overset{\overset{\displaystyle H}{\diagup}}{\underset{\underset{\displaystyle O}{\diagdown}}{}} \longrightarrow H^\cdot + CH_3CH_2C\equiv\overset{+}{O}$$

As well as this simple type of bond-breaking reaction, ketones and aldehydes exhibit loss of H_2O from their molecular ions. The process has been examined by deuterium labelling and found to be complex.

The α-cleavage adjacent to carbonyl is also important in mass spectra of carboxylic acids, esters and amides, and these functional groups may be recognized through loss of appropriate units from the molecular ions. In each of the examples shown in scheme (10.23), simple fission at either side of the carbonyl group may occur, but the abundances of the fragment ions so formed can vary a great deal. Thus, aliphatic acids and amides generally

$$\left. \begin{array}{l} R{-}\underset{\underset{\displaystyle O}{\|}}{C}{-}OH\Big]^{+\cdot} \longrightarrow R^+ \text{ and } \overset{+}{C}O_2H \quad (a) \\ \hphantom{xxxxxxxxxx}\searrow RCO^+ \\[2ex] R{-}\underset{\underset{\displaystyle O}{\|}}{C}{-}OR'\Big]^{+\cdot} \longrightarrow R^+ \text{ and } \overset{+}{C}O_2R', \overset{+}{O}R' \ (b) \\ \hphantom{xxxxxxxxxx}\searrow RCO^+ \\[2ex] R{-}\underset{\underset{\displaystyle O}{\|}}{C}{-}NR'_2\Big]^{+\cdot} \longrightarrow R^+ \text{ and } \overset{+}{C}ONR'_2 \quad (c) \\ \hphantom{xxxxxxxxxx}\searrow RCO^+ \end{array} \right\} \quad (10.23)$$

produce only small peaks in their mass spectra corresponding to the loss of $OH^\cdot$ and $NR'_2{}^\cdot$ (scheme (10.23a,c)), respectively, whereas the ejection of $OR'^\cdot$ from esters (scheme (10.23b)) is much more noticeable. However, aromatic acids and amides do eject $OH^\cdot$ and $NR'_2{}^\cdot$ from their respective molecular ions. Usually, cleavage between the alkyl chain and the carbonyl function is more prominent and $[M - {}^\cdot COOH]^+$, $[M - {}^\cdot COOR']^+$ and $[M - {}^\cdot CONR'_2]^+$ ions are observed. Also, the ejected groups COOH, COOR' and $CONR'_2$ appear as charged species at the relevant m/z values. For example, the methyl ester of an aliphatic acid, $RCOOCH_3$, affords ions at $R^+, RCO^+, COOCH_3{}^+$ and $OCH_3{}^+$.

Similar α-cleavages are found in the mass spectra of sulphoxides and sulphones, but not in nitriles. Although metal carbonyls, $Met(CO)_n$, are not really analogous to organic carbonyl compounds, it is convenient to note here that α-cleavage of the carbon monoxide ligand is extremely easy in these organometallic compounds. When several carbonyl ligands are present, fragment ions are observed corresponding to successive losses of CO:

$$Met(CO)_n^{+\cdot} \longrightarrow Met(CO)_{n-1}^{+\cdot} \longrightarrow Met(CO)_{n-2}^{+\cdot} \longrightarrow etc. \longrightarrow Met^{+\cdot}$$

The α-cleavage in amides is very important for sequence analysis of amino-acid residues in peptide chains. Bond breaking adjacent to carbonyl yields type A and type B ions as shown in scheme (10.24):

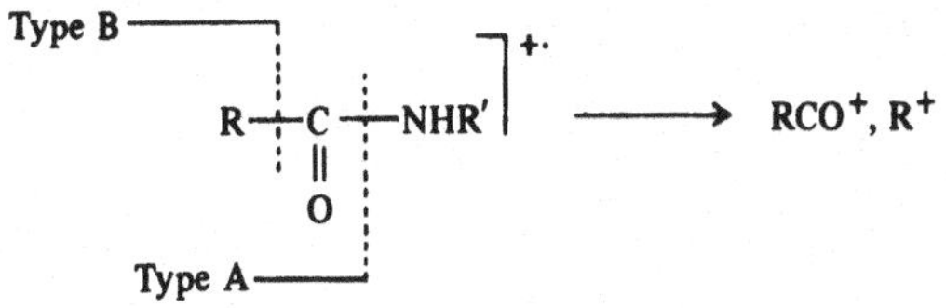

$$\text{Type B} \qquad \left[R \!-\! C \!-\! NHR' \right]^{+\cdot} \longrightarrow RCO^+, R^+ \qquad (10.24)$$

Type A

Type A ions may eject CO to give type B ions. A peptide chain normally cleaves at each CO–NH bond to yield type A fragment ions and, by examining these ions, its sequence of amino acid residues may be determined from the mass spectrum. For EI when the sample must be vaporized, it is usual to acetylate the free amino end of the peptide and to convert the acid terminal to its methyl ester in order to increase volatility. The tripeptide, Leu-Ala-Gly, methylated and acetylated in this way, gives fragment ions at m/z 156 ($C_8H_{14}NO_2$), 227 ($C_{11}H_{19}N_2O_3$), 284 ($C_{13}H_{22}N_3O_4$) and 315 ($C_{14}H_{25}N_3O_5$; the molecular ion) from which the sequence is immediately obtained (scheme 10.25):

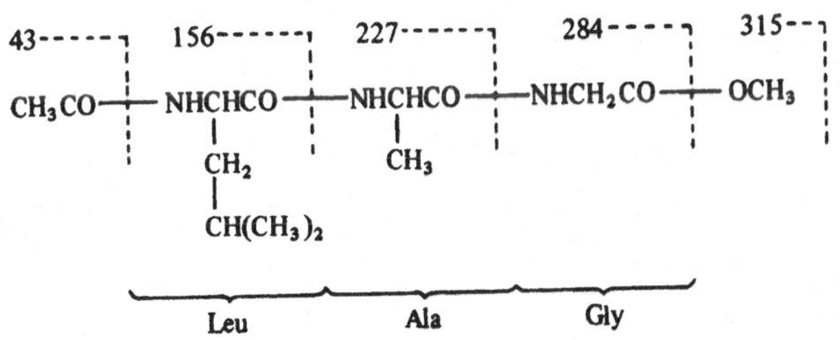

$$\text{43} \quad \text{156} \quad \text{227} \quad \text{284} \quad \text{315}$$

$$CH_3CO \!-\! NHCHCO \!-\! NHCHCO \!-\! NHCH_2CO \!-\! OCH_3 \qquad (10.25)$$

with side groups CH_2–$CH(CH_3)_2$ on Leu and CH_3 on Ala.

Leu   Ala   Gly

For simple peptides it may be unnecessary to determine the elemental compositions of the fragment ions except as confirmation of the sequence. Larger peptides often give complex mass spectra because of competing side-chain fragmentation, and it may then be necessary to determine the elemental compositions of all ions in the spectrum by accurate mass measurement. From the elemental compositions the sequence may be determined as above and computer programmes are available to deal with the sorting involved.

A further parallel between compounds with saturated and those with unsaturated substituents is found in the six-centre transfer of hydrogen in carbonyl compounds, comparable with the six-centre elimination of H_2O from alcohols. The reaction is often called the McLafferty rearrangement and is exemplified in scheme (10.26) for pentan-2-one. The elimination of H_2O from pentan-1-ol is shown for comparison. Again, the rearrangement reaction of the odd-electron molecular ion involves more than one bond cleavage and yields an odd-electron fragment ion. In comparing the two reactions of scheme (10.26), notice that, for the ketone, the charge generally remains with the heteroatom fragment and a neutral aliphatic species is

$$(10.26)$$

ejected but, for alcohols, the charge is retained on the aliphatic part and the heteroatom is ejected as neutral H_2O.

Just as the elimination of H_2O from alcohols is analogous to the elimination of H_2S from thiols, the six-centre rearrangement of carbonyl compounds may be regarded as a particular instance of the general and ubiquitous phenomenon shown schematically in reaction (10.27). The atoms X and Y may be carbon, oxygen, sulphur, phosphorus or nitrogen in various combinations. Sometimes, the charge is retained by the alkene fragment (containing the carbon atoms labelled 4 and 5 in scheme (10.27)) rather than by the fragment containing X and Y. Often a mass spectrum will exhibit

$$(10.27)$$

peaks for both ions. Deuterium labelling has demonstrated that hydrogen transfer takes place mainly from position 5 but transfer also appears to arise from positions 3 and 4. The rearrangement is observed in the mass spectra of amides and sulphoxides as, for example, in reaction schemes (10.28):

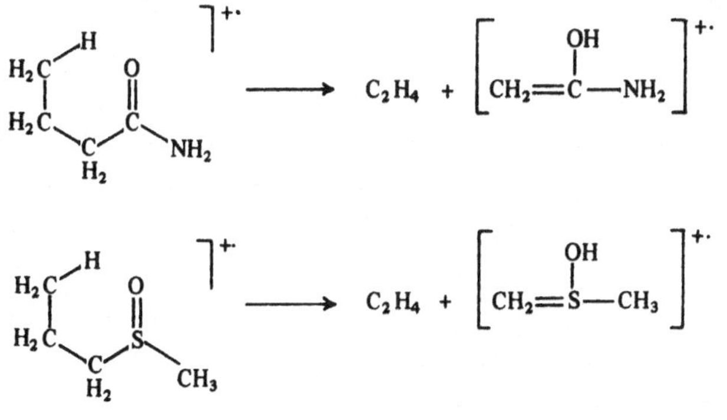

$$\left[\begin{array}{c}H_2C\overset{H}{\diagdown}\;O\\H_2C\diagdown_{C}\diagdown_{C}\diagdown NH_2\\\quad H_2\end{array}\right]^{+\cdot}\longrightarrow C_2H_4 + \left[\begin{array}{c}OH\\CH_2{=}C{-}NH_2\end{array}\right]^{+\cdot}$$

(10.28)

$$\left[\begin{array}{c}H_2C\overset{H}{\diagdown}\;O\\H_2C\diagdown_{C}\diagdown_{S}\diagdown CH_3\\\quad H_2\end{array}\right]^{+}\longrightarrow C_2H_4 + \left[\begin{array}{c}OH\\CH_2{=}S{-}CH_3\end{array}\right]^{+\cdot}$$

Esters that have an alkoxy group longer than methoxy also show the rearrangement and give a peak corresponding to the ionized free acid:

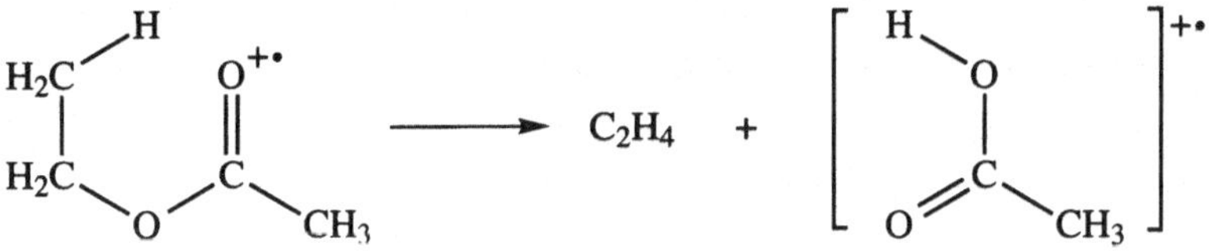

$$\begin{array}{c}H_2C\overset{H}{\diagdown}\;O^{+\cdot}\\H_2C\diagdown_{O}\diagdown_{C}\diagdown CH_3\end{array}\longrightarrow C_2H_4 + \left[\begin{array}{c}H\diagdown O\\O{=}C\diagdown CH_3\end{array}\right]^{+\cdot}$$

Sometimes this single hydrogen transfer in esters is masked by a double hydrogen transfer to give what is thought to be the protonated acid. The 1-propyl ester of ethanoic acid (figure 10.19) gives a large peak at $m/z\,61$ due to this double rearrangement (reaction (10.29)). Deuterium labelling suggests

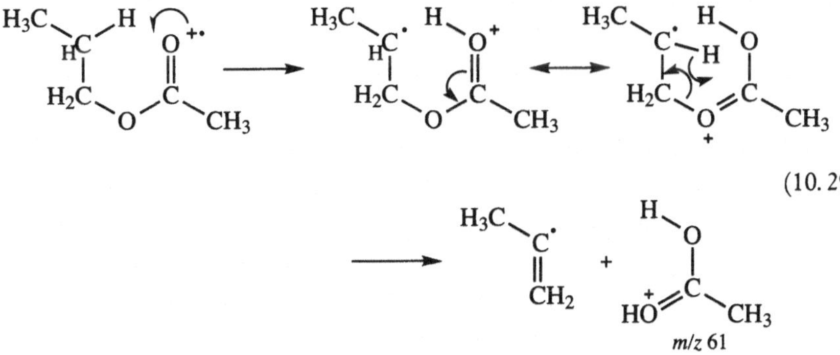

(10.29)

$$\longrightarrow \begin{array}{c}H_3C\diagdown_{C}\!^{\cdot}\\\quad\|\\\quad CH_2\end{array} + \begin{array}{c}H\diagdown O\\HO\overset{+}{=}C\diagdown CH_3\end{array}$$

$m/z\,61$

that two hydrogen atoms are transferred mainly from the β-carbon atoms of the alkoxy group. A similar double hydrogen rearrangement with formation of a protonated species is found in the mass spectra of many types of esters, including carbonates and those of dicarboxylic acids.

The generalized simple cleavage of aliphatic chains in reaction type (iv) of saturated substituents is found in a modified form with unsaturated substituents in alkanes. Thus, esters of general formula, $C_nH_{2n+1}COOR$, give

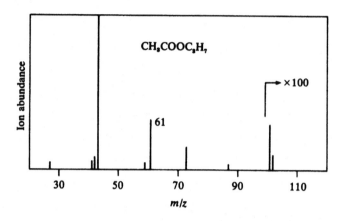

Figure 10.19. A mass spectrum of 1-propyl ethanoate showing a peak at m/z 61 due to double hydrogen rearrangement. The peaks above m/z 100 have been scaled up 100 times.

rise to fragment ions of compositions, $C_nH_{2n}COOR$. However, unlike halide compounds, for which the most abundant fragment ions have a composition $C_nH_{2n}X$ with $n = 4$, the ester fragment ions are most abundant at the compositions corresponding to $n = 2, 6, 10$, etc. The effect has been explained in terms of transfer of hydrogen and fragmentation as shown in the reaction sequence (10.30):

$$\tag{10.30}$$

10.6.1. Summary

Many simple fragmentation reactions of molecular ions containing hetero-atoms may be summarized by reference to scheme (10.31) in which X represents the heteroatom and R^1, R^2 and R^3 are substituents; when X is doubly bonded to the adjacent carbon atom R^3 is not present:

$$\tag{10.31}$$

Type C1 cleavage is breaking of the C–X bond and is observed with saturated heteroatom substituents (X = Cl, Br, I, OH, SH, NH$_2$). The C1 cleavage is not observed with unsaturated heteroatom substituents. Type C2
cleavage is commonly observed for both saturated and unsaturated heteroatom substituents of alkanes. The extent to which C1 or C2 bond breaking
occurs is very dependent on the nature of the heteroatom and the structure
of the molecule. Either or both fragments from C1 and C2 reactions may
appear as even-electron charged species in the mass spectrum as shown in
scheme (10.31).

The other common process observed in aliphatic heteroatom compounds
is rearrangement of hydrogen accompanying elimination of a neutral
species to give odd-electron fragment ions. Again the process can be represented in a common formalism (scheme (10.32)) by two reaction types (RE1
and RE2). With saturated substituents (X), the rearrangement–elimination
reaction (RE1) is general except for amines, but, for unsaturated substitu-

$$\tag{10.32}$$

ents, the alternative process (RE2) is general. The processes shown in
scheme (10.32) have been drawn with the six-membered ring transition
state, but it should be recalled that the results of deuterium labelling are not
unequivocal and some hydrogen transfer occurs through smaller and larger
cyclic transition states. For halogen compounds, a similar five-membered
state appears to be more important.

A double bond or an aromatic ring can be considered here as an unsaturated substituent in alkanes. The 'allylic' or 'benzylic' types of cleavage adjacent to these unsaturated centres are comparable with, for example, the
α-cleavage (C2) of carbonyl compounds if the unsaturated centres take the
place of the heteroatom (scheme (10.33)):

$$CH_3 \!-\!\! \big] \!-\! CH\!\!=\!\!\overset{\cdot\cdot}{O} \big]^{+\cdot} \longrightarrow CH_3^{\cdot} + H\overset{+}{C}\!\!=\!\!O \longleftrightarrow HC\!\!\equiv\!\!\overset{+}{O}$$
$$(10.33)$$

$$CH_3 \!-\!\! \big] \!-\! CH_2\!-\!CH\!\!=\!\!CH_2 \big]^{+\cdot} \longrightarrow CH_3^{\cdot} + \overset{+}{C}H_2\!-\!CH\!\!=\!\!CH_2 \longleftrightarrow CH_2\!\!=\!\!CH\!-\!\overset{+}{C}H_2$$

The allyl cation is isoelectronic with the formyl cation in the resonance formalism; in the first case, a positive charge is delocalized over three carbon atoms and, in the other, over one carbon and one oxygen atom. However, the comparison is not a particularly good one because the olefinic double bond itself may be considered isoelectronic with the chloro-function, for example. As with cleavage adjacent to a chlorine atom, cleavage adjacent to the olefin group occurs with hydrogen transfer and is an important process (scheme (10.34)). In many ways, double bonds and aromatic rings have

$$R\!-\!Cl^{+\cdot} \rightarrow HCl + [R - H]^{+\cdot}$$
$$R\!-\!CH\!\!=\!\!CH_2^{+\cdot} \rightarrow H_2C\!\!=\!\!CH_2 + [R - H]^{+\cdot} \qquad (10.34)$$

characteristics in common with both saturated and unsaturated substituents.

It must be stressed that, so far, only primary fragmentation of the molecular ion has been considered. The effects of subsequent fragmentation may modify the mass spectrum to such an extent that the primary processes appear to have little prominence. For example, if the rate of fragmentation of M^+ to give A^+ is slower than the fragmentation of A^+ to B^+ (scheme (10.35)), then the size of the peak corresponding to A^+ in the mass spectrum

$$M^+ \rightarrow A^+ \rightarrow B^+ \qquad (10.35)$$

will be very much reduced (section 1.1). Also, subsequent fragmentation is not so easy to classify and will be discussed later. If only for these two reasons there is little to substitute for experience gained in the interpretation of mass spectra. Attempts to programme computers to elucidate structures from mass spectra have met with some success (section 4.4.3). In this respect, the exceptional memory characteristics of a computer are invaluable for storing accumulated knowledge and experience.

10.7. PRIMARY FRAGMENTATIONS OF AROMATIC HETEROATOMIC COMPOUNDS

The mass spectra of aromatic compounds may be subdivided into three classes: those in which the aromatic ring is separated from the heteroatom by an aliphatic part (aralkyl compounds), those in which the aromatic ring is attached directly to the heteroatom and those in which the heteroatom forms part of the aromatic ring. Since the presence of the aromatic ring may modify the mass spectrum considerably compared with what is observed with aliphatic compounds, these three classes are considered separately.

Aralkyl compounds. Such compounds have mass spectra with characteristics common to aliphatic and aromatic compounds separately. The presence of an aromatic ring appears to confer greater stability on the molecular ions, which are much more prominent than for simple aliphatic materials. The aromatic ring may itself act as the substituent X in formula (10.31), as mentioned above, and give corresponding fragmentations. Before going on to consider heteroatomic compounds, this point is illustrated by the relatively simple example of butylbenzene. In this compound, cleavage and rearrangement processes C1, C2 and RE2 (scheme (10.36a,b,c)) are observed.

Fragmentation at other positions in the aliphatic chain is also found and, for long chains, this supersedes the above C1, C2 and RE2 reactions in importance. Similar behaviour was discussed earlier for changes observed in the mass spectra of aliphatic heteroatom compounds as chain length increases. In effect, the presence of an aromatic ring in hetero-compounds introduces a

second substituent (X) and, in the absence of interaction between the substituents through space, fragmentation of molecular ions is influenced by
each of the substituents. Thus, on electron ionization, methyl 4-phenylbu-
tanoate gives some of its fragment ions as shown (scheme (10.37)):

$$
C_6H_5 \text{—} CH_2 \text{—} CH_2 \text{—} CH_2 \text{—} \underset{\underset{O}{\parallel}}{C} \text{—} OCH_3 \Bigg]^{+\cdot}
\begin{array}{l}
\nearrow \overset{+}{C}H_2CH_2CH_2CO_2CH_3 \\
\nearrow C_7H_7^+ \\
\searrow C_6H_5CH_2CH_2CH_2^+ \\
\searrow C_6H_5CH_2CH_2CH_2\overset{+}{C}O
\end{array}
\qquad (10.37)
$$

The closer the two substituents are to each other, the more they may interact
through space or through bonding so as to modify their separate effects.

Compounds with the substituent heteroatom attached directly to a benzene ring. In
these cases, simple C1 cleavage may still occur, as with chlorobenzene and
nitrobenzene:

$$
\left[\bigcirc\!\!-\!\text{Cl} \right]^{+\cdot} \longrightarrow C_6H_5^{+\cdot} + Cl^{\cdot}
$$

$$
\left[\bigcirc\!\!-\!\text{NO}_2 \right]^{+\cdot} \longrightarrow C_6H_5^{+} + NO_2^{\cdot}
$$

$$
\left[\bigcirc\!\!-\!\text{F} \right]^{+\cdot} \longrightarrow C_6H_3F^{+\cdot} + C_2H_2
$$

It is interesting to note that the C–F bond strength is great enough to
prevent simple C1 cleavage as with aliphatic fluoro-compounds and,
instead, C_2H_2 is eliminated from the benzene ring. The elimination of
halogen from chlorobenzene may be an ion pair process:

$$
\bigcirc\!\!-\!\text{Cl} + e \longrightarrow C_6H_5^{+} + Cl^{-} + e
$$

This would be a case of primary cleavage into positive and negative ions on
electron ionization and is observed for halogen compounds at low electron-
beam energies. Since mass spectrometers are usually set to monitor only
positive ions, the negative ones are ignored in much mass spectrometry. The
existence of this process does not affect the observed spectrum except

insofar as the energy of the process is different from that in which a chlorine atom is ejected (section 1.1).

Often, simple C1 cleavage does not occur and rearrangement of the molecular ion takes place before fragmentation. Rearrangement processes are covered in more detail in section 10.9 but those germane to this description of fragmentation of aromatic heteroatom compounds are discussed here.

Phenols, phenyl ethers and amines occur commonly in aromatic chemistry and exhibit complex fragmentation behaviour. Phenol eliminates not OH$^\bullet$, but rather CO and CHO$^\bullet$, from the molecular ion. Phenyl methyl ether ejects CH_2O and CHO$^\bullet$ but not methyl or methoxy radical, and aniline eliminates HCN from the molecular ion. These processes are quite general (scheme (10.38)) although they may be modified through interactive substituent effects (see below).

$$
\left.
\begin{array}{l}
\text{[C}_6\text{H}_5\text{—OH]}^{+\bullet} \longrightarrow \text{C}_5\text{H}_6^{+\bullet} + \text{CO} \\
\qquad\qquad\quad \longrightarrow \text{C}_5\text{H}_5^{+} + \text{H}\dot{\text{C}}\text{O} \\
\qquad\qquad\quad \nrightarrow \text{C}_6\text{H}_5^{+} + \dot{\text{O}}\text{H} \\[2ex]
\text{[C}_6\text{H}_5\text{—OCH}_3\text{]}^{+\bullet} \longrightarrow \text{C}_6\text{H}_6^{+\bullet} + \text{CH}_2\text{O} \\
\qquad\qquad\qquad \longrightarrow \text{C}_6\text{H}_7^{+} + \text{H}\dot{\text{C}}\text{O} \\
\qquad\qquad\qquad \nrightarrow \text{C}_6\text{H}_5\text{O}^{+} + \text{CH}_3^{\bullet} \\[2ex]
\text{[C}_6\text{H}_5\text{—NH}_2\text{]}^{+\bullet} \longrightarrow \text{C}_5\text{H}_6^{+\bullet} + \text{HCN} \\
\qquad\qquad\qquad \nrightarrow \text{C}_6\text{H}_5^{+} + \dot{\text{N}}\text{H}_2
\end{array}
\right\} \qquad (10.38)
$$

In some circumstances, nitro-compounds do not simply eject $NO_2^\bullet$ but a complex reaction sequence occurs in which NO$^\bullet$ and CO are eliminated successively (reaction (10.39)). This process is observable to a slight extent in

$$
[\text{C}_6\text{H}_5\text{NO}_2]^{+\bullet} \longrightarrow [\text{C}_6\text{H}_5\text{—O—N}{=}\text{O}]^{+\bullet} \xrightarrow{-\text{NO}^\bullet} \text{C}_6\text{H}_5\text{O}^{+} \xrightarrow{-\text{CO}} \text{C}_5\text{H}_5^{+}
$$
$$(10.39)$$

the mass spectrum of nitrobenzene itself but it can increase in importance in other compounds so as to make the simple elimination of $NO_2^\bullet$ inconsequential. Table 10.7 lists the commoner substituents exhibiting such rearrangements and the neutral species ejected.

Table 10.7. *More commonly occurring substituents of aromatic rings which eject neutral particles resulting from rearrangement.*

Substituent	Neutral particle(s) ejected after rearrangement[a]
NO_2	NO, CO (NO_2)
NH_2	HCN
$NHCOCH_3$	C_2H_2O, HCN
CN	HCN
F	C_2H_2
OCH_3	CH_2O, CHO, (CH_3)
OH	CO, CHO
SO_2NH_2	SO_2, HCN
SH	CS, CHS (SH)
SCH_3	CS, CH_2S, SH (CH_3)

Notes:

[a] Particles shown in parentheses are ejected without rearrangement. Where these are given, it indicates that these particle losses are usually observed together with those resulting from rearrangement; their relative proportions are very variable.

Often there is more than one heteroatomic substituent in an aromatic compound and they may interact to modify the above fragmentation processes considerably. The relative positions of the substituents around the aromatic ring are also important. Whereas the elimination of a methyl radical from the methoxyl group in phenyl methyl ether is a scarcely detectable process, in 2- and 4-methoxyaniline the process is more important than the usual behaviour of the aromatic methoxyl group (losses of CH_2O and CHO•). The ease of elimination of the methyl radical from 2- and 4-methyoxyaniline has been ascribed to a resonance stabilization of the fragment ion. The behaviour of 3-methoxyaniline, in which such resonance interaction is considered impossible, is more 'normal' in that CH_2O and CHO• are eliminated from the molecular ion (reactions (10.40)):

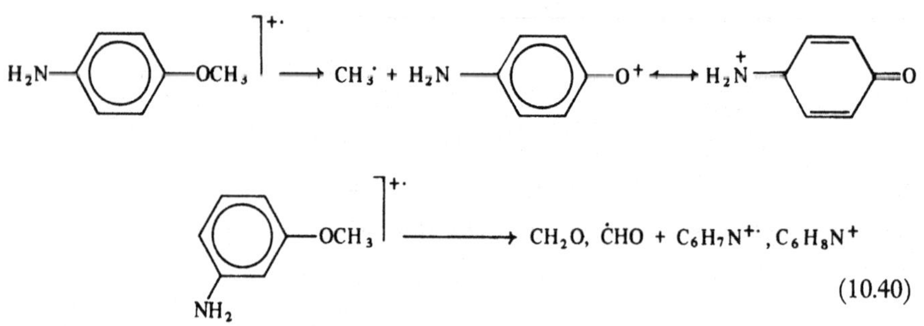

$$CH_2O, \dot{C}HO + C_6H_7N^{+\cdot}, C_6H_8N^+$$

(10.40)

As in solution chemistry, *ortho*-substituents in benzene rings may behave differently from *meta*- and *para*-substituents. The difference in behaviour is often called an *ortho*-effect. The elimination of CH_4O from methyl anthranilate and of OH· from *ortho*-nitrotoluene are examples of this *ortho*-effect (reactions (10.41)). The process is quite general and may be represented by reaction scheme (10.42) in which A, X, Y and Z may be carbon, oxygen, nitrogen and sulphur:

$$(10.41)$$

$$+ \text{ HYZ} \qquad (10.42)$$

It is frequently observed that *ortho*-substituents are eliminated more readily than *meta*- or *para*-substituents and a cyclization process is often invoked as an explanation but there is little evidence to support it. This ejection of *ortho*-substituents is also described as an *ortho*-effect. An example is shown in scheme (10.43):

$$(10.43)$$

$$R = \text{Cl}, \text{NO}_2, \ldots$$

Table 10.8. *The order of ease of fragmentation initiated by the presence of a substituent in a benzene ring.*[a]

Substituent	Neutral particle eliminated
$COCH_3$	CH_3
CO_2CH_3	OCH_3
NO_2	NO_2
I	I
OCH_3 ⎫	⎧ CH_2O, CHO
Br ⎬	⎨ Br
OH ⎭	⎩ CO, CHO
CH_3	H
Cl	Cl
NH_2	HCN
CN	HCN
F	C_2H_2

Notes:
[a] Decreasing ease of fragmentation from top to bottom
of table. The bracketed substituents
are closely similar in ease of fragmentation.

For disubstituted benzenes the order of fragmentation at the substituents may be predicted quite accurately from table 10.8, which gives, alongside each substituent, the neutral particle commonly ejected during its fragmentation. Generally, the higher a substituent in the table, the more readily fragmentation begins there. Thus, in a chloroacetophenone, elimination of methyl radical is observed before that of chlorine and in a fluorobromobenzene loss of bromine is observed but not of fluorine (scheme (10.44)). Table 10.8 is intended as a guide only because strong interaction effects between substituents can modify their behaviour, as shown earlier.

$$(10.44)$$

Benzenes substituted with OCH_3, Br and OH groups require very similar energies for fragmentation so that when these substituents occur together in a benzene compound, table 10.8 serves as only a nominal guide to the order of fragmentation.

The role of substituents in the mass spectra of aromatic compounds has been examined in some detail through the use of Hammett σ-constants (linear free energy relationships), but there is little justification for doing this. The conditions obtaining in electron ionization mass spectrometry do not permit the use of the same thermochemical relationships as are used for solution chemistry (chapter 9). Lack of knowledge of ion structures and the electronic states leading to fragmentation also provides good grounds for reluctance to draw conclusions from observed correlations with σ-constants.

Heteroaromatic compounds. These compounds may be compared with benzene. Benzene yields relatively stable and hence abundant molecular ions. Its most important initial fragmentation leads to the elimination of C_2H_2 and $C_2H_3{}^{\bullet}$. Heteroaromatic compounds also give abundant molecular ions and, apart from side-chain or substituent fragmentations such as those described

$$C_6H_6{}^{+\bullet} \Big\langle \begin{array}{l} C_2H_2 + C_4H_4{}^{+\bullet} \\ C_2H_3{}^{\bullet} + C_4H_3{}^{+} \end{array}$$

in (i) and (ii) above for benzene compounds, their fragmentation leads to the ejection of the heteroatom in a neutral species. Whereas benzene ejects C_2H_2 and $C_2H_3{}^{\bullet}$ from the molecular ion, pyrrole and pyridine eject HCN. Similarly, thiophene ejects $CHS^{\bullet}$ and furan eliminates $CHO^{\bullet}$ from their respective molecular ions:

$$\longrightarrow \quad HCN + C_4H_4{}^{+\bullet}$$

$$\longrightarrow \quad H\dot{C}S + C_3H_3{}^{+}$$

$$\longrightarrow \quad H\dot{C}O + C_3H_3{}^{+}$$

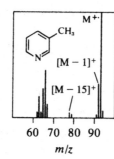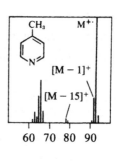

Figure 10.20. Mass spectra of 2-, 3- and 4-methylpyridine showing the enhanced size of the $[M - 1]^+$ peak in the 3-methyl isomer, and the enhanced size of the $[M - 15]^+$ peak in the 2-methyl isomer.

Fragmentations of the heteroaromatic ring itself and of any side chains often compete with each other; the fragmentation of the side chains may be influenced by their positions in the aromatic nucleus. This last property can be invaluable in deducing the position of a substituent in a heteroaromatic ring. The amount of elimination of H• from 2-, 3- and 4-methylpyridine varies with the position of the substituent, the $[M - 1]^+$ peak being largest in the 3-methyl isomer (figure 10.20). Similarly, the 2-methyl isomer has the largest $[M - 15]^+$ peak corresponding to the loss of a methyl radical. The pyridine-2,3-dicarboxylic acid imide (**11**) eliminates both CO and CO_2 from the molecular ion to give ions at m/z 134 and 118 with relative abundances of 17 and 4 per cent compared with the base peak in the spectrum (reaction (10.45)). The 3,4-isomer (**12**) affords ions at the same m/z values but with relative abundances of 1 and 27 per cent.

$$\left[\text{(11)} \right]^{+\cdot} \longrightarrow [M\text{-}CO]^{+\cdot}, \quad [M\text{-}CO_2]^{+\cdot}$$

(**11**) m/z 162 m/z 134 (17%) m/z 118 (4%)

$$\left[\text{(12)} \right]^{+\cdot} \longrightarrow [M\text{-}CO]^{+\cdot}, \quad [M\text{-}CO_2]^{+\cdot}$$

(10.45)

(**12**) m/z 162 m/z 134 (1%) m/z 118 (27%)

Deuterium and ^{13}C labelling strongly suggest that extensive rearrangement of benzene compounds occurs in the molecular ion before fragmentation. It has been suggested that these rearrangements proceed through benzvalene-type intermediates. Sometimes, this sort of rearrangement is apparent from the nature of the fragment ions. For example, the mass spectra of oxazoles contain features very similar to those of isoxazoles and it has been suggested that rearrangement occurs via an azirane (scheme (10.46)) as found in the photochemistry of these compounds:

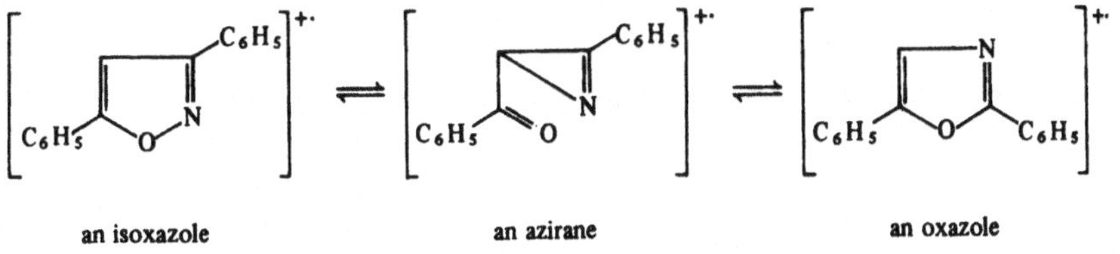

an isoxazole   an azirane   an oxazole

(10.46)

Because C2 type side-chain cleavage of heteroaromatic compounds is often prominent in their fragmentation, it has been suggested that the fragment ions are analogous to the supposed tropylium structure found with some benzyl compounds, i.e, that ring expansion has occurred (reactions (10.47)). However, experimental evidence supporting this contention is not

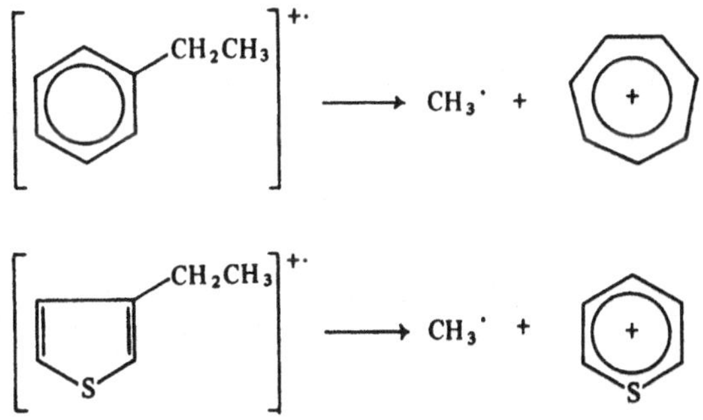

(10.47)

unequivocal in all cases. Nevertheless, in this type of fragmentation behaviour and irrespective of the ring-expansion issue, heteroaromatic compounds are very similar to simple benzene compounds.

The organometallic 'sandwich' compounds like ferrocene exhibit many characteristics of the fragmentation behaviour of aromatic compounds. Abundant molecular ions are observed and any side chains that are present normally fragment like their simple benzene counterparts. Also, just as heteroaromatic compounds decompose by fragmentation of the aromatic ring, so the metallocene molecular ions decompose with destruction of the sandwich structure to leave an ionized metal atom ($Met^{+\cdot}$).

10.7.1. *Summary*

Mass spectra of aromatic compounds exhibit many of the features found in those of aliphatic compounds but are usually considerably modified because of (*a*) the increased stability of the molecular ions, (*b*) the aromatic ring system itself behaving as a substituent (X) in an alkane, (*c*) substituents on the aromatic ring changing fragmentation behaviour and (*d*) heteroatoms in the aromatic ring influencing the fragmentation of the side chain depending on its position.

10.8. SUBSEQUENT DECOMPOSITION OF PRIMARY FRAGMENT IONS

Most molecular ions formed by electron ionization are radical-cations that have one unpaired electron. Radical-cations are known in solution chemistry to be reactive species and the molecular ions formed in mass spectrometry frequently decompose by ejection of a radical to form an even-electron charged species (scheme (10.48*a*)). In even-electron species, the electrons can pair in orbitals to give closed shells; such even-electron cations in solution chemistry, the carbocations, are generally less highly reactive than odd-electron species like radical-cations.

$$[CH_3CH_2CH_3]^{+\cdot} \rightarrow CH_3^{\cdot} + C_2H_5^{+}$$

m/z 44 (even mass number)   radical   *m/z* 29 (odd mass number)
radical–cation   cation
(odd number of electrons)   (even number of electrons)

$$(10.48a)$$

$$[C_6H_5OCO-OCH_3]^{+\cdot} \rightarrow CO_2 + C_7H_8O^{+\cdot}$$

m/z 152 (even mass number)   *m/z* 108 (even mass number)
radical–cation   radical–cation
(odd number of electrons) $\downarrow$ $-\overset{\cdot}{C}HO$

$$C_6H_7^{+}$$

m/z 79 (odd mass number)
cation
(even number of electrons)

$$(10.48b)$$

If the radical-cation molecular ion does not eject a radical to produce an even-electron ion, rearrangement may occur with subsequent elimination of a non-radical fragment to give another odd-electron ion. The rearranged odd-electron fragment can then decompose by ejection of a radical to yield an even-electron ion as in the example of methyl phenyl carbonate (scheme (10.48b)). It is found that many of the ions in a mass spectrum occur at odd-numbered m/z values (even-electron ions). Ions at even-numbered m/z values (odd-electron ions) are produced by rearrangement. For odd numbers of nitrogen atoms in an ion this rule is reversed, i.e., odd-numbered m/z values correspond to odd-electron species and even-numbered m/z values to even-electron species.

The 'even-electron rule' states that odd-electron ions (radical-cations) may eliminate either radicals or neutral molecules but that even-electron ions (cations) may fragment only by loss of neutral molecules and not by ejecting radicals. The situation is summarized in scheme (10.49):

$$
\begin{aligned}
A^{+\cdot} &\longrightarrow C^{+} + N^{1\cdot} \\
A^{+\cdot} &\longrightarrow D^{+\cdot} + N^{2} \\
B^{+} &\longrightarrow E^{+} + N^{3} \\
B^{+} &\xrightarrow{\;\;\;|\!\!|\;\;\;} F^{+\cdot} + N^{4\cdot}
\end{aligned}
\qquad (10.49)
$$

A simpler statement of the rule is that successive losses of radicals are forbidden. The rationale of the rule, in charge localization terms, is that, in an even-electron species, there are no unpaired electrons to 'trigger' radical reactions. Qualitative thermochemical theory argues that product stability largely governs fragmentation pathways and because even-electron species are generally more stable than odd-electron species, loss of a radical from the former to give the latter is energetically unfavourable. For most compounds, the rule is obeyed and may be usefully applied when rationalizing mass spectra. For example, odd-mass ions that contain no nitrogen can fragment to further odd-mass ions but not to even-mass ones. However, the even-electron rule has more than its fair share of exceptions (Karni and Mandelbaum, 1980) and violations appear to be associated with dissociations in which either or both of the odd-electron products (radical and radical-cation) are particularly stable. Polychlorinated, polybrominated and particularly polyiodinated compounds sometimes exhibit successive losses of the relatively stable halogen radicals. Violations are sometimes observed when the loss of two radicals leads to ions for which structures containing

$$\text{(structure)} \xrightarrow{-e} \xrightarrow{-CH_3^{\bullet}} \xrightarrow{-CH_3^{\bullet}} \text{(phenanthrene structure)}^{+\bullet} \qquad (10.50)$$

high degrees of conjugated unsaturation may be postulated (reaction (10.50)). However, arguments based on thermochemical stabilities of assumed structures are not well founded. It is, of course, not necessary to invoke successive loss of two methyl radicals if a compound shows $M^{+\bullet}$, $[M - 15]^+$ and $[M - 30]^{+\bullet}$ ions since losses of $CH_3^{\bullet}$ and C_2H_6 from the molecular ion could explain the results. Unless the elimination of a radical from an even-electron ion is substantiated by observation of metastable $[M - 15]^+$ ions decomposing to $[M - 30]^{+\bullet}$ ions, for example, it should not be proposed. Generally, the even-electron rule does have a place in mass spectrometry as a useful 'rule of thumb.' When even-electron ions are induced to fragment by collisional activation (section 8.7.1), the even-electron rule no longer holds, presumably because the attendant increase of internal energy in the ions overcomes the high activation energy for loss of radicals.

It will be appreciated from the foregoing discussion that subsequent decomposition of primary fragment ions often revolves around the behaviour of even-electron species. It may be recalled that one of the first items to be noted in a mass spectrum is which of the major fragment ions occur at odd-numbered m/z values and which at even-numbered m/z values. It is useful, when studying their subsequent fragmentation, to keep in mind which are even-electron and which are odd-electron ions.

The initial fragmentation reactions of molecular ions can be grouped into a few general schemes, but the decomposition reactions of the fragments themselves are not so easy to classify. The first fragmentations of a molecule in a mass spectrometer are generally those giving most structural information because after several decompositions along any particular fragmentation pathway, the relationship between the ions and the original structure becomes too tenuous to be of much value. Fragment ions (A^+ and B^+) from a molecular ion ($M^{+\bullet}$ in scheme (10.51)) may yield valuable information about $M^{+\bullet}$, e.g. loss of H_2O, $CH_3^{\bullet}$, $CH_3O^{\bullet}$ etc., but by the time fragmentation reaches the ion D^+ much of the original structure of $M^{+\bullet}$ has been destroyed and little structural information may be obtained:

$$M^{+\bullet} \rightarrow A^+ \rightarrow B^+ \rightarrow C^+ \rightarrow D^+ \rightarrow \qquad\qquad (10.51)$$

However, ions at the low mass end of a spectrum should not be ignored. Important clues to the structure of $M^{+\bullet}$ may still be obtained as with $C_2H_3O^+$ ions at m/z 43 from CH_3CO groups and CH_4N^+ ions at m/z 30 from primary amines. (Do note, though, that such useful low-mass ions often arise from primary fragmentation of the molecular ion. Just because they have low mass, they are not necessarily products of a multi-step fragmentation pathway and they are certainly not disqualified from being primary fragment ions.) The loss of $C_2H_3O^\bullet$ from a molecular ion can be recognized by the presence of $[M - C_2H_3O^\bullet]^+$ ions as well as $C_2H_3O^+$ ions at m/z 43. The later stages of a fragmentation pathway frequently only become clear after the structure of a compound has been elucidated. These later fragment ions then serve as confirmation for a proposed structure rather like the use of the 'fingerprint' region in infrared spectroscopy, which may be used for confirmatory evidence for an assignment but is not very useful for initial diagnosis.

Aliphatic fragment ions decompose further by elimination of C_2H_4:

$$R-CH_2-CH_2{}^+ \rightarrow R^+ + C_2H_4$$

Aromatic fragment ions dissociate by ejection of C_2H_2, so that aromatic compounds are often recognized by the series of ions m/z 77 and 51 or m/z 91, 65 and 39:

$$\begin{array}{cc} & -C_2H_2 \\ C_6H_5{}^+ & \longrightarrow \quad C_4H_3{}^+ \\ m/z\ 77 & m/z\ 51 \end{array}$$

$$\begin{array}{ccc} & -C_2H_2 & -C_2H_2 \\ C_7H_7{}^+ & \longrightarrow \ C_5H_5{}^+ & \longrightarrow \ C_3H_3{}^+ \\ m/z\ 91 & m/z\ 65 & m/z\ 39 \end{array}$$

Heteroatomic molecules, after initial fragmentation, often decompose further after hydrogen transfer. The migrating hydrogen atom may arise from a β-, γ- or δ-carbon to the heteroatom. The common C2 cleavage reaction of heteroatomic compounds yields even-electron fragment ions, which, after hydrogen transfer, may fragment along two different pathways to give even-electron ions containing the heteroatom. The processes may be represented generally by scheme (10.52) in which X is oxygen, sulphur or nitrogen (strictly N–R′). Ions at m/z 30 and 58 in the mass spectrum of N-ethylpropylamine (scheme (10.53)) are examples of this type of fragmentation.

$$\left[\begin{array}{c} R\!-\!CHXCH\!-\!R^3 \\ \mid \quad \mid \\ R^1 \quad R^2 \end{array} \right]^{+\cdot} \nearrow^{-R^{1\cdot}} \begin{array}{c} R\!-\!CH\!=\!\overset{+}{X}CH\!-\!R^3 \\ \mid \\ R^2 \end{array} \longrightarrow [R\!-\!H] + \begin{array}{c} CH_2\!=\!\overset{+}{X}CH\!-\!R^3 \\ \mid \\ R^2 \end{array}$$

$$\searrow^{-R^{2\cdot}} \begin{array}{c} R\!-\!\overset{+}{CHX}\!=\!CH\!-\!R^3 \\ \mid \\ R^1 \end{array} \longrightarrow [R^3\!-\!H] + \begin{array}{c} R\overset{+}{CHX}\!=\!CH_2 \\ \mid \\ R^1 \end{array}$$

(10.52)

$$C_3H_7\!-\!\overset{+\cdot}{NH}\!-\!C_2H_5 \nearrow^{-H\cdot} C_2H_5\!-\!CH\!=\!\overset{+}{NH}\!-\!C_2H_5 \xrightarrow{\text{H-transfer}} C_2H_4 + [CH_2\!=\!\overset{+}{NH}\!-\!C_2H_5$$

$$m/z\ 58$$

(10.53)

$$\searrow^{-C_2H_5\cdot} CH_2\!=\!\overset{+}{NHC_2H_5} \xrightarrow{\text{H-transfer}} C_2H_4 + CH_2\!=\!\overset{+}{NH_2}$$

$$m/z\ 30$$

Hydrogen transfer may also occur with elimination of the heteroatom as the neutral species and is quite common for halides and esters. For example, if X = Br $(-CH(R^2)R^3$ absent in scheme (10.52)) then HBr is eliminated as in reaction sequence (10.54):

$$C_2H_5\!-\!CH\!-\!\overset{\cdot+}{Br} \longrightarrow CH_3\cdot + C_2H_5\!-\!CH\!=\!\overset{+}{Br} \xrightarrow{\text{H-transfer}} HBr + C_3H_5^+ \quad (10.54)$$
$$\overset{|}{\underset{CH_3}{}}$$

The elimination of a neutral fragment from even-electron ions is particularly marked in the mass spectra of many simple organometallic compounds. Ejection of a butyl radical from the molecular ion of tetra-1-butyl tin affords an even-electron ion, which then eliminates C_4H_8 with hydrogen transfer to give another even-electron ion; this latter ion then decomposes again by elimination of a second C_4H_8 unit to yield a third even-electron ion:

$$(C_4H_9)_4Sn^{+\cdot} \xrightarrow[-C_4H_9\cdot]{} (C_4H_9)_3Sn^+ \cdot \xrightarrow[-C_4H_8]{\text{H-transfer}} (C_4H_9)_2SnH^+ \cdot \xrightarrow[-C_4H_8]{\text{H-transfer}} (C_4H_9)SnH_2^+$$

When rearrangement of a molecular ion accompanies its fragmentation, the product ion may itself fragment like the molecular ion of the postulated rearranged species. The point may be clarified by two examples. Ejection of CO from the molecular ion of pyranocoumarin (13) yields a fragment ion, which is conveniently postulated to have the furanocoumarin structure (14). The subsequent decompositions of the ion (14) and of the molecular ion produced by a specimen of the furanocoumarin itself were almost identical. Analogously, the methoxyphenazine (15) eliminated CH_2O from its molecular ion to give a fragment ion (16) postulated to be the phenazine molecular ion. The subsequent decomposition of the ion (16) was almost identical to that of the molecular ion of phenazine. The close similarity in behaviour of the respective fragment and molecular ions suggests they may have similar structures although, especially for the decomposing ion, these are not necessarily as postulated. There is thermochemical evidence that at least some of the many observed similarities of ion decompositions are accidental and that the ion structures cannot be as postulated.

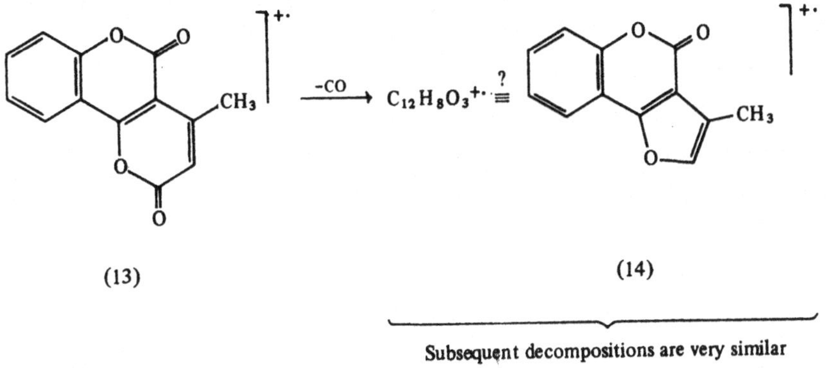

(13)   (14)

Subsequent decompositions are very similar

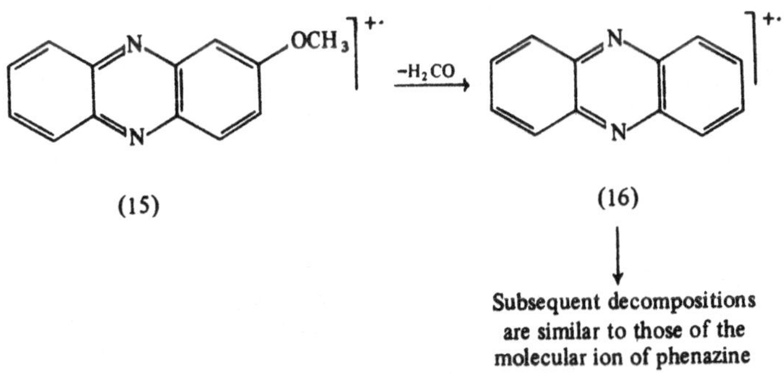

(15)   (16)

Subsequent decompositions
are similar to those of the
molecular ion of phenazine

10.9.   REARRANGEMENT ACCOMPANYING FRAGMENTATION

The formation of new bonds between atoms in an ion leads to rearrangement. Several such processes have been covered in the sections above, as with hydrogen migration, which seems particularly facile. In some instances, mainly with odd-electron ions, the site from which a hydrogen is transferred is fairly specific. A particularly ubiquitous example is the six-centre transfer of hydrogen in carbonyl and related compounds (sometimes called the McLafferty rearrangement) as illustrated in schemes (10.26) and (10.27). In other cases, generally with even-electron ions, the reaction appears to be far less specific. The randomization of hydrogen in ions is widespread and can make the interpretation of deuterium labelling studies difficult or ambiguous.

Skeletal rearrangement is defined as the formation of new bonds between atoms other than hydrogen. The classification of these rearrangements is useful for the interpretation of mass spectra because known structural features giving rise to this phenomenon can be compared with the behaviour of compounds of unknown but postulated structure. Skeletal rearrangements in mass spectrometry are diverse but attempts have been made to classify them.

One of the earliest and most useful classifications was based on (i) the migration of alkyl or aryl groups and (ii) the nature of the neutral particle ejected after rearrangement (Brown and Djerassi, 1967). These compilations are useful and, for structural organic chemistry, the nature of the ejected species may be particularly informative. Typical neutral species eliminated following rearrangement include CO, CO_2, SO_2, CH_2O and HCN, all having favourable negative heats of formation. Two examples are shown in scheme (10.55).

$$\left[\bigcirc\kern-1.2em\bigcirc -SO_2NH_2 \right]^{+\cdot} \longrightarrow SO_2 + \left[\bigcirc\kern-1.2em\bigcirc -NH_2 \right]^{+\cdot}$$

(10.55)

$$\left[\bigcirc \overset{\overset{H}{|}}{\underset{N}{}} \bigcirc \right]^{+\cdot} \longrightarrow HCN + C_{11}H_{10}^{+\cdot}$$

The diagnostic value of the ejected neutral particle may or may not be significant. Thus, elimination of CO_2 from a molecular ion indicates an unsaturated ester, carbonate or cyclic imide group as likely structural features but loss of CO occurs from a wide variety of compounds and yields less immediate information. Elimination of CO_2 from unsaturated esters and carbonates is quite general, although the degree to which it is observed is very variable. The process has been represented by two schemes in which X and Y are doubly or triply bonded in the unsaturated esters and either R or R′ must be aryl in carbonates (reactions (10.56)).

$$
\begin{bmatrix} X \\ \| \\ Y \diagdown_{CO_2^{\diagdown}R} \end{bmatrix}^{+\cdot} \longrightarrow \begin{bmatrix} X_{\diagdown} \\ \| \diagup R \\ Y^{\diagup} \end{bmatrix}^{+\cdot} + CO_2
$$

$$
\begin{bmatrix} R \\ | \\ O_{\diagdown}{}_{CO_2^{\diagdown}R'} \end{bmatrix}^{+\cdot} \longrightarrow \begin{bmatrix} R \\ | \\ O-R' \end{bmatrix}^{+\cdot} + CO_2 \qquad (10.56)
$$

The idea of migration to a positive centre via a nominal four centre process in even-electron ions is embodied in scheme (10.57), two examples of which are shown:

$$
\begin{matrix} A^+ & D \\ | & | \\ B\!-\!C \end{matrix} \longrightarrow \begin{matrix} A\!-\!D \\ | \\ B\!-\!C^+ \end{matrix} \longrightarrow AD^+ + BC
$$

$$
\begin{bmatrix} CH_3O & CH_3 \\ | & | \\ O\!=\!C\!-\!O \end{bmatrix}^{+\cdot} \xrightarrow{-H^\cdot} \begin{matrix} CH_2\!=\!\overset{+}{O}\diagup CH_3 \\ | \diagup| \\ O\!=\!C\!\diagdown O \end{matrix} \longrightarrow CH_2\!=\!\overset{+}{O}\!-\!CH_3 + CO_2 \qquad (10.57)
$$

$$
\begin{bmatrix} ArSO_2NH & Ph \\ | & | \\ N\!=\!CR \end{bmatrix}^{+\cdot} \xrightarrow{-ArSO_2^\cdot} \begin{matrix} \overset{+}{NH}\diagup Ph \\ | \diagup| \\ N\!=\!CR \end{matrix} \longrightarrow \overset{+}{PhNH} + RCN
$$

Whereas these rearrangements appear to be four-centre ones, the corresponding reaction of trimethylsilyl ethers seems amazingly insensitive to the distance between the reacting centres:

$$
\begin{bmatrix} OSi(CH_3)_3 \\ | \\ (CH_2)_n \\ | \\ OPh \end{bmatrix}^{+\cdot} \xrightarrow{-CH_3^\cdot} \begin{matrix} \overset{+}{O}Si(CH_3)_2 \\ \diagdown \\ (CH_2)_n \\ \diagup \\ OPh \end{matrix} \longrightarrow \begin{matrix} Ph\overset{+}{O}Si(CH_3)_2 \\ + \\ (CH_2)_nO \end{matrix}
$$

A variety of formally four-centre rearrangements in odd-electron ions can be collated in one scheme (10.58), which is very similar to that of (10.57) but migration to a positively charged centre is not implied:

$$
\begin{bmatrix} A & D \\ | & | \\ B\!-\!C \end{bmatrix}^{+\cdot} \longrightarrow \begin{bmatrix} A\!-\!D \\ | \\ B\!-\!C \end{bmatrix}^{+\cdot} \diagup^{\displaystyle AD^+ + BC}_{\diagdown\, AD\ +\ BC^+} \qquad (10.58)
$$

The main requirement for this reaction appears to be that atoms A, B, C and D, which may be nitrogen, oxygen, sulphur or carbon, should form contiguous π-centres. It is not suggested that all these rearrangements actually proceed through a four-membered transition state but when four such π-centres are present in a molecule, then rearrangement ions may be expected in the mass spectrum. Two examples from the mass spectra of trifluoroacetyl amides and cyclic thioimides illustrate the general nature of this rearrangement:

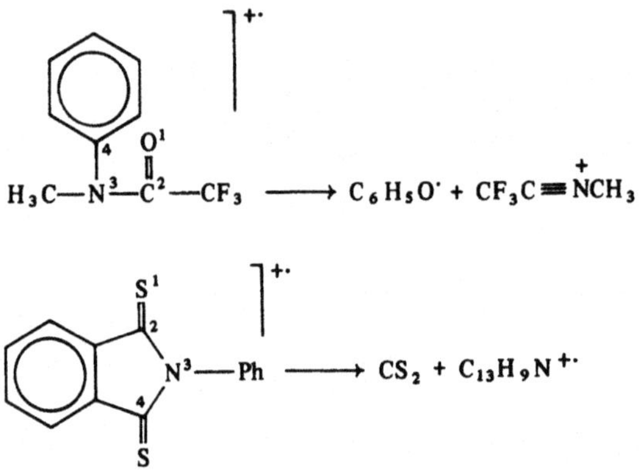

Often rearrangement competes with simple cleavage and the latter may be so easy that peaks representing rearrangement processes are small or not present at all. From this point of view, the closely similar amides (17) and (18) are interesting since the former shows almost no rearrangement whereas the latter has abundant ions corresponding to substantial rearrangement. Unlike the nominal four-centre rearrangement of even-electron ions, migration of alkyl groups seems uncommon in odd-electron ions.

It was pointed out many years ago (Field and Franklin, 1957) that rearrangement processes would have lower frequency factors (see section 9.7) than simple bond-breaking reactions. Subsequent experiments have endorsed this statement. Thus, by gradually reducing the electron-beam energy in electron ionization of molecules, it is observed that rearrangement reactions may increase in importance relative to simple cleavage. Using the simplified form of the quasi-equilibrium rate equation, the rate constant for a rearrangement–cleavage reaction (k_R) may be compared with the rate constant for simple cleavage (k_C) for a molecule with N oscillators. In the equations (10.59), the frequency factors for rearrangement with

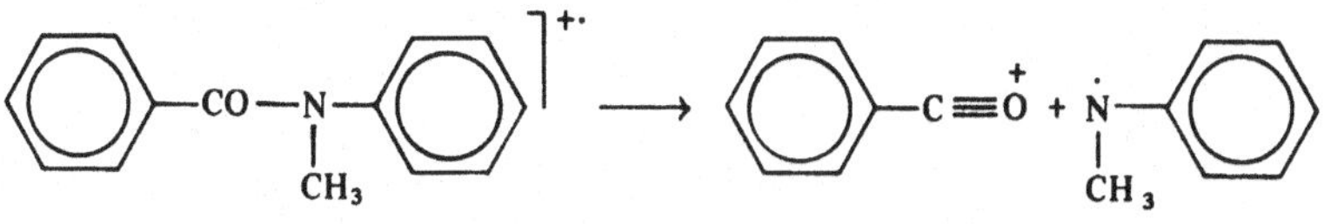

(17)   Simple cleavage

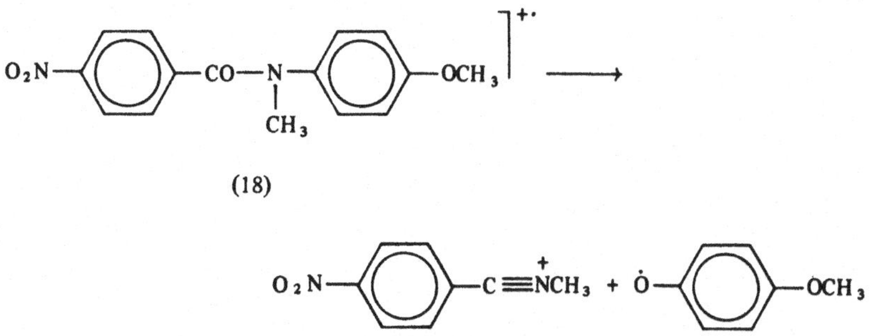

(18)

Rearrangement

cleavage and simple cleavage are v_R and v_C respectively and the energies of activation for the processes, E_R and E_C. The internal energy in the ion is E. Equations (10.59) combine to give equation (10.60), which, to a first approximation, may be written as in equation (10.61), where A is a constant and $\Delta E = E_C{}^0 - E_R{}^0$.

When the energy of activation for simple cleavage is greater than the energy of activation for rearrangement with cleavage, then $\Delta E/E$ in equation (10.61) is positive and $\log(k_R/k_C)$ increases as E is reduced; i.e. reducing

$$k_R = v_R\,(1 - (E_R^0/E))^{N-1}; \quad k_C = v_C\,(1 - (E_C^0/E))^{N-1} \qquad (10.59)$$

$$\log(k_R/k_C) = \log(v_R/v_C) + (N-1)\log(1 - (E_R^0/E))/(1 - (E_C^0/E)) \qquad (10.60)$$

$$\log(k_R/k_C) \approx A + (N-1)(\Delta E/E) \qquad (10.60)$$

the electron-beam energy, and therefore the energy transferred to the molecular ion, causes the rate constant for the rearrangement process to increase relative to the rate constant for the simple bond-breaking reaction. Therefore, as the electron-beam energy decreases, the importance of the rearrangement process increases relative to that of simple bond fission. The

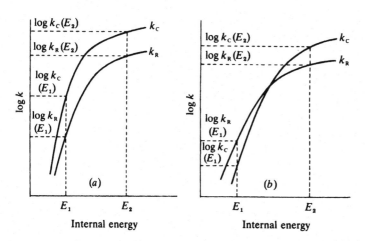

Figure 10.21. The variation of the rate constant for simple bond cleavage (k_C) and for rearrangement with cleavage (k_R) with changes in the internal energy of an ion. In (*a*) the curves do not cross; in (*b*) the curves do cross.

same conclusion may be reached by comparison of the rate constants k_R and k_C at different internal energies of an ion (figure 10.21). Using the simplified rate equations (10.59), graphs of log k_R and log k_C can be plotted for various internal energies as in figure 10.21(*a*). In this case, at all internal energies (e.g. E_1 and E_2), log k_C is greater than log k_R and the rate of simple bond fission always exceeds that for rearrangement followed by fission. With energies of activation and frequency factors different from those obtaining in figure 10.21(*a*), the curves for log k_R and log k_C may cross (figure 10.21(*b*)). Then, at some low excess of internal energy (E_1) the rate of rearrangement with cleavage is faster than the rate of simple bond cleavage but, at higher internal energy (E_2), the order of rates is reversed. Thus, at low electron ionization energies, rearrangement is favoured over ordinary bond cleavage and, by reducing the electron ionization energy below the normal 70 V, the increased importance of the rearrangement process may be observed by increased relative abundances of ions corresponding to it.

Ions formed by field ionization are detected very soon after formation and it has been suggested that there is too little time for much rearrangement to occur. It was expected that comparison of field and electron ionization mass spectra would identify rearrangement processes since they would occur only in the latter. However, there is substantial evidence that rearrangements can occur in field ionization mass spectrometry. The topic is discussed more fully in section 3.7.

10.10. FRAGMENTATION FOLLOWING OTHER METHODS OF
IONIZATION

Scattered references to positive-ion and negative-ion chemical ionization, field ionization and fast atom bombardment occur in this chapter. In this section the information is collated. Generally, the fragmentation of ions is described in empirical terms and rationalization of mass spectra relies upon previous knowledge of the behaviour of large numbers of known compounds in a mass spectrometer. There is ample experience of behaviour following electron ionization, but much less experience of the aforementioned additional methods of ionization. Also, such mild methods of ionization are often employed solely to determine or confirm the relative molecular mass of an unknown compound. In this instance, samples are analysed under conditions that minimize fragmentation. Hence, fragment ions may be absent or, if present, ignored. Therefore, fragmentation of ions generated by mild methods of ionization is treated only briefly here. As a rule, these mild methods of ionization are more likely to identify the relative molecular mass and functional groups of an unknown substance but less likely to show fragmentation of its carbon skeleton than does electron ionization.

Field ionization and field desorption mass spectrometry involve little, if any, fragmentation because molecular or quasi-molecular ions are formed with little excess of internal energy. When molecular ions ($M^{+\cdot}$) do fragment, the product ions are not dissimilar to the primary fragment ions in electron ionization mass spectra although there is more emphasis on direct cleavage of the weaker bonds in ions generated by electric fields. When quasi-molecular ions ($[M + R]^+$, where R = H, Li, Na, etc.) are formed upon ionization in an electric field, their behaviour is similar to that of quasi-molecular ions produced by chemical ionization (see below). Fragmentation in field ionization and desorption mass spectrometry is sometimes due to thermal decomposition of neutral molecules prior to ionization so that 'fragment ions' are really molecular ions of pyrolysis products. In such a case, the behaviour is governed by the thermochemistry of the neutral molecules, but fragmentation frequently parallels that after electron ionization. Collisional activation of otherwise stable ions causes a large increase in their internal energy. The resulting fragmentation is very similar to that in electron ionization mass spectrometry, in which similar amounts of internal energy are imparted to molecular ions.

Chemical ionization and fast atom bombardment afford quasi-molecular ions, most commonly $[M + H]^+$ ions. Being even-electron species and possessing near thermal energies, these ions are less prone to extensive fragmentation than odd-electron molecular ions formed by electron ionization. Protonated ions tend to obey the even-electron rule (see above) inasmuch as they fragment almost exclusively by loss of neutral molecules. Whilst electron ionization of esters, $RCOOCH_3$, affords molecular ions that fragment by loss of $CH_3O^{\bullet}$ and $^{\bullet}COOCH_3$, the same esters form $[M + H]^+$ ions on chemical ionization or FAB and these protonated molecules eliminate CH_3OH and $HCOOCH_3$. The site of proton attachment in $[M + H]^+$ ions is widely assumed to be the functional group associated with highest proton affinity (a heteroatom). In polyfunctional compounds, the proton may be considered as a bridging atom between two heteroatoms. Cleavages are frequently rationalized on the basis of the assumed, specific site of proton attachment. Thus, a common cleavage is ejection of the functional group with the newly bound proton (scheme (10.62)). Since the appearance of chemical ionization mass spectra is highly dependent on the reactant gas used (section 3.3), generalizations are difficult.

$$RX + [H^+] \longrightarrow R{-}\overset{+}{X}{-}H \longrightarrow R^+ + HX$$

$$\underset{O}{RC{-}OCH_3} + [H^+] \longrightarrow \underset{O\ \ H}{RC{-}\overset{+}{O}CH_3} \longrightarrow RC{\equiv}\overset{+}{O} + CH_3OH$$

(10.62)

A variant of chemical ionization is charge exchange (section 3.3.1). An electron is transferred from a sample molecule (M) to reactant gas ion (e.g. $Ar^{+\bullet}$, $N_2^{+\bullet}$, etc.) during collision to yield an odd-electron molecular ion $M^{+\bullet}$. The molecular ions are usually formed with a large excess of internal energy and give mass spectra very similar to those observed by electron ionization, although the amount of internal energy in $M^{+\bullet}$ varies depending on the character of the reactant gas.

Even-electron quasi-molecular ions formed by negative-ion chemical ionization are frequently stable so that the only ions observed are those resulting from reactions of the sample molecule with the negatively charged reactant ion. Such reactions are described in section 3.4. Formation of odd-electron molecular anions, $M^{-\bullet}$, by electron capture may be accompanied by fragmentation. For compounds, RX (X = Cl, Br, F, CN and NO_2), fragment ions, X^-, are common; esters and acids, RCOOR' (R' = H, alkyl), give

RCOO⁻ and R⁻ ions. These simple cleavage reactions are frequently accompanied by hydrogen rearrangements. Skeletal rearrangements observed in negative-ion mass spectrometry are often very complex and *ortho*-effects in polysubstituted aromatic compounds are particularly favoured (reaction scheme (10.63)):

(X = O or NH)

The relatively small body of information on the behaviour of negative ions militates against generalization. The interested reader is referred to more specialized literature on negative ions (Bowie and Williams, 1975; Massey, 1976; Bowie, 1975, 1977, 1979, 1984, 1985, 1987; Bowie, Trenerry and Klass, 1981; O'Hair and Bowie, 1989).

10.11. A SUGGESTED SCHEME FOR INTERPRETATION OF MASS SPECTRA

There is no one right way to interpret a mass spectrum. In fact, any interpretational routine that succeeds in elucidating correct structures in a reasonable amount of time is considered to be correct. The end justifies the means in this case! Despite this uncertainty in procedure, a scheme for examining mass spectra is proposed here by way of a summary. Many mass spectrometrists would follow similar steps, but not slavishly and possibly in a different order.

(i) Determine relative molecular mass via $M^{+\cdot}$ ions or by a mild method of ionization. Note isotopes present. Determine elemental composition if accurate mass measurement is available. Calculate the number of double-bond equivalents.

(ii) Conduct a library search if the facility is available.

(iii) Examine other spectroscopic data alongside the mass spectrometric information.

(iv) Consider the general appearance of the mass spectrum (the degree of fragmentation is diagnostic of aliphatic or aromatic compounds).

(v) Examine odd-electron ions that may arise from very characteristic rearrangements.

(vi) Note low-mass fragments like $H_2C=OH^+$, which often reveal functional groups.

(vii) Consider any remaining important primary losses from $M^{+\cdot}$ ions.

(viii) Pay particular attention to any series of peaks (e.g. ions at m/z 91, 65 and 39). If 14 mass units apart, they indicate a hydrocarbon chain and the m/z values will be characteristic of any functional group attached.

(ix) Propose candidate structure(s). Check that the predicted fragmentation of a given candidate matches the mass spectrum (and that the structure is consistent with any other data).

Many of these points are taken up and reiterated in the next chapter, which concerns some examples of structure elucidation.

11 Examples of structure elucidation by mass spectrometry

INTRODUCTION

Before some worked examples are given to illustrate the elucidation of unknown or partially known structures by mass spectrometry, some general hints are in order. These salient points have been discussed elsewhere in this book and the treatment here is intended as a helpful guide to analysis of samples of various kinds.

A molecular structure is rarely deduced solely from its mass spectrum. What the mass spectrometrist hopes to do in the first instance is to propose a structure that is consistent with the given spectrum. This first structure may need to be modified in the light of subsequent, differently obtained mass spectra or as other information becomes available. Unfortunately, it may even be possible to propose several structures that are compatible with the mass spectrum. Of course, there are degrees of probability of correctness of any structural assignment and, if the elemental composition (high-resolution mass spectrum) and metastable ions are consistent with the proposed structure, then the correctness of an assignment will be much more certain. This is especially true if the spectrum of the unknown structure matches well against a standard spectrum from a library of mass spectra. However, there are many instances of different compounds (in particular, isomers) giving rise to mass spectra having differences that are less than the day-to-day variation in the recording of a mass spectrum, so one should be wary of rash structural assignments. Verification of proposed structures by other techniques is recommended and may be essential. When dealing with very small amounts of material, synthesis, chromatography and UV spectroscopy are widely accepted methods; for larger amounts, nuclear magnetic resonance and infrared spectroscopy are the techniques most likely to yield complementary evidence.

A further word of caution is germane at this point. The sample given for analysis by mass spectrometry may not be pure even if it is claimed to be so. Elemental analyses on less than 0.5 mg of an organic compound are straightforward and well worth having done together with the mass spectrum. Suppose that there were a 50:50 mixture of an inorganic salt and butanol. Simply obtaining an EI spectrum would not reveal the involatile inorganic component but having an elemental analysis would immediately reveal a discrepancy between the elemental and mass spectral analyses. Although this extreme example was chosen to illustrate a point, the principle remains the same for all mixtures. It may even be that a volatile component observed by EI or CI mass spectrometry may only make up a small percentage of the total sample and involatile or not very volatile major components can be missed altogether. For methods of mass spectrometry more suited to involatile materials (fast atom bombardment and so on), any volatile materials may be stripped off so quickly in the vacuum system that their mass spectra are not recorded. Serious discrepancies between a proposed structure suggested by mass spectrometry and one suggested by other methods of analysis are usually indicative of the sample being a mixture. The mixture may actually arise through the process of obtaining a mass spectrum in that, if heat needs to be used to volatilize a sample, then the applied heat may cause its thermal decomposition and only the mass spectra of volatile degradation products are observed.

It is very rare that a totally unknown sample has to be analysed because there is almost always other information. If the compound is a product of a laboratory synthesis, then its likely structure will be known. If the unknown substance is a natural product, its origin will give many clues to its identity or at least type. The method of experimental work-up by which an unknown sample has been obtained often reveals much chemical evidence about its character (e.g., an amine is soluble in aqueous acid but can be extracted into organic solvents from a basic aqueous solution). Information about the substance can vary from its highly diagnostic nuclear magnetic resonance or infrared spectrum to much less specific properties such as colour, smell and physical state, all of which should be kept in mind whilst interpreting the mass spectrum.

Initial examination of a mass spectrum can give much information. Even-mass molecular ions indicate a substance having an even number of nitrogen atoms or none at all. If the molecular ion is of odd number mass then the compound must have an odd number of nitrogen atoms in its structure. Compounds giving an EI spectrum that shows few fragment ions, doubly-charged ions and abundant molecular ions are often aromatic. On the other hand, aliphatic compounds tend to have their most abundant ions at low mass. Any series of ions 14 mass units apart should be noted for these are usually $C_nH_{2n}X^+$ ions and may reveal the character of the functional group, X, and an idea of the length of the carbon chain. Multiple isotope peaks suggest the presence of metals. The above points are illustrated in the examples discussed later in this chapter.

During more detailed examination of a mass spectrum, it should be remembered that primary losses of fragments from the molecular ion are likely to provide the most diagnostic information. In the elucidation of fragmentation pathways, stability of ions is an important consideration but the neutral species ejected must not be ignored just because the mass spectrometer does not detect them directly. The large negative heats of formation of some neutral species, such as CO, CO_2, HCN, N_2, HF and so on, may prove to be the most important factor in making their elimination from ions favourable, rather than the fragment ion itself having any exceptional stability. Also, even-electron ions (cations) rarely eliminate odd-electron neutral species (radicals) but generally dissociate through loss of neutral molecules to give further even-electron ions. In routine mass spectra recorded from a magnetic sector instrument, the ionic products of metastable ion decomposition occur at a mass $m^* = m_2^2/m_1$, where m_1 is the mass of a precursor ion and m_2 that of its product ion. Computerized data acquisition methods and the development of other powerful techniques for examining metastable ion decomposition have seen the demise of this simple method of finding precursor and product ions from a routine mass spectrum recorded on photographic paper. However, the newer approaches to investigation of metastable ion decomposition give more information than did the older method. These linked scanning methods for examining precursor and product ion relationships have been described in chapter 8 and provide important structural information because the relationships of fragment ions with each other (fragmentation pathways) can be unravelled.

For substances ionized using soft or 'mild' methods, it is frequently found that there are few fragment ions and therefore little structural information. In these instances, giving the molecular or quasi-molecular ions more energy (e.g., by collision with a neutral gas during their flight in the analyser of the mass spectrometer) will cause them to fragment; this process has been described in chapter 8 on MS/MS techniques.

As well as the structural information obtainable from a sample presented for mass spectrometry, there are also other uses with impure materials for which information is required. A mass spectrometer is ideal for detecting isotopes and the number of atoms of many elements in a substance can be determined very readily from the isotope patterns (section 12.2.1). Therefore, it is possible to monitor the course of a reaction by taking samples and examining their mass spectra for changes in both molecular mass and isotope patterns. For example, chlorination of a substance can be checked very readily from mass spectra of crude reaction samples taken at intervals. Even though a range of products may be formed and be difficult to separate by the usual chemical or physical means, the chlorination could be stopped at an optimum yield stage by simply recording mass spectra of samples taken during chlorination until the desired degree of chlorination had been achieved. This could be judged by the rate of disappearance of the molecular ion peak and the appearance of new molecular ion peaks of mass and chlorine isotope pattern that were correct for the required chlorination product. In these sorts of examples, the reaction proceeds from a known substance but the structure of the products may be very difficult to unravel without mass spectrometry. Even more information can be gained by subjecting the samples to combined gas or liquid chromatography/mass spectrometry.

The first three of the following examples are relatively simple and are used to illustrate the approach to structure elucidation. The remaining examples have been chosen to illustrate the application of mass spectrometry to different types of compound and the use of various methods of structure elucidation including accurate mass measurement, additional methods of ionization, computerized library searching and analysis of metastable ions. It is shown that mass spectrometry alone provides information that may or may not be sufficient to propose a complete structure. The complementary role of other techniques is emphasized. For the purposes of this chapter, the method of introducing the sample into the mass spectrometer (direct insertion, chromatographic interface, etc.) is irrelevant except for example **F** in which it is shown that the choice of ionization method can be critical. Detailed discussion of an analysis of some medically important lactones by gas chromatography/mass spectrometry is given in section 5.2.2

and quantitative aspects are exemplified in section 7.4. Specialized metastable ion analyses, in particular the differentiation of isomers, are covered in chapter 8.

EXAMPLE A

The EI mass spectrum of compound **A** was recorded on UV-sensitive paper and is reproduced in figure 11.1. Although recording on photographic paper is nowadays unusual, this example has been chosen partly to illustrate how computer processing of raw data leads to compression and reduction of available information and partly to place this early method for examining metastable ion decomposition into its historical perspective. Manual counting of the spectrum is also very rare and so this example is used to show that it can be done and to illustrate how even a computer must be programmed to carry out a similar exercise before it can provide a spectrum. Thus, figure 11.1 represents the raw (amplified, analogue) ion current arising as ions formed in the source reach the detector. This same information (after digitization) is fed into a microprocessor for manipulation by a computerized system.

The upper, most sensitive trace of figure 11.1 is used for counting the spectrum and for finding the broad peaks indicative of metastable ion decomposition and the lower trace is used for structural work. The upper trace was obtained at a sensitivity 100 times greater than that of the lower trace. Just as a computer must be able to identify masses of peaks in a spectrum from information already supplied from a reference compound, before this spectrum can be counted, some peaks at known mass must be identified. The peaks at m/z 28 and 32 (N_2 and O_2 from residual air in the instrument) were identified by comparison with another, previously calibrated mass spectrum. After this the spectrum is counted along the upper trace. Notice how the gap between any two peaks slowly decreases as the m/z value increases. This identifies the analysis as being done on a magnetic sector instrument. During computerized acquisition of these data the decreasing gap is automatically adjusted so that the final spectrum has a constant spacing between successive mass peaks. A sharp change between two successive gaps indicates the presence of doubly charged ions, e.g., see m/z 28.5, which is the doubly charged ion of mass 57 (i.e., $z = 2$ and $m/z = 57/2 = 28.5$). Such ions are very characteristic of aromatic structures but also occur

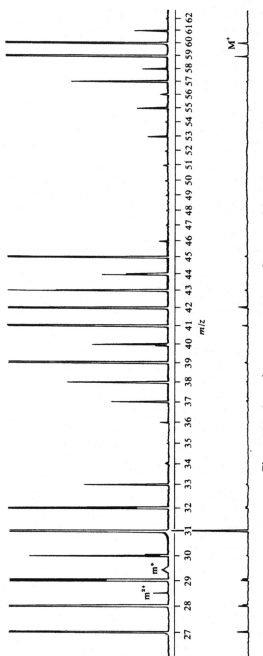

Figure 11.1. An analogue mass spectrum of compound A.

to a lesser extent in the spectra of aliphatic compounds. However, most computerized systems eliminate these multiply charged ions often, unfortunately, by rounding up or down so that m/z 28.5 would be reported as m/z 28 or 29. This constitutes an important loss of information. Counting stops at the last abundant ion, m/z 61, which appears to be the $[M + 1]^{+\cdot}$ ion (^{13}C isotope) and therefore the molecular ion is at m/z 60. There is a broad peak arising from metastable ion decomposition at m/z 29.4. Again, this information on metastable ion peaks from magnetic sector instruments is eliminated in computerized data acquisition and is not available with simple quadrupole instruments but it can be obtained conveniently by any of the linked scanning methods described in chapter 8. Notice that, in figure 11.1, the peak at m/z 42 is equidistant from those of m/z 60 and 29.4. This results from the exponential nature of magnetic scanning and the squared relationship between a peak derived from metastable ion decomposition and the corresponding normal precursor and product ions. Thus, $42^2/60 = 29.4$ and confirms that the metastable molecular ion at m/z 60 undergoes ejection of 18 mass units when it fragments to m/z 42. Computerized data acquisition eliminates all of this information. On the lower trace, many peaks of less abundant ions are not recorded and without the upper trace it would be difficult to count the spectrum accurately. A computerized data system can suffer in the same way so that, with large gaps in a spectrum where there are only small abundances of ions, a miscount can occur often of one but rarely more than two mass units. It is circumspect to check the mass calibration for computerized data systems at regular intervals.

For convenience here, the mass spectrum is converted now into a normalized line diagram and this will lead to a spectrum very similar to that produced by a computerized system. The peak heights on the lower trace are measured (column 2 , table 11.1) Because traces intermediate between highest and lowest sensitivity have been omitted from the mass spectrum for the sake of simplicity, some of the small peak heights measured on the lowest, least sensitive trace cannot be checked (section 1.2). In fact, peaks of less than 1 per cent of the height of the largest have been omitted (e.g., m/z 44), as have peaks below m/z 27, which are generally not diagnostic. Similarly, a computerized data system is programmed to ignore peaks below a certain threshold value; this latter may be 1 per cent of the base peak but the actual threshold can usually be set by the instrument operator. The lower the threshold the more electronic 'noise' appears in the spectrum; the

Table 11.1. *Peak height measurements taken from the mass spectrum of compound A in figure 11.1.*

m/z	Peak height (mm)	Relative abundance (%)
60	6	13
59	8	17
45	2	4
43	1.5	3
42	6	13
41	4	8
40	0.5	1
39	2	4
33	1.5	3
32	1.5	3
31	47	100
29	5.5	11
28	6.5	14
27	7	15

characteristic very flat baseline appearing in computerized systems is attained by setting a threshold high enough that all such 'noise' is eliminated. The largest peak (base peak) in the spectrum is at m/z 31 and this is put equal to 100 arbitrary units (column 3, table 11.1); all other peaks are related to this base peak to give their relative abundances and the normalized line diagram is then drawn (figure 11.2). This representation of the original raw data is almost indistinguishable from a mass spectrum produced on a computerized data system and the remainder of this discussion is independent of whether the data were acquired initially on photographic paper or by computer. Where metastable ions are mentioned, the requisite information can be obtained either from a photographic trace or, which is more likely, from one of the metastable ion scanning techniques covered in chapter 8.

Amongst the larger peaks listed in table 11.1 are m/z 59 (loss of 1 mass unit, H⁺, from the molecular ion), m/z 45 (ejection of 15 mass units, $CH_3^{\cdot}$), and m/z 31 (possible 'characteristic' ion of composition CH_3O; see section 10.1.2). The compound **A** is of even relative molecular mass and therefore cannot contain an odd number of nitrogen atoms and the more abundant fragment ions are at odd m/z values (even-electron ions). One significant

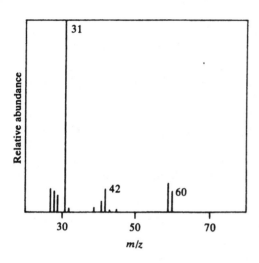

Figure 11.2. A normalized line diagram for the mass spectrum of compound **A** shown in figure 11.1. This display is the same as the digitized mass spectrum produced by a typical data system.

peak occurs at an even-valued m/z 42 and is an odd-electron ion, meaning that rearrangement has occurred with loss of 18 mass units (H_2O) from the molecular ion. The metastable ion decomposition confirms this elimination of H_2O, with the molecular ion being the precursor (m/z 60) and the product ion being at m/z 42 (for the photographic chart, the product of metastable ion decomposition gives a peak at m/z 29.4 and, numerically, $42^2/60 = 29.4$, as required). This last confirmation of the elimination of H_2O from the molecular ion indicates that compound **A** contains a hydroxyl group and is almost certainly an alcohol. From figure 11.1, in which the upper trace is recorded at a sensitivity 100 times that of the lower, the ratio of m/z 60 to m/z 61 peaks can be measured to be 100:3.5. Notice that, in the computerized system, the ions at m/z 61 have not been recorded in the routine spectrum because of their low abundance. It is usually possible to 'magnify' desired region of a mass spectrum acquired by a data system and this would need to be done in the present case to find the abundance of $[M + 1]^{+\bullet}$ ions. Because for each carbon atom of a molecule there is a 1.1 per cent chance that it is ^{13}C, a ratio of abundances for $M^{+\bullet}$ to $[M + 1]^{+\bullet}$ of 100:3.3 is expected for a molecule containing three carbon atoms. This is close enough to the observed ratio to indicate that compound **A** has three carbon atoms. Therefore, it is now known that compound **A** contains the elements C_3H_2O, amounting to a mass of 54. Since the molecular mass is 60, the total composition must be C_3H_8O, indicative of a saturated alcohol, C_3H_7OH. The loss of $H^\bullet$ from the molecular ion indicates that there is at least one hydrogen atom on the α-carbon atom (reaction (11.1); C2 cleavage; see section 10.6). The

$$\left[-\overset{|}{\underset{\overset{\mid}{H}}{C}}\text{-OH}\right]^{+\bullet} \longrightarrow {>}C=\overset{+}{O}H + H^{\bullet} (11.1)$$

loss of H_2O also suggests a hydrogen atom on a γ- or δ-carbon (section 10.6; RE1 reaction). A δ-hydrogen atom implies a chain of four carbon atoms from the OH group, representing $4 \times 12 = 48$ mass units but, if this were correct, there would not be enough mass left over to accommodate an oxygen atom in the molecule. This result concurs with the deduction made above that there are only three carbon atoms in the molecule. A γ-hydrogen atom is ejected with the hydroxyl group on fragmentation of the molecular ion; the remaining hydrogens must fit as shown (scheme (11.2)) and compound **A** is 1-propanol. The ion at m/z 31 is highly characteristic of a primary alcohol and helps to confirm the structure for compound A.

$$\left[\begin{matrix} H_2C\overset{H}{\diagdown}\\ H_2C\diagdown OH\\ \underset{H_2}{C}\diagup \end{matrix}\right]^{+\bullet} \longrightarrow H_2O + C_3H_6^{+\bullet}$$

$$m/z\ 60$$

$$(11.2)$$

$$CH_3CH_2\overset{}{\underset{}{-}}\!\!\overset{}{\diagdown}\!\!-CH_2OH\ \Big]^{+\bullet} \longrightarrow CH_3CH_2^{\bullet} + CH_2=\overset{+}{O}H$$

$$m/z\ 31$$

For comparison, it may be noted that 2-propanol, which does not have a γ-carbon, ejects $OH^{\bullet}$, and not H_2O from its molecular ion. In addition, its base peak occurs at m/z 45, corresponding to C2 cleavage or α-cleavage $(CH_3CH=OH^+)$.

11.3.   EXAMPLE B

The mass spectrum of compound **B** obtained with a computerized data system (figure 11.3) shows a molecular ion at m/z 100 for which accurate mass measurement gives 100.0889. Reference to tables or calculation (see table 1.5) affords the molecular composition, $C_6H_{12}O$ ($C_6 = 72.0000$; $H_{12} = 12.0938$; $O = 15.9949$; total $= 100.0887$; error in measurement $= 2$ parts in 1 000 887, i.e., < 2 ppm, which is acceptable). The composition $C_6H_{12}O$ has one double bond equivalent (section 10.2) which is accounted for by the

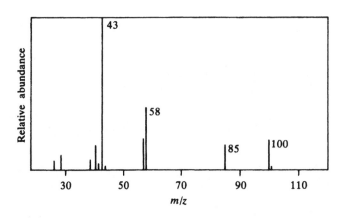

Figure 11.3. A mass spectrum of compound **B**. The accurate mass of the molecular ion at m/z 100 is 100.0887. The infrared spectrum shows a carbonyl absorption band at 1712 cm^{-1}.

carbonyl absorption band in the infrared spectrum. Compound **B** must be an aliphatic aldehyde or ketone. An aldehyde normally shows a loss of hydrogen (C2 cleavage) from a molecular radical-cation but there is no ion at m/z 99 so compound **B** is probably a ketone, RCOR′.

Abundant ions at m/z 85 must be due to loss of CH$_3$· (15 mass units) from the molecular ion and this indicates C2 cleavage adjacent to carbonyl (R = CH$_3$). The alternative C2 cleavage on the other side of the carbonyl group would be expected to give an ion at m/z 43 which can be seen as the base

$$\left[CH_3 - \overset{\xi}{\underset{\xi}{}} - CO - \overset{\xi}{\underset{\xi}{}} - R' \right]^{+\bullet}$$

$$CH_3^\bullet \ + \ R'CO^+ \\ m/z\ 85$$

$$R'^\bullet \ + \ CH_3CO^+ \\ m/z\ 43$$

peak in the mass spectrum of compound **B**. As the total composition of **B** is C$_6$H$_{12}$O, the group R′ must be C$_4$H$_9$ and the possible structures of the unknown compound have been reduced to four, i.e., R = CH$_3$ and R′ = one of CH$_2$CH$_2$CH$_2$CH$_3$, CH$_2$CH(CH$_3$)$_2$, CH(CH$_3$)CH$_2$CH$_3$ or C(CH$_3$)$_3$. To distinguish the possibilities the mass spectrum must be examined further; an odd-electron (even-mass) ion at m/z 58 may be noted. This ion arises by the six-centre rearrangement (RE2) of ketones if a γ-hydrogen is available:

$$\left[\overset{1}{CH_3} \underset{O}{\overset{2}{C}} \overset{H_2}{\underset{H}{\overset{3}{C}}} \overset{CHR^1}{\underset{CHR^2}{}} \right]^{+\bullet} \longrightarrow \left[CH_3 \underset{O_{\diagdown H}}{C} {=} CH_2 \right]^{+\bullet} + \ C_2H_2R^1R^2$$

R^1, R^2 are H and CH$_3$

Table 11.2. *Electron ionization spectrum[a] of compound C.*

m/z	Relative abundance (%)	m/z	Relative abundance (%)
87	5.6	71	0.3
86	100.0	70	1.6
75	3.6	69	0.7
74	18.1	57	2.8

[a] All ions over m/z 50 and greater than 0.2 per cent relative abundance are shown. Metastable ion decomposition reveals the fragmentation m/z 131 → 86. Chemical ionization of compound C with isobutane results in abundant ions at m/z 132.

Hence, a tertiary butyl group (no γ-hydrogen) is ruled out. To afford an ion at m/z 58, there must be an unsubstituted CH_2 at C-3, which eliminates R' $= CH(CH_3)CH_2CH_3$ as a possibility because the corresponding six-centre rearrangement would yield an ion at m/z 72. The two remaining candidate structures for the ketone are $CH_3COCH_2CH_2CH_2CH_3$ and $CH_3COCH_2CH(CH_3)_2$. The sample **B** is, in fact, 2-hexanone but distinction between the two potential structures would be unwise without reference to authentic spectra, as in library searching, or recourse to comparison of metastable ion monitoring of sample **B** with two authentic samples of the candidate structures.

11.4.   EXAMPLE C

The electron ionization mass spectrum of a solid natural product **C** is presented in table 11.2. The base peak at m/z 86 might at first be considered as a candidate for the molecular ion, but this assignment would require loss of 12 mass units to form the fragment ion at m/z 74. Such a loss is most unlikely as it would have to be a carbon atom or a BH• radical and so it is concluded that the molecular ion peak would need to correspond to a mass greater than m/z 86; this molecular ion peak is absent or has such low abundance that it has been missed by the data system. There are no particularly characteristic peaks at low mass and elucidation of the structure from the routine electron ionization spectrum alone is not feasible. Linked scanning of the magnetic field (B) and the electric sector voltage (E) such that B^2/E remains constant was used to search for precursors of metastable ion decomposition occurring just outside the ion source to give m/z 86 (section 8.3.1). Ions of low abundance were found, corresponding to a process in which the ions at m/z 86 were formed from precursor ions at m/z 131 and these are possibly, but not proven to be, the 'missing' molecular ions. To proceed further with the analysis, more information regarding the molecular mass is needed. This is best achieved by means of a mild ('soft') method of ionization.

Chemical ionization of compound **C** with isobutane resulted in only two significant mass peaks at m/z 132 (100 per cent relative abundance) and 86 (72 per cent relative abundance), which together accounted for 80 per cent of the total ion current. Field ionization and field desorption both afforded abundant ions at m/z 132. It is concluded that the molecular mass of compound **C** is 131, these additional methods of ionization having yielded $[M + H]^+$ ions (chapter 3). Because the molecular mass is an odd number, the compound must contain an odd number of nitrogen atoms.

The formation of ions at m/z 86 in the electron ionization spectrum must be due to very easy (rapid) loss of 45 mass units from an initially formed molecular ion at m/z 131. Loss of 45 mass units is often due to ejection of $C_2H_5O^{\bullet}$ from ethyl esters or of $^{\bullet}CO_2H$ from carboxylic acids. If the substance were an ester, $RCO_2C_2H_5$, the ions RCO^+ (m/z 86) resulting from ejection of $C_2H_5O^{\bullet}$ would be expected to lose CO to give further fragment ions at m/z 58 (R^+), but these last are not observed (table 11.2). Thus, the compound is probably a carboxylic acid. This deduction is supported by the fact that the compound is a solid with a high melting point; an ester of this low molecular mass is likely to be a liquid. Ions at m/z 74 must arise through loss of 57 mass units (usually $C_2H_5CO^{\bullet}$ or $C_4H_9^{\bullet}$) from the molecular ions. The loss of $C_2H_5CO^{\bullet}$ would imply that compound **C** is an ethyl ketone, but such a compound would give $[M - C_2H_5^{\bullet}]^+$ ions at m/z 102 and abundant ions at m/z 57 ($C_2H_5CO^+$; C2 cleavage; section 10.6). The former peak is absent and the latter is rather small, so that loss of a butyl radical is indicated, with m/z 57 being partly due to $C_4H_9^+$.

The groups C_4H_9 and CO_2H account for a mass of 103, leaving 29 mass units still to be explained. As the compound contains an odd number of nitrogen atoms, the outstanding elemental composition must be CH_3N. The losses of $C_4H_9^{\bullet}$ and $^{\bullet}CO_2H$ are so favourable in the electron ionization spectrum that C2 cleavage is suspected, the stabilities of the fragment ions being ascribed to formation of immonium species as shown below; this requires both groups to be attached to an α-carbon:

$$H_2N \overset{+}{=} CH—COOH \quad m/z\ 74$$

$$H_2N—\underset{|}{\overset{\overset{\displaystyle C_4H_9}{|}}{CH}}—COOH \xrightarrow[-C_4H_9{}^\bullet]{-e}$$

$$\xrightarrow[-{}^\bullet COOH]{-e} H_2N \overset{+}{=} CH—C_4H_9 \quad m/z\ 86$$

Ions at m/z 74 (relative abundance 18.1 per cent) contain only two carbon atoms so that the relative abundance of ions at m/z 75 expected from ^{13}C isotopic content is $18.1 \times 2 \times 1.1/100 = 0.4$ per cent. Ions at m/z 75 are much more abundant than this and require an alternative explanation. It may be proposed that the peak at m/z 75 arises mostly from a six-centre rearrangement (for compounds with an odd number of nitrogen atoms and therefore an odd-numbered molecular mass, the normal rule that rearrangement ions have even masses becomes one in which rearrangement ions have odd masses (section 10.2):

$$\left[\, H_2N\underset{\underset{\displaystyle C}{\overset{|}{CH}}}{\overset{OH}{\underset{|}{\overset{\bullet}{C}}}{\overset{\displaystyle .}{\cdot}}O \,\right]^{+\bullet} \longrightarrow \left[\, H_2N\underset{\underset{\displaystyle H}{|}}{C}\overset{OH}{\underset{|}{\overset{.}{C}}}OH \,\right]^{+\bullet} \quad + \quad \nearrow\!\!=\!\!\searrow$$

$$m/z\ 75$$

This fragmentation is in accord with the proposed structure, but does not shed any light on the character of the butyl group.

The mass spectra are consistent with the amino acids leucine, isoleucine and norleucine, which have butyl groups $(CH_3)_2CHCH_2$, $C_2H_5(CH_3)CH$ and $CH_3CH_2CH_2CH_2$, respectively. These isomers would be distinguished readily by ^1H nuclear magnetic resonance spectroscopy or, if the sample size were limited, by chromatography. In fact, compound **C** is leucine. Note that the analysis would have been much simpler had the amphoteric nature of the compound been known from the method of isolation. An alternative approach to structure elucidation that becomes apparent with hindsight would be to use a mild method of ionization to obtain protonated molecular ions and then to carry out MS/MS by colliding these ions with a neutral gas (section 8.7.1). Not only would this approach give molecular mass information more readily but, by varying the energy of collision, it is likely that the three amino acids could be distinguished; comparison of collision-induced mass spectra often reveals small differences in mass spectral behaviour of isomers.

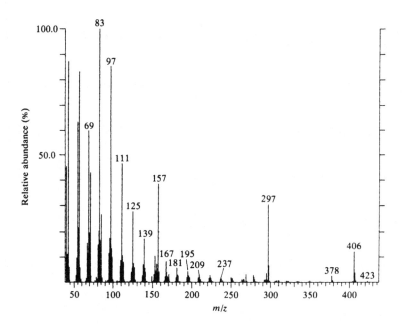

Figure 11.4. A digitized mass spectrum of compound **D**.

11.5.   EXAMPLE D

The electron ionization mass spectrum (figure 11.4) of an unknown white solid (**D**) extracted from a plant was acquired and processed by a data system (chapter 4). The spectrum shows features typical of an aliphatic compound. A long hydrocarbon chain is indicated by the series of ions at m/z 57, 71, 85, 99 and 113 ($C_nH_{2n+1}^+$) and m/z 55, 69, 83, ..., 209, 223, 237 and 251 ($C_nH_{2n-1}^+$). Because the ions of the two series decrease steadily in abundance as m/z increases, an unbranched chain is indicated (branched chains afford greater ion abundances at branching points; section 10.5). Further structural information is not readily ascertained from the spectrum; for instance, a molecular ion is difficult to assign because of the low abundances at large m/z values. The data system was used to compare the measured mass spectrum with several thousand reference spectra (the library), stored in computer memory. The five library spectra showing closest resemblance to the spectrum of the unknown were identified (figure 11.5) and all were found to be long-chain alcohols. The resemblances or 'matches' were assessed on a scale of 0 for a complete mismatch and 1000 for a perfect match (details of library searching appear in section 4.4.3). For the highest ranked entry according to matching, 1-nonacosanol (n-$C_{29}H_{59}OH$; relative molecular mass 424), the resemblance between measured and library spectrum (the

LIBRARY SEARCH OF COMPOUND D
25409 SPECTRA IN LIBRARY SEARCHED FOR MAXIMUM PURITY
52 MATCHED AT LEAST 7 OF THE 16 LARGEST PEAKS IN THE UNKNOWN

RANK	NAME
1	1-NONACOSANOL
2	1-DOCOSANOL
3	1-HEPTACOSANOL
4	1-HEXACOSANOL
5	1-TETRACOSANOL

RANK	FORMULA	MOL. WT	PURITY	FIT
1	C29.H60.O	424	585	971
2	C22.H46.O	326	554	960
3	C27.H56.O	396	546	971
4	C26.H54.O	382	537	950
5	C24.H50.O	354	523	950

Figure 11.5. A computer printout from the data system showing the result of a library search for a mass spectrum comparable to that of compound **D**.

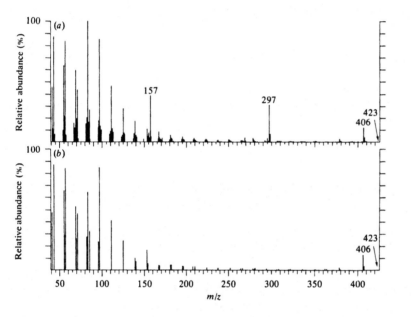

Figure 11.6. A comparison of (*a*) the mass spectrum of compound **D** and (*b*) the reference mass spectrum for 1-nonacosanol from a library of mass spectra stored in the data system.

fit figure) is high because all the peaks in the reference spectrum occur also in the unknown spectrum with about the same heights. However, the purity of fit is much lower because there are some peaks in the unknown spectrum that are not present in the reference spectrum (see figure 11.6 and section 4.4.3).

This situation would obtain if the sample were a mixture containing 1-nonacosanol as a component, or if the unknown were a compound with much structural similarity to 1-nonacosanol. It is seen (figure 11.6) that the reference spectrum shows no molecular ion (at m/z 424) but does have an ion at m/z 423 (loss of H$^{\cdot}$) and 406 (loss of H$_2$O). Because the same behaviour is shown by compound **D**, this implies that it is an isomer of 1-$C_{29}H_{59}OH$.

Branching of the chain has already been excluded and compound **D** is not actually 1-$C_{29}H_{59}OH$, so it is concluded that the unknown is a secondary alcohol. A secondary alcohol is expected to undergo C2 cleavage as shown in scheme (11.3) to give ions (**1**) and (**2**), which would not occur in the spectrum of 1-nonacosanol. Inspection of figure 11.6 immediately reveals ions at m/z 157 and 297, which are not present in the reference spectrum. The ion $C_9H_{19}CH=OH^+$ has mass 157 and the ion $C_{19}H_{39}CH=OH^+$ mass 297.

$$
\begin{array}{l}
\underset{R^2}{\overset{R^1}{>}}C=\overset{+}{O}H \quad + \quad H \cdot \\
\quad\quad m/z\ 423 \\[2ex]
\underset{R^2}{\overset{R^1}{>}}\underset{H}{\overset{}{C}}=\overset{+}{O}H \quad + \quad \cdot R^2 \\
\quad\quad\quad \mathbf{1} \\[2ex]
\underset{R^2}{\overset{H}{>}}C=\overset{+}{O}H \quad + \quad \cdot R^1 \\
\quad\quad\quad \mathbf{2}
\end{array}
$$

$$\underset{R^2}{\overset{R^1}{>}}\underset{OH}{\overset{H}{C}} \quad \xrightarrow{-e^-} \quad \tag{11.3}$$

In the 70 eV electron ionization spectrum, elimination of the largest radical is favoured (section 10.6) and it is observed that the ion abundances decrease in the order m/z 157 > 297 > 423 as expected. It may be concluded that compound **D** is 10-nonacosanol, $C_9H_{19}CH(OH)C_{19}H_{39}$, a substance not represented in the mass spectral library. The molecular mass was confirmed by obtaining a chemical ionization spectrum of the unknown in butane as reagent gas. A prominent protonated molecule, $[M + H]^+$, was observed at m/z 425.

The structural assignment may be confirmed by nuclear magnetic resonance spectroscopy and by derivatization (e.g., by oxidation to the corresponding ketone) followed by further mass spectrometry.

11.6. EXAMPLE E

The electron ionization mass spectrum of compound **E** (figure 11.7) is typical of an aromatic compound in general appearance with abundant molecular ions and relatively few fragment ions. In the molecular ion region there are two peaks at m/z 273 and 275, separated by two mass units and with an abundance ratio of 1:1, indicating the presence of bromine. The odd-numbered molecular mass is characteristic of an odd number of nitrogen

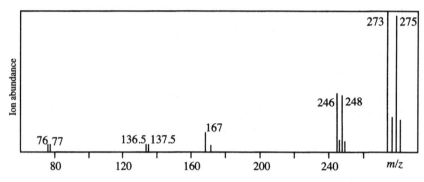

Figure 11.7. An electron ionization mass spectrum of the unknown compound E. Accurate masses of two of the ions were found to be 272.9904 and 245.9782. Peaks from a precursor ion scan obtained with a triple quadrupole system revealed that both m/z 246 and 248 were precursors to the product ion at m/z 167.

atoms in the molecule. This is confirmed by the accurate mass (272.9904) of the molecular ion, which corresponds to a composition of $C_{12}H_8BrN_3$ of expected mass 272.9901. The difference between the measured mass and that required for the composition $C_{12}H_8BrN_3$ is 0.0003 mass units in 272.9901, i.e., a difference of 1.1 ppm (usually, differences or errors in mass measurement of ±10 ppm are considered acceptable). Two fragment ions at m/z 246 and 248 in the ratio of 1:1 correspond to the loss of 27 mass units from the molecular ions and suggest the expulsion of $C_2H_3{}^{\cdot}$ or HCN.

The loss of HCN from aromatic amines and nitrogen heteroaromatic compounds is commonly observed (section 10.7) and is confirmed in this case by the accurate mass 245.9782 corresponding to the composition $C_{11}H_7BrN_2$ ($C_{12}H_8BrN_3 - C_{11}H_7BrN_2 = HCN$). Peaks resulting from metastable ion decomposition correspond to respective losses of ^{79}Br and ^{81}Br from m/z 246 and 248 to yield the ion at m/z 167, which contains no bromine (there is no ion at m/z 169 of equal abundance to that at m/z 167 as would be required for the presence of bromine). The ions at m/z 77 and 76 are usually observed with aromatic compounds containing benzenoid (C_6) rings and the relatively abundant doubly charged molecular ions (M^{2+}) at 136.5 and 137.5 are also indicative of multi-ring aromatic character. Note that the molecular formula $C_{12}H_8BrN_3$ for E corresponds to ten double-bond equivalents and signifies a highly unsaturated or condensed ring system.

Thus, the mass spectrum can be used to propose that the compound is aromatic, possibly a condensed polycyclic compound with one or more benzenoid rings, and that it contains one bromine and three nitrogen atoms. The ejection of one nitrogen as HCN from the molecular ion before bromine is ejected strongly suggests the presence of a primary amino group (section 10.7; table 10.8). Beyond this, further deductions would be speculative unless supported by other experimental information or experience based on similar compounds. For example, the amino group could be confirmed by acetylation or trimethylsilylation followed by mass spectrometric analysis. For suspected condensed polycyclic aromatic compounds, an ultraviolet spectrum would require very little material and would be highly informative. The compound **E** is 1-amino-3-bromophenazine (**3**).

11.7.   EXAMPLE F

After attempted chlorosulphonation of 5,10,15,20-*tetrakis*(2,6-dichlorophenyl)-porphyrin (**4**), the resulting product, thought to have structure (**5**) was hydrolysed to the corresponding sulphonic acid (**6**) and presented for mass spectral analysis (Ashcroft, Johnstone, Lowes and Stocks, 1994). The acid was very soluble in water, very involatile and not easy to derivatize. Electron, chemical and fast atom bombardment ionization did not afford ions corresponding to the expected product in either positive-ion or negative-ion modes. However, an excellent spectrum could be obtained by negative-ion electrospray ionization (figure 11.8(*a*)). Examination of the spectrum revealed that the product (**6**) was not pure but consisted of a mixture of di-, tri- and tetrasulphonated porphyrins and a second attempted synthesis led to over-chlorosulphonation as shown by the presence of penta- and hexa-sulphonic acids (figure 11.8(*b*)). These findings enabled the laboratory researchers to adjust the reaction conditions, aiming for tetrachloro-sulphonation. This time, the $[M - H]^-$ ion expected for the tetrasulphonic acid product (**6**) was found without any evidence for under- or over-chloro-sulphonation (figure 11.8(*c*)).

The peaks observed in the mass spectra of figures 11.8(*a*)–(*c*) are actually unresolved 'envelopes' of $^{35}Cl/^{37}Cl$ isotopes because there are eight chlorine atoms in the molecule (**6**). At higher resolution, the envelopes are resolved and reveal the expected pattern of chlorine isotopes (inset, figure 11.8(*c*)). Also, it will be noticed that there is a prominent peak at m/z 604, corresponding to $[M - 2H]^{2-}$. The negative-ion mode was utilized in this

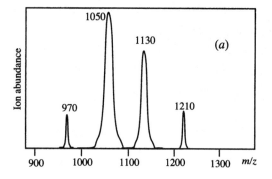
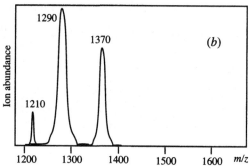
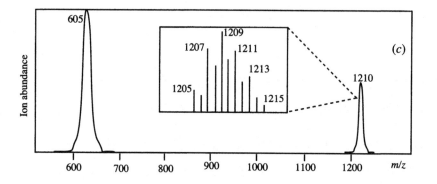

study because sulphonic acids readily accommodate the charge as their con-jugate bases.

This example shows how important the method of ionization can be in obtaining a structure. Electron ionization, requiring vaporization of an involatile sample, simply caused thermal destruction of compound (F). Similarly, the mild chemical ionization method failed because attempts were made to volatilize the sample at an even higher ion source pressure than for EI. Neither positive-ion nor negative-ion fast atom bombardment ionization revealed any ions typically expected of an ionic material such as a sulphonic acid. In this last case, no heating was necessary but, even though a range of liquid matrices was tried, the sulphonic acid could not be volatilized. On the other hand, electrospray could be expected to provide ions from an ionic compound such as the product (6). Whereas much emphasis on electrospray has concerned the examination of very large mol-ecular mass peptides, proteins and carbohydrates, this example shows that the technique is equally effective with compounds of small molecular mass in giving structural information. By adjusting the 'cone' voltage (between

Figure 11.8. Electrospray ionization mass spectra of chlorosulphonated tetraphenyl-porphyrins. (a) The mass spectrum from an under-reacted species, showing $[M - H]^-$ peaks for mono-, di-, tri- and tetra-substituted compounds at m/z 970, 1050, 1130 and 1210. (b) The mass spectrum of an over-reacted species with peaks representing tetra-, penta- and hexa-substituted products at m/z 1210, 1290 and 1370. (c) The mass spectrum of the required tetrasubstituted compound F showing the required singly and doubly charged ions at m/z 605 and 1210. The inset reveals the range of isotope peaks due to chlorine in the envelope of m/z 1210 obtained at lower resolution.

4

5

6

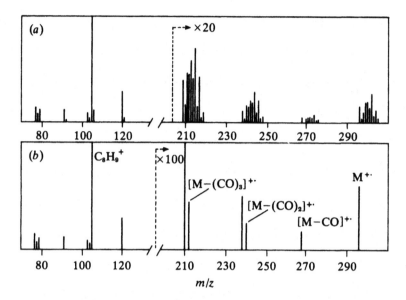

Figure 11.9. (*a*) The mass spectrum of unknown compound **G** and (*b*) the same mass spectrum with isotope contributions removed. Ions of very low abundance also occur at *m*/*z* 92, 94, 95, 96, 97, 98 and 100 but do not appear on the first printout because the computerized data system has eliminated all peaks below a set threshold; they were found by magnifying the region *m*/*z* 80–100. Ion abundances above *m*/*z* 200 are drawn to magnified scales as indicated. MS/MS studies on a neutral loss scan obtained in an ion trap show that H_2 is ejected for pairs of ions at *m*/*z* 238 and 240, 210 and 212, and 103 and 105.

the nebulizer and the skimmer; section 3.9) so as to induce multiple collisions between ions before they enter the analyser of the mass spectrometer, fragmentation could be induced, thereby giving added structural information (successive losses of SO_3 from the $[M - nH]^{n-}$ ions, $n = 1, 2$) to confirm the initial findings. This example shows how, even with impure samples, synthetic chemistry can be aided by examining crude reaction mixtures of substances for which structures could be proposed but not easily verified by other means.

11.8.   EXAMPLE G

The electron ionization mass spectrum of compound **G** (figure 11.9(*a*)) shows closely spaced groups of peaks at high *m*/*z* values but only isolated peaks at low *m*/*z* ratios. There are ten major peaks in the molecular ion region and this immediately suggests either that the sample is grossly impure or that there is an element present, which has several isotopes. The last peak at *m*/*z* 305 must be a ^{13}C isotope, leaving nine other possible isotope peaks in the ratio $1.0:0.2:0.6:1.1:1.2:0.7:1.6:0.3:0.7$.

Examination of isotope tables suggests that molybdenum is present since it has seven isotopes at 92, 94, 95, 96, 97, 98 and 100 in the ratio $1.0:0.6:1.0:1.0:0.6:1.5:0.6$ (the peaks at *m*/*z* 297 and 303 in the molecular ion region for compound **G** would then have to be ^{13}C isotopes of *m*/*z* 296 and 302 respectively). Confirmation of this identification is obtained from the presence of ions in the expected abundance ratios for molybdenum at m/z 92, 94, 95, 96, 97, 98 and 100, albeit in very low overall abundance. If the peak at m/z 297 is a ^{13}C contribution, then the abundance of this ion relative to that of m/z 296 would enable the ^{13}C isotope contributions to be removed from all of the ions between m/z 296 and m/z 305 (section 12.2.3). When this is done, the residual peaks number only seven and are in the abundance ratios 1.0 : 0.6 : 1.0 : 1.1 : 0.7 : 1.5 : 0.7, which are very close to the ratios required for molybdenum. All isotope contributions other than ^{92}Mo and ^{12}C can be removed from the mass spectrum leaving it much simpler to interpret (figure 11.9(b)).

The series of peaks at m/z 296, 268, 240 and 212 suggest successive losses of 28 mass units from the molecular ion (on the evidence of peaks from MS/MS in an ion trap, the ions at m/z 238 and 210 arise from m/z 240 and 212 respectively by loss of H_2). Three successive ejections of 28 mass units from an organometallic compound suggest elimination of three CO molecules from a metal complex (confirmed by accurate mass measurement). Thus, the compound contains molybdenum (^{92}Mo, figure 11.9(b)) and three CO ligands (3 × CO = 84 mass units). Together, these account for 176 mass units, leaving 296 − 176 = 120 mass units to assign. Abundant ions are found at m/z 120 and these decompose through elimination of 15 mass units ($CH_3^{\cdot}$) to give ions at m/z 105 (the peak representing metastable ion decomposition at m/z 91.8).

Accurate mass measurement shows that the ions at m/z 105, which also lose H_2, have the composition C_8H_9. Apart from molybdenum and the CO groups, the remainder (C_9H_{12}) of compound **G** must be very unsaturated (four double bond equivalents) and, together with the presence of ions at m/z 77, 78, 79 and 91, this suggests a benzenoid ring, possibly benzylic (m/z 91; section 10.5), with a C_3H_7 side chain. The prominent loss of a methyl radical from m/z 120 and the relatively small peak at m/z 91 indicate that the side chain is 2-propyl. In this instance, on the basis of the mass spectrum alone, a tentative total structure can be proposed for compound **G**, viz., $Mo(CO)_3$ complexed to 2-propylbenzene. The actual structure of **G** is as shown (**7**).

11.9.   EXAMPLE H

Not infrequently, mass spectrometry is called upon to identify the structure
of an oligomer or polymer. In particular, the method is required to establish
the sequence of monomer units. This task is straightforward if (i) the
monomer units have unique masses, (ii) the polymer fragments reliably
between each monomer unit, leaving the charge on the same end of the
chain to give so-called 'sequence ions' and (iii) competing fragmentation at
other sites in the ions does not obscure the sequence ions. The analysis is
illustrated schematically by a hypothetical polymer of unknown sequence
that contains monomer units A (relative mass 50), B (mass 60), C (mass 75)
and/or D (mass 80). If the mass spectrum of this polymer exhibits clear
peaks at m/z 80, 160, 210, 285, 365, 445, 495 and 555, then its sequence must
be D-D-A-C-D-D-A-B, as shown in scheme (11.4).

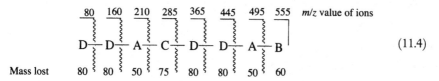

$$(11.4)$$

In real life, with real polymers like polypeptides or polysaccharides, this
ideal is not achieved. Many different monosaccharide units and some different amino acid residues have the same mass. Fragmentation usually does
occur between residues but may be obscured by other fragmentations (for
example, cleavages of the side chains of amino acid residues), and the charge
may be retained on either fragment when a bond cleaves. This latter effect
may give rise to a second set of sequence ions adding both clutter and,
potentially, extra information to the mass spectrum, assuming that it can be
disentangled. In the hypothetical case above, the spectrum could be complicated by a series of sequence ions at m/z 60 (B), 110 (B-A), 190 (B-A-D), 270
(B-A-D-D), 345 (B-A-D-D-C), 395 (B-A-D-D-C-A) and 475 (B-A-D-D-
C-A-D), together with the molecular ion at m/z 555. Even without the
complications of competing fragmentations at positions other than the chain
links, isomeric monomer units, and missing sequence peaks, the initial set of
ions is quite confusing (m/z 555, 495, 475, 445, 395, 365, 345, 285, 270, 210,
190, 160, 110, 80 and 60). The data require careful interpretation. Computer
programmes have been written to solve such sequencing problems.

When addressing a real problem like the sequencing of a polynucleotide, protein or polysaccharide, there is a dilemma. An ionization method is required that is capable of handling large polar molecules. Such methods (electrospray and matrix-assisted laser desorption, predominantly) yield abundant $[M + nH]^{n+}$ ions but little fragmentation. Hence, the application demands a method (mild ionization) that will not immediately provide the information required to deduce a sequence. The dilemma is solved by inducing fragmentation of the protonated molecules, usually in an MS/MS experiment. With electrospray, it is also possible to increase the cone voltage in the ion source, thus collisionally heating the protonated molecules and inducing fragmentation.

Taking the example of a protein, the fundamental strategy for determining its sequence by mass spectrometry typically involves the following steps (Biemann and Papayannopoulos, 1994). (1) The relative molecular mass (M_r) of the protein is measured by electrospray or matrix-assisted laser desorption mass spectrometry. (2) Samples of the protein are digested with proteases like trypsin and chymotrypsin. (3) Each of the resulting digests is fractionated by HPLC or capillary electrophoresis (CE). (4) The M_r value of each of the peptides detected is measured by electrospray or fast atom bombardment and fragmentation is induced (usually by collisional activation in an LC/MS/MS or CE/MS/MS experiment) to obtain information on sequence. (5) The sequences of as many of the oligopeptides as possible are deduced from the data. (6) By looking for identical overlapping sequences, individual peptide sequences are assembled into the full sequence of the protein. (7) If there is insufficient information to complete step 6, further digestions are used to generate a fresh crop of peptides for further sequence analysis. In this brief section, the whole of this procedure cannot be considered but an example of a relatively simple peptide serves to illustrate steps 4 and 5.

Methionyl human growth hormone has been digested with trypsin to form a number of smaller peptides (Covey, Huang and Henion, 1991). The digest was analysed by LC/MS using electrospray mass spectrometry. One such peptide was determined to have an M_r value of 1361 because its electrospray spectrum, recorded during its elution from a liquid chromatograph, showed a base peak at m/z 681.5 due to $[M + 2H]^{2+}$ ions. The sequence of this peptide could only be established by inducing fragmentation of the $[M + 2H]^{2+}$ ions in a triple quadrupole system and acquiring a product ion spectrum (table 11.3). Table 11.3 does not show every peak in the complex spectrum but it lists the main ones. Trypsin specifically

Table 11.3. *The approximate relative ion abundances in the product ion scan of the collisionally induced diprotonated molecules of the unknown peptide H (M_r = 1361).*

m/z	Relative abundance (%)	m/z	Relative abundance (%)
85	12	544	25
102	13	566	13
142	6	577	48
175	7	629	8
201	100	657	14
229	95	705	92
232	28	785	4
259	20	818	29
298	7	858	5
316	20	875	96
358	46	886	4
363	29	987	10
383	10	1004	80
400	14	1116	10
476	27	1133	46
487	16	1246	4

hydrolyses proteins at the carboxyl terminus of all lysine and arginine residues. Therefore the unknown peptide should have lysine (Lys) or arginine (Arg) at its C-terminus. Knowing this fact, it is a sensible place to start the interpretation.

Peptides tend to cleave in a number of ways at each peptide bond. To

$$H\!\!-\!\!\left[\text{NH}\!-\!\text{CH}\!-\!\text{CO}\right]_n\!\!-\!\!\text{OH} \xrightarrow[\substack{\text{electrospray} \\ \text{and} \\ \text{MS/MS}}]{\text{via}} \text{NH}_2\text{-CH-CO-}\overset{+}{\text{O}}\text{H}_2 \; + $$

with R_i and R_n substituents

$$\text{NH}_2\text{-CH-CO-NH-CH-CO-}\overset{+}{\text{O}}\text{H}_2 \; + $$

with $R_{(n-1)}$ and R_n substituents

$$\text{NH}_2\text{-CH-CO-NH-CH-CO-NH-CH-CO-}\overset{+}{\text{O}}\text{H}_2 \; + \; \text{etc.} $$

with $R_{(n-2)}$, $R_{(n-1)}$ and R_n substituents

(11.5)

Table 11.4. *Masses of amino acid residues*
(-NHCHRCO-).

Amino acid residue	Residue mass
Gly	57
Ala	71
Ser	87
Pro	97
Val	99
Thr	101
Cys	103
Leu/Ile	113
Asn	114
Asp	115
Lys/Gln	128
Glu	129
Met	131
His	137
Phe	147
Arg	156
Tyr	163
Trp	186

simplify this particular problem, two modes of fragmentation only will be considered, starting with that shown in scheme (11.5).

The series of cleavages shown contain the C-terminus. The end residue should be Lys or Arg, so it is prudent to check for evidence of either. Lysine would give such a peak at m/z 147 whilst arginine would exhibit m/z 175. The former is absent from the spectrum but the latter is present, albeit at low abundance, so Arg is proposed as the C-terminus. In identifying the next residue, it is helpful to refer to the masses of amino acid residues (table 11.4). The next C-terminal ion should appear at an m/z value that is m/z 175 plus the residue mass of a second amino acid. Adding each of the masses in table 11.4 to m/z 175 in turn provides only one m/z value that actually appears in the spectrum. A glycine (Gly) residue would provide ions at m/z 232 (m/z 175 + 57) and the corresponding peak is significant (about 28 per cent). The next residue is assigned in the same way. Adding each residue mass to m/z 232 provides the candidates. Again, there is only one addition that corresponds to a peak in the spectrum: methionine (Met) adds a mass of 131 to give m/z 363 (about 29 per cent). Hence the C-terminus is assigned to -Met-Gly-Arg. Continuing this process, only a leucine (Leu) or isoleucine (Ile) residue at this point in the chain would afford a peak that appears in the mass spectrum. The additional residue mass of 113 provides a peak at m/z 476 (27 per cent). The next residue must be Thr; only this amino acid contributes a mass that is represented in the spectrum (m/z 476 + 101 = m/z 577). At 48 per cent, this peak is large and clear. Just one residue mass adds to m/z 577 to give another peak in the spectrum: m/z 705 is obtained by adding 128. This value can correspond to either Lys or Gln, which have the same integer mass. A Leu or Ile residue comes next, explaining the peak at m/z 818.

At this point, the interpretation becomes slightly more complicated. A Gly residue would give rise to a peak at m/z 875 (818 + 57) but, alternatively, a tryptophan (Trp) residue would afford a peak at m/z 1004 (818 + 186). Both peaks occur in the spectrum at relative abundances of 96 and 80 per cent respectively, so an unambiguous assignment is not immediately possible. Thus far, the possible sequences are

-Gly-Leu/Ile-Gln/Lys-Thr-Leu/Ile-Met-Gly-Arg

or

-Trp-Leu/Ile-Gln/Lys-Thr-Leu/Ile-Met-Gly-Arg.

Working on the first of these candidates, the next residue must be glutamine (Glu) to give m/z 1004 (875 + 129). This peak is coincident with that of the second candidate above. In other words, Glu-Gly has the same residue mass as Trp and so the two alternative structures are not readily distinguished.

Above m/z 1004, there are only three significant peaks, plus the precursor ion. The next residue has to be Glu (m/z 1004 + 129 = m/z 1133). To produce m/z 1246, there must be another leucine or isoleucine residue. Finally, it is necessary to consider the formation of the ion at m/z 1246 from the diprotonated ion at m/z 681.5 This precursor is equivalent, for interpretational purposes, to an $[M + H]^+$ ion at m/z 1362 and so the mass lost in the first fragmentation is 116. This corresponds to an N-terminus of asparagine (Asp) because its residue mass is

115 [-NH-CH(CH$_2$COOH)-CO-] but, being at the end of the chain, it needs one hydrogen atom to be added to the nitrogen to make a complete molecule [NH$_2$-CH(CH$_2$COOH)-CO--]. The deductions so far suggest strongly one of the following peptides:

Asp-Leu/Ile-Glu-(Glu-Gly or Trp)-Leu/Ile-Gln/Lys-Thr-Leu/Ile-Met-Gly-Arg

The interpretation here has been simplified inasmuch as the type of ion generated by the Arg residue was assumed at the outset. With this type of analysis, a second series of peaks for another type of sequence ion would be anticipated. This second set, if present, may add weight to the pro-

$$
H-\left[NH-CH-CO\right]_n-OH \xrightarrow[\substack{electrospray \\ and \\ MS/MS}]{via} H_2N\text{-}CH-CO^+ \quad +
$$

with R$_i$ on the left structure and R$_1$ on the right structure

$$
H_2N\text{-}CH-CO-NH-CH-CO^+ \quad +
$$
with R$_1$ and R$_2$

$$
H_2N\text{-}CH-CO-NH-CH-CO-NH-CH-CO^+ + \text{etc.}
$$
with R$_1$, R$_2$, R$_3$

(11.6)

posed assignments. The fragmentation arises from the N-terminus of the peptide and involves ions of the type shown in scheme (11.6). It is useful now to calculate the m/z values for ions of this type expected to arise from the candidate peptides and to check for their presence in the spectrum. Four of the expected peaks (m/z 116, 999, 1130, 1187) are not present in the spectrum, but the remainder do occur, at very variable abundances (4–95 per cent). Note that there is a peak at m/z 487 that suggests that the peptide containing Glu-Gly is more likely than the Trp analogue.

116	(95%) 229	(46%) 358	(16%) 487	(25%) 544	(14%) 657	(4%) 785	(4%) 886	999	1130	1187

Asp — Leu/Ile — Glu — Glu — Gly — Leu/Ile — Gln/Lys — Thr — Leu/Ile — Met — Gly — Arg

Summarizing, a full set of sequence ions containing the C-terminus (scheme (11.5)) occurs in the spectrum and, in this instance, the

N-terminal ions (scheme (11.6)) do not provide complete sequence information. The abundances of the key ions are very variable. Interestingly, the base peak at m/z 201 is not a sequence ion. Peak size is not a reliable guide to importance! The true sequence of this peptide is shown in section 5.4.4. Here, the interpretation has produced an answer that is entirely consistent with the true structure and the only ambiguities involved those residues with common integer masses.

12   Further discussion of selected topics

12.1.   IONIZATION AND APPEARANCE ENERGIES

The least energy required to remove an electron from an atom or molecule is the *ionization energy*. Strictly, this should be the first ionization energy and correspond to the first adiabatic ionizing transition in a molecule. Potential energy (Morse) curves for a simple diatomic molecule, M, are illustrated in figure 12.1. Except at high temperatures, most molecules populate the zeroth vibrational state ($\nu = 0$) and this is the only vibrational state of the

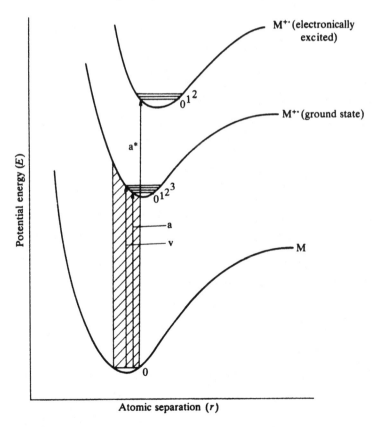

Figure 12.1. Potential energy (Morse) curves showing ionizing transitions from molecule (M) to ground state and electronically excited states of the ion M$^{+\cdot}$. The transition v is a 'vertical' one and transitions a and a* are adiabatic ones.

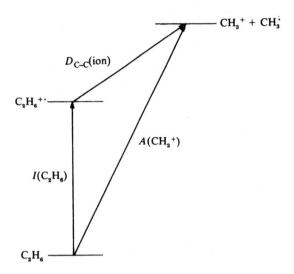

Figure 12.2. An energy diagram showing the relationship between the appearance energy of CH_3^+ from C_2H_6 and the ionization energy of C_2H_6. For a ground-state reaction, the C–C bond energy (D_{C-C}) in the ion is $A(CH_3^+)$ − $I(C_2H_6)$.

molecule considered here. A vertical transition such as v (figure 12.1) representing ionization to the ground-state ion, $M^{+\cdot}$, produces the ion in an excited *vibrational* state $(\nu = 3)$; this transition is $0 \rightarrow 3$. The shaded area in figure 12.1 defines the range within which vertical transitions are possible from the $\nu = 0$ state of the molecule, M, to different vibrational states of the ion, $M^{+\cdot}$. One of these transitions is called adiabatic (a) since ionization to the ground-state ion, $M^{+\cdot}$, produces the ion in its lowest vibrational state $(\nu = 0)$; this is a $0 \rightarrow 0$ transition. The $0 \rightarrow 0$ transition is that corresponding to the first ionization energy. The $0 \rightarrow 1, 0 \rightarrow 2$, etc., transitions are closely separated in energy and are due to vibrational energy levels in the ground state of the ion; these transitions follow each other by a few tenths of an electron-volt. A $0 \rightarrow 0$ ionizing transition (a* in figure 12.1) to form the ion $(M^{+\cdot})$ in its first *electronically* excited state (ionization to the second molecular electronic orbital level) corresponds to the second ionization energy of the molecule (M). Similarly, there are third, fourth and higher ionization energies, each of which is accompanied by a series of closely spaced vibrational levels. Rotational energy is frequently neglected in mass spectrometry because the amount of energy involved is small compared with the total excess of energy usually present in a newly formed ion.

The first ionization energy (I) of ethane is represented on the energy level diagram of figure 12.2, which illustrates the energy change in reaction (12.1):

$$C_2H_6 \rightarrow C_2H_6^{+\cdot} + e^- \tag{12.1}$$

If more energy is gradually imparted in forming the $C_2H_6^{+\bullet}$ ion, a point will be reached at which the ion contains enough vibrational energy for it to fragment as in reaction (12.2).

$$C_2H_6^{+\bullet} \rightarrow CH_3^+ + CH_3^\bullet \qquad (12.2)$$

The energy at which the fragment ion, CH_3^+, first appears is the *appearance energy* for that ion (the appearance energy of a molecular ion is its first ionization energy). These ionization and appearance energies can provide fundamental thermochemical information in chemistry, and mass spectrometers may be used to measure them. In the example for ethane given here, the C–C bond dissociation energy, D_{C-C}, may be obtained from the appearance energy of the CH_3^+ ion and the ionization energy of the $CH_3^\bullet$ radical. Equations (12.3) illustrate how the dissociation energy, D_{C-C}, is calculated, assuming that there are no excess of energy terms.

$$
\begin{array}{lll}
C_2H_6 \longrightarrow CH_3^+ + CH_3^\bullet + e^- & A & \\
CH_3^\bullet \longrightarrow CH_3^+ + e^- & I & (12.3) \\
\hline
\therefore C_2H_6 \longrightarrow CH_3^\bullet + CH_3^\bullet & D_{C-C} = A - I &
\end{array}
$$

Since a mass spectrometer is a detector of ions, it is only necessary to increase the energy applied to the molecules until molecular or selected fragment ions appear at the detector. The incident energy may be from photons from a vacuum monochromator, or from electrons. A graph of ion current versus incident radiation energy is an *ionization efficiency curve*, as illustrated in figure 10.4 for electron ionization. This graph is a convolution of total ion current with change in the wavelength of the incident radiation. When photons are used together with separation of individual m/z values, the ionization efficiency curve becomes, in effect, an 'action spectrum' (section 9.6.2). The onset of ionization is the ionization energy but it can be difficult to assign an accurate value to it because the position of onset of ion formation may be somewhat difficult to observe. The latter remark applies particularly to values of ionization energies derived through electron ionization and there seems little doubt that many of the older values determined in this way will need to be corrected. More recently introduced methods of mathematical smoothing and analysis of raw data promise greater accuracy and consistency. Also, by use of closely defined photon or electron energies and repeated scanning of the ionization efficiency curve, accurate data have become available.

Appearance energies determined by mass spectrometry may have dubious significance. In the example given above (equations (12.3)), it was assumed there were no excess of energy terms or that these cancelled. Errors arise in several ways but probably the most important are (i) the molecular and electronic structures of the ions produced are assumed ones, (ii) the energy of activation for the reverse of the reaction that leads to fragmentation is assumed to be zero, i.e. $A = I + D_{C-C}$ in equations (12.3) only if the activation energy for the reverse reaction is zero (section 9.10.1) and (iii) for CH_3^+ ions to be detected in the mass spectrometer, the $C_2H_6^{+\bullet}$ ions in equations (12.3) must decompose at a measurable rate and must therefore contain *some* excess of internal energy. If results obtained by mass spectrometric procedures agree with those found by other methods it may be supposed that the assumptions are satisfactory. When no check is available, the assumptions need to be kept firmly in mind and results must be regarded as open to question. There are many mass-spectrometric results that do not agree with thermochemical arguments based on other methods of evaluating thermochemical data. On the other hand, mass spectrometric data are sometimes the only ones available. Modern methods of calculating heats of formation of ions are sufficiently well advanced to provide an alternative check on mass spectrometric data obtained through electron ionization and appearance energies. (Curtiss, Raghavachari, Trucks and Pople, 1991; Smith and Radom, 1993; Szulejko and McMahon, 1993). Indeed, calculated heats of formation are generally so reliable that they can be used to check experimental ionization and appearance energies or to compare theoretical and observed values so as to gain knowledge of ion structures. For example, the ion $C_3H_6^{+\bullet}$ formed from cyclopropane (C_3H_6) could conceivably have a closed structure **1** or an open (linear) structure **2** as shown in scheme (12.4). The ionization energies (I_1 and I_2) for these two possibilities can be calculated from equations (12.5):

$$(12.4)$$

$$I_1 = \Delta H_f(\text{cyclo-C}_3\text{H}_6^{+\bullet}) - \Delta H_f(\text{cyclo-C}_3\text{H}_6)$$

$$I_2 = \Delta H_f(\text{linear-C}_3\text{H}_6^{+\bullet}) - \Delta H_f(\text{linear-C}_3\text{H}_6)$$

(12.5)

Now, all of the terms on the right-hand side of these equations are heats of formation (ΔH_f), which can be estimated with high accuracy. Therefore, I_1 and I_2 may be predicted and compared with measured values. Alternatively, the estimated ionization energies may be compared with the measured values to decide which agrees best with structure **1** or **2** (or neither!).

12.2. ISOTOPE ANALYSIS AND LABELLING

The measurement of isotope abundances by mass spectrometry can be performed readily to a fair degree of accuracy. If relative abundances are required to better than about 3 per cent accuracy, special precautions must be taken and mass spectrometers have been designed specifically for isotope analysis. These spectrometers compare simultaneously two beams of ions for any pair of isotopes and provide very accurate isotope ratios by null methods. The methods involved have been extensively discussed (Brenna, 1994). A more recently developed technique uses an inductively coupled plasma to generate ions, which are then examined by similar analysers to those used for 'standard' isotope ratio measurements. Because of the method of ionization, these ICP/MS instruments are capable of being used to support investigations into complex substances in environmental science, geochemistry, medicine, forensic science and so on that are normally difficult to analyse by other methods (Gray and Date, 1983; Houk, 1994).

12.2.1. *Natural isotopes*

A mass spectrometer measures mass-to-charge ratios and the effect of elemental isotopes on the appearance of a mass spectrum has been described in section 1.4. Ions containing carbon (^{12}C) are always accompanied by ions of one mass unit greater because of the ^{13}C isotope. The natural abundance of ^{13}C is quite low so that it does not complicate the mass spectrum too much except when there are relatively large numbers of carbon atoms in the compound under investigation (section 12.2.3). Other elements such as nitrogen and oxygen also contain only low percentages of isotopes other than a main one so that the abundances of the isotope peaks are not of much significance unless there are large numbers of these atoms present in the molecule or ion.

However, some elements (listed in table 1.4) have abundant isotopes, for example, chlorine, bromine and many of the metals present in organometallic compounds. When elements with abundant isotopes are found in mass spectrometry, it is necessary to have some idea of how the appearance of the mass spectrum will depend on their number, i.e., to know what the *isotope pattern* should be.

Ions, M^+, containing one chlorine atom will be accompanied by ions, $[M + 2]^+$, of approximately one-third the abundance because the natural relative abundances of ^{35}Cl and ^{37}Cl are in the ratio 3:1. When there are two chlorine atoms in the ion, RCl_2^+, the calculation of the relative abundances of the isotope peaks (isotope pattern) becomes more complicated. In such a case, three peaks, due to M^+, $[M + 2]^+$ and $[M + 4]^+$ ions, occur corresponding to the possible combinations of ^{35}Cl and ^{37}Cl (formula (12.6)):

$$R \underbrace{^{35}Cl \, ^{35}Cl}_{M^+} \qquad R \underbrace{^{35}Cl \, ^{37}Cl}_{[M + 2]^+} \qquad R \underbrace{^{37}Cl \, ^{37}Cl}_{[M + 4]^+} \qquad (12.6)$$

Simple calculation of probabilities shows that the ions M^+, $[M + 2]^+$ and $[M + 4]^+$ will have relative abundances of 9:6:1 (figure 1.6). For more than two chlorine atoms, this sort of calculation becomes more tedious, but a short cut is to expand the simple formula (12.7), which yields immediately the number and relative abundances of the isotope peaks:

$$(a + b)^n = a^n + na^{n-1}b + n(n-1)a^{n-2}b^2/2! + \text{etc} \qquad (12.7)$$

In the binomial expansion (12.7), n is the number of isotopic atoms to be considered and a and b are the relative abundances of the isotopes. For two chlorine atoms in an ion, $n = 2$, $a = 3$ and $b = 1$ and hence $(a + b)^n = (3 + 1)^2 = 9 + 6 + 1$, and these terms are the relative abundances of the chlorine isotope peaks for ions containing two chlorine atoms. For three bromine atoms, $n = 3$, $a = 1$ and $b = 1$ ($^{79}Br:^{81}Br \approx 1:1$; see table 1.4) and so $(a + b)^n = (1 + 1)^3 = 1 + 3 + 3 + 1$. Thus, ions with three bromine atoms would give peaks in the mass spectrum for M^+, $[M + 2]^+$, $[M + 4]^+$ and $[M + 6]^+$ with relative abundances of 1:3:3:1. The reader familiar with nuclear magnetic resonance spectroscopy will note how the same formula (12.7) is used to obtain spin 'splitting' or coupling patterns for protons.

When ions contain two or more different elements with isotopes, the calculation *ab initio* of relative abundances becomes laborious and time-consuming and it is better to write a computer programme to effect the calculation. Usually, this facility is provided as part of the software package on modern mass spectrometric data systems. In simpler cases, there is a quick way of obtaining the isotope pattern, which is best exemplified by two specimen calculations (12.8 and 12.9).

Firstly, consider ions with two chlorine and two bromine atoms:

$$\left.\begin{array}{l} \text{For } Cl_2, (a+b)^n = (3+1)^2 = 9:6:1 \\ \text{For } Br_2, (a+b)^n = (1+1)^2 = 1:2:1 \end{array}\right\} \begin{array}{l} \text{relative} \\ \text{abundances} \end{array} \tag{12.8}$$

These relative abundances are combined,

$$
\begin{array}{rl}
(9\ 6\ 1)(1\ 2\ 1) \text{ and } \therefore (9\ 6\ 1) \times 1 = & 9:\ 6:\ 1 \\
(9\ 6\ 1) \times 2 = & 18:12:2 \\
(9\ 6\ 1) \times 1 = & \underline{\hphantom{18:}9:6:1} \\
\text{Total} & 9:24:22:8:1
\end{array}
$$

Each of the terms relating to the relative abundances for two chlorine atoms (9 6 1) is multiplied by each term relating to the two bromine abundances (1 2 1). After each multiplication, the resulting terms are moved one column to the right and finally totalled as in scheme (12.8). Thus, the isotope pattern for Cl_2Br_2 is one consisting of five ions, M^+, $[M + 2]^+$, $[M + 4]^+$, $[M + 6]^+$ and $[M + 8]^+$, with relative abundances of $9:24:22:8:1$.

Secondly, consider ions having two sulphur atoms and one chlorine:

$$
\begin{array}{l}
\text{For } S_2, (a+b)^n = (95+5)^2 = 9025:950:25 \\
(^{32}S : {}^{34}S \cong 95:5\ ;\ \text{see table 1.4}) \\
\text{For } Cl, (a+b)^n = (3+1)^1 = 3:1
\end{array} \tag{12.9}
$$

Combining these relative abundances (equations (12.7)) gives

$$
\begin{array}{rl}
(9025\ 950\ 25)(3\ 1) \text{ and } \therefore (9025\ 950\ 25) \times 3 = & 27075:\ 2850:\ 75 \\
(9025\ 950\ 25) \times 1 = & \underline{\hphantom{27075:}9025:950:25} \\
\text{Total} & 27075:11875:1025:25 \\
\text{or} & 1083:\ 475:\ 41:\ 1
\end{array}
$$

Therefore, the isotope pattern consists of four peaks for M^+, $[M + 2]^+$, $[M + 4]^+$ and $[M + 6]^+$ in the ratios $1083:475:41:1$. The peak due to $[M + 6]^+$ ions is very much smaller than the other three and might be ignored in the mass spectrum. Notice that the ratio (2.3:1) of the abundances of the M^+ and $[M + 2]^+$ ions is quite different from the $3:1$ that it would be if only chlorine and no sulphur were present. Although the natural abundance of ^{34}S is quite modest, its contribution to the spectrum would not be missed by an experienced mass spectrometrist.

Mechanisms of chemical reactions may be investigated by changes in natural isotope abundances as caused by kinetic isotope effects. The technique is best illustrated by reference to an example. The pyrolytic *cis*-elimination of xanthate esters (the Chugaev reaction) may proceed by either of two mechanisms, schemes (12.10*a*, *b*):

$$(12.10a)$$

$$(12.10b)$$

$$CH_3S\text{-}CO\text{-}SH \xrightarrow{\text{fast}} COS + CH_3SH$$

In mechanism (*a*) the bond between the thioether sulphur atom and the carbonyl carbon is broken whereas, in mechanism (*b*), the same bond is not significantly altered in the rate-determining step but the double bond between the thione sulphur and the attached carbon becomes a single bond. In each case, the rate-determining step is the formation of the cyclic transition state. Therefore, since the $C-^{32}S$ bond is weaker than the $C-^{34}S$ bond, a $^{32}S/^{34}S$ kinetic isotope effect will be observed for either the thioether (*a*) or the thione sulphur atom (*b*). Also, a $^{12}C/^{13}C$ isotope effect of the carbonyl carbon atom is anticipated. Theory predicts a relatively large effect for mechanism (*a*) in which dissociation of the C–S bond occurs, and a negligible effect for mechanism (*b*). An isotope effect manifests itself as a change in the natural isotope content for the particular atom as the reaction progresses from starting material to products with different relative rates for the different isotopes. This change can be monitored readily by mass spectrometry once the carbonyl carbon atom and the thione and thioether sulphur atoms have been separated. The gaseous products, COS and CH_3SH, were swept out of the reaction vessel, chemically separated and subjected to degradative processes such that the sulphur atoms were converted into SO_2 and the carbonyl carbon into CO_2. No significant change in the natural $^{12}C/^{13}C$ ratio was observed and there was only a small change in the natural isotopic composition of the thioether sulphur atom, indicating that the bond between them is not cleaved in the rate-determining step. On the other hand, a large change in the isotope ratio and hence a large isotope effect was observed for the thione sulphur atom, proving that the bonding at this atom is altered during formation of the transition state. It was concluded that mechanism (12.10b) is operative in the Chugaev reaction. This and other applications of isotope effects have been discussed (Shiner and Buddenbaum, 1975).

Measurement of the natural isotopic content of substances has other, diverse applications. For example, archaeologists have long used the number of tree rings as a method of dating (dendrochronology) and the thickness of each ring as an indication of the climate of the year in which it formed. Mass spectrometry has been used to quantify climatic changes by measuring extremely accurately changes in the $^{12}C/^{13}C$ ratio in the rings. The measurements can differentiate alterations in average temperature for each year of about 0.2°C. The variations in stable isotope ratios are attributed either to a temperature-dependent isotope effect during photosynthesis in the tree or to variation in the composition of the atmosphere. Insight into thermal history and chronology may also be obtained by measurement of the $^{16}O/^{18}O$ ratio in carbonates in limestone. As this rock is formed on the sea-bed, the ratio of ^{16}O to ^{18}O varies in successive layers of the limestone, depending on the temperature of the sea at the time the limestone deposits formed.

In fields more familiar to the chemist, metastable ion analysis can be used in conjunction with natural isotopes to aid structure elucidation. Consider the unknown substance in which a metastable molecular ion (m/z 282) ejected 87 mass units and which exhibited an isotope pattern in the molecular ion region indicating two sulphur atoms and about 15 carbon atoms. This suggested a molecular composition of $C_{15}H_{22}OS_2$ or $C_{15}H_6O_2S_2$. To a good approximation, the $[M + 1]^{+\bullet}$ ions at m/z 283 are derived from molecules with one ^{13}C atom randomly distributed through-out the structure. Using one of the techniques that detects selectively product ions of a given precursor ion (chapter 8), it was possible to focus on $[M + 1]^{+\bullet}$ ions and to observe losses of 87 mass units (^{13}C retained in the ion) and 88 mass units (^{13}C in the ejected neutral particle) without interference from ions derived from other fragmentations. The ratio of abundances of the resulting metastable product ion peaks for loss of 87 and 88 mass units was found to be $2:1$ so that, out of every three $[M + 1]^{+\bullet}$ ions undergoing this fragmentation, only one on average ejected its ^{13}C atom. Therefore, the number of carbon atoms in the neutral species must have been one third of that in the precursor ion. This fact necessitated that the number of C atoms in the original molecule was a multiple of three and was consistent with a compound having 15 carbon atoms. It followed that the ejected neutral species contained five carbon atoms (60 of the 87 mass units) and that its composition must have been $C_5H_{11}O$, there being no nitrogen atoms in the molecule. Therefore, the original molecule could not have been $C_{15}H_6O_2S_2$ but must have been $C_{15}H_{22}OS_2$ (Beynon, Morgan and Brenton, 1979). This low-resolution mass spectrometric analysis contributed to the elucidation of structure (**3**) for the substance under investigation.

The reasoning used with the metastable ion analysis could not have been used with normal ions because of interfering fragmentation of molecular ions. Note that this same information could have been gained from high-resolution accurate mass measurement on the ions at m/z 282 and 195 (sections 1.6 and 4.4).

12.2.2.   *Labelling with stable isotopes*

One of the simplest uses of isotope labelling is the determination of the number of 'active' hydrogens in a molecule. Treatment with D_2O or CH_3OD of a ketone, $R_2CHCOCH_2R$, that has three enolizable hydrogens adjacent to the carbonyl group in the presence of a small amount of a base will yield a deuteriated ketone, $R_2CDCOCD_2R$, which has a molecular mass three units greater than the original ketone. This increase in mass is easily detectable by mass spectrometry and it is even unnecessary for there to be complete incorporation of deuterium. Indeed, even if incorporation of the label were chemically complete, it could well happen that, on putting the compound into a mass spectrometer, some back exchange with protons from water adsorbed on surfaces inside the instrument would occur; the incorporation of the label would then seem incomplete in the first place but, generally, molecular ions corresponding to incorporation of one, two and three deuteriums would be observed. The use of stable isotopes as an aid to structural studies, by which many ambiguities may be resolved, has been described in section 6.3.3 and exemplified in chapter 11 (see examples E, F and G).

Labelling of compounds specifically for mass spectrometric work is usually done either to clarify a fragmentation sequence or to investigate mechanisms of mass spectrometric reactions. The 'shift' technique is used to help unravel fragmentation pathways. For example, the mass spectrum of a methyl ester may be compared with that of its trideuteriomethyl analogue; all peaks that shift by 3 mass units (CD_3, 18 m.u.; CH_3, 15 m.u.) must incorporate the methyl group.

Most labelling has been done to gain some insight into mass spectrometric fragmentation reactions and may be illustrated for the case of toluene. The fragmentation of toluene in the mass spectrometer has been investigated extensively; its major features are shown in scheme (12.11). Elimination of a hydrogen atom from the molecular ion of toluene was first believed to occur from the side-chain methyl group to give the benzyl cation, $C_7H_7^+$, at m/z 91. However, various deuterium labelling experiments showed that the ejected hydrogen was almost randomly derived from any of the original eight hydrogens in the toluene molecule. Further, ^{13}C-labelling of toluene in the side chain or in the ring revealed that ejection of C_2H_2 from the fragment ion, $C_7H_7^+$, was again a random process in that the labelled carbon atom was not specifically retained in or ejected from the decomposing ion. These observations, together with measurements of the heat of formation of the ion, $C_7H_7^+$, suggested that a tropylium structure (**4**; scheme (12.11)) was plausible for it:

$$\text{CH}_3 \text{ (toluene)}^{+\bullet} \xrightarrow{-\text{H}^\bullet} \text{C}_7\text{H}_7^+ \xrightarrow{-\text{C}_2\text{H}_2} \text{C}_5\text{H}_5^+ \xrightarrow{-\text{C}_2\text{H}_2} \text{C}_3\text{H}_3^+ \tag{12.11}$$

$m/z\ 92$ $m/z\ 91$ (**4**) $m/z\ 65$ $m/z\ 39$

If the toluene molecular ion at $m/z\ 92$ first rearranged to a hydrotropylium ion, the random loss of hydrogen to give the tropylium ion at $m/z\ 91$ could be explained and also the random loss of the ^{13}C label as C_2H_2 from this last ion was understandable (scheme (12.11)).

Experiments with the doubly ^{13}C-labelled toluene (**5**; scheme (12.12)) have shown that, although the two ^{13}C labels are adjacent in the molecule before ionization, when the ion at $m/z\ 93$ fragments, the loss of ^{13}C label in the ejected C_2H_2 corresponds to complete randomization of all the carbon atoms as well as hydrogen in the ion at $m/z\ 93$ (scheme (12.12)).

$$^{13}\text{CH}_3 \text{ (labelled toluene, } \mathbf{5}) \xrightarrow{-\text{e}^-} {}^{13}\text{C}_2{}^{12}\text{C}_5\text{H}_8^{+\bullet} \xrightarrow{-\text{H}^\bullet} {}^{13}\text{C}_2{}^{12}\text{C}_5\text{H}_7^+$$

$m/z\ 94$ $m/z\ 93$

$$\xrightarrow{\text{random}} \left\{ \begin{array}{l} -\,{}^{12}\text{C}_2\text{H}_2 \\ -\,{}^{13}\text{C}^{12}\text{CH}_2 \\ -\,{}^{13}\text{C}_2\text{H}_2 \end{array} \right\} \longrightarrow \left\{ \begin{array}{l} {}^{13}\text{C}_2{}^{12}\text{C}_3\text{H}_5 \\ {}^{13}\text{C}^{12}\text{C}_4\text{H}_5 \\ {}^{12}\text{C}_5\text{H}_5 \end{array} \right\}^+ \tag{12.12}$$

$m/z\ 65\text{--}67$

The above results show that the randomization of the carbon atoms may occur in the molecular ion, which could not then have the toluene structure. Alternatively, it could occur in the $C_7H_7^+$ ion before it fragments; it could also be the result of the $C_7H_7^+$ ion not having a tropylium structure. More recent experiments with techniques such as field ionization kinetics, collisional activation and ion cyclotron resonance (section 12.3), which allow ions with specific energies and lifetimes to be studied, reveal a complex picture. Carbon and hydrogen scrambling in the molecular ion of toluene ($C_7H_8^{+\bullet}$) occur by at least four mechanisms and there may be dynamic equilibrium between the toluene and cycloheptatriene structures. Low-energy $C_7H_7^+$ ions from toluene may have either of two structures, possibly tropylium and benzyl ions. The two can be distinguished because the former is unreactive in ion/molecule reactions with neutral toluene whereas the latter is reactive. In the absence of ion/molecule reactions, the benzyl ion, $C_6H_5CH_2^+$, is long-lived and stable. However, substantial rearrangement by several mechanisms precedes fragmentation of high-energy $C_7H_7^+$ ions. Isotope labelling for determination of fragmentation mechanisms has been reviewed (Holmes, 1975).

To be unequivocal, labelling experiments in mass spectrometry need to give either 100 per cent positive or 100 per cent negative results. Intermediate values may be interpreted as being due to scrambling of the label, the operation of alternative mechanisms or the occurrence of rearrangement. Hydrogen is particularly prone to scrambling in the mass spectrometer and deuterium experiments are frequently invalidated by this effect.

For good quantitative results and less ambiguity in isotope studies, it is necessary to have as great an enrichment of the isotopic label as possible. The methods used to introduce labels are legion, but often some ingenuity is required in the synthesis of labelled compounds from expensive isotopically labelled starting materials, the available range of which is limited. Labelled ethanoic acid, $CH_3{}^{13}CO_2H$, enriched to 99 per cent with ^{13}C is expensive but is in fact one of the cheapest commercially available ^{13}C-labelled compounds. By working on a small scale, the cost of the labelling experiments may be reduced considerably. The following synthesis of ^{13}C-labelled 1,4-diphenylbutadiene required a total of about 100 mg of ethanoic acid (scheme (12.13)).

The incorporation of deuterium labels is usually easier and the required chemicals cheaper than those for ^{13}C labels. Thus, D_2O or CH_3OD may be used to exchange deuterium for hydrogen in amines, amides, alcohols and enolizable carbonyl groups. Similarly, lithium aluminium deuteride is readily available and can be used to reduce many compounds. The use of an ^{18}O isotope is relatively expensive and fewer ^{18}O-labelled compounds are available compared with deuterium and ^{13}C.

In the fields of biochemistry, medicine and toxicology, before the advent

$$(12.13)$$

of mass spectrometry, the fate of a substance in biological tissue was often investigated by employing radioactive substrates. Mass spectrometry has largely superseded such studies because it does not require that the isotope be radioactive, thus removing radiation hazards both to experimental organism and to investigator. The metabolism of exogenous compounds such as drugs and pollutants as well as endogenous substances may be studied. A substance is 'labelled' with a stable isotope at a site that does not affect and is not affected by the natural processes under investigation and, after incubation, any products of the labelled compound will have the same isotopes in the same ratio as the original substrate, providing unequivocal evidence of origin. For example, a sample of progesterone, containing some unlabelled, some mono-, di- and mainly tri-deuteriated molecules (6), was incubated with frog ovaries. After appropriate work-up, several compounds with the same isotope pattern in their molecular ion regions were identified by gas chromatography/mass spectrometry. Figure 12.3 shows the high-mass regions of the mass spectra of the labelled progesterone substrate (*a*) and one extracted product (*b*) together with a reference spectrum of pregnanolone (*c*, structure **7**). The extracted product was identified as pregnanolone by its retention time and mass spectrum, its high-mass ions being 3 mass units greater than those of natural pregnanolone (figure 12.3(*b*) and (*c*)). Note also that the extracted pregnanolone must have been a product of the

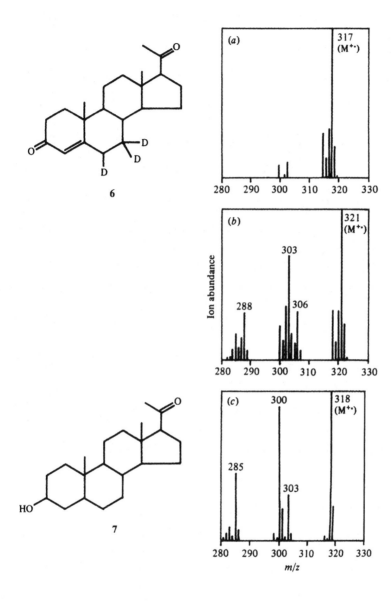

Figure 12.3. Partial mass spectra of (*a*) deuteriated progesterone (**6**) used for incubation with biological tissue, (*b*) a product of the incubation and (*c*) natural pregnanolone (**7**). Note that the sizes of peaks in the molecular ion regions of spectra (*a*) and (*b*) are the same and that the major peaks of spectrum (*b*) are the same as those in (*c*) but displaced to higher mass by 3 mass units.

progesterone because the mass spectra of the two compounds show the same isotope pattern for deuterium content in the molecular ion region (figure 12.3(*a*) and (*b*)). The proportion of unlabelled product is not taken into account because this will change unpredictably if the organism has 'natural pools' of substrate and/or products.

The above example illustrates a principle that is very widely applied for tracing the metabolic fate of compounds (Baillie, 1978; Klein and Klein, 1979, Caprioli and Bier, 1980).

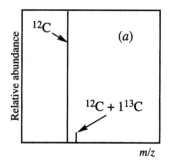 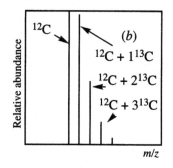 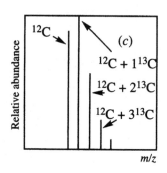

12.2.3. *Isotopes in molecules having large numbers of carbon atoms*

With the advent of mass spectrometric systems capable of measuring very large relative molecular masses, often of large organic molecules, it is of some importance to consider the appearance of the isotope pattern to be expected from ^{13}C isotopes. At its natural abundance level of 1.1 per cent, the chances of there being a ^{13}C atom in an organic compound containing only one carbon is 98.9:1.1 (figure 12.4(a)). With two carbon atoms in a molecule, the chance that *one* of them will be a ^{13}C isotope is doubled to 2.2 in 98.9. Thus, with n carbon atoms in a molecule, the chance that *one* of them is a ^{13}C isotope is $1.1 \times n : 98.9 = 1.1n : 98.9$. As n nears 100 there arises the, at first curious, result that the first ^{13}C isotope peak after the ^{12}C molecular ion peak becomes larger than the ^{12}C isotope peak. For 90 carbon atoms, the chance of finding *one* ^{13}C isotope in any one molecule is $1.1 \times 90 : 98.9 = 99 : 98.9$, *viz.*, the first ^{13}C isotope peak, $[M + 1]^{+\cdot}$, is almost the same height as the ^{12}C isotope peak, $M^{+\cdot}$ (figure 12.4(b)); for more carbon atoms, say 100, the isotope peak becomes even larger than the ^{12}C isotope peak (figure 12.4(c)). Similarly, the isotope peaks corresponding to molecules containing two, three or more ^{13}C atoms also increase so that, for a compound containing 100 carbon atoms, the isotope distribution pattern at the molecular ion is like that shown in figure 12.4(c).

In section 12.2.1 above, it was shown that the isotope pattern to be expected from a compound having a number of isotopic atoms present can be calculated from the binomial expansion (equation (12.7)). If the binomial expansion is applied to the ^{12}C and ^{13}C isotopes of carbon, it becomes that shown in equation (12.15), for n carbon atoms. Because the ^{12}C isotope is far more abundant than the ^{13}C isotope (98.9:1.1), terms of the expansion beyond the fourth are small compared with the first ones (equation (12.15)). Therefore, even with 100 carbon atoms in the molecule, the chance of

Figure 12.4. Diagrams showing the pattern of peaks expected in the molecular ion region due to natural abundances of ^{12}C and ^{13}C isotopes in organic compounds having increasing numbers of carbon atoms. (a) A compound with only one carbon atom, (b) a compound with 90 carbon atoms and (c) a compound with 100 carbon atoms.

finding five or more ^{13}C atoms in any one molecule is very small and may be safely ignored (equation (12.16)).

$$(a + b)^n = a^n + na^{n-1}b + \frac{n(n-1)a^{n-2}b^2}{2!} + \frac{n(n-1)(n-2)a^{n-3}b^3}{3!} + \ldots \tag{12.14}$$

$$(98.9 + 1.1)^n = 98.9^n + n98.9^{n-1}1.1$$
$$+ \frac{n(n-1)98.9^{n-2}1.1^2}{2!} + \frac{n(n-1)(n-2)98.9^{n-3}1.1^3}{3!} + \ldots \tag{12.15}$$

For $n = 100$,

$$(98.9 + 1.1)^{100} = 98.9^{100} + 100 \times 98.9^{99} \times 1.1$$
$$+ 100 \times 99 \times 98.9^{98} \times 1.1^2 / 2 + 100 \times 99 \times 98 \times 98.9^{97} \times 1.1^3 / 6 + \ldots$$

or, when normalized with respect to the first term ($= 100$),

$$100 \times (98.9 + 1.1)^{100} / 98.9^{100} \approx 100 + 111 + 61 + 22 + 6 + \ldots \tag{12.16}$$

A further consequence of the presence of ^{13}C isotopes in a compound having many carbon atoms lies in the mean relative molecular mass. For the natural abundances of the ^{12}C and ^{13}C isotopes, the mean atomic mass is 12.011 and, for 100 carbon atoms, the total mass is 1201.1, *viz.*, the average mass is 1.1 mass units greater than if only ^{12}C ($= 12.000$) had been used. As the chance of incorporating one or more ^{13}C atoms increases, there is a steady increase in the average atomic mass over that calculated on the basis of C = 12.000 mass units.

It might also be noted that a substance having many carbon atoms in its molecule will usually also have many hydrogen atoms. As the atomic mass for ^1H is 1.0078, a compound with 200 hydrogen atoms will have a hydrogen component of 201.6 rather than 200 if ^1H be counted as 1.000. Thus, a hydrocarbon, $C_{100}H_{200}$, will have a mass of 1402.7, which is almost 3 mass units greater than the figure that would have been arrived at by considering carbon and hydrogen to have masses 12 and 1 respectively. The 'additional' mass becomes important in the examination of mass spectra of large polypeptides, proteins, carbohydrates, nucleotides and artificial polymers. For such molecules, the possibility of there being more than 90 carbon atoms is high. A computerized data handling system, which is usually programmed to place a number against the largest peak in a bunch of peaks, will indicate a peak other than the ^{12}C isotope peak as being most significant (figure 12.4(d)) and it is easy to be misled as to the true relative molecular mass of the molecule. The true molecular mass for most chemical purposes is of course the average of all the isotope peaks but if 12.0000 has been used for the atomic mass of carbon in calculations then the apparent mass picked out by the computer will appear to be wrong. On the basis of 12.0000 as the atomic mass of carbon, the first peak in the molecular ion region (figure 12.4(d)) would have to be taken as the molecular ion despite the fact that isotope peaks immediately following are bigger. If the analyser of a mass spectrometer does not provide sufficient resolution to separate all of the peaks in the isotope pattern for molecular ions, then a broad unresolved envelope will be observed. An example of this effect can be seen in figure 11.8. If a protein of mass 300 000 is examined by MALDI, for example, it is inevitable that the molecular ion region will conglomerate into a broad envelope. The maximum of such an envelope of peaks, assigned by computer, will be the average of all of the contributing isotopes.

When normalized by putting the first term (^{12}C isotope) equal to 100, equation (12.15) gives an isotope pattern of $100 : 111 : 61 : 22 : 6$ for the first five terms (equation (12.16)). Thus, the first four terms set the appearance of the isotope pattern for a compound containing 100 carbon atoms. This pattern would be modified by any isotopes of other elements that were present in the same compound.

12.3.   ION CYCLOTRON RESONANCE SPECTROSCOPY

This method of mass spectroscopy, together with ion traps already described in sections 2.4.4 and 2.4.5, has features that are different from the usual sector and quadrupolar analysers and merits extra discussion. Two major differences between ion cyclotron resonance (ICR) and the commoner instruments lie in the production of long-lived ions and in the ability to examine ion/molecule reactions more readily. Important changes in ICR methodology have taken place since the last edition of this book. Nevertheless, the basic instrumentation and many uses have not changed and, therefore, the account given in the previous edition is largely repeated here but a new section has been added to accommodate some of the more important changes.

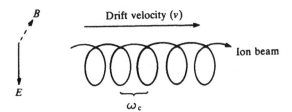

Figure 12.5. A side view of the path of an ion 'beam' in ion cyclotron resonance spectroscopy (see figure 12.6 for a top view). The ion cyclotron frequency (ω_c) and drift velocity (v) are shown. The electric (E) and magnetic field (B) are crossed (E is in the plane of the paper but B is vertical to it).

Under the influence of crossed magnetic and electric fields, ions follow a cycloidal path of frequency, $\omega_c = zeE/m$, and have a drift velocity, $v = E/B$, where E is the electric field strength, B the magnetic field strength, m the mass of the ion and z its charge (figure 12.5). Both positive and negative ions obey the same laws and are distinguished only by the direction of motion along the cycloid. In ion cyclotron resonance spectroscopy, ions are formed in an 'open' ion source (having no formal exit slits) and then drift into the analyser, where they are subjected to the effects of the crossed magnetic and electric fields. After travelling through the analyser the ions reach a detector (figure 12.6). Normally, space-charge effects would lead to the ions drifting to the walls of the analyser and a 'trapping' potential is used to prevent this. In the region of the detector there is no trapping potential and the ions strike the walls and are recorded as a total ion current. Because the ions follow a long cycloidal path at a small drift velocity, the actual time taken to traverse the analyser is often 5–10 ms although the analyser itself is only a few centimetres long. This time is in marked contrast to the few microseconds flight-time in sector, quadrupole and time-of-flight instruments. The relatively long flight-time in the ICR instrument leads, in combination with the gas pressure, to a high probability that ion/molecule collisions will occur before the ions reach the collector. Therefore, it is possible to investigate specific ion/molecule reactions by varying the background gas in the instrument (Lehman and Bursey, 1976; Nibbering, 1990).

Figure 12.6. A top view of the path of an ion 'beam' in ion cyclotron resonance spectroscopy showing the side-to-side oscillatory motion of ions impressed on their cycloidal path (see figure 12.5). The ions produced in the source pass through an analyser to a detector. A trapping potential prevents the ions from striking the sides of the analyser. The electric field (E) is perpendicular to, and the magnetic field (B) is in the plane of, the paper.

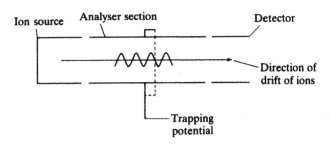

Since all ions proceed through the analyser with the same drift velocity, individual ion species (different m/z values) are examined by applying a radiofrequency electro-magnetic field to the ion beam in the analyser section. Although all ions have the same drift velocity, their individual cyclotron frequencies (ω_c) are proportional to their m/z values. When the frequency of the applied radiofrequency field equals the ion cyclotron frequency, absorption of electromagnetic radiation energy occurs; this energy absorption is measured to assign m/z values. Through application of radiofrequency energy of sufficient intensity, ions at particular m/z values can be given more kinetic energy so as to eject them to the walls of the analyser, where they are discharged. Such a process can be used to select ions of any one or several m/z values so that they are ejected from the analyser, leaving only ions that are of interest. More details of these techniques are described below.

It is outside the scope of this book to discuss ion/molecule reactions in detail but it may be mentioned that, apart from their intrinsic interest, they can yield valuable thermochemical data such as proton and electron affinities through estimation or comparison of equilibrium constants (Knewstubb, 1969; Freiser, 1988). For example, from a knowledge of the pressures of substituted pyridines in an ion cyclotron resonance cell and measurement of the abundances of the corresponding protonated pyridines (structures 8 and 9; reaction (12.17)), the free energy changes in the proton-transfer equilibria and the relative basicities of the pyridines can be mea-

$$\text{8} \qquad + \qquad \rightleftharpoons \qquad + \qquad \text{9} \qquad \qquad (12.17)$$

sured in the gas phase (Taagepera, Henderson, Brownlee, Beauchamp, Holtz and Taft, 1972). Many other relative basicities or acidities have been obtained from similar gas-phase equilibria (Szulejko and McMahon, 1993) and have been compared with pK_a values determined in solution.

Similarly, rate constants for gas-phase reactions can be measured (Lehman and Bursey, 1976) and have been compared with collision rate constants calculated from 'Langevin' equations (for some leading references, see Johnstone (1979)). For systems in which reaction occurs each time an ion collides with a neutral molecule, the measured rate constant and the collision (momentum-transfer) rate constants are equal so that the rate constant can be estimated from a knowledge of the collision rate for ions and neutral molecules; the latter rate may be obtained by variation of the neutral gas pressure in the ICR cell and its effect on any particular ion/molecule reaction.

Investigation of ion/molecule reactions by ion cyclotron resonance spectroscopy can provide considerable information on ion structures. The ion resulting from six-centre rearrangement of 2-hexanone with elimination of C_3H_6 is considered to be the enol (10) of acetone rather than the molecular ion of acetone itself (11), as shown in scheme (12.18); the ion resulting from simple fragmentation of the methyl cyclobutanol (12) is also thought to possess the enol structure (10):

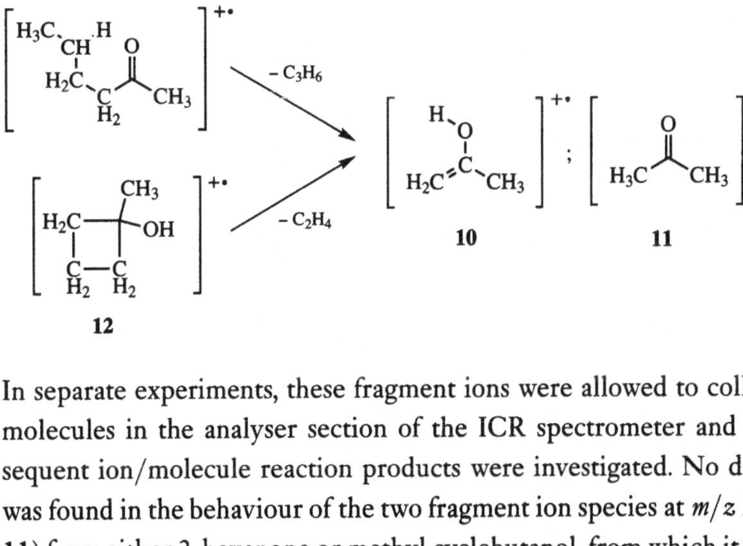

In separate experiments, these fragment ions were allowed to collide with molecules in the analyser section of the ICR spectrometer and the subsequent ion/molecule reaction products were investigated. No difference was found in the behaviour of the two fragment ion species at m/z 58 (10 or 11) from either 2-hexanone or methyl cyclobutanol, from which it was concluded that the ions had the same structure. One typical ion/molecule reaction of the ion at m/z 58 that illustrates this conclusion is shown in scheme (12.19). The molecular ions of methyl cyclobutanol at m/z 86 fragment to give the ion (10 or 11) at m/z 58 ($C_3H_6O^{+\cdot}$), which, after ion/molecule collision with a neutral molecule of the cyclobutanol (12), then affords an ion, $C_6H_{10}O^{+\cdot}$, at m/z 98 (scheme (12.19)).

To differentiate the ion, ($C_3H_6O^{+\cdot}$), at m/z 58 produced by the cyclobutanol from the ion of the same composition produced by fragmentation of 2-hexanone, the latter was labelled with deuterium in the 5-position. The

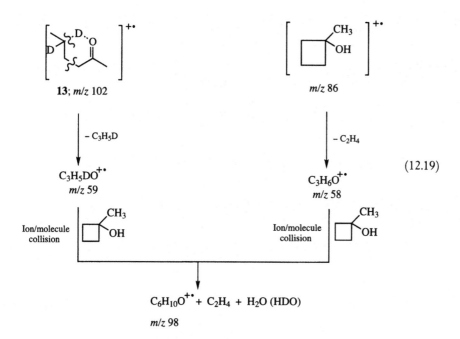

13; *m/z* 102

m/z 86

(12.19)

$C_3H_5DO^{+\cdot}$
m/z 59

$C_3H_6O^{+\cdot}$
m/z 58

$C_6H_{10}O^{+\cdot} + C_2H_4 + H_2O$ (HDO)

m/z 98

5, 5-d_2-hexanone (**13**) fragmented like 2-hexanone itself to give the ion (**10** or **11**) but now of composition, C_3H_5DO, at *m/z* 59. In the ICR spectrometer, this ion at *m/z* 59 also underwent an ion/molecule reaction with neutral methyl cyclobutanol (**12**) to give the ion, $C_6H_{10}O^{+\cdot}$, at *m/z* 98 (scheme (12.19)). Thus, the fragment ions, $C_3H_6O^{+\cdot}$, from methyl cyclobutanol and $C_3H_6O^{+\cdot}$ from the hexanone, when produced together in the spectrometer, were both found to react with neutral methyl cyclobutanol molecules to give the ion, $C_6H_{10}O^{+\cdot}$, at *m/z* 98. The molecular ion of acetone (**11**) was investigated in the same way but its ion/molecule reactions were found to be different from those of the ion at *m/z* 58 from either 2-hexanone or methylcyclobutanol. It may be concluded that the fragment ions of composition C_3H_6O derived from both methyl cyclobutanol and from 2-hexanone (scheme (12.18)) have the same structure, which is different from that of the molecular ion of acetone (**11**) and may be that of the enol (**10**). Although structures of ions may be compared in this way by ion cyclotron spectroscopy, it should be borne in mind that the actual structure may not be as postulated since rearrangement may occur as a result of energy transfer following ion/molecule collision.

Negative-ion spectra may be measured as readily as positive ones by ion cyclotron resonance spectroscopy. An interesting application of this facility is the measurement of electronic excitations in molecules resulting from impact with low-energy electrons. In the reaction sequence (12.20), electrons of energy E_1 are captured by SF_6 and the abundance of the resulting SF_6^- ions is determined by ICR spectroscopy:

$$M \qquad + e(E_1) \rightarrow \qquad M^* \qquad + e(E_2); (E_2 < E_1)$$
$$\text{(neutral molecule)} \qquad \text{(electronically excited)}$$

$$SF_6 + e(E_1 \text{ or } E_2) \rightarrow SF_6^- \tag{12.20}$$

In the presence of other gas molecules, energy may be transferred from the electrons with energy E_1 through electronic excitation of the gas molecules; as a result of this process, the electrons after 'collision' have less energy (E_2). The electronic excitation energy of the gas molecule must then equal $E_1 - E_2$. The change in the energy of the colliding electron leads to a change in the abundance of SF_6^- ions. Thus, by monitoring the ion current of SF_6^- ions as the energy of the colliding electrons is gradually increased, electronic excitation processes in the other neutral gas molecules can be revealed as changes in the SF_6^- ion current.

It is also possible to carry out a technique of double resonance spectroscopy by irradiating ions of a particular ion cyclotron resonance frequency (effectively an m/z value) whilst observing the effect of this on the abundance of other ions at some other cyclotron frequency (some other m/z value). If an ion/molecule reaction (12.21) is suspected, ions A^+ can be irradiated in the ICR cell at their resonance frequency, ω_A.

$$A^+ + B \rightarrow C^+ + D \tag{12.21}$$

The irradiation speeds up the motion of the ions, thereby increasing the energy of their collision with neutral gas molecules, B. If C^+ ions are derived from A^+, then changing the translational collision energy will lead to a change in the yield of C^+ ions and therefore changes the peak size at the m/z value corresponding to C^+. Thus, the relationship between A^+ and C^+ can be demonstrated. For example, the ion at m/z 59 resulting from a double McLafferty rearrangement in the molecular ion of 4-nonanone-1,1,1-d_3 has been considered to have either structure (14) or (15), as illustrated in scheme (12.22). The molecule of a ketone such as 4-nonanone is readily protonated by collision with radical-cations that have a hydrogen or deuterium atom (reaction (12.23)).

$$
\left[
\begin{array}{c}
\text{H}_3\text{C}-\overset{\text{H}_2}{\text{C}}-\text{CH} \quad \overset{\text{H}}{\text{O}}\cdots\overset{\text{D}}{\text{O}}-\text{CD}_2 \\
\text{H}_2\text{C}-\overset{}{\underset{\text{H}_2}{\text{C}}}-\overset{\text{O}}{\underset{\text{H}_2}{\text{C}}}-\overset{}{\text{C}}-\text{CH}_2
\end{array}
\right]^{+\cdot}
\longrightarrow
\begin{array}{c}
\text{H}-\overset{+\cdot}{\text{O}} \\
\text{H}_2\text{C}=\overset{\text{C}}{}\text{CH}_2\text{D} \\
\mathbf{14}
\end{array}
\quad \text{or} \quad
\begin{array}{c}
\text{H}\diagdown\overset{+}{\text{O}}\diagup\text{D} \\
\text{H}_2\text{C}\cdots\overset{\text{C}}{}\cdots\text{CH}_2 \\
\mathbf{15}
\end{array}
$$

$$\text{C}_3\text{H}_5\text{DO}^{+\bullet}\,;\, m/z\ 59 \qquad (12.22)$$

$$
\text{CH}_3(\text{CH}_2)_4\overset{\text{O}}{\overset{\|}{\text{C}}}(\text{CH}_2)_2\text{CD}_3 + \underset{m/z\ 59}{\text{C}_3\text{H}_5\text{DO}^{+\bullet}} \longrightarrow \text{CH}_3(\text{CH}_2)_4\overset{\overset{+}{\text{O}}\text{H(D)}}{\overset{\|}{\text{C}}}(\text{CH}_2)_2\text{CD}_3 + \text{C}_3\text{H}_4\text{D(H)O}^{\bullet}
$$

$$
\begin{array}{l}
[\text{M} + \text{H}]^+\,;\, m/z\ 146 \\
[\text{M} + \text{D}]^+\,;\, m/z\ 147
\end{array} \qquad (12.23)
$$

Therefore, collision of the ion at m/z 59 with 4-nonanone-1,1,1-d_3 (relative molecular mass $= 145$) could lead to transfer of H^+ or D^+ with formation of protonated 4-nonanone at m/z 146 ($[\text{M} + \text{H}]^+$) or m/z 147 ($[\text{M} + \text{D}]^+$) depending on the structure of the ion at m/z 59 (structure **14** or **15** in reaction (12.22)). Thus, in the experiment, the ion at m/z 147 was observed at its cyclotron frequency and the ion at m/z 59 was irradiated at its cyclotron frequency. No change was found in the abundance of the ion at m/z 147 and it could not have been formed by collision with the ion at m/z 59. On the other hand, the abundance of the ion at m/z 146 was strongly affected when the ion at m/z 59 was irradiated. It was concluded from this *double resonance* experiment that the ions at m/z 59 could not have had the structure (**15**) in scheme (12.22) because transfer of deuterium from the ion to 4-nonanone was not detectable but should have been almost as equally likely as transfer of hydrogen (both hydrogen and deuterium coming from the same oxygen atom in structure **15**). Useful discussions of ion/molecule reactions have appeared (Freiser, 1988; Nibbering, 1990).

As well as recording a mass spectrum by measuring the energy loss of the irradiating field at each ion cyclotron frequency corresponding to the various m/z values of the ions in the spectrum, the latter can be recorded by measuring the total ion current at high field strengths. Under such conditions, when resonance occurs, an ion is not merely accelerated somewhat but is ejected onto the sides of the cell before reaching the collector and therefore there is a decrease in the total ion current. These changes in the total ion current at each m/z value yield a mass spectrum. The technique is used for obtaining Fourier transform ICR mass spectra, as is discussed in more detail below. A simple corollary of ejecting all ions to the cell walls through use of a band of frequencies at high field strength is that individual ions can be ejected from the cell by use of a specific frequency at high field strength. This mode of operation leaves behind in the cell any other ions of different m/z values. An application of this approach is described here.

Thus, it is possible to assess the relative contributions of different ion/molecule reactions leading to a *common* ion. Consider the two ion/molecule reactions (scheme (12.24)):

$$A^+ + B \rightarrow C^+ + D$$
$$X^+ + B \rightarrow C^+ + Z \tag{12.24}$$

If the mass spectrum could be recorded without A^+ (or X^+), the relative importance of the two reactions leading to the formation of C^+ could be assessed. In ion cyclotron resonance spectroscopy, this assessment can be made by 'ejecting' A^+ (or X^+) from the cell by application of an oscillating electric field to the trapping plates (figure 12.5). Since the ions are oscillating between the trapping plates as well as following a cycloidal path, then at resonance the oscillations increase and the ions are ejected to the walls of the cell. Ejection of A^+ would reduce the abundance of C^+ ions to that due solely to their formation from X^+ ions. The technique can be used with any number of reactant ions so long as they can be ejected selectively and has been utilized for application of FTICR to MS^n ($n = 2$–5; see later).

In older ICR instruments, mass resolution was not particularly good at higher m/z values and much of the research was carried out at relatively low mass. Generally, resolving power varies inversely with m/z values for ICR. If the ions in the ICR cell are ejected sequentially to obtain a spectrum or the transient signal induced in an external circuit by the oscillatory motions of the ions is monitored, the mass spectrum is recorded in the 'time' domain. Fourier transformation of the latter time domain signal to a 'frequency' domain gives the individual ion cyclotron frequencies and hence their masses. This is the Fourier transform mass spectrum. The advantage of gathering m/z information in this way is that the short irradiating pulse used to obtain the frequency domain spectrum can be done once or repeated many times in a short period of time. Because the ions are retained in the cell, the technique is non-destructive and many measurements of the mass spectrum can be obtained from the same set of trapped ions. Successive measurements of the mass spectrum can be added together to give a final spectrum, which has the advantage that background electronic noise, being random, tends to cancel out but any specific signals (m/z values) accumulate. A much better resolved spectrum is attained and the effective resolution of the instrument is considerably enhanced; resolution is still inversely proportional to m/z value for any one ion/molecule collision frequency, *viz.*, it varies with gas pressure in the ICR cell and, as discussed in section 8.9.2, ionization and activation of ions outside the ICR cell is used to maintain a low pressure within the cell itself. The transform technique is similar to that used with infrared and nuclear magnetic resonance spectroscopy. This advance led to major development of ICR as FTICR (Fourier transform ion cyclotron resonance) spectroscopy (Hanson, Kerley and Russell, 1989; Jacoby, Holliman and Gross, 1992).

Whereas, for a variety of reasons, ICR was not a favoured technique for analytical purposes, the later developments in FTICR and in mass spectrometry generally have provided an impetus for the use of these 'ion traps' in such areas. In particular, improvements in resolution and detection limits, the ability to effect ionization outside the FTICR cell and the growth of MS/MS have played a major role in the application of FTICR to analytical problems (Asamoto, 1988; Asamoto and Dunbar, 1991; Smith, Smith, Pasztor, McKelvey, Meunier, Frohlicher and Ellaboudy, 1993). Application of FTICR to measurement of the proton-to-electron mass ratio and mass spectrometric investigations of antimatter positrons and antiprotons are striking examples of the advances made with this technique (Marshall and Schweikhard, 1992), as is the observation of single trapped ions from poly(ethylene glycol) with masses up to about 7 000 000 (Smith, Cheng, Bruce, Hofstadler and Anderson, 1994).

It has been realized that best resolution enhancements in FTICR are only possible with very low pressures in the cell. Otherwise, ion/molecule collisions with residual gas in the cell are enough to reduce overall resolution severely. For best resolution, the cell must be operated below 10^{-7} Torr. These sorts of pressures have been achieved by ionizing the sample outside the cell and transferring the resulting ions into the cell. Differential pumping between the external ion source and the analysis cell is necessary. With these sorts of operating conditions it is possible to achieve ultra-high resolving powers. At m/z 4, a resolution of 200×10^6 can be attained, enabling the complete separation of $^4He^+$ and D_2^+. A less esoteric but still impressive use of this high-resolution capability can be found in a separation of isobaric product ions. The isobaric molecules, acetophenone and mesitylene, yield molecular ions on electron ionization that lose methyl radicals (reaction (12.25)). Thus, the two product ions at m/z 105 ($C_7H_5O^+$ and $C_8H_9^+$) were produced by colliding the molecular ions with a pulse of inert gas and then separating the resulting two isobaric (induced) fragment ions at a mass resolution of 250 000. At a lower signal-to-noise ratio, the resolution could be increased to 500 000 (Wise, 1987; Cody, 1988). This sort of resolution performance for *product* ions from collisional activation of precursors easily exceeds that for MS/MS experiments on other types of instrument.

$$\text{acetophenone} \rightarrow \underset{m/z\,120}{C_8H_8O^{+\bullet}} \rightarrow CH_3^{\bullet} + \underset{m/z\,105}{C_7H_5O^+} \qquad (12.25a)$$

$$\text{mesitylene} \rightarrow \underset{m/z\,120}{C_9H_{12}^{+\bullet}} \rightarrow CH_3^{\bullet} + \underset{m/z\,105}{C_8H_9^+} \qquad (12.25b)$$

With the development of ion sources external to the analysis cell of the instrument, it has been possible to use a wide variety of ionization methods in FTICR. Most of the ionization sources and combined inlet/ion sources described in chapter 3 have been used with FTICR. External laser desorption is widely used because it yields mainly $[M + K]^+$ ions (depending on the matrix used), which can be examined by MS/MS techniques. This application of lasers is particularly useful for analysis of polymers because all of the molecular ions for a polymer series, M_n, appear as a series of $[M_n + K]^+$ molecular ions so that the molecular mass distribution of the polymer can be read off from the mass spectrum. Atmospheric pressure ionization methods, including electrospray, have been utilized as well as field ionization, chemical ionization, fast atom bombardment and so on. One problem with the external sources is the 'time-of-flight' effect of ions being produced in one place and then being transferred by electric fields to the FTICR cell. If the transfer line is not short then the times of arrival of ions in the cell from the source vary with mass and, over a wide mass range, this can mean that reactions of lower mass ions can be occurring in the cell before the heavier ones have arrived.

New operating modes have been developed for the investigation of ions trapped in the FTICR cell. A significant advance in this area has been the application of a *stored waveform inverse Fourier transforms* (SWIFT) method, which provides optimum waveforms for excitation of the ions in the cell (Marshall and Schweikhard, 1992). These SWIFT excitation profiles allow

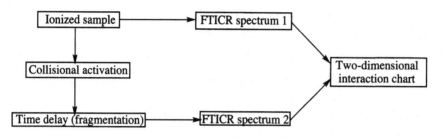

Figure 12.7. An illustration of the application of SWIFT to two-dimensional mass spectrometry. By comparing mass spectra before and after collisional activation (MS/MS), abundances of precursor ions that decrease can be correlated with those of product ions that increase. This cross-correlation is used to produce a two-dimensional chart showing the connections between precursor and product ions as off-diagonal elements.

multiple selected ion monitoring, multiple ion ejection and simultaneous removal of ions at specified m/z values whilst retaining and detecting ions at other specified values. The same excitation transform provides more accurate ion abundances and enables MS/MS to be performed, all at high resolution. In one application of the SWIFT method it was possible to de-excite previously excited ions and return them to their original locations in the cell and this enabled the development of two-dimensional MS/MS. Figure 12.7 shows a schematic outline of a two-dimensional (2D) FTICR MS/MS experiment. Ions of m/z ratios p, q, r, etc., are irradiated in the presence of a collision gas such as helium. After a specified short delay during which some of the initial ions will have collided with the helium and will have decomposed to give fragment ions, a second waveform profile returns the excited ions back to their original state except that some of them have disappeared through fragmentation and consequently their abundance has decreased; some newly formed fragment ions will have increased in abundance. Comparison of the two such sets of spectra reveals specifically the various ion fragmentations, i.e., which ions have given which others. The various ion/molecule reactions are connected by the off-diagonal matrix elements from the two sets of spectra. Again, this approach is similar to the two-dimensional methods of nuclear magnetic resonance spectroscopy and provide a powerful tool for examination of the fragmentation of single compounds or for investigation of mixtures. The ions at m/z p, q, r, etc. can be molecular ions, perhaps produced by laser desorption, and several MS/MS investigations can be carried out on them in one experiment. This is potentially valuable for analysis by MS/MS of complex mixtures that by 'standard' methods (chapter 8) require several successive MS/MS examinations of a series of ions to unravel a complex mass spectrum of a mixture (Marshall and Schweikhard, 1992).

Finally, a novel investigation of electronic levels in ions has been carried out. This produces information similar to that from the excitation of SF_6

discussed early in this section. By use of a tunable laser, ions trapped in an FTICR cell have been irradiated and the resulting emission of fluorescence has been observed (Wise, Buchanan and Guerin, 1990). In suitable cases, the fluorescence spectrum of an ion can be determined. The advance of FTICR has placed it in a very competitive position with regard to the range of analytical procedures that can be tackled by 'standard' mass spectrometry using sector or quadrupole instruments. The major fundamental difference between FTICR and sector or quadrupole instruments lies in two time and position domains. Sector instruments provide (linear) successive ionization, analysis and detection of ions and the ions are 'used' once only as they pass through the sectors. In FTICR (as with ion traps generally), after ionization, analysis and detection of ions is carried out in one 'sector' as and when required, thereby giving a greater time dimension to a mass spectrometric experiment. Thus, FTICR, quadrupole and sector instruments have their own individual advantages and disadvantages but all largely use similar ionization methods, similar approaches to analysis of samples and the same ion or ion/molecule chemistry and have the same problems of connection to other analytical methods for combined techniques. Thus, this section of this book has concentrated on aspects of FTICR (or ICR) that are not readily achievable with sector or quadrupole instruments or that can be carried out better than with the latter. Many of the other topics dealt with earlier in this book have used sector instruments as exemplars because these are usually simpler to understand and this is, after all, a teaching text. However, once the principles of ICR have been understood, it is easier to understand how the topics considered earlier in the book are applied similarly in ICR mass spectrometry.

12.4. PYROLYSIS/MASS SPECTROMETRY AND PYROLYSIS/CHROMATOGRAPHY/MASS SPECTROMETRY

12.4.1. *Introduction*

The action of heat on most pure compounds has a number of possible outcomes. The substance being heated may simply volatilize, it may rearrange to something else, it may decompose in a specific fashion or it may degrade to form a variety of other substances. For mass spectrometry, which often involves heating a sample to a high enough temperature to volatilize it, it is preferable that no thermal changes should occur if a proper structural analysis is to be carried out. This is especially true for mixtures and several techniques have been combined with mass spectrometry to provide an initial separation of a mixture followed by mass spectral analysis of the individual separated components. Such combined methods as GC/MS, LC/MS, CE/MS, TLC/MS and SFC/MS have been covered in earlier chapters. Many mixtures can be separated into individual components through use of several chromatographic techniques used in tandem and, in theory, even extremely complex mixtures can be dealt with in this way. However, practically, it may not be feasible to carry out long complicated multiple separations because of the time and cost entailed. Additionally, there are some mixtures that are simply not amenable to separation by dissolution in solvent. Ceramics, bone, cement and many biological materials are examples of mixtures that are generally too complex for straightforward chromatography and some other means is needed for dealing with them either for the purpose of identifying components or for comparing one sample with another. Intentional application of heat or light in a short time interval (pyrolysis or laser desorption) is one way that has been developed to deal with awkward mixtures; the pyrolysate can be examined by mass spectrometry, which, with pyrolysis, gives a new combined method of pyrolysis/mass spectrometry (Py/MS, sometimes written PY/MS) or, if also combined with chromatography, it gives the triple combined approach of pyrolysis/chromatography/mass spectrometry. The diversity of analytically 'difficult' materials that can be examined by pyrolytic methods is impressively large and covers synthetic polymers, biopolymers, complex inorganic substances, micro-organisms such as bacteria and fungi, coconut shells, paints, ceramics, works of art, wood, fossils, parts of plants and so on. All of these materials present severe difficulties for 'standard' methods of analysis because they are complex mixtures of complex molecular structures, often difficult to dissolve or totally insoluble in all reasonable solvents. Therefore, pyrolysis/mass spectrometry provides a powerful technique that fills a niche in a wide range of 'difficult' substances that need to be analysed for their thermal properties (e.g., cured polymers) or for their compositions (e.g., bacteria).

12.4.2. *General pyrolysis of complex substances*

Thermogravimetric analysis, by which the change in weight of a sample is monitored as its temperature is raised, uses a relatively slow change in temperature covering tens of seconds or even minutes. Different volatile substances may be detected at different temperatures as the sample is heated. This may be referred to as *time-resolved* pyrolysis. As discussed more fully below, much pyrolysis is effected extremely rapidly (10^{-9}–5 s) such that individual processes become indistinguishable and equilibrium conditions do not hold. The actual rate of heating leads to different compositions of the resulting pyrolysates (Plage and Schulten, 1991) and can be used to gain insights into the dependence of the pyrolytic process itself on the rate of heating and on the volatilization rates of components formed on pyrolysis (Boon, 1992).

Apart from such specialized applications, pyrolysis can be applied to many substances or complex mixtures so as to provide a pyrolysate that may be easier to analyse than the original intractable substance. The ways in which the pyrolysate is treated before mass spectral analysis are manifold but a few of these approaches are shown in scheme (12.26):

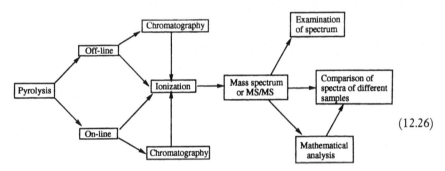

(12.26)

The pyrolysate may be obtained *off-line* from the mass spectrometer, as would be the case for a forensic examination of the cause of a fire (Ardrey, 1993). At the other extreme, the pyrolysate may be passed immediately after production *on-line* into the ion source of a mass spectrometer or may even be produced actually inside the ion source. Often the pyrolysate may be subjected to some sort of separation as with gas chromatography and this leads to the development of the triple combined technique of pyrolysis/gas chromatography/mass spectrometry (Py/GC/MS). Which analytical treatment of the pyrolysate is used will be determined by the end-use for the resulting information. If it is required to relate complex samples through comparison of their pyrolysates then it is only necessary, in principle, to examine the total mass spectrum obtained by passing all of the pyrolysate, without any separation, into the ion source of a mass spectrometer. This would apply to characterization and quantification of unknown microbial samples for which a rapid comparison with known samples was required; individual components within the pyrolysate do not need to be identified but the appearance of the mass spectrum of the *total* pyrolysate becomes a 'fingerprint' for each micro-organism. By comparison of these fingerprints, identification and determination of the amount of a micro-organism can be carried out. At the other extreme, it may be necessary to identify at least some of the components resulting from pyrolysis and this will involve either separation (gas or liquid chromatography) followed by mass spectrometric examination of the individual separated components or application of MS/MS techniques as described earlier in section 8.8. For example, in food chemistry, the smell or flavour components of roasting meat may need to be identified and this will involve pyrolysing (roasting) the meat followed by separation and analysis only of the volatile substances, as these are the ones that reach olfactory and taste organs. For these last methods in which combined techniques are used for chromatography and mass spectrometry, many earlier sections of this book give all the necessary information, the only speciality in this section being the pyrolysis step used for obtaining the sample to be examined. Therefore these types of pyrolysis/chromatography/mass spectrometry are not explored in detail here. There are many analytical applications for which it is not necessary to separate and identify components, only the total mass spectral fingerprint being used in the analysis. This sort of pyrolysis/mass spectrometry has some unique features not covered in earlier sections of this book and is discussed in more detail.

12.4.3. *Chemical aspects of pyrolysis*

The application of heat to substances may be totally uncontrolled, as in cases of arson, approximately controlled, such as with the smoking of tobacco, or closely regulated as in high-temperature manufacturing processes like the catalytic cracking of heavy oils to give lighter ones. With most organic compounds serious thermal instability begins at quite low temperatures and, as the temperature of the sample being heated increases, bonds begin to break homolytically to give radicals or heterolytically to give ions. If the heating is carried out in air, oxygen begins to react with the material being heated so as to produce combustion products, the simplest being water, carbon dioxide and carbon monoxide. As this burning (oxidation) complicates the basic pyrolytic reactions, most intentional pyrolyses are carried out in an inert gas atmosphere of helium or nitrogen, or in a vacuum. The primary pyrolytic reactions are important for analysis or for structure determination and not those resulting from combustion because these obscure structural details.

The primary species formed on pyrolysis are mainly either simple elimination products of molecular species or reactive radicals formed by homolytic bond cleavage. If the radicals leave the heating zone rapidly from thin layers of sample, they have no time to react with unpyrolysed material. Thus the result of very fast heating, as with laser ablation (see below) is to give information on elemental compositions and functional groups, together with a lot of relatively small molecules. With 'slow' heating and large samples, there is ample opportunity for the initial pyrolysis materials (radicals) to react with themselves or with unpyrolysed material as they diffuse out from the body of the sample; this type of heating tends to give more structural and molecular mass information about the original sample. If the heating parameters (rate, initial and final temperatures) and sample sizes are very closely controlled then it might be expected that the resulting complex pyrolysate would be similar from one sample to another for the same initial substance but would be different for different initial substances. This is the basis for analytical examination or comparison of complex substances by pyrolytic methods. These same experimental parameters may be varied to give different kinds of information, as with slow and fast heating, but for valid comparisons to be made from sample to sample the parameters must be reproducible. Some of the methods of heating given below are highly reproducible even from one laboratory to another but, with other methods of pyrolysis, close reproducibility is difficult and may not even be required.

As heating rates may be controlled fairly accurately, pyrolysis/mass spectrometry can be carried out in a *temperature-resolved* mode or a *time-resolved* mode depending on whether the pyrolysis is controlled more by the actual temperature range through which the sample is heated or by the speed with which the temperature is ramped.

12.4.4. *Thermogravimetric analysis*

One simple application of this approach lies in thermogravimetric analysis. In this method, a substance is heated relatively slowly, usually to temperatures of 200–800°C, and the weight of the sample substance is monitored as the temperature increases. The resultant graph of sample weight versus its temperature composes the thermogravimetric analysis. As materials are heated, phase or chemical changes may occur without formation of volatile material and these are recorded as exo- or endotherms in differential thermal analysis or differential scanning calorimetry. Degradation usually leads to volatilization of lower molecular mass materials from the heated substance. There is an accompanying weight loss, which is recorded as a change in the slope of the thermogravimetric analysis curve. There is no way of determining exactly what has volatilized simply from the weight loss itself but, if the volatile substance (or substances) is passed into the ion source of a mass spectrometer, it can be identified from its mass spectrum and, thereby, can indirectly provide additional information on the substance being heated. This is the basis of thermogravimetric analysis/mass spectrometry (TGA/MS or TG/MS), which provides much more information than just the thermogravimetric procedure by itself (Redfern, 1991).

12.4.5.   *Pyrolysis through Curie point heating*

If pyrolysis is to be a consistently repeatable technique for comparison of 'fingerprint' mass spectra of total pyrolysates, it is important that the parameters leading to the formation of the pyrolysate should be controlled as closely as possible. Variables such as the rate of heating, the temperature reached during heating, the presence or absence of air (oxygen) and possible catalytic effects from the support holding the sample must all be considered and held as constant as possible. Usually, a small amount of sample is heated rapidly to a high temperature in the absence of air. The volatile substances that come off during rapid heating will be a mixture of (i) components that simply vaporize without decomposition and (ii) components formed by the act of heating (rearranged or degraded or built up from the original). As the rates and extents of reaction that occur vary with both the *rate* and *length of time* of heating, a small number of methods has been devised for carrying out the heating in a controlled, reproducible manner. One of these is 'Curie-point' inductive heating.

The 'Curie point' of any ferromagnetic material is the temperature at which its magnetic permeability (its magnetic property) disappears. Inductive heating occurs when a metal of magnetic permeability greater than unity is placed in a radiofrequency electric field; there is no physical contact between the metal being heated and the coil or solenoid forming the electric field. Thus, if a magnetic metal wire or foil is placed in a suitable radiofrequency field it becomes inductively heated and its temperature increases up to its Curie point, at which inductive heating stops even with the radiofrequency field still switched on. At this point, the temperature of the wire or foil will fall until, just below the Curie point, heating begins again. This 'hunting' or 'feedback' control means that a magnetic substance can be heated to its Curie temperature, at which point the temperature stabilizes, remaining constant to within about 1°C. Another effect of radiofrequency heating lies in the fact that *primary* heating lies within a small region close to the surface of the metal (a 'skin-effect'), *viz.*, usually within about 0.1 mm of the surface of the metal. For analytical pyrolysis methods, in which it is important to heat samples quickly and repetitively to one fixed temperature, this *Curie point heating* in a radiofrequency field allows the temperature of a sample to be increased rapidly, accurately and repetitively to a fixed temperature. For pyrolysis/MS, the sample is placed on a thin wire, or more probably, ribbon or foil, which is about the same thickness as the depth of the skin effect. The wire or ribbon is then heated inductively to its Curie point in about 0.1–0.5 s, i.e., a rate of heating of about $10^3–10^4$ K s^{-1}. Under these conditions the temperature of the sample can be brought rapidly from ambient to a pre-set fixed temperature that can range up to about 1200°C, but which is predetermined by the chemical composition of the ribbon. For example, pure iron reaches 770°C, nickel 358°C and a 1:1 alloy of nickel and iron has a Curie point of 500°C. The wires or ribbons are so cheap that they are often used for only one sample so that there can be no 'memory' or carry-over effects from one sample to the next. In this way, it is easy to control the conditions of a pyrolysis experiment and results are accurately reproducible from one sample of a substance to the next, a feature that is highly desirable for analytical work. After pyrolysis, the volatile components pass through an expansion chamber and then diffuse down a so-called 'molecular beam' tube into the ion source of the mass spectrometer, where some of the pyrolysate is ionized by low-energy electrons (typically 25 eV; see sections 3.2 and 9.3). Un-ionized material is condensed out by a trap cooled in liquid nitrogen (Goodacre, Kell and Bianchi, 1993).

12.4.6. *Resistive (filament) heating*

In resistive heating, an electric DC or AC current is passed through a wire or filament. The heating effect depends on the resistivity of the metal composing the filament. The current or voltage is either applied at set values so that the filament heats up rapidly to the maximum attainable temperature or the current/voltage may be increased in a controlled manner (pre-set rate) up to a pre-set limit. Small samples may be heated directly from being coated onto the filament but larger samples are heated inside quartz or similar tubes, which are heated externally by a filament. For tight control, reproducible heating rates and longevity in an oxidizing (air) atmosphere with wires or filaments, pure metals are best if their resistivity is to remain constant over time. Because the filaments have to withstand repeated heating to high temperatures and to have no or little direct effect on the sample itself, they are usually made from pure platinum or rhenium and are quite expensive. Unlike Curie point heating, in which the heated foil does not come into direct contact with the applied RF electric field, with resistive heating direct from a filament, the sample is in direct contact with the electric current. Because filaments are made from precious metals they are too expensive to throw away after each sample analysis and must be cleaned and re-used.

Unlike Curie point heating and laser pyrolysis, which entail almost instantaneous heating to the required maximum pyrolysis temperature, resistive heating (certainly from a crucible) is slower because of the increased thermal masses involved. Often, sample size is quite large. With extremely rapid heating the pyrolysate is produced faster than or at similar times to a standard mass spectrometer scan time and maybe only one or a few scans of the total pyrolysate can be obtained. With slower heating rates, scans of the mass spectra of the pyrolysate may be taken over a long time period as the pyrolysis develops. This is particularly true of crucible heating or with thermogravimetric analysis. In these instances, temperature ramps of 0.5–20 K s^{-1} are common in contrast to the 10^3–10^6 K s^{-1} of laser and Curie point heating. The slower heating of a sample allows successive new pyrolytic effects to be observed as their onset is passed (time-resolved pyrolysis). For some applications at least, there are substrates for which filament heating gives better results than do some of the other methods. A time-resolved study of styrene/methyl methacrylate co-polymers with a filament pyrolyser and thin samples was found to yield the most reliable results in comparison with other means of pyrolysis (Atkinson and Lehrle, 1991).

As with Curie point and laser heating, few or no ions are formed directly in the pyrolysate during resistive heating and it is necessary to attach an ion source (EI, CI, etc.) to which the pyrolysate must be passed before a mass spectrum can be obtained, or the pyrolysis is performed inside the ion source itself close to the ionizing agency.

12.4.7.   *Laser pyrolysis*

Lasers provide a variety of ways for pyrolysing samples prior to mass spectrometric investigation of the pyrolysate. These various ways are governed by whether the laser beam is highly focussed or not, the wavelength(s) of the radiation, the power delivered by the beam and the continuous or pulsed nature of the irradiation step. At one extreme there can be exceedingly high rates of heating with intense laser pulses, which cause rapid (*flash*) or even almost explosive pyrolysis (*ablation*). At the other extreme with *continuous wave* (CW) heating using a defocussed beam impinging on a relatively large area, heating rates are comparatively low (but still fast compared with standard thermal methods) and the results of heating are more like those of standard fast thermal pyrolyses. The types of information given by the two extremes of operating conditions are usually very different and, for many substances, mixed 'slow' and 'fast' laser pyrolysis (or desorption) occurs, with the results depending on a mixture of the two regimes.

Ablation. Typically, with a pulsed laser operating at, say, 266 nm and focussed on a spot size of diameter 10–25 μm for 10 ns, peak power is approximately 10^8 W cm^{-2} and a heating rate of 10^6 K s^{-1} is attained. Absorption of so much energy into a small sample area leads, through internal conversion into vibrational and kinetic energy, to the irradiated material being ejected or ablated. Indeed, by the time a second laser pulse appears or even before the first has finished, a plasma develops above the irradiation zone and leads to more efficient absorption of light energy (less reflected energy). Because photon density is very high, there are significant chances of more than one photon being absorbed by a molecule before internal energy conversion can return it to the ground electronic state. At a wavelength of 266 nm, absorption of two photons deposits enough energy into a molecule to transform it to an electronic excited state above its ionization energy and an electron is ejected with concomitant formation of an ion. Thus, high-intensity laser irradiation at short wavelength leads to some direct ionization of the pyrolysate as well as to pyrolytic decomposition. Generally, the yield of ions by this approach is not particularly high and most laser ablation methods employ a secondary ionization step to increase ion yield. Once the pyrolysate has been obtained, almost any secondary ionization method can be used and this is often electron ionization, although CI, API, FI and so on have all been utilized.

A further feature of the small area of sample that is heated by the focussed laser beam is the ability to obtain pyrolysis mass spectra from different areas over even a small total sample surface. This allows changes in composition to be examined as the laser spot is moved over the surface; this is surface profiling. However, by irradiating one spot on the surface with successive pulses of laser light, the beam gradually works its way down through the depth of the sample. Therefore, each successive set of mass spectra following the laser pulses examines deeper and deeper into the specimen; this is known as 'depth profiling' and is valuable for determining changes in the composition of a sample throughout its depth. A combination of the two modes of surface and depth profiling is a powerful technique for examining otherwise intractable materials. For example, in forensic science, the total composition of a small paint sample from the scene of an accident can be compared with a small paint chip from a suspect vehicle; the layers of paint can be compared and even their thicknesses can be gauged by the number of laser pulses needed to pass from one layer to the next.

Laser continuous wave (CW) heating. If, instead of the pulsed laser, a continuous wave laser (e.g. an argon ion laser) is used, operating at visible or infrared wavelengths, the sample is still heated over a small spot size but much more slowly because usually less energy is absorbed at longer wavelengths and much is reflected. The spot size is normally somewhat larger (150–200 μm diameter) and the energy input much smaller at about 10^5 W cm^{-2}, giving a heating rate of about 10^3 K s^{-1} somewhat similar to that of fast Curie point heating.

Because these lasers often operate at visible wavelengths, the *colour* of a sample can have a marked effect on the appearance of the pyrolysate, a coloured substance being able to absorb more of the incoming light than a clear one. Thus, two samples of plastic, one coloured and one colourless but otherwise identical, will give different fingerprint mass spectra and could mislead the investigator into thinking that they were different plastics or polymers. This effect of a coloured substance by which visible light is more efficiently absorbed by a material having an absorption band matching the incoming radiation wavelength is the same as that of adding a matrix material to promote desorption and ionization in matrix-assisted laser desorption ionization (MALDI; section 3.11).

Laser desorption is also discussed with further references in section 3.11 on ionization methods. These and many other aspects of laser ablation and pyrolysis, together with details of instrumentation have been reviewed fully, particularly with regard to systems of large molecules such as synthetic and natural polymers (Creasy, 1992; Boon, 1992).

12.4.8. *In-line, off-line and in-source methods of analysis*

Condensation of the pyrolysate is usually a problem in pyrolysis/mass spectrometry, so that regular cleaning of ion sources becomes essential. As material is vaporized during pyrolysis the less volatile and more polar components will tend to separate out and condense onto cooler surfaces. If allowed to remain, this condensation will give rise to memory effects whereby components of a previous sample analysis turn up in the next because traces of this previous sample are swept along as the pyrolysate from the next sample passes through the apparatus. Therefore, various means have been devised to minimize the effects of condensation. Any transfer lines between a pyrolysis chamber and the mass spectrometer ion source are normally heated, as are any connections between the pyrolysis area and any chromatographic column that might be used on-line. The transfer lines are made from material that is as inert as possible, such as quartz or gold-plated metal. Often, an inert carrier gas is used to help sweep the pyrolysate into the mass spectrometer. Once in the ion source of the mass spectrometer, only a fraction of the pyrolysate will actually be ionized and much neutral material will be pumped away by the pumps maintaining the vacuum in the ion source. Also, attached to the source, there may be a trap cooled with liquid nitrogen to condense some of the un-ionized pyrolysate. Despite these precautions, collections of unwanted condensate are an occupational problem for most pyrolysis/mass spectrometrists, who can expect to have to clean out the pyrolysis chamber, transfer lines, traps, ion source and even the analyser of the mass spectrometer much more frequently than they would in 'straightforward' mass spectrometry. When a pyrolysate is produced off-line and then introduced at a later stage into the mass spectrometer or chromatograph/mass spectrometer system, these effects of condensation are much reduced. Another advantage of off-line working is that pyrolysate components can be derivatized before mass spectrometry so that, as demonstrated in chapter 6, advantage can be taken of better chromatographic and/or mass spectrometric performance resulting from the use of a derivative. However, a major disadvantage is that the additional variables introduced by any off-line stage(s) may cause irreproducibility in the data, thereby making subsequent mathematical analysis unreliable (section 12.4.9).

It has been stated that the best place for pyrolysis is right inside the ion source. Thus, non-isothermal, fast heating of small samples within the vacuum of the ion source allows immediate ionization of the pyrolysate as it is produced and, at the same time, the ions so formed are immediately extracted from un-ionized material by the electric fields in the source, which transmit the ions to the mass analyser (Boon, 1992). However, in practice, ion sources for most mass spectrometers are not designed for in-source pyrolysis and condensation onto the walls of the source cannot be prevented; these not only give rise to memory effects but can cause shorting of any high-voltage electrical insulators in or near the source, leading to breakdown in its operation. When in-source pyrolysis is carried out, there is usually a cold trap to prevent too rapid a build-up of condensate on vital parts of the source but, frequently, cleaning of the source is a regular feature of its operation. When chromatographic separation of pyrolysate components is needed, in-source operation is ruled out but MS/MS can still be used for analysis, much as it is used for routine examination of mixtures without prior chromatographic separation (sections 8.8.1 and 8.9). It might be noted that direct chemical ionization (DCI, section 3.6) is regarded as a form of in-source pyrolysis since the sample is heated directly into an ion plasma or may be heated by part of the electron beam striking the sample, causing simultaneous desorption and ionization.

12.4.9.   *Mathematical analysis of the total mass spectrum of a pyrolysate*

For those means of heating, such as the Curie point method, that are so tightly controlled as to give a reproducible total mass spectrum (fingerprint) from sample to sample of any one substance, it is possible to analyse mathematically this fingerprint to provide even more definitive information. For this mathematical analysis to be successful for those cases in which any differences in samples are slight, as with similar strains of micro-organisms, it is essential that there should be an absolute minimum of external differences introduced by the instrumentation or treatment of the samples. A major possible

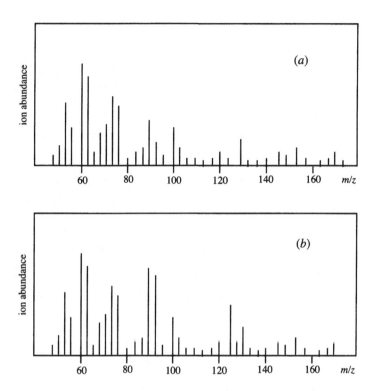

Figure 12.8. Pyrolysis mass spectra of two malt whiskies, (*a*) Glenfiddich and (*b*) Tullibardine. The spectra have been simplified for the purposes of this book but note that all *m/z* values in one spectrum occur in the other but their relative abundances vary, usually not by much but occasionally significantly. The small differences are reproducible through the use of Curie point heating. (Adapted from diagrams kindly supplied by Horizon Instruments.)

difference lies in the means of pyrolysis, for which it is necessary to ensure reproducible heating rates, maximum temperatures and so on. Curie point heating with no prior separation or treatment of the sample has proved to be a very successful approach for subsequent mathematical analysis of the data.

During pyrolysis, volatile compounds with low molecular masses, such as water, hydrogen halides, methane and ammonia, are frequently formed and it is a convenient simplification to ignore *m/z* values of less than 50. On the other hand, materials with high molecular masses are not very abundant compared with those at lower masses and, again, for many analytical purposes it is convenient to ignore *m/z* values above 200. Depending on the interests of the individual investigator, this upper limit is variable. In the example considered here, the fingerprint that is examined is limited to the mass range 51–200 since, with low-energy electron ionization, $z = 1$ and $m/z = m$. Figure 12.8 shows parts of typical *m/z* 51–200 range fingerprints for two different samples of malt whisky.

At the simplest level of application, visual examination or comparison of two total pyrolysis mass spectra may be all that is required to confirm the identity of or difference between two samples. For accurate work it is

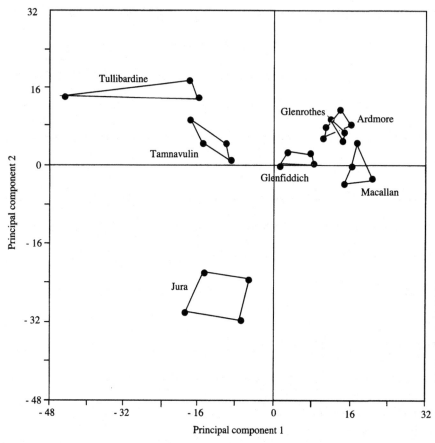

Figure 12.9. A principal components analysis of pyrolysis mass spectra from seven different malt whiskies. This statistical technique identifies correlations within the large data sets resulting from all of the m/z values and abundances of ions in the mass spectra. The plot of the first two principal components is the best two-dimensional representation of the variance within the mass spectra. The seven whiskies are separated into groups but some of the groups are close, as with the Glenrothes and Ardmore types, which actually overlap. (Adapted from diagrams kindly supplied by Horizon Instruments.)

necessary to analyse these raw results mathematically. In one widely used approach, data from multiple scans of a pyrolysate sample are normalized and then subjected to *principal components analysis*, which picks out broad differences or similarities between them (figure 12.9). The initial 150 dimensions (peaks in the range m/z 51–200) are reduced to a relationship in two dimensions whilst at the same time retaining the major (principal) differences in the sample sets (Gutteridge, Vallis and McFie, 1985). Figure 12.9 shows that the various whiskies are separated into groups but some of the groups are not well separated from others. To reveal finer differences, the principal component analysis can be reworked if there is an added piece of information on the likely number of groupings (*canonical variates analysis*). This last bit of information may be known (the number of different specimen samples is known in advance) or may be deduced from an examination of the initial results. Figure 12.10 illustrates a typical canonical variates analysis resulting from the principal components analysis of figure 12.9 (Berkeley, Goodacre, Helyer and Kelley, 1990). Now it can be seen that the malt whiskies are segregated into clearly separated groups.

Whilst the above mathematical deductive analyses are very powerful for revealing first-order (linear) differences and similarities amongst a range of samples, they do not pick out small second-order (non-linear) differences. A second heuristic approach through *artificial neural network* processing may be of benefit. Essentially, artificial neural networking examines data for similarities and differences as does canonical variates analysis but by quite different methods. By utilization of a computer programme that simulates a neural network (a number of computation centres working in unison), the data from a number of standard samples are introduced as a training set, i.e., a set that is used to train the computer in what to look for in the available training spectra. Once the network has been trained, it can be asked to examine mass spectra from pyrolysates of other materials to find out which match and which do not. Importantly, the neural network can estimate the amounts of impurity in the various samples even though these be only small in proportion to the main component (Goodacre and Kell, 1993). These methods have been used for example to identify micro-organisms for biological work, discriminate between paints in forensic science and analyse subtrates, metabolites and products of biochemical processes.

12.4.10. *Some selected applications of pyrolysis/mass spectrometry*
Results and methods of Py/MS are widely scattered in the literature, the journals used often being concerned with the substances being examined rather than with the technique itself. However, there is now a journal devoted entirely to pyrolysis (see section 12.5) and there are regular international conferences. The following examples have been chosen to give some idea of the range of applications, using the techniques discussed above.

Thermogravimetry. To examine a mixture of kaolinite and calcite, a first sample was heated at 15°C min^{-1} from room temperature to about 800°C. The resulting thermogravimetric analysis curve is shown in figure 12.11. The gases given off were swept into a mass spectrometer set for continuous scanning. At the end of the analysis, examination of the resulting mass spectra showed that only two major components at m/z 18 and 44 had been evolved. A second sample was then heated under identical conditions and

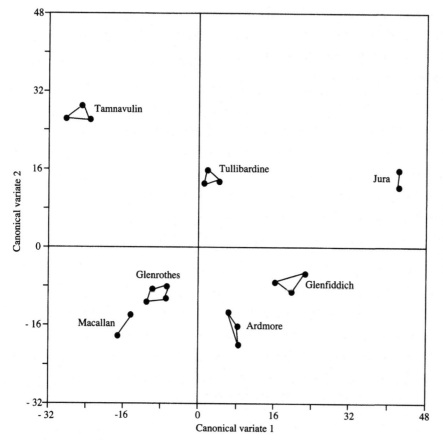

Figure 12.10. A canonical variates analysis from the mass spectra discussed in figure 12.9. This statistical technique aims to maximize the ratio of variation between groups as against within groups, viz., it attempts to differentiate between groups of similar substances. The malt whisky samples are now clearly separated into well-defined sets, which do not overlap. (Adapted from diagrams kindly supplied by Horizon Instruments.)

the mass spectrometer was set to monitor the abundance of ions at m/z 18 (water) and 44 (carbon dioxide); the ion current curves for these two ions are shown in figure 12.11. It can be seen that the first loss in mass (weight reduction) at about 550°C was due to water being evolved from the kaolinite–calcite mixture whereas the second loss at about 730°C was caused by evolution of carbon dioxide. This is a fairly simple example chosen to show how the information from Py/MS is obtained and utilized but exactly the same technique can be employed to examine the curing of polymers or the thermal stability of cured polymers (Redfern, 1991). Thermogravimetry/mass spectrometry (TG/MS) has been discussed (Yun and Meuzelaar, 1991).

In a somewhat different approach, the connection between a thermogravimetric analysis instrument and a mass spectrometer was provided by a short GC capillary column, which allowed rapid transfer of volatile

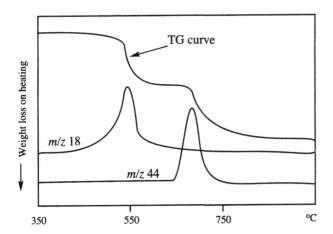

Figure 12.11. Three curves are shown. The thermal gravimetric analysis curve reveals significant losses of mass, recorded as a changes in weight, from a sample of mixed kaolinite and calcite at 551 and 732°C. A second curve gives the abundance of ions at m/z 18 (water), the evolution of which maximizes at 551°C. Similarly, the third curve shows that carbon dioxide (ions at m/z 44) is evolved near 732°C.

components when a styrene/isoprene co-polymer was pyrolysed under helium. The volatile material was swept through the capillary column by the helium and each separated component was ionized in a mass spectrometer by electron ionization. Mass spectra of the individual components allowed them to be identified and the total ion current provided a record of the course of the pyrolysis. The Py/GC/MS separation into individual components was done so as to enhance the degree of information available from an otherwise complex pyrolysate. This approach allows both quantitative and qualitative determination of volatile components, thereby providing kinetic information about specific pyrolytic reaction mechanisms (McClennen, Buchanan, Arnold and Dworzanski, 1993).

Pyrolysis/mass spectrometry and mathematical analysis. (a) In an attempt to distinguish virgin olive oil samples from adulterated oils, 0.5 μl samples of the oils were heated to 530°C on an iron/nickel alloy foil within 0.5 s by Curie point heating. The pyrolysate was allowed to diffuse down a narrow ceramic tube, forming a molecular beam across the electron ionization source of a quadrupole mass spectrometer in which electron ionization was effected at 25 V. Samples from eight virgin olive oils were examined together with the same oils adulterated with soya, sunflower, peanut, corn or rectified olive oil. From the normalized mass spectral results of each sample, a canonical variates analysis was carried out to differentiate them. This statistical analysis revealed differences between the eight types of oil both in virgin and in adulterated samples but was unable to show which samples had been adulterated. However, by application of an artificial neural network analysis, in which the net was trained to distinguish virgin from adulterated oils, it was then possible to find out which olive oil samples were virgin and which adulterated (Goodacre, Kell and Bianchi, 1993). Once the net had been trained, the analysis became routine, accurate and quick.

(b) A similar artificial neural network analytical advance has been applied to the differentiation of micro-organisms (Goodacre, 1994) although, earlier, multivariate analysis alone was shown to be capable of distinguishing even closely related organisms. Thus, Py/MS was used to discriminate the four bacterial species, *B. subtilis, B. pumilus, B. licheniformis* and *B. amyloliquefaciens*. This was a significant success for the pyrolysis approach because few classical methods of microbiology are able to separate these bacteria (Shute, Gutteridge, Norris and Berkeley, 1984). The same type of approach using multivariate analysis following Py/GC/MS was used as a kind of pattern recognition for nucleosides (chromatographic fingerprinting). The nucleosides were heated in a quartz tube by a platinum coil pyrolyser for 5 s at 750°C in helium, passed through a GC column and then into a mass spectrometer for 70 V electron ionization of individual components. The mass spectra allowed recognition of nucleoside bases and the GC patterns could then be used diagnostically for recognizing nucleosides (Sahota and Morgan, 1993).

Laser desorption and ionization. In a novel approach to the examination of lignins by pyrolysis/mass spectrometry, lasers were used first to desorb components from lignin and then to ionize the components by a second tunable laser, using multiphoton absorption (section 9.6.2) to characterize them without formal separation. Thus, an infrared CO_2 laser operating at a wavelength of 10.6 μm was focussed onto a 1.5 mm² area of a lignin sample. The desorbed pyrolysate was transferred to an ion source *via* a supersonic jet, which cooled the components so as to increase subsequent optical resolution and sensitivity. This 'freezing out' also prevents thermal decomposition after laser desorption. In the ion source, a second laser was used to irradiate the cooled pyrolysate over a wavelength range of 272–282 nm. Each of the main molecular lignin components, such as coniferyl alcohol, absorbed photons according to its ultraviolet spectral absorption bands to give electronically excited molecules. With the intense laser irradiation, before the excited molecules could release this excess of energy, a second photon was absorbed. Successive absorption of two photons carried the total excess of internal energy in the molecules above their ionization energies and they transformed into ions by ejection of an electron. This resonance-enhanced multiphoton ionization (REMPI) has been discussed in section 9.6.2. As the laser radiation was tuned through the range 272–282 nm, the molecules responded by ionizing and fragmenting because of the excess of energy absorbed. Therefore, as the molecular ions fragmented, the abundance of the molecular and fragment ions changed because of the induced fragmentation, which varied as a function of irradiating wavelength. The measured variations in the abundances of the different molecular ions as a function of wavelength of irradiation are, in fact, excitation or action spectra (section 9.6.2), which can be used to identify each component. In effect, this is an ultraviolet spectroscopic absorption method of identification in which the absorption of energy is measured not by a photocell but by the fragmentation of ions produced after absorption of light.

12.5.   FURTHER LITERATURE ON MASS SPECTROMETRY

To close this introductory text, pointers to more specialized information are provided to help interested readers extend their knowledge of mass spectrometry. There are four major categories of literature: books, periodicals, journals and compilations of mass spectra, which are discussed in turn. In addition, information is increasingly becoming available as software. Some of these packages are also mentioned here.

Many books on general and specialist mass spectrometry have been referred to at the appropriate points in the text. Other than these, several books may be recommended for the following topics: MS/MS (McLafferty, 1983; Busch, Glish and McLuckey, 1988); time-of-flight mass spectrometry (Cotter, 1994); chemical ionization mass spectrometry (Harrison, 1992); artifacts that turn up with alarming regularity in mass spectra (Middleditch, 1989); ion/molecule reactions (Franklin, 1979); collision processes (Cooks, 1978); mass spectrometry of inorganic and organometallic compounds (Litzow and Spalding, 1973; Jarvis, Gray, Williams and Jarvis, 1990; Holland and Eaton, 1991); characterization of micro-organisms (Fox, Morgan, Larsson and Odham, 1990; Fenselau, 1994); biomedical, clinical, environmental and forensic applications (Waller, 1972; Watson,1976; Waller and Dermer, 1980; Lawson, 1989; Desiderio, 1992, 1994); analysis of organic pollutants in water (Keith, 1976); large molecules (Standing and Ens, 1991);

peptides (Desiderio, 1990); lipids (Murphy, 1993); a brief introduction to bench-top GC/MS (Halket and Rose, 1990); and practical mass spectrometry (Middleditch, 1979; Chapman, 1993). Complementing such books, there is an audio-cassette available from the American Chemical Society on the interpretation of mass spectra, and video tapes on basic instrumentation and interpretation from both the Royal Society of Chemistry and the American Chemical Society.

There are too many periodicals that cover mass spectrometry to be comprehensive here; only the most important ones are mentioned. Three particularly useful sources of information are *Specialist Periodical Reports on Mass Spectrometry* (published by the Royal Society of Chemistry), *Mass Spectrometry Reviews* and biennial reviews in *Analytical Chemistry* (published by the American Chemical Society in each even numbered year). The first publication contains regular chapters on the main fields of mass spectrometry, extending from theory and energetics to methods and applications, and occasional chapters critically appraising specialized areas such as food science, quantification of metals, gas-phase ion clusters and environmental aspects. The second offers thorough and timely reviews of both applied and fundamental mass spectrometric topics. Several references have been made to both of these sources in the preceding pages. The reviews in *Analytical Chemistry* (for instance, Burlingame, Boyd and Gaskell (1994)) cover all areas of mass spectrometry in a highly condensed fashion but form a useful summary of the primary literature. Occasional easy-to-read reviews of selected topics are published by Fisons Instruments as *VG Monographs in Mass Spectrometry*, for example a basic account of LC/MS (Mellon, 1991). Many other periodicals are the published proceedings of conferences and symposia. All aspects of mass spectrometry are covered at the International Conferences on Mass Spectrometry, the proceedings of which are published triennially as *Advances in Mass Spectrometry* (Heyden, Wiley and currently Elsevier), and the *Annual Conferences on Mass Spectrometry and Allied Topics* (American Society for Mass Spectrometry). The invited lectures presented in 1991 at the former (Kistemaker and Nibbering, 1992) were also reproduced in special issues of the *International Journal of Mass Spectrometry and Ion Processes*. There are many other worthwhile and more specialized conferences, such as the International Symposium on Mass Spectrometry in the Health and Life Sciences, the International Conference on Biological Mass Spectrometry, the Montreux Conference on LC/MS (which covers considerably more than LC/MS!), and biennial International Conferences on Pyrolysis, which are reported in the *Journal of Analytical Applied Pyrolysis*. The proceedings of some such conferences are published, as with the first two of these (Burlingame and McCloskey, 1990; Matsuo, Caprioli, Gross and Seyama, 1994). Many published proceedings of symposia are useful for assessing the state of a particular subject, but are not necessarily up-to-date because there may be a considerable delay between the conference and publications resulting from it. It is recommended that the reader takes notice of the date of the conference as well as that of the publication. Many of the research papers presented at such conferences appear in greater detail at a later date in the primary journals, and not infrequently before the conference has begun!

Several excellent journals in the English language are devoted solely to mass spectrometry. The *International Journal of Mass Spectrometry and Ion Processes* (Elsevier) concentrates on theory, energetics, instrumentation and methodology whilst the main themes of *Organic Mass Spectrometry* (Wiley) and *Biological Mass Spectrometry* (formerly *Biomedical and Environmental Mass Spectrometry*, Wiley) are self-evident. From 1995, the latter two journals were replaced by the *Journal of Mass Spectrometry* (Wiley). *Rapid Communications in Mass Spectrometry* (Wiley), *European Mass Spectrometry* (IM Publications) and the *Journal of the American Society for Mass Spectrometry* (Elsevier) are devoted to research papers in all aspects of the science of gas-phase ions. A glance at the bibliography at the end of this book will show that several references have been made to such journals. Some journals containing substantial numbers of research papers on mass spectrometry are the *Journal of Chemical Physics* (American Institute of Physics) concerned with theoretical aspects, *Analytical Chemistry* (American Chemical Society) concentrating on methodology, applications and computers, *Analytical Biochemistry* (Academic Press), the *Journal of Analysis and Applied Pyrolysis* devoted to its eponymous subject, the *Journal of Chromatography* (Elsevier), which contains many reports of gas and liquid chromatography/mass spectrometry as well as capillary electrophoresis/mass spectrometry, and many other journals, too numerous to list here. Frequently, such publications include useful reviews of specialized areas within mass spectrometry.

Most manufacturers of mass spectrometers and allied equipment issue brief publications describing new or interesting analytical methods or developments in instrumentation. Whilst there is clearly an element of self-advertising in these issues, they are nevertheless frequently of excellent value in drawing the reader's attention to advances in mass spectrometric instrumentation and applications. These publications are normally freely available on enquiry to the relevant manufacturer.

Some abstracting journals are devoted to mass spectrometry. Published monthly by the Mass Spectrometry Data Centre of the Royal Society of Chemistry, the *Mass Spectrometry Bulletin* contains a comprehensive list of references, selected from current research literature, that are relevant to mass spectrometry. Each abstract in the list contains key words defining the subject area of the original paper. The key words form the basis of an extensive subject index, making it easy for the reader to find abstracts of interest; a useful author index is also included. More specialized abstracting services are provided by *GC/MS Updates* and *LC/MS Update* (published by HD Science Ltd), which form a valuable aid to literature searching in the combined analytical techniques. Each entry in the *Updates* contains bibliographic information and an abstract that emphasizes the chromatographic and mass spectrometric aspects of the original paper. Entries in the Updates are also collated into a few application areas, such as pesticides, for publication as the annual *Mass Spectrometry Analytes* (HD Science). All of the above abstracting services are available in printed versions and on diskettes. It should be borne in mind that abstracting journals, whilst being very convenient, are necessarily retrospective.

One route towards interpretation of an 'unknown' mass spectrum is to compare it with many reference spectra of known compounds for a possible match. Many data systems contain mass spectral libraries for comparison with mass spectra of unknown compounds (section 4.4.3). Such libraries, supplied by the computer manufacturers, may run to many tens of thousands of reference spectra. If a computer is not available, there are many different sources of published compilations of mass spectra for manual comparisons. Usually these collections contain abridged rather than whole spectra for the sake of brevity. One of the best known collections is the *Eight-Peak Index of Mass Spectra* from the Royal Society of Chemistry. For each compound in the collection, the largest eight peaks and the origin of the mass spectrum, the relative molecular mass, elemental formula, name and CAS Registry Number are listed. The compilation of over 65 000 spectra is arranged in several different ways for easy reference. The mass spectra of over 112 000 compounds are presented in the *Wiley/NBS Registry of Mass*

Spectral Data (McLafferty and Stauffer, 1989). A very large compilation of mass spectra (400 000 entries) is available as a derivative of the *Wiley/NBS Registry of Mass Spectral Data* (McLafferty and Stauffer, 1991). In this latter printed library, mass spectral peaks are listed according to abundance and statistical importance. Its main advantage, other than its sheer size, is that it includes peaks that are deemed to be the most diagnostic for each substance; these are not always the largest. All of the above compilations are general in the sense that there is no bias towards any particular type of compound. Smaller, specialized libraries of mass spectra are also available from various sources. A selection is provided here: environmental contaminants (Hites, 1992); negative-ion electron-capture mass spectra of environmental contaminants (Stemmler and Hites, 1988); drugs, poisons, pesticides, pollutants and their metabolites (Pfleger, Maurer and Weber, 1992); essential oils by ion-trap mass spectrometry (Adams, 1989); derivatized organic acids and body fluid components by GC/MS (Chalmers, Halket and Mills, 1994); derivatized amino acids by GC/MS (Halket, Blau and Down, 1990); and prostaglandins, thromboxanes and leukotrienes (Pace-Asciak, 1989).

Finally, the reader is referred to the International Union of Pure and Applied Chemistry (IUPAC) recommendations on nomenclature for mass spectrometry (Zerbi and Beynon, 1977; Todd, 1991*b*). Unfortunately, in the past, practitioners of mass spectrometry, unrestrained by such standardization, produced much unnecessary symbolism and jargon to confound newcomers and experts alike. Happily, as the expertise of an individual in mass spectrometry increases, it becomes easier to decipher the jargon. We hope that this book will enable the reader to appreciate better the valuable techniques hiding behind such obfuscations.

References

Adams, F., Gijbels, R. and Van Grieken, R. (1988). *Inorganic Mass Spectrometry*. New York: Wiley.

Adams, J. and Gross, M.L. (1987). *Analytical Chemistry*, **59**, 1576–1582.

Adams, R.P. (1989). *Identification of Essential Oils by Ion Trap Mass Spectrometry*. San Diego: Academic.

Adams, R.P., Granat, M., Hogge, L.R. and von Rudloff, E. (1979). *Journal of Chromatographic Science*, **17**, 75–81.

Anderegg, R.J. (1988). *Mass Spectrometry Reviews*, **7**, 395–424.

Ardrey, R.E. (1993). *VG Monographs on Mass Spectrometry*, No. 5. Manchester: VG Instruments.

Arpino, P. (1989). *Mass Spectrometry Reviews*, **8**, 35–55.

Arpino, P. (1990a). *Fresenius' Zeitschrift für analytische Chemie*, **337**, 667–6 85.

Arpino, P. (1990b). *Mass Spectrometry Reviews*, **9**, 631–669.

Asamoto, B. (1988). *Spectroscopy International*, **1**, 28–36.

Asamoto, B. and Dunbar, R.C. (1991). *Analytical Applications of Fourier Transform Ion Cyclotron Resonance Spectroscopy*. New York: VCH.

Ashcroft, A.E, Johnstone, R.A.W., Lowes, S. and Stocks, P.A., (1994). *Abstracts of the 13th International Mass Spectrometry Conference, Budapest, Hungary*, Th C10.

Atkinson, D.J. and Lehrle, R.S. (1991). *Journal of Analysis and Applied Pyrolysis*, **19**, 319–331.

Baillie, T.A. (1978). *Stable Isotopes. Application in Pharmacology, Toxicology and Clinical Research*. London: Macmillan.

Baldwin, M.A. (1995). *Natural Products Reports*, **12**, 33–44.

Barber, M., Bordoli, R.S., Elliott, G.J., Sedgwick, R.D. and Tyler, A.N. (1982). *Analytical Chemistry*, **54**, 645A–657A.

Barber, M., Bordoli, R.S. Sedgwick, R.D. and Tyler, A.N. (1981). *Nature*, **293**, 270–275.

Barber, M. and Elliott, R.M. (1964). 12th ASTM E-14 Meeting on Mass Spectrometry, Montréal.

Bayer, E. (1991). *Angewandte Chemie, International Edition*, **30**, 113–129.

Beauchemin, D. (1991). *Trends in Analytical Chemistry*, **10**, 71–77.

Beckey, H.D. (1961). *Zeitschrift für Naturforschung*, **16a**, 505–510.

Beckey, H.D. (1977). *Principles of Field Ionisation and Field Desorption Mass Spectrometry*. Oxford: Pergamon.

Beckey, H.D. (1979). *Organic Mass Spectrometry*, **14**, 292.

Beckey, H.D. and Röllgen, F.W. (1979). *Organic Mass Spectrometry*, **14**, 188–190.

Beckey, H.D. and Schulten, H.-R. (1975). *Angewandte Chemie, International Edition*, **14**, 403–415.

Benninghoven, A. and Bispinck, H. (1979). *Modern Physics in Chemistry*, **2**, 391–421.

Benninghoven, A., Evans, C.A., McKeegan, K., Storms, H.A. and Werner, H.W. (1990). *Secondary Ion Mass Spectrometry*. New York: Wiley.

Benninghoven, A., Rudenauer, F.G. and Werner, H.W. (1987). *Secondary Ion Mass Spectrometry: Basic Concepts, Instrumental Aspects, Applications and Trends*. New York: Wiley.

Benninghoven, A. and Sichtermann, W.K. (1978). *Analytical Chemistry*, **50**, 1180–1184.

Bentley, T.W. (1979). In *Mass Spectrometry*, vol. 5, ed. R.A.W. Johnstone, pp. 71–74. London: The Chemical Society.

Berkeley, R.C.W., Goodacre, R., Helyer, R. and Kelley, T. (1990). *Laboratory Practice*, **39**, 81–83.

Beynon, J.H. (1960), *Mass Spectrometry and its Application to Organic Chemistry*. Amsterdam: Elsevier.

Beynon, J.H. and Caprioli, R.M. (1980). In *Biochemical Applications of Mass Spectrometry, First Supplementary Volume*, ed. G.R. Waller and O.C. Dermer, pp. 89–102. New York: Wiley-Interscience.

Beynon, J.H., Caprioli, R.M., Baitinger, W.E. and Amy, J.W. (1970). *Organic Mass Spectrometry*, **3**, 455–477.

Beynon, J.H. and Cooks, R.G. (1975). In *MTP International Review of Science, Physical Chemistry Series Two*, vol. 5, ed. A. Maccoll, pp. 159–205. London: Butterworths.

Beynon, J.H., Morgan, R.P. and Brenton, A.G. (1979). *Philosophical Transactions of the Royal Society of London*, Series A, **293**, 157–166.

Beynon, J.H., Saunders, R.A. and Williams, A.E. (1968). *The Mass Spectra of Organic Molecules*. Amsterdam: Elsevier.

Beynon, J.H. and Williams, A.E. (1963). *Mass and Abundance Tables for Use in Mass Spectrometry*. Amsterdam: Elsevier.

Biemann, K. (1962). *Mass Spectrometry, Organic Chemical Applications*. New York: McGraw-Hill.

Biemann, K. (1979). *Journal of Molecular Evolution*, **14**, 65–70.

Biemann, K. and Papayannopoulos, I.A. (1994). *Accounts of Chemical Research*, **27**, 370–378.

Biller, J.E. and Biemann, K. (1974). *Analytical Letters*, 7, 515–528.

Bjoerkhem, I. (1979). *CRC Critical Reviews in Clinical Laboratory Sciences*, 11, 53–105.

Blackley, C.R., Carmody, J.C. and Vestal, M.L. (1980). *Clinical Chemistry*, 26, 1467–1473.

Blakley, C.R. and Vestal, M.L. (1983). *Analytical Chemistry*, 55, 750–754.

Blau, K. and Halket, J. (1993). *Handbook of Derivatives for Chromatography* (2nd edition). London: Wiley.

Blease, T.G., Scrivens, J.H. and Morden, W.E. (1989). *Biomedical and Environmental Mass Spectrometry*, 18, 775–779.

Boerboom, A.J.H. (1990). *Rapid Communications in Mass Spectrometry*, 4, 385–395.

Boerboom, A.J.H. (1991). *Organic Mass Spectrometry*, 26, 929–935.

Boon, J.J. (1992). *International Journal of Mass Spectrometry and Ion Processes*, 118, 755–787.

Bordoli, R.S. and Bateman, R.H. (1992). *International Journal of Mass Spectrometry and Ion Processes*, 122, 243–254.

Bowers, W.D., Delbert, S. and McIver, R.T. (1986). *Analytical Chemistry*, 58, 969.

Bowie, J.H. (1971). In *Mass Spectrometry*, vol. 1, ed. D.H. Williams, pp. 91–138. London: The Chemical Society.

Bowie, J.H. (1973). In *Mass Spectrometry*, vol. 2, ed. D.H. Williams, pp. 90–142. London: The Chemical Society.

Bowie, J.H. (1975). In *Mass Spectrometry*, vol. 3, ed. R.A.W. Johnstone, pp. 288–295. London: The Chemical Society.

Bowie, J.H. (1977). In *Mass Spectrometry*, vol. 4, ed. R.A.W. Johnstone, pp. 237–241. London: The Chemical Society.

Bowie, J.H. (1979). In *Mass Spectrometry*, vol. 5, ed. R.A.W. Johnstone, pp. 279–284. London: The Chemical Society.

Bowie, J.H. (1984). In *Mass Spectrometry*, vol. 7, ed. R.A.W. Johnstone, pp. 151–167. London: The Royal Society of Chemistry.

Bowie, J.H. (1985). In *Mass Spectrometry*, vol. 8, ed. M.E. Rose, pp. 161–183. London: The Royal Society of Chemistry.

Bowie, J.H. (1987). In *Mass Spectrometry*, vol. 9, ed. M.E. Rose, pp. 172–195. London: The Royal Society of Chemistry.

Bowie, J.H., Trenerry, V.C. and Klass, G. (1981). In *Mass Spectrometry*, vol. 6, ed. R.A.W. Johnstone, pp. 233–240. London: The Royal Society of Chemistry.

Bowie, J.H. and Williams, B.D. (1975). In *MTP International Review of Science, Physical Chemistry Series Two*, vol. 5, ed. A. Maccoll, pp. 89–127. London: Butterworths.

Boyd, R.K. and Beynon, J.H. (1977). *Organic Mass Spectrometry*, 12, 163–165.

Brenna, J.T. (1994). *Accounts of Chemical Research*, 27, 340–346.

Bricker, D.L., Adams, T.A. and Russell, D.H. (1983). *Analytical Chemistry*, 55, 2417–2418.

Brooks, C.J.W. Edmonds, C.G., Gaskell, S.J. and Smith, A.G. (1978). *Chemistry and Physics of Lipids*, **21**, 403–416.

Brooks, C.J.W. and Middleditch, B.S. (1975). In *Mass Spectrometry*, vol.3, ed. R.A.W. Johnstone, pp. 296–338. London: The Chemical Society.

Brooks, C.J.W. and Middleditch, B.S. (1977). In *Mass Spectrometry*, vol. 4, ed. R.A.W. Johnstone, pp. 146–185. London: The Chemical Society.

Brooks, C.J.W. and Middleditch, B.S. (1979). In *Mass Spectrometry*, vol. 5, ed. R.A.W. Johnstone, pp. 142–185. London: The Chemical Society.

Brown, M.A. (1990). *Liquid Chromatography/Mass Spectrometry. Applications in Agricultural, Pharmaceutical and Environmental Chemistry*. Washington: American Chemical Society.

Brown, P. and Djerassi, C. (1967). *Angewandte Chemie, International Edition*, **6**, 477–496.

Bruins, A.P. (1991). *Mass Spectrometry Reviews*, **10**, 53–77.

Bruins, A.P., Covey, T.R. and Henion, J.D. (1987). *Analytical Chemistry*, **59**, 2642–2646.

Bruins, A.P., Jennings, K.R. and Evans, S. (1978). *International Journal of Mass Spectrometry and Ion Physics*, **26**, 395–404.

Budzikiewicz, H. (1986). *Mass Spectrometry Reviews*, **5**, 345–380.

Budzikiewicz, H., Djerassi, C. and Williams, D.H. (1964*a*). *Interpretation of Mass Spectra of Organic Compounds*. San Francisco: Holden-Day.

Budzikiewicz, H., Djerassi, C. and Williams, D.H. (1964*b*). *Structure Elucidation of Natural Products*, vols. 1 and 2. San Francisco: Holden-Day.

Budzikiewicz, H., Djerassi, C. and Williams, D.H. (1967). *Mass Spectrometry of Organic Compounds*, San Fransisco: Holden-Day.

Burlingame, A.L., Boyd, R.K. and Gaskell, S.J. (1994). *Analytical Chemistry*, **66**, 634R–683R.

Burlingame, A.L. and McCloskey, J.A. (1990). *Biological Mass Spectrometry*. Amsterdam: Elsevier.

Busch, K.L., (1987). *Trends in Analytical Chemistry*, **6**, 95–100.

Busch, K.L., (1990). *Chromatographic Science (Handbook of Thin-layer Chromatography)*, **55**, 183–209.

Busch, K.L., Brown, S.M., Doherty, S.J., Dunphy, J.C. and Buchanan, M.V. (1990). *Journal of Liquid Chromatography*, **13**, 2841–2869.

Busch, K.L., Glish, G.L. and McLuckey, S.A. (1988). *Mass Spectrometry/Mass Spectrometry: Techniques and Applications of Tandem Mass Spectrometry*. New York: VCH.

Busch, K.L., Unger, S.E., Vincze, A., Cooks, R.G. and Keough, T. (1982). *Journal of the American Chemical Society*, **104**, 1507–1511.

Campana, J.E. (1980). *International Journal of Mass Spectrometry and Ion Physics*, **33**, 101–117.

Caprioli, R.M. (1990*a*). *Analytical Chemistry*, **62**, 477A–485A.

Caprioli, R.M. (1990*b*). *Continuous-flow Fast Atom Bombardment Mass Spectrometry*. New York: Wiley.

Caprioli, R.M. and Bier, D.M. (1980). In *Biochemical Applications of Mass Spectrometry, First Supplementary Volume*, ed. G.R. Waller and O.C. Dermer, pp. 895–925. New York: Wiley-Interscience.

Carroll, D.I., Dzidic, I., Stillwell, R.N., Haegele, K.D. and Horning, E.C. (1975). *Analytical Chemistry*, **47**, 1308–1312.

Catlow, D.A. and Rose, M.E. (1989). In *Mass Spectrometry*, vol. 10, ed. M.E. Rose, pp. 222–252. London: The Royal Society of Chemistry.

Chalmers, R.A., Halket, J.M. and Mills, G.A. (1994). *GC/MS Companion – Derivatized Organic Acids and Body Fluid Components*. Nottingham: HD Science.

Chapman, J.R. (1978). *Computers in Mass Spectrometry*. London: Academic Press.

Chapman, J.R. (1985). In *Mass Spectrometry*, vol. 8, ed. M.E. Rose, pp. 123–140. London: The Royal Society of Chemistry.

Chapman, J.R. (1987). In *Mass Spectrometry*, vol. 9, ed. M.E. Rose, pp. 143–171. London: The Royal Society of Chemstry.

Chapman, J.R. (1989). In *Mass Spectrometry*, vol. 10, ed. M.E. Rose, pp. 118–144. London: The Royal Society of Chemistry.

Chapman, J.R. (1993). *Practical Organic Mass Spectrometry*. Chichester: Wiley.

Chesnavich, W.J. and Bowers, M.T. (1978). *Journal of Chemical Physics*, **68**, 901–910.

Claeys, M., Muscettola, G. and Markey, S.P. (1976). *Biomedical Mass Spectrometry*, **3**, 110–116.

Cochran, R.L. (1986). *Applied Spectroscopy Reviews*, **22**, 137–187.

Cody, R.B. (1988). *Analytical Chemistry*, **60**, 917–923

Cole, M.J. and Enke, C.G. (1991). *Analytical Chemistry*, **63**, 1032–1038.

Conzemius, R.J. and Capellen, J.M. (1980). *International Journal of Mass Spectrometry and Ion Physics*, **34**, 197–271.

Cooks, R.G. (1978). *Collision Spectroscopy*. New York: Plenum Press.

Cooks, R.G., Beynon, J.H., Caprioli, R.M. and Lester, G.R. (1973). *Metastable Ions*. Amsterdam: Elsevier.

Cotter, R.J. (1979). *Analytical Chemistry*, **51**, 317–318.

Cotter, R.J. (1988). *Analytical Chemistry*, **60**, 781A–793A.

Cotter, R.J. (1994). *Time-of-flight Mass Spectrometry*. ACS Symposium Series, vol. 549. Washington: American Chemical Society.

Cottrell, T.L. (1965). *Dynamic Aspects of Molecular Energy States*, pp. 32–44. Edinburgh: Oliver and Boyd.

Covey, T.R., Huang, E.C. and Henion, J.D. (1991). *Analytical Chemistry*, **63**, 1193–1200.

Craig, A.G. and Derrick, P.J. (1986). *Australian Journal of Chemistry*, **29**, 1421–1434.

Creaser, C.S. (1988). *European Spectroscopy News*, **80**, 13–18.

Creaser, C.S., McCoustra, M.R.S. and O'Neill, K.E. (1991). *Organic Mass Spectrometry*, **26**, 335–338.

Creasy, W.R. (1992). *Polymer*, **33**, 4486–4492.

Curtiss, L.A., Raghavachari, K., Trucks, G.W. and Pople, J.A. (1991). *Journal of Chemical Physics*, **94**, 7221–7230.

Danis, P.O. (1991). *Abstracts of the 12th International Mass Spectrometry Conference*, Amsterdam, The Netherlands, p. 211.

Daves, G.D., Jr (1979). *Accounts of Chemical Research*, **12**, 359–365.

Dawson, P.H. (1976). *Quadruple Mass Spectrometry and its Applications*. Amsterdam: Elsevier.

Dawson, P.H. (1986). *Mass Spectrometry Reviews*, **5**, 1–37.

Dekrey, M.J., Mabud, Md. A., Cooks, R.G. and Syka, J.E.P. (1985). *International Journal of Mass Spectrometry and Ion Processes*, **67**, 295–303.

de Leenheer, A.P. and Cruyl, A.A. (1980). In *Biochemical Applications of Mass Spectrometry, First Supplementary Volume*, ed. G.R. Waller and O.C. Dermer, pp. 1169–1207. New York: Wiley-Interscience.

Dell, A. and Rogers, M.E. (1989). *Trends in Analytical Chemistry*, **8**, 375–378.

De Pauw, E. (1986). *Mass Spectrometry Reviews*, **5**, 191–212.

De Pauw, E. (1989). *Advances in Mass Spectrometry*, **11**, 383–398.

Derrick, P.J. (1977). In *Mass Spectrometry*, vol. 4, ed. R.A.W. Johnstone, pp. 132–145. London: The Chemical Society.

Desiderio, D.M. (1990). *Mass Spectrometry of Peptides*. Boca Raton, FL: CRC Press.

Desiderio, D.M. (1992). *Mass Spectrometry: Clinical and Biomedical Applications*, vol. 1. New York: Plenum.

Desiderio, D.M. (1994). *Mass Spectrometry: Clinical and Biomedical Applications*, vol. 2. New York: Plenum.

Despeyroux, D., Wright, A.D., Jennings, K.R., Evans, S. and Riddoch, A. (1992). *Proceedings of the Second European Meeting on Tandem Mass Spectrometry*, July, Warwick, U.K., p. 50.

Detter, L.D., Hand, O.W., Cooks, R.G. and Walton, R.A. (1988). *Mass Spectrometry Reviews*, 7, 465–502.

Dewar, M.J.S. and Worley, S.D. (1969). *Journal of Chemical Physics*, **50**, 654–667.

Dorey, R.C. (1992). In *Mass Spectrometry in the Biological Sciences*, ed. M.L. Gross, pp. 79–92. Dordrecht: Kluwer Academic.

Dromey, R.G., Buchanan, B.G., Smith, D.H. Lederberg, J. and Djerassi, C. (1975). *Journal of Organic Chemistry*, **40**, 770–774.

Dromey, R.G. and Foyster, G.T. (1980). *Analytical Chemistry*, **52**, 394–398.

Dromey, R.G., Stefik, M.J., Rindfleisch, T.C. and Duffield, A.M. (1976). *Analytical Chemistry*, **48**, 1368–1375.

Esteban, N.V., Liberato, D.J., Sidbury, J.B. and Yergey, A.L. (1987). *Analytical Chemistry*, **59**, 1674–1677.

Evershed, R.P. (1987). In *Mass Spectrometry*, vol. 9, ed. M.E. Rose, pp. 196–263. London: The Royal Society of Chemistry.

Evershed, R.P. (1989). In *Mass Spectrometry*, vol. 10, ed. M.E. Rose, pp. 181–221. London: The Royal Society of Chemistry.

Fenn, J.B., Mann, M., Meng, C.K., Wong, S.F. and Whitehouse, C.M. (1989). *Science*, **246**, 64–70.

Fenn, J.B., Mann, M., Meng, C.K., Wong, S.F. and Whitehouse, C.M. (1990). *Mass Spectrometry Reviews*, **9**, 37–70.

Fenselau, C. (1994). *Mass Spectrometry for the Characterization of Microorganisms*. ACS Symposium Series, vol. 541. Washington: American Chemical Society.

Fenselau, C. and Cotter, R.J. (1987). *Chemical Reviews*, **87**, 501–512.

Field, F.H. and Franklin, J.L. (1957). *Electron Impact Phenomena*. New York: Academic Press.

Fox, A., Morgan, S.L., Larsson, L. and Odham, G. (1990). *Analytical Microbiology Methods: Chromatography and Mass Spectrometry*. New York: Plenum.

Franklin, J.L. (1979). *Ion–Molecule Reactions, Parts I and II*. Stroudsburg: Hutchinson & Ross.

Franzen, J., Küper, H. and Riepe, W. (1974). *Analytical Chemistry*, **46**, 1683–1690.

Freiser, B.S. (1988). In *Techniques for the Study of Ion/Molecule Reactions*, Vol. 20, ed. J.M. Farrar and W.H. Saunders, pp. 61–118. New York: Wiley.

Futrell, J.H., Ryan, K. and Siek, L.W. (1965). *Journal of Chemical Physics*, **43**, 1832–1833.

Garland, W.A. and Barbalas, M.P. (1986). *Journal of Clinical Pharmacology*, **26**, 412–418.

Garland, W.A. and Min, B.H. (1979). *Journal of Chromatography*, **172**, 279–286.

Gaskell, S.J. and Finlay, E.M.H. (1988). *Trends in Analytical Chemistry*, **7**, 202–208.

Gaskell, S.J., Finney, R.W. and Harper, M.E. (1979). *Biomedical Mass Spectrometry*, **6**, 113–116.

Gaskell, S.J., Pike, A.W. and Millington, D.S. (1979). *Biomedical Mass Spectrometry*, **6**, 78–81.

Gelpi, E. (1986). *Advances in Mass Spectrometry 1985*, 397–415.

Geno, P.W. (1992). In *Mass Spectrometry in the Biological Sciences*, ed. M.L. Gross, pp. 133–142. Dordrecht: Kluwer Academic.

Gioia, B., Franzoi, L. and Arlandini, E. (1991). *Abstracts of the 12th International Mass Spectrometry Conference*, Amsterdam, The Netherlands, August, poster B05.

Glish, G.L. and Goeringer, D.E. (1984). *Analytical Chemistry*, **56**, 2291–2295.

Glish, G.L., McLuckey, S.A., McBay, E. and Bertram, L.K. (1986). *International Journal of Mass Spectrometry and Ion Processes*, **70**, 321–338.

Glish, G.L., McLuckey, S.A. and McKown, H.S. (1987). *Analytical Instrumentation*, **16**, 191–206.

Goodacre, R. (1994). *Microbiology Europe*, **2**, 16–22.

Goodacre, R. and Kell, D.B. (1993). *Analytica Chimica Acta*, **279**, 17–26.

Goodacre, R., Kell, D.B. and Bianchi, G. (1993). *Journal of Science of Food and Agriculture*, **63**, 297–307.

Gower, J.L. (1985). *Biomedical Mass Spectrometry*, **12**, 191–196.

Gray, A.L. and Date, A.R. (1983). *Analyst*, **108**, 1033–1050.

Grob, R.L. (1985). *Modern Practice of Gas Chromatography*. New York: Wiley.

Grönneberg, T.O., Gray, N.A.B. and Eglington, G. (1975). *Analytical Chemistry*, **47**, 415–419.

Gross, M.L. (1994). *Accounts of Chemical Research*, **27**, 361–369.

Gutteridge, C.S., Vallis, L. and MacFie, H.J.H. (1985). In *Computer-Assisted Bacterial Systematics*, ed. M. Goodfellow, D. Jones and F. Priest, pp. 369–401. London: Academic Press.

Halket, J.M., Blau, K. and Down, S. (1990). *GC/MS Companion – Derivatized Amino Acids*. Nottingham: HD Science.

Halket, J.M. and Rose, M.E. (1990). *Introduction to Bench-top GC/MS*. Nottingham: HD Science.

Hammond, G.S. (1955). *Journal of the American Chemical Society*, **77**, 334–338.

Hanson, C.D., Kerley, E.L. and Russell, D.H. (1989). In *Treatise on Analytical Chemistry*, Part 1, vol. 11, ed. I.M. Kolthoff, J.D. Winefordner and M.M. Bursey, pp. 117–187. New York: Wiley; idem *Analytical Chemistry*, **61**, 83–85.

Harrison, A.G. (1989). *Advances in Mass Spectrometry*, **11**, 582–595.

Harrison, A.G. (1992). *Chemical Ionization Mass Spectrometry*. Boca Raton, FL: CRC Press.

Harrison, W.W., Barshick, C.M., Klingler, J.A., Ratliff, P.H. and Mei, Y. (1990). *Analytical Chemistry*, **62**, 943A–949A.

Harrison, W.W. and Donohue, D.L. (1989). In *Treatise on Analytical Chemistry*, Part 1, vol. 11, ed. I.M. Kolthoff, J.D. Winefordner and M.M. Bursey, pp. 189–235. New York: Wiley.

Hertz, H.S., Hites, R. and Biemann, K. (1971). *Analytical Chemistry*, **43**, 681–691.

Hill, H.C. (1972). *Introduction to Mass Spectrometry*. London: Heyden.

Hillenkamp, F. (1989). *Advances in Mass Spectrometry*, **11**, 354–362.

Hillenkamp, F. and Ehring, H. (1992). In *Mass Spectrometry in the Biological Sciences*, ed. M.L. Gross, pp. 165–179. Dordrecht: Kluwer Academic.

Hillenkamp, F. and Karas, M. (1990). *Methods in Enzymology*, **193**, 280–295.

Hiraoka, K. and Kudaka, I. (1990). *Rapid Communications in Mass Spectrometry*, **4**, 519–526.

Hites, R.A. (1992). *Handbook of Mass Spectra of Environmental Contaminants*. Boca Raton, FL: Lewis.

Holland, J.F. (1979). *Organic Mass Spectrometry*, **14**, 291.

Holland, J.F., Soltmann, B. and Sweeley, C.C. (1976). *Biomedical Mass Spectrometry*, **3**, 340–345.

Holland, J.G. and Eaton, A. (1991). *Applications of Plasma Source Mass Spectrometry.* Cambridge: The Royal Society of Chemistry.

Holmes, J.L. (1975). In *MTP International Review of Science, Physical Chemistry Series Two*, vol. 5, ed. A. Maccoll, pp. 207–287. London: Butterworths.

Holmes, J.L. and Terlouw, J.K. (1980). *Organic Mass Spectrometry*, **15**, 383–396.

Horning, E.C., Carroll, D.I., Dzidic, I., Haegele, K.D., Horning, M.D. and Stillwell, R.N. (1974). *Journal of Chromatography*, **99**, 13–21.

Horning, E.C., Mitchell, J.R., Horning, M.G., Stillwell, W.G., Stillwell, R.N., Nowlin, J.G. and Carroll, D.I. (1979). *Trends in Pharmacological Science*, **1**, 76–81.

Houk, R.S. (1994). *Accounts of Chemical Research*, **27**, 333–339.

Howe, I. (1973). In *Mass Spectrometry*, vol. 2, ed. D.H. Williams, pp. 34–42. London: The Chemical Society.

Hsu, J-P. (1993). *VG Monographs in Mass Spectrometry*, No. 7. Manchester: VG Organic (now Micromass UK Ltd).

Huang, E.C., Wachs, T., Conboy, J.J. and Henion, J.D. (1990). *Analytical Chemistry*, **62**, 713A–725A.

Hunt, D.F. (1982). *International Journal of Mass Spectrometry and Ion Physics*, **45**, 111–123.

Hunt, D.F. and Crow, F.W. (1978). *Analytical Chemistry*, **50**, 1781–1784.

Hunt, D.F., Stafford, Jr, G.C., Crow, F.W. and Russell, J.W. (1976). *Analytical Chemistry*, **48**, 2098–2105.

Iribarne, J.V., Dziedzic, P.J. and Thompson, B.A. (1983). *International Journal of Mass Spectrometry and Ion Physics*, **50**, 331–347.

Iribarne, J.V. and Thompson, B.A. (1976). *Journal of Chemical Physics*, **64**, 2287–2284.

Isenhour, T.L., Kowalski, B.R. and Jurs, P.C. (1974). *CRC Critical Reviews in Analytical Chemistry*, **4**, 1–44.

Ishii, D. and Takeuchi, T. (1989). *Trends in Analytical Chemistry*, **8**, 25–29.

Ito, Y., Takeuchi, T., Ishii, D., Goto, M. and Mizuno, T. (1986). *Journal of Chromatography*, **358**, 201–207.

Jacoby, C.B., Holliman, C.L. and Gross, M.L. (1992). In *Mass Spectrometry in the Biological Sciences*, ed. M.L. Gross, pp. 93–116. Dordrecht: Kluwer Academic; idem, *NATO ASI Series, Series C*, **353**, 93–116.

Jaeger, H. (1987). *Capillary Gas Chromatography Mass Spectrometry in Medicine and Pharmacology.* New York: Huethig.

Jarvis, K.E., Gray, A.L., Williams, J.G. and Jarvis, I. (1990). *Plasma Source Mass Spectrometry.* Cambridge: The Royal Society of Chemistry.

Jennings, K.R. (1965). *Journal of Chemical Physics*, **43**, 4176–4177.

Jennings, K.R. (1971). In *Mass Spectrometry. Techniques and Applications*, ed. G.W.A. Milne, pp. 419–458. New York: Wiley-Interscience.

Jennings, K.R. (1977). In *Mass Spectrometry*, vol. 4, ed. R.A.W. Johnstone, pp. 203–216. London: The Chemical Society.

Jennings, K.R. (1979). *Philosophical Transactions of the Royal Society of London*, Series A, **293**, 125–133.

Jinno, K. (1992). *Hyphenated Techniques in Supercritical Fluid Chromatography and Extraction*. Amsterdam: Elsevier.

Johnson, J.V., Yost, R.A., Kelley, P.E. and Bradford, D.C. (1990). *Analytical Chemistry*, **62**, 2162–2172.

Johnstone, R.A.W. (1979). In *Mass Spectrometry*, vol. 5, ed. R.A.W. Johnstone, pp. 1–63. London: The Chemical Society.

Jolly, W.L. and Gin, C. (1977). *International Journal of Mass Spectrometry and Ion Physics*, **25**, 27–37.

Jonsson, G.P., Hedin, A.B., Hakansson, P.L., Sundqvist, B.U.R., Save, B.G.S., Nielsen, P.F., Roepstorff, P., Johansson, K.-E., Kamensky, I. and Lindberg, M.S.L. (1986). *Analytical Chemistry*, **58**, 1084–1087.

Justice, J.B. and Isenhour, T.L. (1974). *Analytical Chemistry*, **46**, 223–226.

Karas, M., Bahr, U., Ingendoh, A. and Hillenkamp, F. (1989). *Angewandte Chemie, International Edition*, **28**, 760–761.

Karni, M. and Mandelbaum, A. (1980). *Organic Mass Spectrometry*, **15**, 53–64.

Keith, L.H. (1976). *Identification and Analysis of Organic Pollutants in Water*. Ann Arbor: Ann Arbor Science.

Kimble, B.J. (1978). In *High Performance Mass Spectrometry*, ed. M.L. Gross, pp. 120–149. Washington: American Chemical Society.

Kiser, R.W. (1965). *Introduction to Mass Spectrometry and its Applications*. New Jersey: Prentice-Hall.

Kistemaker, P.G. and Nibbering, N.M.M. (1992). *Advances in Mass Spectrometry*, vol. 12. Amsterdam: Elsevier.

Klein, E.R. and Klein, P.D. (1979). *Stable Isotopes – Proceedings of the Third International Conference*. New York: Academic Press.

Klots, C.E. (1976). *Journal of Chemical Physics*, **64**, 4269–4275.

Knapp, D.R. (1979). *Handbook of Analytical Derivatization Reactions*. New York: Wiley and Sons.

Knewstubb, P.F. (1969). *Mass Spectrometry and Ion–Molecule Reactions*. London: Cambridge University Press.

Kwok, K.-S., Venkataraghavan, R. and McLafferty, F.W. (1973). *Journal of the American Chemical Society*, **95**, 4185–4194.

Lacey, M.J. and Macdonald, C.G. (1979). *Analytical Chemistry*, **51**, 691–695.

Larsen, B.S. (1990). In *Mass Spectrometry of Biological Materials*, ed. C.N. McEwen and B.S. Larsen, pp. 197–214. New York: Marcel Dekker.

Lattimer, R.P., Muenster, H. and Budzikiewicz, H. (1989). *International Journal of Mass Spectrometry and Ion Processes*, **90**, 119–129.

Lattimer, R.P. and Schulten, H.-R. (1989). *Analytical Chemistry*, **61**, 1201A–1215A.

Lawson, A.M. (1989). *Clinical Biochemistry. Principles, Methods and Applications. Volume 1. Mass Spectrometry*. Berlin: De Gruyter.

Le Beyec, Y. (1989). *Advances in Mass Spectrometry*, **11**, 126–145.

Lederberg, J. (1964). *Compilation of Molecular Formulas for Mass Spectrometry*. San Francisco: Holden-Day.

Lee, E.D., Mueck, W., Henion, J.D. and Covey, T.R. (1988). *Journal of Chromatography*, **458**, 313–321.

Lee, E.D., Mueck, W., Henion, J.D. and Covey, T.R. (1989). *Biomedical and Environmental Mass Spectrometry*, **18**, 844–850.

Lehman, T.A. and Bursey, M.M. (1976). *Ion Cyclotron Resonance Spectroscopy*. New York: Wiley.

Light, J.C. (1967). *Discussion of the Faraday Society*, No. 44, 12–29.

Litzow, M.R. and Spalding, T.R. (1973). *Mass Spectrometry of Inorganic and Organometallic Compounds*. Amsterdam: Elsevier.

Louris, J.N., Wright, L.G., Cooks, R.G. and Schoen, A.E. (1985). *Analytical Chemistry*, **57**, 2918–2924.

Lubman, D.M. (1990). *Lasers and Mass Spectrometry*. Oxford: Oxford University Press.

Lyon, P.A. (1985). *Desorption Mass Spectrometry: Are SIMS and FAB the Same?* ACS Symposium Series, vol. 291. Washington: American Chemical Society.

Mabud, Md. A., Dekrey, M.J. and Cooks, R.G. (1985). *International Journal of Mass Spectrometry and Ion Processes*, **67**, 285–294.

McClennen, W.H., Buchanan, R.M., Arnold, N.S. and Dworzanski, J.P. (1993). *Analytical Chemistry*, **65**, 2819–2823.

McFadden, W.H. (1973). *Techniques of Combined Gas Chromatography/Mass Spectrometry*. New York: Wiley-Interscience.

Macfarlane, R.D. (1988). *Trends in Analytical Chemistry*, **7**, 179–183.

Macfarlane, R.D., Hill, J.C., Jacobs, D.L. and Geno, P.W. (1989). *Advances in Mass Spectrometry*, **11**, 3–21.

Macfarlane, R.D. and Torgerson, D.F. (1976). *International Journal of Mass Spectrometry and Ion Physics*, **21**, 81–92.

McIver, R.T., Hunter, R.L. and Bowers, W.D. (1985). *International Journal of Mass Spectrometry and Ion Processes*, **64**, 66–77.

McKelvey, J.M., Alexandratos, S., Streitwieser, A. Abboud, J.L.M. and Hehre, W.J. (1976). *Journal of the American Chemical Society*, **98**, 244–246.

McLafferty, F.W. (1963). *Mass Spectral Correlations.* Washington: American Chemical Society.

McLafferty, F.W. (1977). *Analytical Chemistry,* **49,** 1441–1443.

McLafferty, F.W. (1980). *Accounts of Chemical Research,* **13,** 33–39.

McLafferty, F.W. (1983). *Tandem Mass Spectrometry.* New York: Wiley.

McLafferty, F.W. and Stauffer, D.B. (1989). *Wiley/NBS Registry of Mass Spectral Data.* New York: Wiley and Son.

McLafferty, F.W. and Stauffer, D.B. (1991). *Important Peak Index of the Registry of Mass Spectral Data.* New York: Wiley and Son.

McLafferty, F.W. and Turecek, F. (1993). *Interpretation of Mass Spectra.* Mill Valley: University Science Books.

McLuckey, S.A. (1984). *Organic Mass Spectrometry,* **19,** 545–550.

McLuckey, S.A., Ouwerkerk, C.E.D., Boerboom, A.J.H. and Kistemaker, P.G. (1984). *International Journal of Mass Spectrometry and Ion Processes,* **59,** 85–101.

Majer, J.R. and Boulton, A.A. (1970). *Nature,* **225,** 658–660.

Mann, M. (1990). *Organic Mass Spectrometry,* **25,** 575–587.

Mann, M. and Fenn, J.B. (1992). In *Mass Spectrometry: Clinical and Biochemical Applications,* vol. 1, ed. D.M. Desiderio, pp. 1–35. New York: Plenum Press.

March, R.E. (1991). *Organic Mass Spectrometry,* **26,** 627–632.

March, R.E. and Hughes, R.J. (1989). *Quadrupole Storage Mass Spectrometry.* New York: Wiley.

Marcus, R.A. (1952). *Journal of Chemical Physics,* **20,** 359–364.

Marcus, R.A. (1975). *Journal of Chemical Physics,* **62,** 1372–1384.

Marshall, A.G. and Grosshans, P.B. (1991). *Analytical Chemistry,* **63,** 215A–229A.

Marshall, A.G. and Schweikhard, L. (1992). *International Journal of Mass Spectrometry and Ion Processes,* **118/119,** 37–70.

Massey, H.S.W. (1976). *Negative ions.* London: Cambridge University Press.

Mather, R.E. and Todd, J.F.J. (1979). *International Journal of Mass Spectrometry and Ion Physics,* **30,** 1–37.

Matsuo, T., Caprioli, R.M., Gross, M.L. and Seyama, Y. (1994). *Biological Mass Spectrometry: Present and Future.* New York: Wiley.

Meili, J., Walls, F.C., McPherron, R. and Burlingame, A.L. (1979). *Journal of Chromatographic Science,* **17,** 29–42.

Meisel, W.S. (1972). *Computer Orientated Approaches to Pattern Recognition.* New York: Academic Press.

Mellon, F.A. (1975). In *Mass Spectrometry,* vol. 3, ed. R.A.W. Johnstone, pp. 117–142. London: The Chemical Society.

Mellon, F.A. (1977). In *Mass Spectrometry,* vol. 4, ed. R.A.W. Johnstone, pp. 59–84. London: The Chemical Society.

Mellon, F.A. (1979). In *Mass Spectrometry*, vol. 5, ed. R.A.W. Johnstone, pp. 100–120. London: The Chemical Society.

Mellon, F.A. (1981). In *Mass Spectrometry*, vol. 6, ed. R.A.W. Johnstone, pp. 196–232. London: The Royal Society of Chemistry.

Mellon, F.A. (1991). *VG Monographs in Mass Spectrometry*, No. 2. Manchester: VG Organics.

Mellon, F.A., Chapman, J.R. and Pratt, J.A.E. (1987). *Journal of Chromatography*, **394**, 209–222.

Message, G.M. (1984). *Practical Aspects of GC/MS*. New York: Wiley.

Middleditch, B.S. (1979). *Practical Mass Spectrometry*. New York: Plenum Press.

Middleditch, B.S. (1989). *Analytical Artifacts*. Amsterdam: Elsevier.

Millard, B.J. (1978a). *Quantitative Mass Spectrometry*. London: Heyden.

Millard, B.J. (1978b). In *Quantitative Mass Spectrometry in Life Sciences*, vol. 2, ed. A.P. de Leenheer, R.R. Roncucci and C. van Peteghem, pp. 83–102. Amsterdam: Elsevier.

Miller, J.M. (1988). *Chromatography: Concepts and Contrasts*. New York: Wiley.

Miller, J.M. (1990). *Mass Spectrometry Reviews*, **9**, 319–347.

Millington, D.S., Norwood, D.L., Kodo, N., Moore, R., Green, M.D. and Berman, J. (1991). *Journal of Chromatography*, **562**, 47–58.

Millington, D.S. and Smith, J.A. (1977). *Organic Mass Spectrometry*, **12**, 264–265.

Milne, G.W.A. and Lacey, M.J. (1974). *CRC Critical Reviews in Analytical Chemistry*, **4**, 45–104.

Moseley, M.A., Deterding, L.J., Tomer, K.B. and Jorgenson, J.W. (1991a). *Analytical Chemistry*, **63**, 109–114.

Moseley, M.A., Deterding, L.J., Tomer, K.B. and Jorgenson, J.W. (1991b). *Analytical Chemistry*, **63**, 1467–1473.

Munson, M.S.B. and Field, F.H. (1966). *Journal of the American Chemical Society*, **88**, 2621–2630.

Murphy, R.C. (1993). *Mass Spectrometry of Lipids*. New York: Plenum.

Nibbering, N.M.M. (1984). *Mass Spectrometry Reviews*, **3**, 445–477.

Nibbering, N.M.M. (1985). In *Mass Spectrometry*, vol. 8, ed. M.E. Rose, pp. 141–160. London: The Royal Society of Chemistry.

Nibbering, N.M.M. (1990). *Accounts of Chemical Research*, **23**, 279–285.

Niessen, W. and van der Greef, J. (1992). *Liquid Chromatography-Mass Spectrometry*. New York: Marcel Dekker.

Nourse, B.D. and Cooks, R.G. (1990). *Analytica Chimica Acta*, **228**, 1–21.

O'Hair, R.A.J. and Bowie, J.H. (1989). In *Mass Spectrometry*, vol. 10, ed. M.E. Rose, pp. 145–180. Cambridge: The Royal Society of Chemistry.

Olesik, S.V. (1991). *Journal of High Resolution Chromatography*, **14**, 5–9.

Overberg, A., Hassenburger, A. and Hillenkamp, F. (1992). In *Mass Spectrometry in the Biological Sciences*, ed. M.L. Gross, pp. 181–197. Dordrecht: Kluwer Academic.

Pace-Asciak, C.R. (1989). *Advances in Prostaglandin, Thromboxane and Leukotriene Research. Volume 18. Mass Spectrometry of Prostaglandins and Related Products*. New York: Raven.

Paradisi, C., Todd, J.F.J., Traldi, P. and Vettori, U. (1992). *Organic Mass Spectrometry*, **27**, 251–254.

Pesyna, G.M., Venkataraghavan, R., Dayringer, H.E. and McLafferty, F.W. (1976). *Analytical Chemistry*, **48**, 1362–1368.

Pfleger, K., Maurer, H.H. and Weber, A. (1992). *Mass Spectral and GC Data of Drugs, Poisons, Pesticides, Pollutants and Their Metabolites*. Weinheim: VCH.

Pierce, A.E. (1968). *Silylation of Organic Compounds*. Illinois: Pierce Chemical Company.

Pihlaja, K., Rossi, K. and Vainiotalo, P. (1985). *Journal of Chemical Engineering Data*, **30**, 387–394.

Plage, B. and Schulten, H.-R. (1991). *Journal of Analysis and Applied Pyrolysis*, **19**, 285–299.

Poole, C.F. and Poole, S.K. (1991). *Chromatography Today*. Amsterdam: Elsevier.

Poole, C.F. and Schuette, S.A. (1984). *Contemporary Practice of Cromatography*. Amsterdam: Elsevier.

Pratt, D.E., Eagles, J. and Self, R. (1987). In *Mass Spectrometry*, vol. 9, ed. M.E. Rose, pp. 407–430. London: The Royal Society of Chemistry.

Price, D. (1990). *Trends in Analytical Chemistry*, **9**, 21–25.

Prokai, L. (1990). *Field Desorption Mass Spectrometry*. New York: Marcel Dekker.

Radom, L. (1992). *International Journal of Mass Spectrometry and Ion Processes*, **118/119**, 339–368.

Redfern, J.P. (1991). *Polymer International*, **26**, 51–58.

Reed, R.I. (1962). *Ion Production by Electron Impact*. New York: Academic Press.

Reed, R.I. (1966). *Applications of Mass Spectrometry to Organic Chemistry*. New York: Academic Press.

Richter, W.J., Blum, W., Schlunegger, U.P. and Senn, M. (1983). In *Tandem Mass Spectrometry*, ed. F.W. McLafferty, p. 417–434. New York: Wiley.

Rinehart, K.L. (1982). *Science*, **218**, 254–60.

Roellgen, F.W., Bramer-Weger, E. and Buetfering, L. (1987). *Journal of Physics, Colloquia*, **48**, 253–256.

Roepstorff, P. (1987). *European Spectroscopy News*, **73**, 18–23.

Roepstorff, P. (1989). *Accounts of Chemical Research*, **22**, 421–427.

Roepstorff, P. (1992). In *Mass Spectrometry in the Biological Sciences*, ed. M.L. Gross, pp. 213–227. Dordrecht: Kluwer Academic.

Rose, M.E. (1981). *Organic Mass Spectrometry*, **16**, 323–324.

Rose, M.E. (1982). In *Carotenoids 6. Proceedings of the Sixth International Symposium of Carotenoids*, ed. G. Britton and T.W. Goodwin, pp. 167–174. Oxford: Pergamon Press.

Rose, M.E. (1984). In *Mass Spectrometry*, vol. 7, ed. R.A.W. Johnstone, pp. 196–292. London: The Royal Society of Chemistry.

Rose, M.E. (1985). In *Mass Spectrometry*, vol. 8, ed. M.E. Rose, pp. 210–283. London: The Royal Society of Chemistry.

Rose, M.E. (1987). In *Mass Spectrometry*, vol. 9, ed. M.E. Rose, pp. 264–284. London: The Royal Society of Chemistry.

Rose, M.E., Longstaff, C. and Dean, P.D.G. (1983). *Biomedical Mass Spectrometry*, **10**, 512–527.

Rosenstock, H.M., Draxl, K., Steiner, B.W. and Herron, J.J. (1977). *Journal of Physical and Chemical Reference Data*, vol. 6, Supplement no. 1, 786 pp.

Rosenstock, H.M., Wallenstein, M.B., Wahrhaftig, A.L. and Eyring, H. (1952). *Proceedings of the National Academy of Science of the U.S.A.*, **38**, 667–678.

Russell, D.H., McBay, E.H. and Mueller, T.R. (1980). *International Laboratory*, 49–61.

Safron, S.A., Weinstein, N.D., Herschblach, D.R. and Tully, J.C. (1972). *Chemical Physics Letters*, **12**, 564–568.

Sahota, R.S. and Morgan, S.L. (1993). *Analytical Chemistry*, **65**, 70–77.

Sanderson, R.T. (1982). *Journal of Organic Chemistry*, **47**, 3835–3839.

Sargent, M. and Webb, K. (1993). *Spectroscopy Europe*, **5** (3), 21–28.

Schmelzeisen-Redeker, G., Buetfering, L. and Roellgen, F.W. (1989). *International Journal of Mass Spectrometry and Ion Processes*, **90**, 139–150.

Schmitz, B. and Klein, R.A. (1986). *Chemistry and Physics of Lipids*, **39**, 285–311.

Schulten, H.-R. (1977). *Methods of Biochemical Analysis*, **24**, 313–448.

Schulten, H.-R. (1979). *International Journal of Mass Spectrometry and Ion Physics*, **32**, 97–283.

Scrivens, J.H. and Rollins, K. (1992). *VG Monographs in Mass Spectrometry*, No. **4**. Manchester: VG Organic (now Micromass UK Ltd).

Scrivens, J.H., Rollins, K., Jennings, R.C.K., Bordoli, R.S. and Bateman, R.H. (1992). *Rapid Communications in Mass Spectrometry*, **6**, 272–277.

Sedgwick, R.D. (1981). In *Mass Spectrometry*, vol. 6, ed. R.A.W. Johnstone, pp. 174–195. London: The Royal Society of Chemistry.

Sen, N.P., Miles, W.F., Seaman, S and Lawrence, J.F. (1976). *Journal of Chromatography*, **128**, 169–173.

Shackleton, C.H.L., Gaskell, S.J. and Liberato, D.J. (1987). In *Chromatographic and Non-chromatographic Mass Spectrometric Techniques for Clinical Steroid Analysis*, ed. H. Jaeger, pp. 185–203. Heidelberg: Huethig.

Shiner, V.J., Jr and Buddenbaum, W.E. (1975). In *MTP International Review of Science, Physical Chemistry Series Two*, vol. 5, ed. A. Maccoll, pp. 129–158. London: Butterworths.

Shute, L.A., Gutteridge, C.S., Norris, J.R. and Berkeley, R.C.W. (1984). *Journal of General Microbiology*, **130**, 343–355.

Simmons, D.S., Colby, B.N. and Evans, C.A. Jr (1974). *International Journal of Mass Spectrometry and Ion Physics*, **15**, 291–302.

Slagle, J.R. (1971). *Artificial Intelligence, the Heuristic Programming Approach*. New York: McGraw-Hill.

Smith, B.J. and Radom, L. (1993). *Journal of the American Chemical Society*, **115**, 4885–4888.

Smith, C.G., Smith, P.B., Pasztor, A.J., McKelvey, M.L., Meunier, D.M., Frohlicher, S.W. and Ellaboudy, A.S. (1993). *Analytical Chemistry*, **65**, 217R–243R.

Smith, D.H., Olsen, R.W., Walls, E.C. and Burlingame, A.L. (1971). *Analytical Chemistry*, **43**, 1796–1806.

Smith, R.D., Cheng, X., Bruce, J.E., Hofstadler, S.A. and Anderson, G.A. (1994). *Nature*, **369**, 137–139.

Smith, R.D., Kalinoski, H.T. and Udseth, H.R. (1987). *Mass Spectrometry Reviews*, **6**, 445–496.

Smith, R.D., Loo, J.A., Edmonds, C.G., Barinaga, C.J. and Udseth, H.R. (1990). *Analytical Chemistry*, **62**, 882–899.

Smith, R.D., Olivares, J.A., Nguyen, N.T. and Udseth, H.R. (1988). *Analytical Chemistry*, **60**, 436–441.

Smith, R.D. and Udseth, H.R. (1987). *Analytical Chemistry*, **59**, 13–22.

Smith, R.M. (1988*a*). *Supercritical Fluid Chromatography*. London: The Royal Society of Chemistry.

Smith, R.M. (1988*b*). *Gas and Liquid Chromatography in Anlaytical Chemistry*. Chichester: Wiley.

Snedden, W. and Parker, R.B. (1976). *Biomedical Mass Spectrometry*, **3**, 295–298.

Spalding, T.R. (1979). In *Mass Spectrometry*, vol. 5, ed. R.A.W. Johnstone, pp. 312–346. London: The Chemical Society.

Speir, J.P., Gorman, G.S. and Amster, I.J. (1992). In *Mass Spectrometry in the Biological Sciences*, ed. M.L. Gross, pp. 199–212. Dordrecht: Kluwer Academic.

Standing, K.G. and Ens, W. (1991). *Methods and Mechanisms for Producing Ions from Large Molecules*. New York: Plenum.

Stemmler, E.A. and Hites, R.A. (1988). *Electron Capture Negative Ion Mass Spectra of Environmental Contaminants and Related Compounds*. Weinheim: VCH.

Stimpson, B.P. and Evans, C.A., Jr (1978). *Biomedical Mass Spectrometry*, **5**, 52–63.

Sundqvist, B. and Macfarlane, R.D. (1985). *Mass Spectrometry Reviews*, **4**, 421–460.

Szulejko, J.E. and McMahon, T.B. (1993). *Journal of the American Chemical Society,* **115**, 7839–7840.

Taagepera, M., Henderson, W.G., Brownlee, R.T.C., Beauchamp. J.L., Holtz, D. and Taft, R.W. (1972). *Journal of the American Chemical Society,* **94**, 1369–1370.

Todd, J.F.J. (1984). *International Journal of Mass Spectrometry and Ion Processes,* **60**, 3–10.

Todd, J.F.J. (1991*a*). *Mass Spectrometry Reviews,* **10**, 3–52.

Todd, J.F.J. (1991*b*). *Pure and Applied Chemistry,* **63**, 1541–1566.

Todd, J.F.J. and Lawson, G. (1975). In *MTP International Review of Science, Physical Chemistry Series Two,* vol. 5, ed. A. Maccoll, pp. 289–348. London: Butterworths.

Todd, J.F.J. and Penman, A.D. (1991). *International Journal of Mass Spectrometry and Ion Processes,* **106**, 1–20.

Tomer, K.B., Guenat, C.R. and Deterding, L.J. (1988). *Analytical Chemistry,* **60**, 2232–2236.

Tong, H.Y., Giblin, D.E., Lapp, R.L., Monson, S.J. and Gross, M.L. (1991). *Analytical Chemistry,* **63**, 1772–1780.

Torgerson, D.F., Skowronski, R.P. and Macfarlane, R.D. (1974). *Biochemical and Biophysical Research Communications,* **60**, 616.

Van Berkel, G.J., McLuckey, S.A. and Glish, G.L. (1991). *Analytical Chemistry,* **63**, 1098–1110.

van Marlen, G. and Dijkstra, A. (1976). *Analytical Chemistry,* **48**, 595–598.

Vekey, K. and Zerilli, L.F. (1991). *Organic Mass Spectrometry,* **26**, 939–944.

Vestal, M.L. (1983). *Mass Spectrometry Reviews,* **2**, 447–480.

Vestal, M., Wahrhaftig, A.L. and Johnston, W.H. (1962). *Theoretical Studies in Basic Radiation Chemistry,* Aeronautical Research Laboratory Report, pp. 62–426.

von Ardenne, M., Steinnfelder, K. and Tümmler, R. (1971). *Electronenanlagerungs-Massen-spektrographic organischer Substanzen.* Berlin: Springer Verlag.

Voyksner, R.D. (1989). *Chemical Analysis (New York),* **100**, 173–202.

Waller, G.R. (1972). *Biochemical Applications of Mass Spectrometry.* New York: Wiley-Interscience.

Waller, G.R. and Dermer, O.C. (1980). *Biochemical Applications of Mass Spectrometry, First Supplementary Volume.* New York: Wiley-Interscience.

Wangen, L.E., Woodward, W.S. and Isenhour, T.L. (1971). *Analytical Chemistry,* **43**, 1605–1614.

Ward, S.D. (1971). In *Mass Spectrometry,* vol. 1, ed. D.H. Williams, pp. 253–287. London: The Chemical Society.

Ward, S.D. (1973). In *Mass Spectrometry,* vol. 2, ed. D.H. Williams, pp. 264–301. London: The Chemical Society.

Watson, J.T. (1976). *Introduction to Mass Spectrometry, Biomedical, Environmental and Forensic Applications.* New York: Raven Press.

Weston, A.F., Jennings, K.R., Evans, S. and Elliott, R.M. (1976). *International Journal of Mass Spectrometry and Ion Physics*, **20**, 317–327.

Willoughby, R.C. and Browner, R.F. (1984). *Analytical Chemistry*, **56**, 2626–2631.

Wilson, J.M. (1971). In *Mass Spectrometry*, vol. 1, ed. D.H. Williams, pp. 1–30. London: The Chemical Society.

Wilson, J.M. (1973). In *Mass Spectrometry*, vol. 2, D.H. Williams, pp. 1–32. London: The Chemical Society.

Wilson, J.M. (1975). In *Mass Spectrometry*, vol. 3, ed. R.A.W. Johnstone, pp. 86–116. London: The Chemical Society.

Wilson, J.M. (1977). In *Mass Spectrometry*, vol. 4, ed. R.A.W. Johnstone, pp. 102–131. London: The Chemical Society.

Winkler, P.C., Perkins, D.D., Williams, W.K. and Browner, R.F. (1988). *Analytical Chemistry*, **60**, 489–493.

Wise, M.B. (1987). *Analytical Chemistry*, **59**, 2289–2293.

Wise, M.B., Buchanan, M.V. and Guerin, M.R. (1990). *Proceedings of the 38th ASMS Conference on Mass Spectrometry and Allied Topics*, p. 619, Tucson, AZ, U.S.A.

Wolkoff, P., van der Greef, J. and Nibbering, N.M.M. (1978). *Journal of the American Chemical Society*, **100**, 541–545.

Yergey, A.L., Edmonds, C.G., Lewis, I.A.S. and Vestal, M.L. (1990). *Liquid Chromatography/Mass Spectrometry. Techniques and Applications*. New York: Plenum.

Yun, Y. and Meuzelaar, H.L.C. (1991). *Energy Fuels*, **5**, 22–29.

Zerbi, G. and Beynon, J.H. (1977). *Organic Mass Spectrometry*, **12**, 115–118.

Index

accurate mass measurement 15–17,
124–8, 157, 193, 218–19, 228–9, 332
and metastable ions 237
by computers 125–7
manual 124–5
acids 186
using negative ions 247
derivatization of 186
action spectrum 305–6, 473
activation of normal ions 261–70
acylcarnitines 147–53, 168–9, 194
adduct ions 64, 68, 79, 90, 108
alcohols 186–7, 192, 197, 259, 355, 357,
368, 401–6, 411–13
derivatization of 186–7, 192, 197,
200–1
aldehydes 333, 365–6
alkaloids 163, 216, 259
alkanes 210, 244, 346–9
alkenes 189–90, 349–50
localization of double bonds in 189–90,
350
amides 187, 367, *and see* peptides, proteins
derivatization of 187
amines 104, 186–7, 192, 197, 223–4
aliphatic 355, 357
aromatic 375, 377, 379
derivatization of 186–7, 192, 197
amino acids 80, 187
by gas chromatography/mass
spectrometry 131–3
derivatization of 187
analogue-to-digital converters 114–15
analysers 37–54
hybrid instruments 46–7, 278–9

ion cyclotron resonance instruments –
see ion cyclotron resonance and
Fourier transform ion cyclotron
resonance spectroscopy, and 51–3,
123, 127, 270, 283–8
ion traps 47–51, 123, 270, 283–8
magnetic sector instruments 38–41
multiple sector instruments 37–8, 271
quadrupole mass filters 42–6
reverse geometry 41
single focussing 41
time-of-flight instruments 53–4, 109,
235, 279
APCI – *see* atmospheric pressure
chemical ionization
API – *see* atmospheric pressure ionization
appearance energy 313–14, 318–19, 429
archaeological applications 157, 435
aromatic compounds
interaction of substituents in 377–8,
395
linear free energy relationships 379
order of fragmentation of substituents
378
ortho-effects in 377–8, 395
array detectors 39–40, 55–7, 34-6, 106,
282–3
artificial intelligence 139, 467–79
atmospheric pressure chemical ionization
30–1, 75–6, 165
atmospheric pressure ionization 30–1,
75–6, 164–5, 227–8, 300–1

background subtraction 8–9, 128–9, 149
base peak 3

bath gas – *see* buffer gas
Biller–Biemann enhancement 132
bond dissociation energies in ions
291–8, 313–14, 318–19, 355–63,
429–31
buffer (bath) gases 30, 78, 288, 445–54

CA – *see* collisional activation
calibration compounds 127–8
calibration graphs 208–213
capillary electrophoresis/mass
spectrometry 142, 172–6
carbohydrates 86, 163, 200, 228, 261
by liquid chromatography/mass
spectrometry 228
derivatization of 200–1
carbonates 369, 383, 388
carboxylic acids 104, 186–7, 189, 247,
366, 369
carotenoid compounds 237
carrier effect in gas chromatography
212–13
cationization 30, 90, 104, 108, 196, 228
CE/MS – *see* capillary
electrophoresis/mass spectrometry
centre of mass collision energy 263
^{252}Cf plasma desorption 35, 110–12
channel electron multiplier 39–40, 55–7,
282–3
channeltron – *see* channel electron
multiplier and array detectors
charge exchange 74–5, 82, 100, 178,
308–9
charge localization theory 321–2
charge permutation processes 264

chemical ionization 18–21, 30–1, 60, 66–77, 152–3, 160, 178–80, 189, 216–17, 220–1, 307–8
    atmospheric pressure 75–6
    negative ions 78–81, 223–4, 247
    pulsed positive ion/negative ion 82
CI – *see* chemical ionization
CIA – *see* collision induced activation
CID – *see* collision induced dissociation (or decomposition)
cluster ions 64, 69, 109
collisional activation 50–1, 90, 123, 230–1, 235, 261–70, 275, 301–2, 447–55
    of multiply charged ions 265–8
collision induced activation – *see* collisional activation
collision induced dissociation (or decomposition) – *see* collisional activation
computer control of mass spectrometers 113–14, 213
computers – *see* data systems
cone voltage 97, 416–18
continuous flow FAB 26, 100–6, 163–4, 167–8, 181

Daly detector 35–6, 55, 254
data acquisition by computer 114–21
data display units 141
data processing by computers 123–40, 154–5, 170
    accurate mass measurement 124–8
    analogue-to-digital converters 114–15
    digital-to-analogue converters 114–15
    multiscan analyses 128–34
    single scan analyses 134–41
    transputers 120–1
data systems 113–141
daughter ions – *see* product ions
defocussing 239–47
deprotonated molecules 31, 79
derivatization 147, 183–204
    for enantiomeric separation 197–8
    for gas chromatography/mass spectrometry 183–8, 197

for isotope studies 193–5
for negative ion mass spectrometry 196–7
metastable ions and MS/MS 195–6
of inorganic compounds 203–4
of organic compounds 183–202
detectors 55–8, 282
    channeltron arrays 39–40, 55–7
    electron multiplier arrays 39–40, 55–7
    electron multipliers 55
    Faraday cups 55
    photographic plates 57, 207
    scintillation counting 55
digital-to-analogue converters 114–15
dioxins 221
directed fragmentation 189–91
double-bond equivalents 332
double-bond location 189–90
double-focussing mass spectrometers 38–41
drugs 157, 181–2, 221, 223–4, 228

EI – *see* electron ionization
electrohydrodynamic ionization 32
electron affinity 80
electron capture 31, 65, 78–9, 82, 303
electron multiplier 39–40, 55–7
electron ionization 1–9, 24, 28–30, 60–6, 303–4
electrospray ionization 10, 26, 32–4, 60, 91–9, 164–6, 171–5, 180, 199, 226–8, 258, 261, 264, 266, 268, 281, 287, 300
elemental compositions of ions 15–17, 125–6
elemental maps 335–7
enantiomers 197–8
energy states of ions 289–98
    at low pressures 289–98
    at medium gas pressures 298–300
    in the condensed liquid state 300–1
    resulting from activation 301–2
environmental analysis 76, 138, 154–5, 180, 205, 207, 220–1, 281, 439–41
equations of motion of ions 38–43, 53, 239, 241–6, 251–2, 255, 444–55

esters 366, 369
ethers 358–9, 375–6, 389
even electron rule and even/odd mass 331, 334, 398–9
excess of energy in ions 289–98, 347, 353, 363
    effect on mass spectrum of 1–2, 289–98, 337–44

FAB – *see* fast atom bombardment
Faraday cup 55
fast atom bombardment 26, 34, 60, 100–8, 163–4, 166–9, 173, 181, 194, 225, 261, 275, 281, 300
    dynamic or continuous flow 34
fatty acids 189, 196, *and see* carboxylic acids
field desorption 31–2, 60, 89–91, 225, 258, 264, 173, 306–7
field free regions 238, 250, 254–5, 257
field ionization 31–2, 85–9, 306–7
    kinetics 87–8
    negative ions 87
fit 137, 411
focal plane detector 39–40, 55–7, 106, 207
food flavour and odour research by mass spectrometry 157, 169–71, 180, 218–19, 221, 275
forensic science 157, 431
Fourier-transform ion cyclotron resonance mass spectrometry 51–3, 121–3, 263, 270, 279, 284–5, 444–5
four-sector mass spectrometer 46
fragmentation 1–2, 310–16, 334, 346–92
    odd/even mass ions 331, 334
    primary 353–87, 398–400
    secondary 382–7
fragmentation pathways 1–3, 324–5, 327, 344–6, 384–5
fragmentation pattern 2
fragmentation rates 315–16
fragment ions 286, 313–19
    formation of 1
    fragmentation of – *see* fragmentation, secondary

gas chromatography/mass spectrometry 25, 44, 128–34, 142, 144–58, 183–8, 212–24

    computers with 128–34

    interfaces 145–7

GC/MS – *see* gas chromatography/mass spectrometry

geochemical applications 157, 431, 435

glow discharge mass spectrometry 37

glucosinolates 169–71

halides 329, 355, 361, 364

Hall effect sensor 121

heats of formation 77, 313–15

heteroatomic compounds

    aliphatic 353–72

    aromatic 373–82, 413–15, 415–18

hybrid analysers 46, 278–9

in-beam electron or chemical ionization 24, 60, 76, 82–5, 466

inductively coupled plasma mass spectrometry 36–7, 431

inlet systems 23–7, 453

    cold 23

    direct probe 24

    for solutions 25–7

    gas chromatographic 25

    hot 23–4

    moving belt 26

inorganic compounds 203–4, 225–6, 327, 357–8, 366, 381, 386

    quantitative measurements 225–6

integrated ion current technique 224

internal standard 208–9

inverse library searching 137, 147–55

'involatile' substances 183–8

    derivatization of 186–8, 196–7, 203–4

    liquid chromatography/mass spectrometry of 198–203

ion abundance 2–8

ion bombardment 34, 107–108

ion cyclotron resonance spectroscopy 51–3, 121–3, 207, 263, 270, 279, 284–5, 306–16, 444–55

double resonance experiments 449–51

ion evaporation 33, 91–9, 162–3

ionization 289–302, 427–8

ionization efficiency 198–9, 205, 223, 225, 227, 427–31

ionization energy 64, 290–3, 313–15, 337–41, 427–9

ionization methods 60–109

    at atmospheric pressure 75–6

    charge exchange 74–5, 82, 308–9

    chemical 30, 60, 66–77, 307–8

    combined 35–6

    electric field 31–4, 85–9, 306–8

    electron 28–30, 60–6, 303–4

    fast atoms, ions 34, 60, 100–8

    field desorption 60, 89–91, 308

    inductively coupled plasma 36–7

    modified, chemical 82–5

    modified, electron 82–5

    nuclear fission 35, 110–12

    photons 34–5, 108–10

    spark 37

    spray 32–3, 162–6

ion kinetic energy spectroscopy (IKES) 241

ion lifetime 87–8, 315–16

ion/molecule reactions 30–1, 66–75, 79, 85, 90, 94, 100, 123, 262, 265, 301–2

ion sources 27

    residence times of ions in 31–2

ion spray 32–4, 91–9, 162–6

ion structures 309, 344–6

ion trap 47–51, 121–3, 141, 156, 207, 264, 266, 283–8, 306, 419

isomers, differentiation of 69, 75, 87, 237, 261

isotope abundance patterns 10–14, 134, 220–1, 331, 405, 413–15, 418–19, 431–44

    in large molecules 331–2, 442–4

    simplification of 13–14

isotope ratios, measurement of 331, 431

isotopes 10–14, 134, 209–12

    accurate masses of 15–16

    and metastable ions 436

    natural abundances of 10–12

use for investigation of reaction mechanism 436–40

use in internal standards 209–12

use in labelling experiments 134, 436–41

jet separator 147

ketones 365, 368, 406–8

kinetic isotope effects, measurement by mass spectrometry 434–5

laser desorption 108–10, 461–5

    ionization 108–10

LC/MS – *see* liquid chromatography/mass spectrometry

learning machines 139

library searching 134–41, 149, 154–5, 325

linked scanning 243–53

    with gas chromatography/mass spectrometry 250

lipids 217, 276–7

liquid chromatography/mass spectrometry 25–7, 32–3, 75–6, 142–3, 158–72, 197–8, 208, 211, 217, 227–8

    computers for 128–40, 159

    interfaces 143–4

    atmospheric pressure 164–6

    continuous flow fast atom bombardment 163–4

    moving belt 159–60

    particle beam 160–2

    thermospray 162–3, 166, 169

    microbore columns 158–9

literature on mass spectrometry 473–7

magnetic sector mass spectrometers 38–41, 156, 232–3, 238–9, 241–3, 250–3, 254–5

    computer control of 121–2

    double-focussing 38

    of reverse geometry 41, 256–7

    quadruple-focussing 40, 279

    single-focussing 41

    triple-focussing 260–1

MALDI (matrix-assisted laser desorption ionization) 35, 108–9, 465
mass-analysed ion kinetic energy spectroscopy (MIKES) 256–9
mass calibration 27, 114–20, 250
  for accurate mass measurement 114–20
mass chromatograms 129, 156–7, 206
mass fragmentography 218
mass spectra
  appearance of 17–18, 329–35
  classification of 325–91
  examination of 17–18, 325–35
  formation of 1–9
  libraries of 9, 134–9
  modification of, by derivatization 188–202
  modification of, by instrumental changes 337–44
  normalization of 3–9
mass spectrometry/mass spectrometry 46–53, 123, 172, 176, 195–6, 238–41, 270–88, 399–400
match factor 137
mathematical analysis 471–2
Mattauch–Herzog geometry 41
metastable ions 14–15, 87, 89, 120, 122, 128, 139, 195–6, 228–31, 232–55, 316, 436
  in conventional mass spectrometers 232–5, 399
  in field ionization kinetics 87–8
  in reversed geometry mass spectrometers 256–9
  in triple-focussing mass spectrometers 277–8
  in triple quadrupole mass spectrometers 259–61
  release of internal energy 252–3
  shapes of product ion peaks 252–3
  uses of 236–8
microdensitometer 40, 57
mixtures, qualitative analysis of
  by gas chromatography/mass spectrometry 144–58
  by liquid chromatography/mass spectrometry 158–80

without prior separation of components 270–82
mixtures, quantitative analysis of
  by gas chromatography/mass spectrometry 183–8
  by liquid chromatography/mass spectrometry 227–8
  with non-chromatographic inlets 270–82
molecular ions 1–2
  formation of 1–10, 18–20, 289–302
  fragmentation of
    by rearrangement with cleavage 353, 357, 367–71, 377–82, 395
    by simple cleavage 354–5, 363–7, 373
molecular separators 147
moving belt interface 26, 159–60
MS/MS – see mass spectrometry/mass spectrometry
  two dimensional 154
multiphoton fragmentation 285, 304–6
multiphoton ionization 35, 270, 473
multiple analysers 46, and see mass spectrometry/mass spectrometry
multiply charged ions 9–10, 91–9, 109
multiscan analyses 128–34
multivariate analysis 140, 466–9, 471–2, and see pyrolysis

negative-ion chemical ionization – see negative ions by chemical ionization
negative-ion field ionization – see negative ions by field ionization
negative ions 31, 35–6, 66, 78–80, 84, 87, 94, 96, 104, 165–6, 196–7, 223, 258, 264, 268–9, 277, 289, 303, 308–9, 394
  by chemical ionization 30–1, 75, 78–81, 220, 258
  by charge exchange 308–9
  by electron capture 31
  by electrospray 97, 415–18
  by field ionization 85–9, 308
  by ion pair processes 308–9
  multiply charged 265–6
Nier–Johnson geometry 39

neutral loss analyses 272
nitriles 364
nitro-compounds 375, 377
nitrosamines 221
normalization of mass spectra 3–9
normalized relative abundance 3
nucleosides
  by liquid chromatography/mass spectrometry 163
  by pyrolysis/mass spectrometry 472

organometallic compounds 13, 203–4, 327, 356–8, 366–7, 381, 386, 418–19
ortho effect 377

parent ions – see precursor ions
particle beam interface 160–2
pattern recognition 140
peptides 163, 187, 191, 200, 247, 264, 281–2, 285, 367, 420–5
  by capillary electrophoresis/mass spectrometry 172–6
  by gas chromatography/mass spectrometry 190–1
  by liquid chromatography/mass spectrometry 171, 200
  chemical ionization of 72
  derivatization of 187, 191, 200
percentage of reconstructed ion current presentation 3
percentage of total ion current presentation 3–5
percentage relative abundance 3
phase space 310
phenols 186, 192, 375–6, 378
  derivatization of 186, 192
photographic plate 39, 41, 56–7, 207, 283, 334
photoionization 34, 304–5
photodissociation (photon-induced decomposition) 269–70, 285, 305–7, 473
plasma desorption 35, 60, 110–12
plasmaspray 33
pollutants in the environment – see environmental science

polymers 157, 178, 180, 420–1
polysaccharides 420–1
porphyrins 90, 287–8, 415–18
precursor ions 260–1, 272, 277, 280, 284–5
product ions 260–1, 272, 277, 280, 284–5
prostaglandins 191–2
proteins 94–5, 109, 420–2
proton affinity 71, 77
protonated molecules 18–21, 30, 60, 63, 67–8, 76, 78, 85–6, 90–6, 108, 111, 200, 219–20, 289, 307–8, 393–4
pulsed positive-ion/negative-ion chemical ionization 82
purge and trap devices 154–5
purity of fit – see fit
pyrolysis
    chemical aspects of 183, 458–9
    using Curie point heating 460–2
    using lasers 463–5, 472–3
    using resistive heating 461–3
pyrolysis/gas chromatography/mass spectrometry 140, 158, 456
pyrolysis/mass spectrometry 140, 158, 259, 455–73

quadrupole mass filters 28, 42–6, 121–2, 156, 213, 220, 235, 259–60, 276–7
    accurate mass measurement with 128
    computerized 124–8
quantitative mass spectrometry 205–13
    by gas chromatography/mass spectrometry 212–24
    by liquid chromatography/mass spectrometry 208, 211, 217, 227–8
    by non-chromatographic inlets 224–6
    by soft methods of ionization 227
    calibration graphs 209
    internal standards for 208–13
quasi-equilibrium theory 310–12
    simplified 310–12, 360–3, 390–1
quasi-molecular ions 1, 18–21, 393–5

radionuclide ionization 35, 110–12
rapid heating methods 83, and see pyrolysis and/or lasers

reactant gases for chemical ionization 30, 66–81, 178–9
reactant ion chemical ionization – see reactant gases for chemical ionization
rearrangement accompanying fragmentation 87–8, 292, 309, 311, 316, 353, 357, 367–71, 387–92, 395
rearrangements 87–8, 292, 309, 311, 316
reconstructed ion current 7, 133
recording of mass spectra 18
    as tabulations 5–7
    on charts 3–5, 9, 15, 19, 57
    on photographic plates – see photographic plates
    via computers 57–8, and see data systems
reference compounds for mass calibration 24, 27, 127–8
resolution 15–17, 451–3
resonance-enhanced multiphoton ionization (REMPI) 35, 304–5, 473 – see also multiphoton ionization and fragmentation
reverse searching 137
reverse-geometry mass spectrometers 41, 256
RRKM theory 310

sampling 114–15
scintillation counting 55–7
secondary-ion mass spectrometry 101, 107–8, 163, 181–2, 200, 182, 300
selected current profile 218
selected ion monitoring 60, 122–3, 128, 156, 157, 191–3, 207, 213–27, 220–1
    at high resolution 157
selected metastable ion monitoring 229
selected (or single) reaction monitoring 195, 228–9, 279
separation of ions
    according to times of flight 41
    by ion cyclotron resonance frequencies 51
    in magnetic sector instruments 38–41
    in quadrupole mass filters 42–6

SFC/MS – see supercritical fluid chromatography/mass spectrometry
similarity index 137
single-focussing mass spectrometers 41
single ion monitoring 206, 210–11
single ion retrieval 129
single scan analyses 134
soft methods of ionization 60–4
space research by mass spectrometry 22
spark source mass spectrometry 37, 207
    quantitative 207
stability diagrams 43, 50
steroid conjugates 84, 166
steroids 69, 129–30, 133, 138, 157, 166, 188, 191, 195, 225, 228–31, 245, 258, 261, 440–1
    derivatization of 228–31
    quantitative measurements 228–31
structure elucidation, examples of 397–426
sugars 186, 200–1, 228 and see carbohydrates
sulphones 364
sulphonic acids 415–18
sulphoxides 364, 366, 368
supercritical fluid chromatography/mass spectrometry 27, 176–80
suppression 104, 108
surface-induced decomposition 268–79
surface ionization 106
surfactants 199–200

tandem mass spectrometry – see mass spectrometry/mass spectrometry 46–53
target analysis 149, 154
temperature, effects on mass spectra 341–4
terpenes 136, 250
theory of mass spectrometry 289–324
    qualitative aspects 316–24
    quantitative aspects 310–16
thermochemical considerations 313–15, 345, 355–6, 363
thermogravimetry/mass spectrometry 459–60, 469–70

thermospray ionization 33, 60, 91–4, 162–3, 166, 169, 227–8

thin layer chromatography/mass spectrometry 107, 142, 180–2

thioethers 359

thiols 356–7

time-of-flight analyser – *see* analysers

TLC/MS – *see* thin layer chromatography/mass spectrometry

TG/MS – *see* thermogravimetry/mass spectrometry

total ion current 6–7, 192, 215–16

transputers 120–1

trimethylsilylation 184–5, 187

triple-focussing mass spectrometers 259–61

triple quadrupole mass spectrometers 46, 167–8, 178, 180, 230, 259–61, 273–7

visual display unit 141,

volatility of samples 159, 162, 186–8 imparting 186–8

Printed in the United States
By Bookmasters